Computer Integrated Manufacturing

Current Status and Challenges

NATO ASI Series

Advanced Science Institutes Series

A series presenting the results of activities sponsored by the NATO Science Committee, which aims at the dissemination of advanced scientific and technological knowledge, with a view to strengthening links between scientific communities.

The Series is published by an international board of publishers in conjunction with the NATO Scientific Affairs Division

A Life Sciences **B Physics**	Plenum Publishing Corporation London and New York
C Mathematical and Physical Sciences **D Behavioural and Social Sciences** **E Applied Sciences**	Kluwer Academic Publishers Dordrecht, Boston and London
F Computer and Systems Sciences **G Ecological Sciences** **H Cell Biology**	Springer-Verlag Berlin Heidelberg New York London Paris Tokyo

Series F: Computer and Systems Sciences Vol. 49

Computer Integrated Manufacturing

Current Status and Challenges

Edited by

I. Burhan Turksen

University of Toronto
Department of Industrial Engineering
Toronto, Ontario M5S 1A4, Canada

Co-editors:

Kiyoji Asai

University of Osaka Prefecture
Department of Industrial Engineering
College of Engineering
Sakai, Osaka 591, Japan

Gunduz Ulusoy

Bogazici University
Department of Industrial Engineering
P.k.2 Bebek, Istanbul, Turkey

Springer-Verlag
Berlin Heidelberg New York London Paris Tokyo
Published in cooperation with NATO Scientific Affairs Division

Proceedings of the NATO Advanced Study Institute on Computer Integrated Manufacturing: Current Status and Challenges, held in Istanbul, Turkey, August 30–September 12, 1987.

ISBN-13: 978-3-642-83592-6 e-ISBN-13: 978-3-642-83590-2

DOI: 10.1007/978-3-642-83590-2

Softcover reprint of the hardcover 1st edition 1988

2145/3140-543210 – printed on acid-free paper

Preface

The Current state of expectations is that Computer Integrated Manufacturing (CIM) will ultimately determine the industrial growth of world nations within the next few decades. Computer Aided Design (CAD), Computer Aided Manufacturing (CAM), Flexible Manufacturing Systems (FMS), Robotics together with Knowledge and Information Based Systems (KIBS) and Communication Networks are expected to develop to a mature state to respond effectively to the managerial requirements of the factories of the future that are becoming highly integrated and complex.

CIM represents a new production approach which will allow the factories to deliver a high variety of products at a low cost and with short production cycles. The new technologies for CIM are needed to develop manufacturing environments that are smarter, faster, close-coupled, integrated, optimized, and flexible. Sophistication and a high degree of specialization in materials science, artificial intelligence, communications technology and knowledge-information science techniques are needed among others for the development of realizable and workable CIM systems that are capable of adjusting to volatile markets. CIM factories are to allow the production of a wide variety of similar products in small batches through standard but multimission oriented designs that accommodate flexibility with specialized software.

CIM factories are to operate in accordance with such characteristics as decentralization, disaggregation, flexibility, rapid conversion of product lines, response to innovation, production tied to demand, and multiple functions. Other requirements for CIM are minimal downtime and maintenance, maximum product family range, ability to adapt to variability in materials and process conditions, the ability to handle increasingly complex product designs and technologies within the existing systems with minimum disruption and minimum cost.

At present, a full CIM implementation does not exist in any factory. The main issue to overcome is that of integration. Integration is considered as consisting of three areas: (1) data integration which is to ensure that a single definition of the information in the organization is considered and updated by all activities within the organization; (2) transactional integration which is to ensure that a single transaction made in any department of the organization is employed to record and update an event; and (3) operational integration which is to allow users to initiate transactions easily and more efficiently between applications by having access to standardized computer menus and security systems.

The NATO Advanced Study Institute entitled "Computer Integrated Manufacturing: Current Status and Challenges" was held between August 30 and September 12, 1987 at The Grand Tarabya Hotel in Istanbul, Turkey to address some of the contemporary issues in this regard.

Two speakers from industry, one from IBM Europe, the other from the Robotics Institute of Japan, together with two speakers from research institutions, one from Europe, the other from Japan, addressed the "Current Status" of CIM from their individual perspectives. New Directions in CIM were treated by four speakers from the USA, Germany, and Portugal, addressing various aspects of decision support systems, applied artificial intelligence, and information systems.

Many aspects of CIM technologies, from strategic planning to robot design, necessitate natural language understanding of vagueness and management of uncertainty. Fuzzy set theories and related research offers frontier knowledge to cope with various aspects of imprecise, ill-defined or uncertain problem domains of CIM. Four speakers from the USA, Canada, and Japan have presented fuzzy set and fuzzy logic methodologies potentially useful for CIM as well as other areas of future technologies. The last paper in this section incorporates both the fuzzy and precise aspects of scheduling problems. In the last section, flexible manufacturing problems, group technology, plant layout, scheduling and assembly, shape models, image acquisition, and processing techniques were addressed by nine speakers from Europe and North America.

We acknowledge with gratitude the cooperation and support we have received from all the speakers and the participants during NATO-ASI-CIM. In addition, we acknowledge the generous support received from a number of industrial corporations to fulfil our requirement for the "Double Jump" program. These corporations are: IBM, NETAS, NIXDORF, SISE-CAM, TOFAS, KORDSA, ARCELIK, GENERAL ELECTRIC. Our special thanks go to Miss H. Santos for her secretarial help, and to graduate students for their assistance in the preparation of the indices: Dr. A. Dagnino, Mr. Z. Leskowsky, Mr. Z. Zhong, Miss D. Ulguray, Mr. K. Jagadisan, and Mr. R. Bahatnagar.

Toronto, July 1988 I.B. Turksen

Table of Contents

Part 4. Models Toward Integration

PART 1: CURRENT STATUS

(A) EUROPEAN VIEWS

(I) IBM - CIM
(II) TNO - Process Planning

(B) JAPANESE VIEWS

(I) CIM
(II) Robotics

Computer Integrated Manufacturing

Horst W. Artmann
CIM Center Europe
Mgr. Mfg. Systems & Integration

IBM's corporate goal in the industrial sector is to be the leading CIM product supplier. Therefore, our products will provide a comprehensive and well proven integrated spectrum of offerings for a smooth installation and migration towards a CIM enterprise.

The IBM offerings for CIM include

> IBM Hardware products
> IBM Software products
> IBM Communication products
> IBM Workstations
> IBM education facilities
> Packaged solutions for an enterprise.

This presentation demonstrates IBM's commitment to CIM as

> CIM Developer
> CIM Manufacturer
> CIM Supplier.

After this presentation, you will understand, what IBM is doing about CIM, how CIM can serve you in accomplishing your objectives, and how you could make CIM a reality. We will see how IBM supports its customers on their way towards CIM:

- Computer Integrated Manufacturing -

NATO ASI Series, Vol. F49
Computer Integrated Manufacturing
Edited by I. Burhan Turksen

IBM **CIM**

Industry Trend

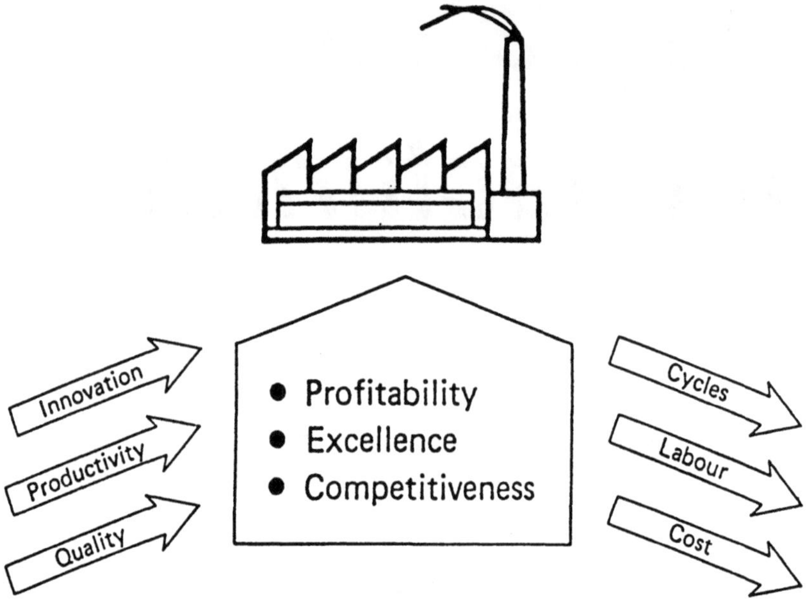

In the last decade, the pressure for greater efficiency has brought many changes in the manufacturing environment.

Trade press, trade conferences and your own experiences show that the key objectives for an Industrial enterprise are:

PROFITABILITY - EXCELLENCE - COMPETITIVENESS.

Innovation is key to new markets and new products. Improvements in productivity and quality are continuously required to meet these objectives: Profitability, Excellence and Competitiveness.

Conflicting trends are shorter product cycles, less direct labour and lower cost.

Shorter cycles reduce inventories and allow higher flexibility. Fast reaction to customer or market demand changes are needed. All major cost factors need careful attention. Direct Labour certainly has to be reduced, especially in an automated, process driven environment.

The conflict between these trends is evident. Only a solid and well balanced enterprise strategy could ultimately resolve this issue.

IBM's solutions for CIM will help you to manage this challenge.

IBM ***CIM***

Changing Environment

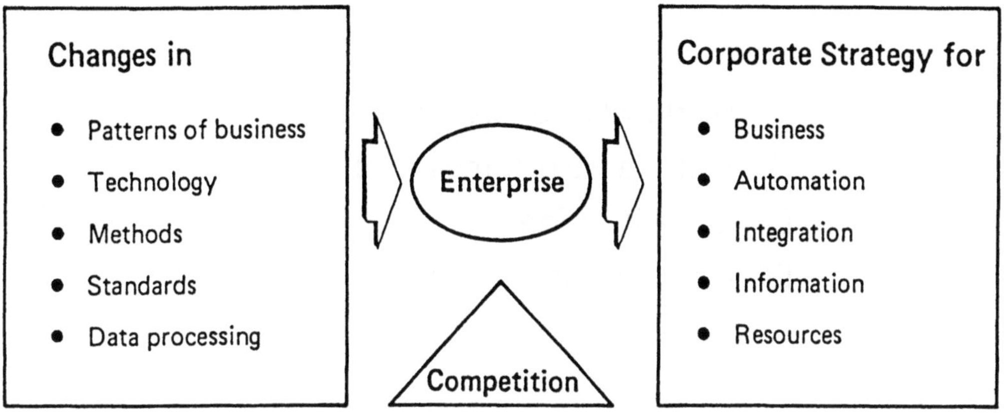

What has changed?

It's the total environment of the business.

Think about the changes of

the pattern of the business
the technology
the methods and standards
the DP technology.

Changes have occurred faster than ever before in these areas.

If you don't cope with these changes, - your competitors will.

Therefore, we need a corporate strategy for the business.
This strategy must cover the aspects of

- Automation
- Integration
- Information and last not least
- Resources.

A Corporate CIM strategy will help you to protect your investments and make the best use of them.

Let's look more closely at the changes.

Changes in Patterns of Business

- ☐ Worldwide co-operation
- ☐ Multi faceted company relationship
- ☐ Diversification
- ☐ Government sponsorships
- ☐ Customization

What has changed in the pattern of the business?

New means of transportation and and communication make today's world appear smaller. The speed of exchange of goods or information has drastically improved. Therefore, world-wide co-operation is feasible - and in some segments, it has become a MUST.

Many companies also see a changing Business partner relationship. Vendors and suppliers play more important roles than ever before. The business of today asks for services that hardly any company could provide with its own resources.

There is an increase in product and customer diversification. Goods and services are required by many new markets, demanding also broader product spectrum and shorter cycles. Think about the changes in your company. Are you satisfied with your ability to cope with your customers' demands?

Sometimes, governmental support of industrial activities, e.g. funding of research, development or recommending or regulating certain product use, unbalance the market situation.

Your Corporate CIM strategy must address these Changes of the Business Patterns.

Changes in Technology

Manufacturing Technology

- Materials
  - Plastics
  - Fiber
  - ETC
- Material shaping
  - Laser
  - Electronic discharge machine
- Manufacturing process
  - Micro
  - Image
  - Automation
  - Transportation
  - Storage
- Information
  - Communication
  - Data
  - Application
- Biotechnology

The changes in technology are evident. Drastic changes have occurred in

Use of new material
Methods of shaping
Manufacturing processes
Information processing

This foil highlights some of the most recent technological changes, everybody is faced with.

The speed of innovation is ever increasing, thus cutting the time available for getting acquainted with the new technological environment.

Changes in Methods

- ☐ Simultaneous engineering
  - Design for manufacturing
  - Design for serviceability
- ☐ Continuous flow manufacturing
- ☐ Just in time
- ☐ Modelling/Simulation
- ☐ Group technology
- ☐ Closed loop processes
- ☐ Remote control
- ☐ Zero defect
- ☐ Inline quality
- ☐ Inhouse logistics

With the advent of new technologies, the use of new methods became possible and they now form the new manufacturing world.

Product development is not a sequential process anymore, e. g. the engineering activities overlap with those on the manufacturing floor for the same product. In the manufacturing process itself, new methods of production are introduced.

"Lot size one", a magic philosophy allows each product to be different from the previous lot. Economic order quantities or lot sizes are terms of the past.

"Continuous flow manufacturing" with a "Just in time" component supply are key words to the new manufacturing community. These new concepts in turn ask for higher flexibility of the processes and of the equipment used.

To obtain maximum benefits of the high investments in machinery and equipment, modelling and simulation must replace physical prototyping. The simulation and modelling capabilities in a CIM environment also allow the representation of a line's behaviour well in advance.

The changes in methods provide the evidence for the need of CIM. CIM allows you to "close the loop", to control the process from a remote place, to guarantee high quality thru a comprehensive and transparent logistic system.

Changes in Standards

- ☐ Social standards
- ☐ Environmental regulations
- ☐ Inter/Intra standards
- ☐ Engineering standards
- ☐ DP standards
- ☐ ISO / OSI / MAP
- ☐ NBS / ANSI
- ☐ CEN / CENELEC / ESPRIT

Standards, often the representation of a traditional way of doing things right, are under thorough review.

While people (employees and workers) change their standard of living and their view of work and engagement, governments and the society express their environmental concerns and come up with new regulations. Computer controls allow adherance to the new values.

But changes also occur to the standards of your internal processes. It is vital to follow external and internal changes of the standards.

There is a growing interest in standardizing key portions of CIM, to allow communication of Hardware and Software from different sources. Some of the abbreviations, you might have met already listed here:

ISO	- International Standards Organisation
OSI	- Open Systems Interconnection
MAP	- Manufacturing Automation Protocol
NBS	- National Bureau of Standardisation
ANSI	- American National Standards Institute
CEPT	- Conference Europeen des Administrations des Postes et des Communications
CEN	- Committee Europeen Normalisation
CENELEC	- Committee Europeen de Normalisation Electronique
ESPRIT	- European Strategic Program for Research in Information Technology (Pilot Project)

Changes in Data Processing

- ☐ Price/Performance
- ☐ Distributed concepts
- ☐ Networking/Connectivity
- ☐ Application enabling/Portability
- ☐ Data transparency
- ☐ Personal computing
- ☐ Expert system
- ☐ "Third party" involvement
- ☐ Open system concept

An industry, which has changed significantly and has demonstrated the dramatic impact of innovative changes is the DP Industry.

Within 2 decades, the 'punched card' tabulating scenario of the 1960's has completely changed to a full on-line, paperless, display driven environment.

Cuts in prices and increases of capabilities by the thousands led to a dramatic change in price/performance. This now allows distributed concepts, intelligent workstations and personal computing to the vast majority of the field users.

These offerings require networking and connectivity capabilities, which ask for standards and comprehensive product line commitments from the DP equipment supplier.

Traditional programming effort is reduced more and more by the use of 'enabling packages'. These enablers allow also the transport of the logical function and the data responsibility back to/from its source.

Expert Systems assist as multipliers of knowledge in many domains of the DP environment.

There is an open system concept defined, which allows many independent suppliers to provide Software and/or Hardware extensions.

Competition

- ☐ Worldwide competition
- ☐ Responsiveness to customer demand
- ☐ On-time delivery
- ☐ Lowest price
- ☐ Best quality
- ☐ Innovation

Competition has grown to a new dimension.

It is world-wide competition now. The marketplace has grown - and so did your competitors. Worldwide markets require responsiveness to worldwide needs and the ability to cope with different manufacturing environments in other areas.

The acceptance of a product often depends on the capability to respond to individual customer demands. These demands could be in functionality, price, performance or appearance. The product spectrum has enlarged.

Only products, that are delivered on time with the lowest possible price and the best achievable quality will survive.

All this underlines the need for an innovative strategy in all areas of the business - engineering, design, manufacturing, marketing and sales.

And last not least, don't forget your responsibility to maintain your diversified products for optimal customer satisfaction. The ability for follow up support of your product motivates your customer to select you as a reliable supplier.

CIM is key to be leading in the marketplace.

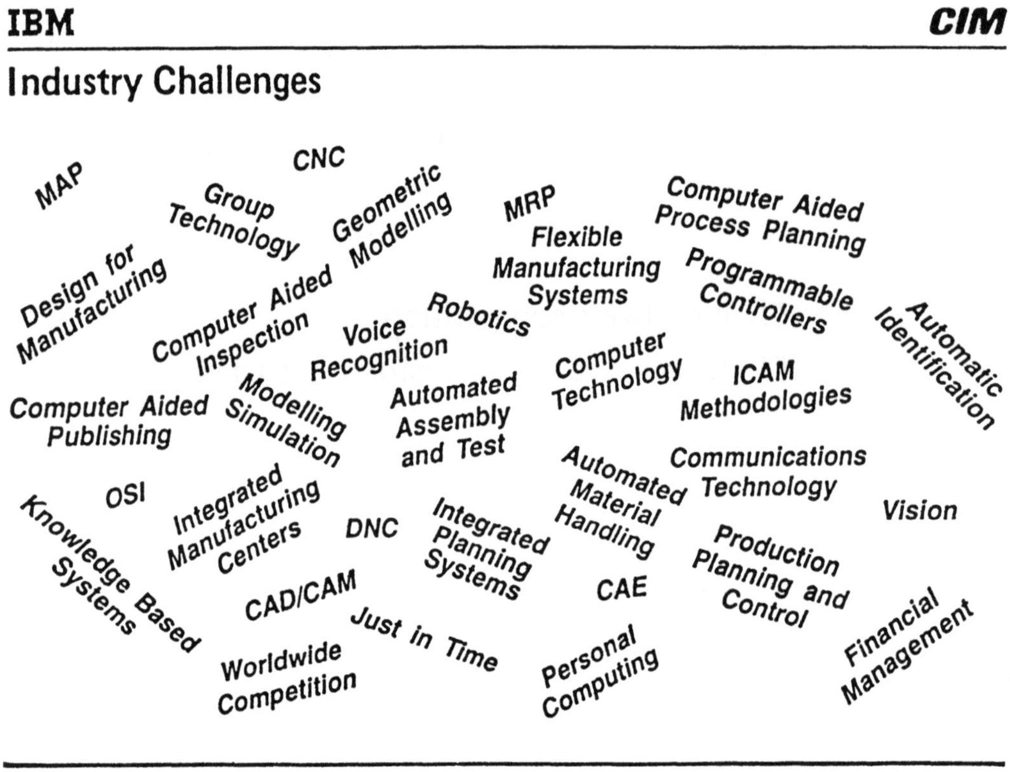

This chart gives an impression on the many recipies, from which the Industry must select from to survive.

Which one is most significant?

Where should you start?

Which one ensures your company remains in the marketplace with an extended share?

There is no single answer to these important questions. But there is help available.

You need a reliable and competent partner.
A partner, who not only understands these challenges theoretically, but a partner, who also works himself on integrated solutions and uses the concepts in its own manufacturing environment. Ideally, he also follows and offers a comprehensive CIM product strategy.

For most challenges, there are different ways to succeed.

Only well tested and well proven concepts will survive, protect your investments and assure you your expected benefits.

IBM has experience in this challenging task. IBM can assist you in finding your way thru this jungle.

CIM Approach

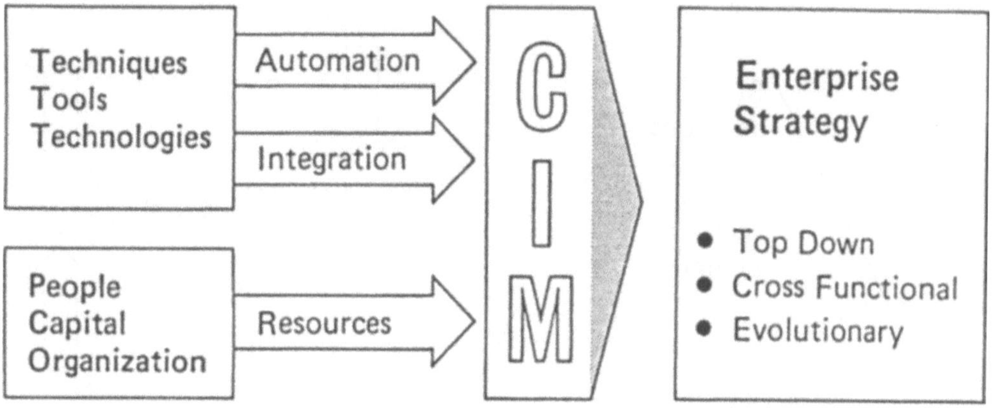

Your approach to CIM requires a clearly defined enterprise strategy. CIM will not happen by accident - or by migrating individual concepts on a low level, low key activity.

Your enterprise strategy towards CIM must be TOP DOWN, it must be a cross functional strategy with strong management backing and control.

The implementation then will occur bottom up in an evolutionary process. But it is important to understand, that all CIM components must be approached in a co-ordinated, in a well orchestrated, planned way.

Automation alone is not CIM! Nor is it the integration of islands or only the better control of resources. All these topics have their individual justifications. But CIM is more.

Only an integrated, cross functional approach, well planned in its entirety will lead to a successful CIM solution.

A CIM enterprise will result in higher effectivity, maximum flexibility, will assure prudent and intelligent use of all resources.

CIM Enterprise Structure

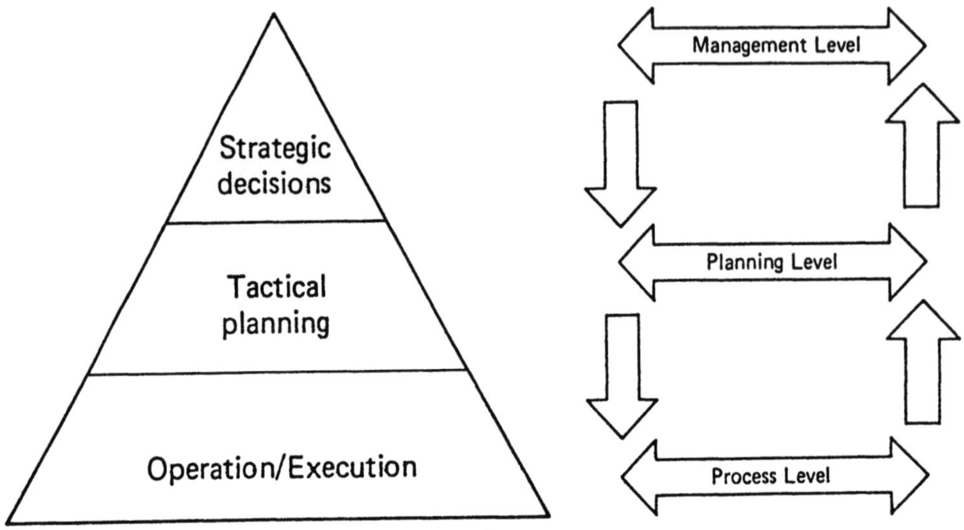

Have a look at your enterprise structure and how it basically operates.

There is the management level on top. Management has to establish strategies, decisions and directions.

The planning level converts the management directions into a plan. This conversion usually consists of various iterative processes, until all parameters are well balanced and the plan becomes the operating plan for the operational level.

The process level then executes the plan - it operates according to the plan - and in doing so, detects planning deficiencies.

All levels must have a good communication not only within the level but also to the other levels of the enterprise. Directions and instructions by management must be communicated fast to the appropriate planning functions and the operative level.
This is the top-down information flow.

Feedback information, coming from bottom up is thoroughly needed to maintain the strength and identify the weaknesses of your plan.

A CIM environment assures you of the information flow. It guarantees a controlled, closed loop type information exchange. CIM identifies the responsibilities for creating information, and often automatically creates the data, and it defines the routing of the information thru the infrastructure of your enterprise.

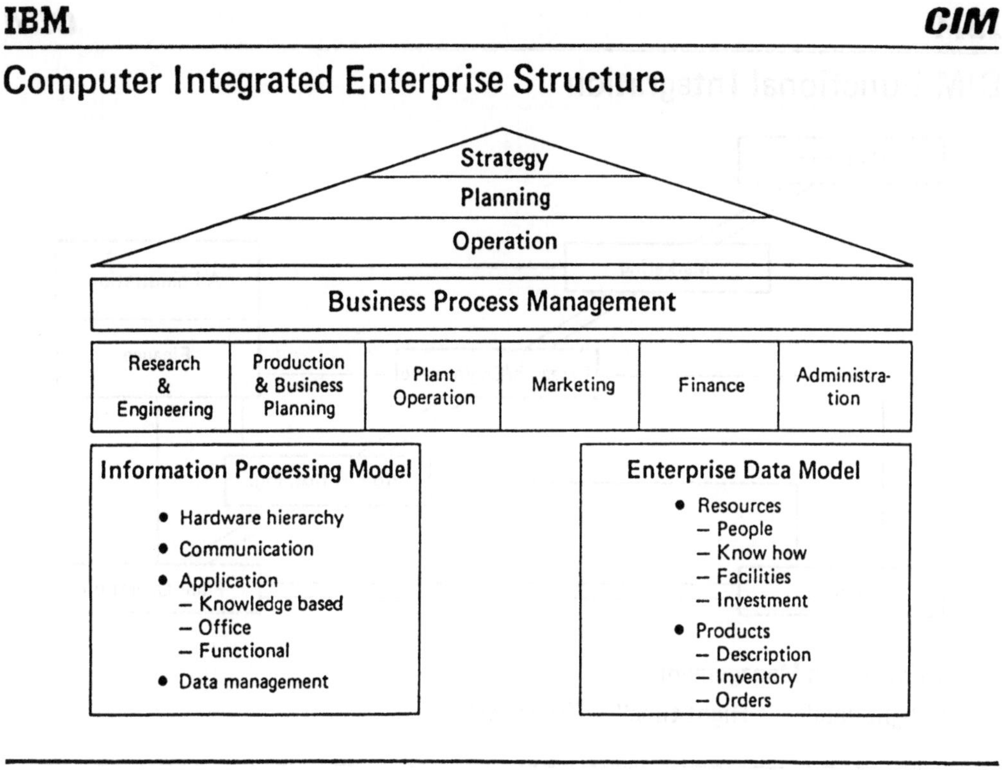

This picture shows in more detail the CIM relevant structure of an enterprise and the CIM components.

On the top, you recognize again the 3 levels

Management level - to provide Strategies and Directions
Planning level - to develop the Operating Plan
Process/Operational level - to execute the Plan

We describe the communication model across the functions as Business Process Management.

Within a CIM environment, Business Process Management is based on an

Information Processing Model and an
Enterprise Data Model.

The Data Model contains the description of the products, processes and resources (in its widest sense) of your enterprise.

The Information Processing Model connects this information, ties together the relevant data, defines the rules for decisions and actions. It comprises hardware, software, communication facilities, applications and the physical data management into an integrated environment - which in its entirety means CIM.

IBM **CIM**

CIM Functional Integration

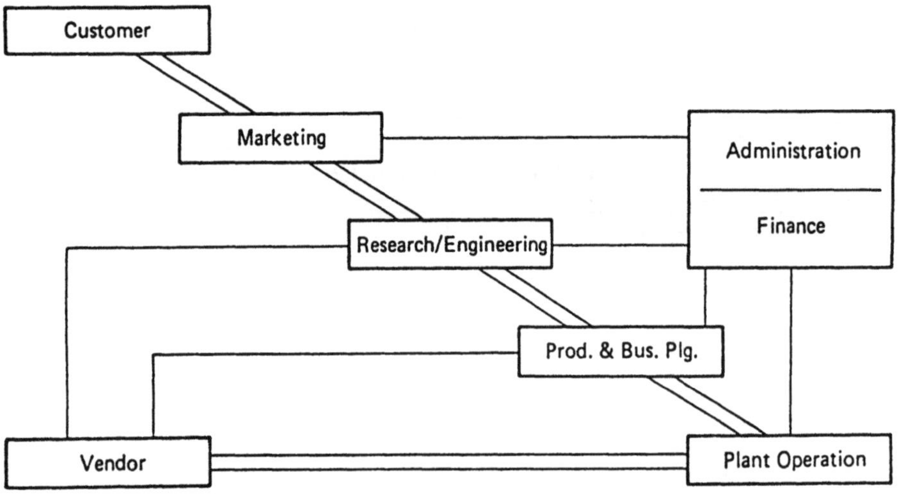

Information Management
"Right data" – "Right time" – "Right place"

So far, we mainly concentrated on CIM w i t h i n your enterprise.

But CIM goes beyond the functional integration of your 'in house' departments.

CIM assures an integral understanding of your customers' request thru all functions. This starts at your Marketing organisation, thru the Research and Development Function all the way down to the Plant Floor. It synchronizes your internal activities with the contributions of your vendors.

Your administrative and financial data is created as a by-product keeping your records up-to-date and so allowing for effective and timely control of all activities.

IBM CIM

Business Events (Overview)

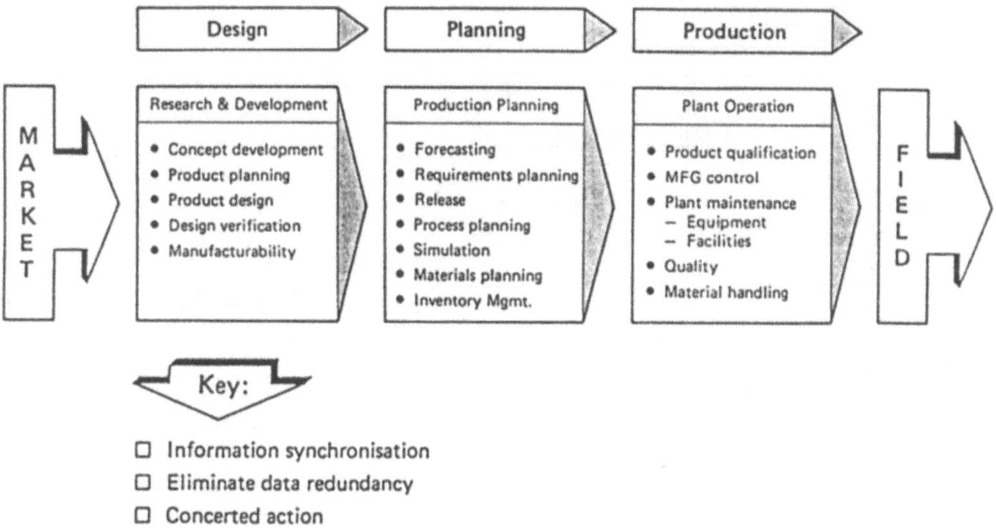

Let's look closer into our business events.

CIM supports a straight forward approach. All phases of the product require the synchronisation of activities and the elimination of redundant work. Each function involved should work from the same set of data, putting added value to the product or the process, eventually guaranteeing field satisfaction and fulfillment of your enterprise objectives.

Data is a vital resource. This resource needs careful attention. With CIM, all participating functions are concerted, contributing on time, their share to the end product.

The next foils show the functional share in more detail.

IBM **CIM**

Research & Development

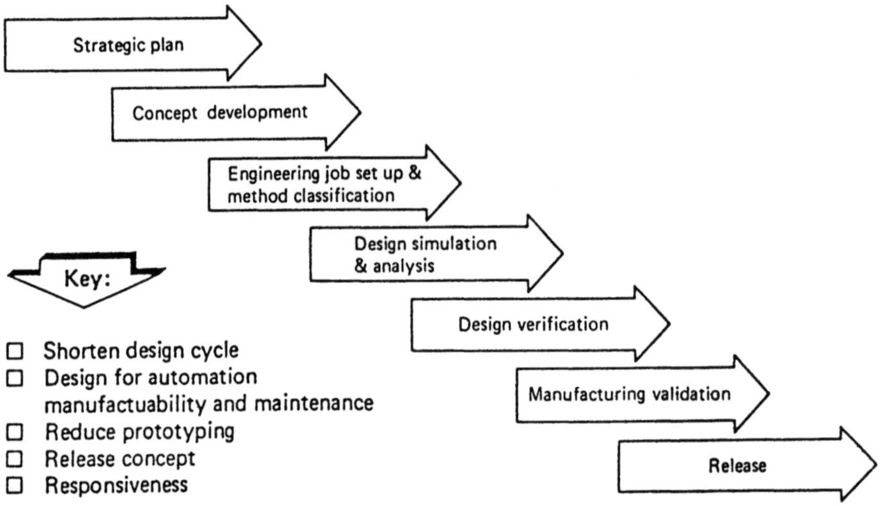

- ☐ Shorten design cycle
- ☐ Design for automation manufactuability and maintenance
- ☐ Reduce prototyping
- ☐ Release concept
- ☐ Responsiveness

The Research and Development function has the objective to generate the product description for further use in the manufacturing process.

A product concept must be detailed to a final release. The better this process is supported by CIM, the better the key factors on the lower left hand side of the picture could be achieved.

These key factors are directly derived from the overall industry challenges, we looked at in some of our introductory foils of this presentation.

What are these factors - and more important, how do they impact the business? - which of the business challenge do they address?

Shortening the design cycle	- guarantees your responsiveness to changing customer and market demands
Design for automation manufacturability and maintenance	- allows for higher competitiveness and lower cost
Reduced prototyping	- leads to lower cost, shorter lead time, better quality thru simulation/modelling
Release concept	- assures the integrity of design, responsiveness to changes and so customer satisfaction
Responsiveness	- means immediate reaction on demand and flexibility

IBM CIM

Production Planning (Logistic)

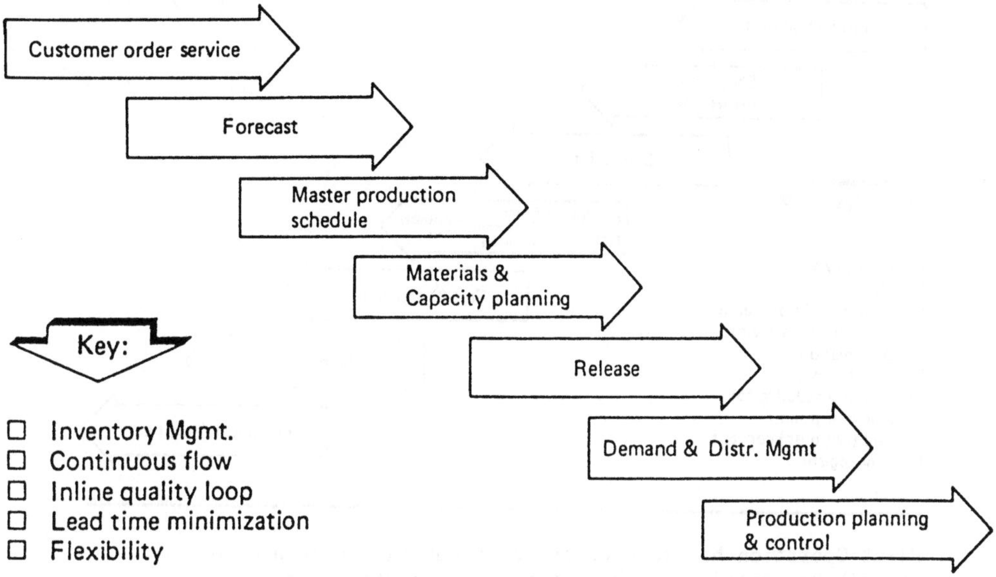

Production planning has to balance between minimal inventory and maximum capability to deliver on time, cosidering line capacity, quality and cost objectives.

Customer orders and product forecasts have to be converted into a Master Production Schedule, allowing changes and modifications of products according to the demand, supporting the 'responsiveness' objective.

Some of the key words in achieving this critical goal are:

- continuous flow manufacturing
- just-in-time supply management
- KANBAN
- lot size 'one'.

CIM allows to balance the conflicting objectives of the Production Planning function.

IBM ***CIM***

Plant Operations

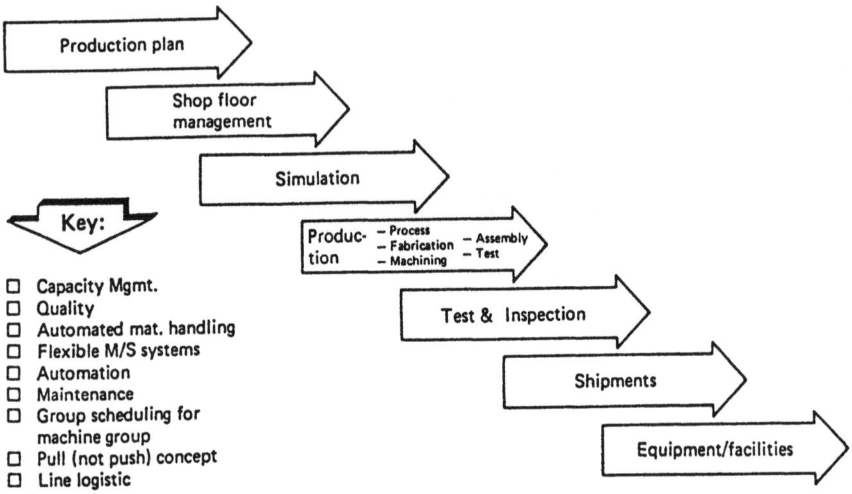

Plant Operation has to execute the production plan and so contributes finally to the enterprise objective, to

deliver	the right product
with	the right cost
and	the right quality
on time.	

The plant operation could prove or disaprove all the planning activities of the other enterprise functions. With the increasing requirement to react more flexibly on any situation, like a customer or demand change, a product change, an equipment change or a change in operator skill, an integrated CIM data model is vital for the plant operation.

Planning Data must be downloaded to the line to assure the lines' performance to plan. The same is needed for the product data. The product has to be a solid representation of the engineering data belonging to it.

Simulation and modelling techniques need life data from the line to see well in advance the behaviour of the environment.

Parameter and statistical data from the plant operation must be fed back to the planning level.

Equipment and facilities must operate also according to plan. Unscheduled maintenance or repair activities must be avoided thru preventive, scheduled actions. While an equipment failure in a non integrated, buffered and independent environment causes normally only 'minor' damages, in a CIM environment, under 'just in time' concepts, the exposure to the total floor is much more serious. CIM again collects all relevant equipment data, compares it to performance objectives and considers these new parameters for the next planning cycle.

CIM is the base for an optimal Plant Operation.

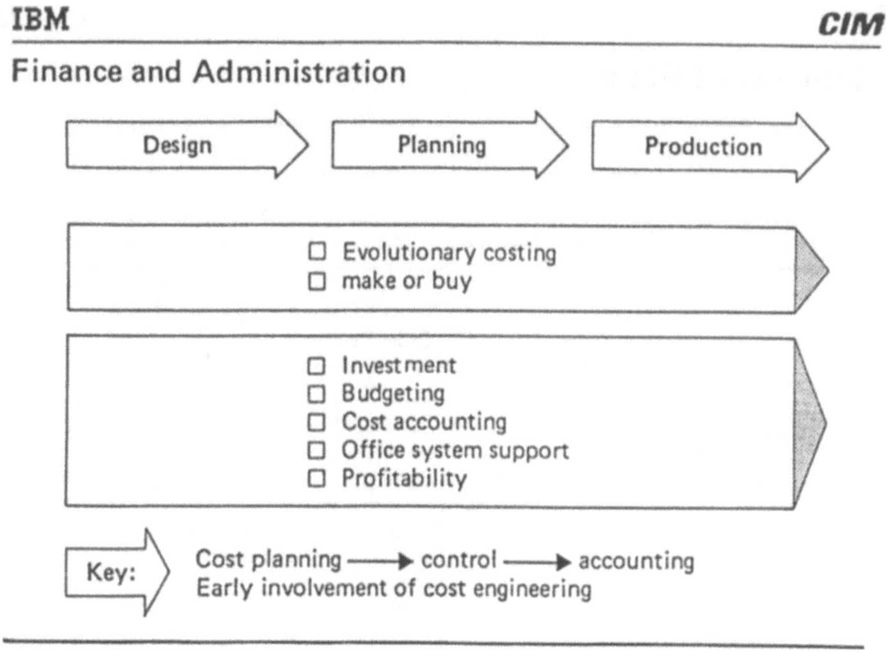

Some scholars exclude Finance and Administration from the CIM scope. We don't.

Finance and Administration must be part in the CIM process.

In a highly automated and integrated environment, the majority of product cost is defined during early phases of the product and process design. Once the product and its associated processes are firm, there is very little potential to impact the cost in the line. The material, the process steps and the equipment to be used are fixed. And because these are quantifiable cost factors, the product cost are firm as well.

Cost engineering must be involved very early in the product design activities. The controller has to participate in the design process to follow the product cost from the very beginning and to control profitability of the project. As more details become available, more accurate cost projections are needed. We call this process 'evolutionary costing'.

'Make or buy' decisions need to be made. This involves further functions like Procurement, Engineering and Quality.

Investments, capacities, load and coverage are some budgeting factors in your finance and administration functions. They all need thorough attention. CIM assures integrity of the decision base.

A CIM concept, assuring fast communication between the engineering, planning and administrative functions is vital in today's industry.

IBM supplies not only applications, hardware and software and communication tools to the engineer and the floor people, but also to the financial and administrative people in your enterprise.

There is a comprehensive offering for office automation, administation in general, procurement and the evolving area of expert systems.

IBM **CIM**

CIM Functional Flow

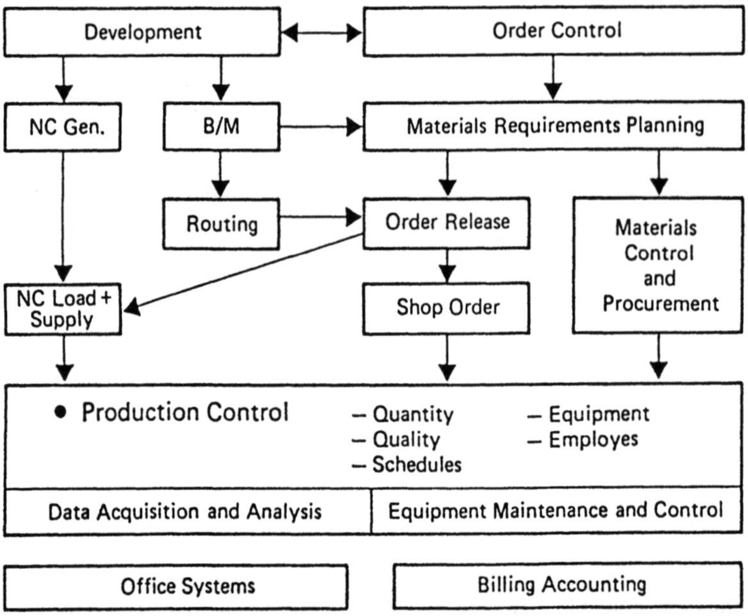

This picture - very generalized - represents the functional flow in your enterprise.

Technical Data is provided by the Development function. This contains product descriptions (e. g. drawings or Bill of Material) and process descriptions (like routings or numerical control data).

Under the topic order entry, customer orders and other relevant qantitative planning figures are related to and then converted to a production schedule, and finally to instructions to the floor.

Production Control balances and assigns the resources, which make up the production, the manufacturing process.

We discussed the role of Finance and Administration on the previous chart.

Think of the lot of information which is necessary to effectively and successfully execute all these functions. Without CIM, an almost impossible attempt.

IBM **CIM**

CIM Information Flow

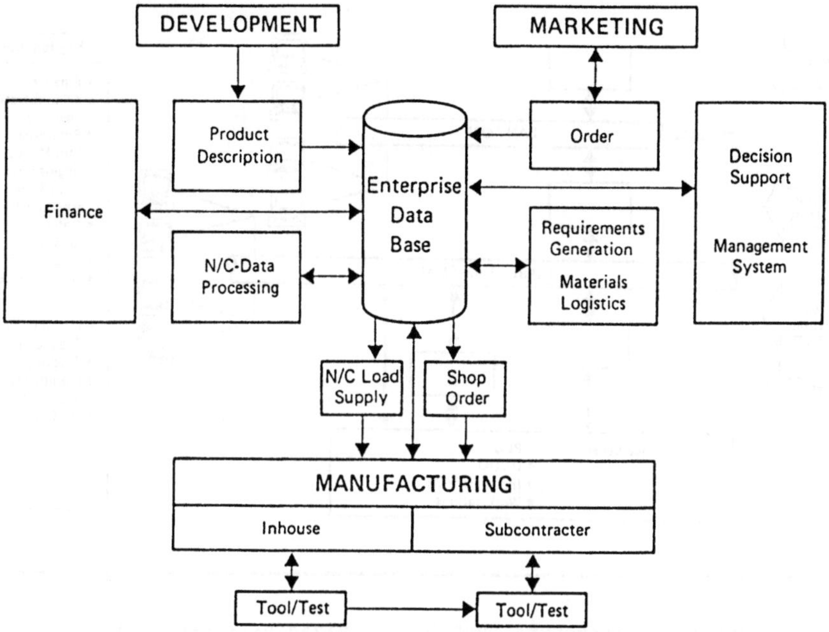

This picture - very generalized - represents the functional flow in your enterprise.

Technical Data is provided by the Development function. This contains product descriptions (e. g. drawings or Bill of Material) and process descriptions (like routings or numerical control data).

Under the topic order entry, customer orders and other relevant qantitative planning figures are related to and then converted to a production schedule, and finally to instructions to the floor.

Production Control balances and assigns the resources, which make up the production, the manufacturing process.

We discussed the role of Finance and Administration on the previous chart.

Think of the lot of information which is necessary to effectively and successfully execute all these functions. Without CIM, an almost impossible attempt.

IBM **CIM**

Enterprise System Model

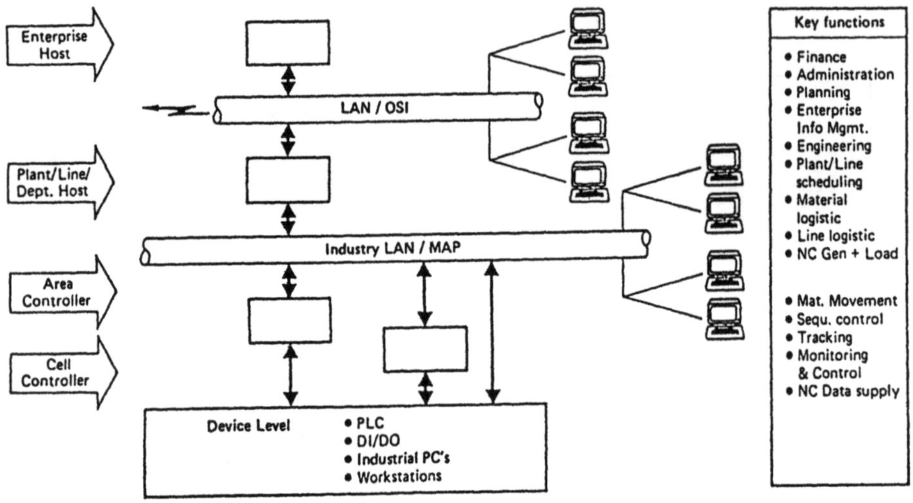

The CIM Systems architecture must be organized hierarchically.

First of all, there is the enterprise Host processor. It could control one or more plant or departmental Host system(s). Further down, Area Controllers, Cell controllers and Device controllers are assigned to their respective level of control.

Each level has its dedicated task to fulfil and has its own characteristics in terms of data availability, responsiveness to demand and reaction capabilities.

A Cell Controller needs the data for a specific cell. It is not necessary, that the cell controller contains long range planning data or data for another production line. Accordingly, the enterprise Host processor does not necessarily handle the Numerical Control data for a specific piece of equipment within the line. It is sufficient to keep this data in the Area or Cell controller and advice the processors higher in the hierarchy of its availability.

In turn, the Cell controller only needs the data for the active operations (e.g. with a 1-day or 1-shift outlook). In this example the cell controller works autonomously, that means independent from its next higher processor level for 1 day or 1 shift. A connection to the higher processor is not required, unless an unplanned situation occurs.

Networks connect the processors with each other. LAN stands for Local Area Network, OSI for Open Systems Interconnection. MAP is the acronym for Manufacturing Automation Protocol.

While the upper processor levels are connected in an LAN (e. g. SNA) mode, in the floor itself, an Industry Local Area Network is used. We will further discuss the techniques in the technical section of this presentation.

IBM CIM Strategy

- ☐ Deliver building blocks & integrated systems
  - IBM standard products
  - Unique industry products
  - Open architecture
  - Application & enablers
  - Industry standards
- ☐ Provide cross systems consistency
- ☐ Utilize alternate channels to supplement and complement IBM direct marketing
- ☐ Synergy with IBM internals

Now, what is IBM doing about CIM? What is IBM's strategy for CIM?

First of all, IBM is committed to CIM. IBM has delivered and will continue to supply its customers with a wide range of products, all of them supporting the CIM concepts for any industry environment.

The 'Open Architecture' of the IBM products allow also 3rd parties to provide additional components towards a comprehensive CIM offering.

IBM has a proven track record of developing systems which meet the broad range of requirements of our customers; and we equally assure protection of our customers investments in future developments.

It is important in CIM to allow for cross systems consistency - a means to inter-connect various CIM components to a single CIM environment, not only within the operational level on the floor, but also with the tactical planning groups and embedded in the overall information systems processing strategy.

IBM also has the unique capability to test its own CIM products internally, in the development laboratories, in the sales and service organisations and headquarters' functions, and in more than 40 IBM plants worldwide, so benefitting from the experiences of thousands of 'internal customers' before giving a product to the market. This is what IBM understands under synergy with IBM internals.

IBM's CIM Commitment

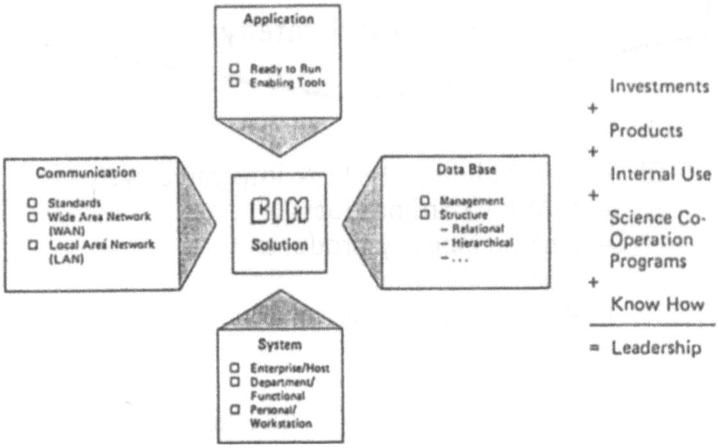

IBM's commitment to CIM is to deliver over time a comprehensive range of offerings to ease CIM implementation.

In the application area, IBM offers 'ready to run' applications. Typical examples are the CAD/CAM program offerings or offerings in the Production and Business Planning or Financial Systems area. Other functions benefit from the support of a wide range of tools that ease application development. These tools are called enablers. Under this topic, you could find different languages, software for communication, line and cell controls, robot programming, tools for structuring floor communication and expert systems shells.

The Data Base is key to a successful CIM operation. IBM's Data Base Systems support all requirements for effective Data Base Management. There are various techniques for data representation. All of them support the integrity of the data and allow extension according to the maturity of a CIM installation.

The IBM hardware offerings cover all levels of an enterprise CIM environment. Beginning at the top with high performance large enterprise host systems down to the Personal Computers and/or workstation level. In between, you find a wide range of departmental or functional computers, tailorable exactly to the needs of your problem or your environment.

But it's the software, which makes the system work for you. Depending on your situation, the IBM coverage is the broadest possible, supporting the well established /370 or System 3X architecture, as well as the UNIX/RSC architecture.
The requirements to CIM differ from enterprise to enterprise, and so the solutions must be tailorable accordingly. The IBM product line offers a full range of hardware architectures, that cover the needs of small business as well as those of large corporations, acting on a world-wide basis.

CIM means also communication. Data must be shared by multiple functions. An accurate and timely information exchange asks for standardisation of communication. Internally, we talk about the LAN, a local area network; external communication must be supported by a wide area network (WAN). IBM has facilities to support all major networking standards and IBM continues to standardize communication protocols.

IBM's commitment to CIM is manifested in IBM's investments in CIM products (hardware and software), the internal use of CIM, the scientific co-operation programs with universities, which help to recruit the univeristy graduates wiht the necessary theoretical know-how for future offerings. All together adds to IBM's leadership in CIM.

IBM CIM Emphasis Products

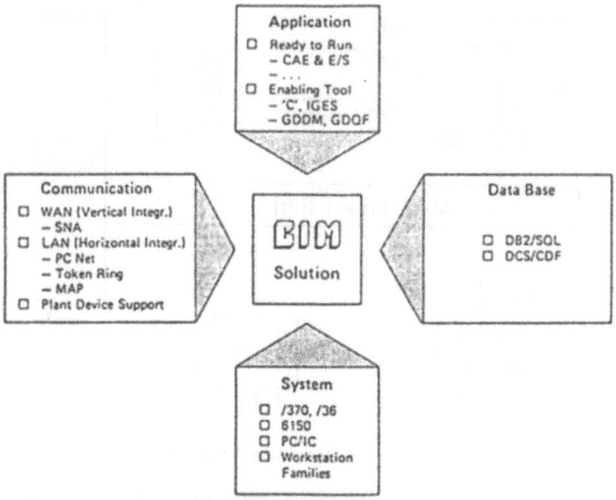

Here are some samples of IBM products in the CIM segment.

These offerings are key to making your investment in CIM pay back what you expected.

The integrity of a CIM environment could be thoroughly exposed, if 1 component of the total installation does not perform in harmony with the other building blocks. CIM is integration, so the strength of your architecture only comes across, when all products are fully harmonized.

As we saw already earlier, your Data Base plays a key role. Traditional Data Base concepts are not sufficient. The power of IBM's relational products are needed. DB2 and SQL have proven their ability in CIM environments. DCS and CDF are Data Base enablers for sharing data from different applicastions.

DCS (Data Communication System) and CDF (Consolidated Design File) are typical examples, of how IBM internally developed packages have proven their ability in IBM's lab/manufacturing environment and are now made available for customer usage.

This type of systems development will continue to provide solutions which have proven their usefulness internally to the benefit of the customers.

This is true also especially for the applications and communications area.
IBM's unique concept of single Laboratory responsibility, and our manufacturing concept in many countries, require a high level of applications and communications facilities. These areas of CIM have a long tradition, and the expertise of IBM engineering and manufacturing people are reflected in many of the products available to everybody now.

On the system side, you have a wide choice of hardware, each box fitting a specific segment of the business, depending on your requirements. And there is a growth path. The 9370 architecture allows departmental computing with the concepts of a typical mainframe processor. Workstations and RISC processors support the use of many commonly available applications from many different sources. Key to an effective use of all these components is cross systems consistency and the definite commitment for supporting the strategy long term.

System Infrastructure Model (IBM Offerings)

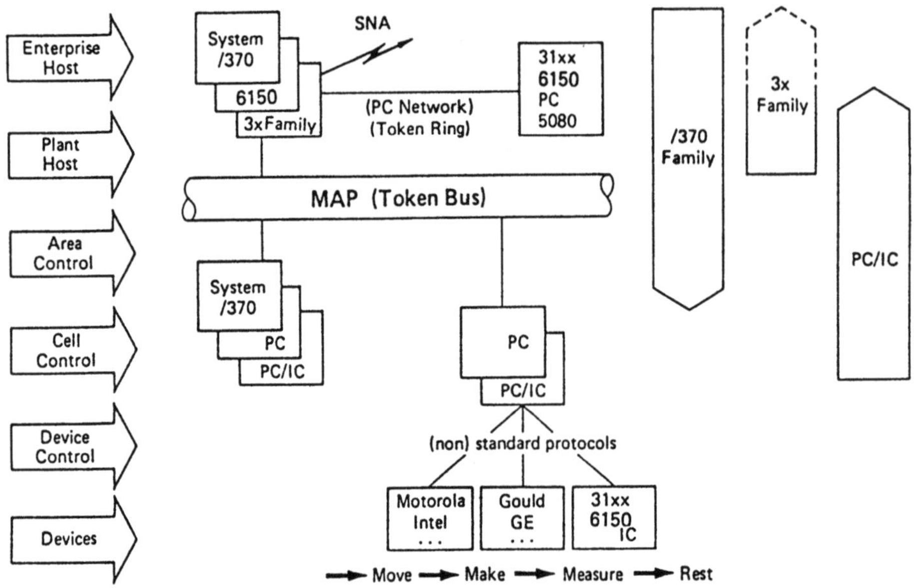

Here, you see an infrastructure Model for the IBM Hardware offerings.

Let's start on the very left hand side of the picture, where you see the level hierarchy. At the top, you find the enterprise host, usually a /370 processor which supports the total enterprise administration, including e. g. corporate financial systems and the office automation, text, business graphics and communication to external organisations.

Plant / Site host processors might be located at the plant, or site level in a multi-site environment. All site relevant activities are controlled by the site host, synchronizing internally the site information in both ways, upwards to the enterprise host and downwards to the floor controlling systems. Typically, an Engineering system (e.g. CAD) or a PPS System (COPICS or MAPICS) resides on a site host.

The communication to/from the lines is controlled by the industrial Local Area Network (LAN), here represented with MAP, the strategic solution for a standardized communication between host and floor processors and among the controllers in the floor.

At the Area, Cell and Device controller levels, dedicated high performance controllers are required to guarantee the flexibility and reliability of the processes. The IBM Industrial Computer is a typical example for a floor controller. Ruggedized design and the experience of our own manufacturing floor applications make it perfectly fit to the unique conditions in the work centers.

Some of the areas covered by the IBM Industrial Controller are Process Control, Energy management, Quality control, Logistics, and the maintenance of plant components.

IBM hardware families overlap, as you recognize on the right hand side of the picture. Depending on the specific requirements of many different customers, a tailored IBM hardware selection with the possiblity to further extend and grow gives a maximum guarantee that your investments will be by IBM's CIM strategy.

Communications Model (IBM Offerings)

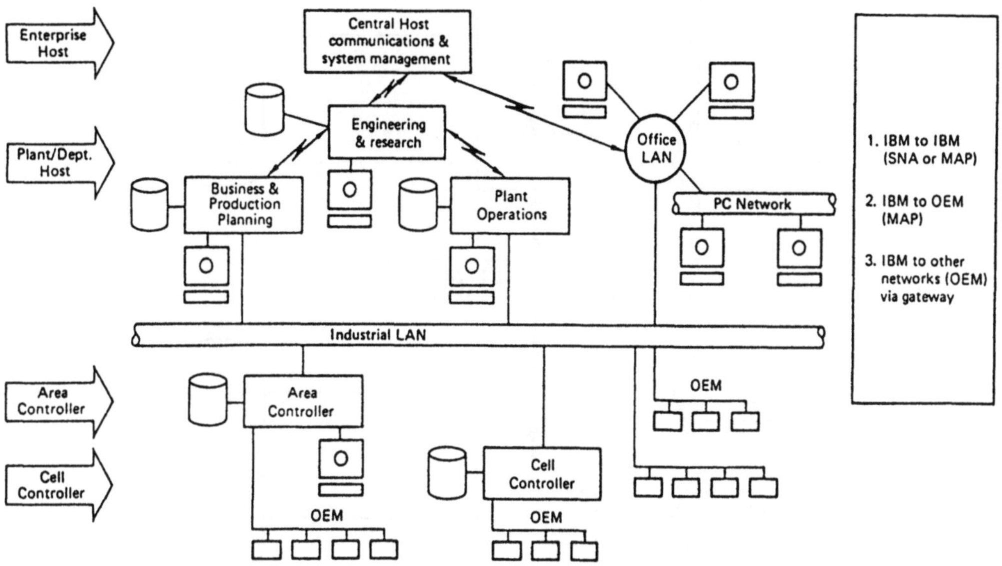

Here, you can see, how your CIM strategy could assure the communication among the various processer levels and between the different applications.

The communication requirements vary widely. IBM has a comprehensive offering for proprietory communications plus the necessary gateways to other networks.

The IBM SNA (Systems Network Architecture) and IBM Token Ring Network offerings are used for IBM to IBM communication. MAP is an IBM offering, proposed for use in the floor communication on your premises. Another LAN (Local Area Network) has to control the Office communication (Office LAN). Here, the PC network has been very successfully installed to interconnect Office Work-stations.

If you have OEM controllers installed in your floor equipment, an IBM PC based gateway should synchronize the device controllers with MAP, so ensuring the long-term strategic CIM investments on your driving systems.

Data Base (IBM Offerings)

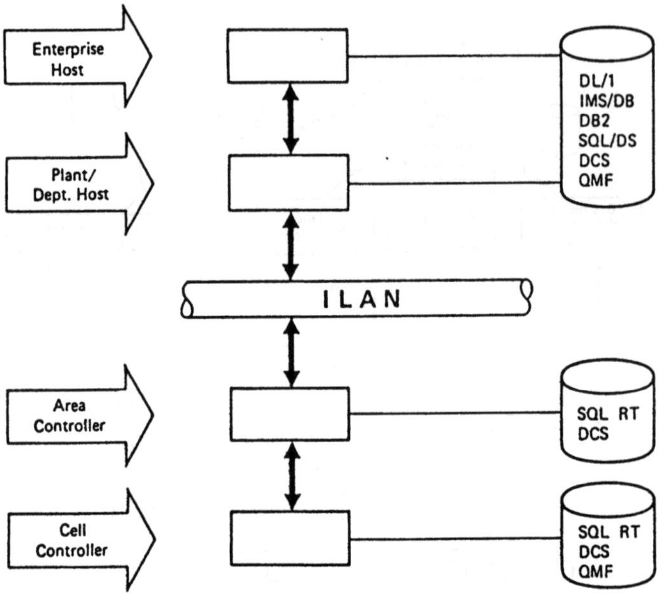

The Data Base Systems must support a reliable storage of the data and, depending on the business needs, fast and accurate retrieval and update capabilities. The data must be accessible on line, redundant data should be minimized and the system DB software must assure long term protection of this vital resource to your enterprise.

Relational DB Systems are a fact. Although current applications still might refer to DL/1, IMS DB or even flat files, a CIM strategy must define the evolution of your data and the responsibilities for the data in you processor hierarchy.

The key objective for the Data Base system is the vertical and horizontal connectivity of functions and information, tying all elements together to an integer and up-to-date information library. IBM has the products to support this objective and to provide the power to manage large data volumes.

IBM **CIM**

Data Integration Model

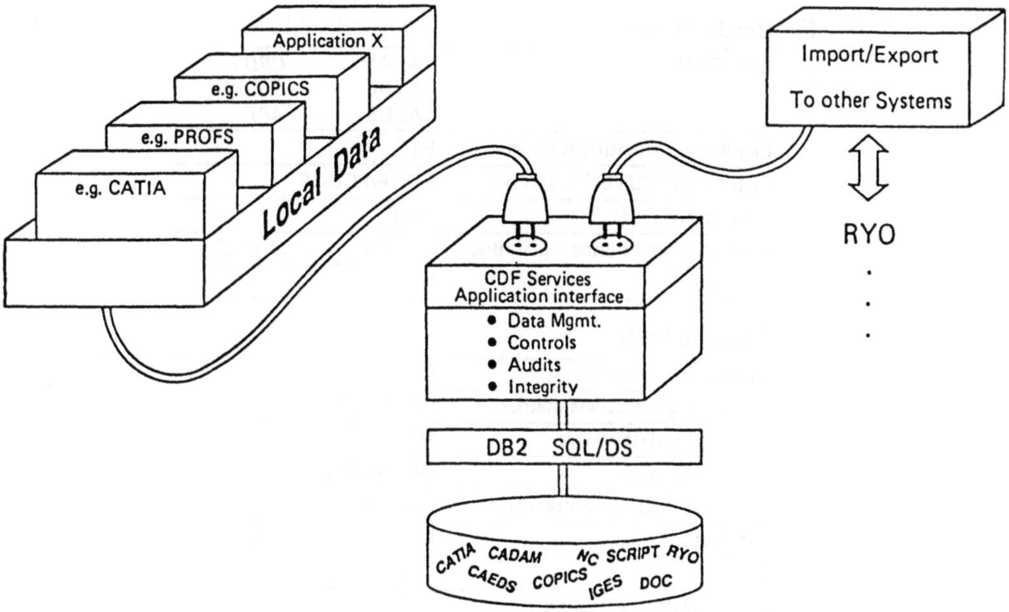

Many of us are still used to think in applications. The application is a combination of data with certain logical connections.

Under CIM, integration of data is much more important, than the individual combination of data and logic. Certainly, every application needs a specific set of data elements, but these elements are not proprietary to the specific application. In a CIM environment, the Data is a common resource, open to every application or user, who qualifies to use the data. And this view includes not just the traditional data as numbers, quantities, vectors or codes, but the text and voice data as well.

And these data need to be shared by the various in-house applications, as well as with our vendor and/or supplier systems. CDF Services, the software for a Consolidated Data File provides all applications with their view to the Data, supporting the Data Management aspect and the necessary services to assure the integrity, correctness and logical connection of the data.

Necessary routines for security and safety are included to protect the data from any unauthorized access.

Applications (IBM Offerings)

- Ready to run

CAD/CAM:	CADAM CATIA CAEDS	*CBDS* *CIEDS* *APT*
Engineering scientific support:	ES 3	
Distributed GX/TEXT/DATA:	GDQF/PGDQF	
GX exchange std:	IGES	
Production and business planning:	COPICS	*MAPICS*
Finance:	IFPS	

- Enabling tools

Languages/Generators	C	*CSP*
Engineering GX access method	GRAPHIGS	
Engineering data service:	DCS	
Presentation GX:	GDMM/GKS	
Plant floor application services:	PACS/PCS	
Data base:	DB2	*SQL/DS*
Query:	QMF	
Logic and decision support:	ESE	

It is not possible to show all the available IBM application offerings on one (or even several charts). It would blow the frame of this presentation by far, if we would attempt to give an impression of the available packages.

IBM basically provides two sets of helpful application code. The one is set is 'ready to run' applications, which in essence only need tailoring and interfacing to your existing environment, while others, so called 'enabling tools' assist you in building your own set of applications.

'Ready to run' applications usually cover a complete set of functions, for a common problem area. The systems have sometimes to be customized or the necessary functions have to be selected from various options. For very unique problems, the 'enablers' allow fast implementation providing basic functions or function blocks, which sometimes the end user needs only to combine to built its own solution.

Justifying CIM

Henry Ford: "If You need a new Machine
and don't buy it,
You pay for it without getting it"

- ☐ Plan and justify CIM investment against enterprise strategy
- ☐ Implement in stages and control single projects

There is no doubt, CIM requires investments.

These investments need to be planned and justified. You need to know, how much CIM will cost you, and where you can expect the benefits - and when.

It is very common, to concentrate only on topics with an obvious return of investment. But in doing so, you might miss even higher opportunities, which might become evident, as soon as an enterprise CIM strategy is established and followed.

There is a danger, to forget the overall enterprise strategy in behalf of 'minor' automation projects.

Your CIM Business Case must view the total effect of CIM to all functions of your business. It must comprise the tangible benefits, like lower inventory thru shorter lead time and less direct labour thru higher degree of automation. But you also must see factors hard to quantify, like responsiveness to customer demands, flexibility, synchronisation of your enterprise departments etc.

CIM needs to be an enterprise - top down - strategy, with high management approval and control, which then needs to be implemented in stages, controlled on an individual project level.

IBM **CIM**

Productivity Effects
Conventional versus CIM

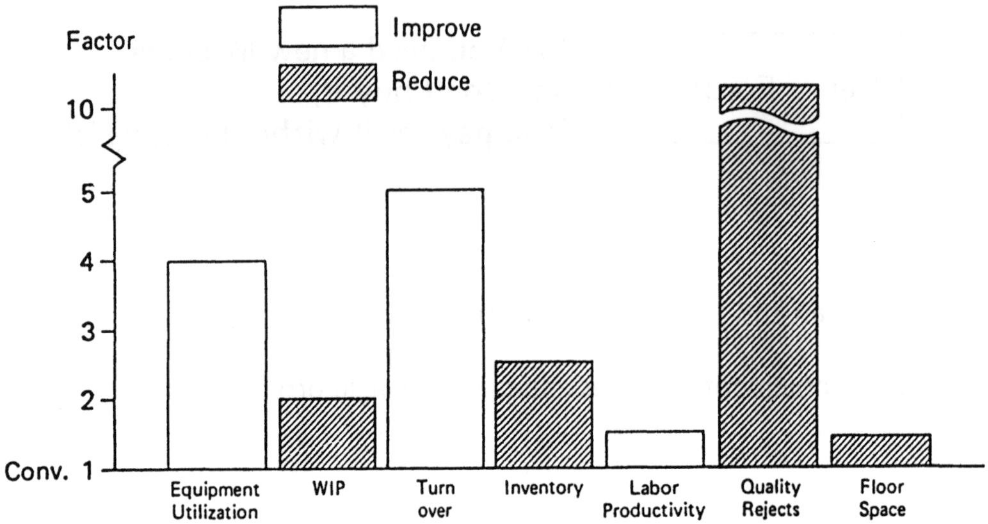

Source: Harvard Business School

What is on the benefit side of such a business case?

Some benefits have been calculated by business schools and have been proven by installations in various areas:

Among these are:

- A 15 to 30 percent reduction in engineering design cost.
- A 30 to 60 percent reduction in lead time.
- A 40 to 70 percent increase in productivity.
- A 30 to 60 percent reduction of work in process.
- A 5 to 20 percent reduction in labour.
- Quality improvements by a factor of 2 to 5 have been reported.
- The engineering productivity is claimed to improve 3 to 35 times.
- And you could expect an up to 4 times better utilisation of your equipment.

Not all factors might turn out to be applicable to your environment. Based on today's characteristics of your manufacturing process, your CIM strategy must identify the expected improvements.

Summary

Getting started

- ☐ Enterprise strategy
  - – Top down planning
  - – Bottom up implementation
  - – Flexibility in organization
  - – Shift in skill profile
  - – Treat information as ASSET
- ☐ Focus on enterprise automation and integration
- ☐ Staged approach
- ☐ Competent partners
- ☐ Personnel programs

Start n o w to develop your CIM strategy.

IBM sales organisation will be delighted to help you with advice and experience. It might well pay off, to visit an advanced IBM plant to see, what CIM could do in a manufacturing environment. Not all concepts of IBM plants might fit your situation, but IBM plants have been working towards CIM since years, and in many instances, advanced solutions have been developped and applied internally, before they became available as marketing offerings.

As soon as we realize, that information is an asset, like capital, material or resource, this asset gets the proper attention and we make sure, that this asset is used to the maximum benefit of our enterprise.

Automation and Integration must not be separated. They need to be developed simultaneously.

Next is the definition of the implementation approach. It might well take several years to implement, and over time, a review of the original strategy should certify the approach, or a revision of the strategy, considering the latest information might become necessary.

For all activities, work closely with competent partners. Not everybody talking CIM really understands the challenge, and, even more important, has experience and proven offerings.

As emphasized, CIM has a serious impact - not only on your operation but also on your people. They must understand and carry the strategy. As CIM brings new concepts, you might also need a personnel program to re-educate your people.

AUTOMATED PROCESS PLANNING

Ir. B. Furth

Metaalinstituut TNO
Laan van Westenenk 501 Pob 541
7300 AM APELDOORN
The Netherlands

ABSTRACT

Automation of process planning is developing via administrative support, the group technology supported variant systems and the generative system possibly supported by the use of artificial intelligence in expert systems. Problems not yet completely solved are CAD-process planning integration, extraction of logic or heuristic process planning rules, composing of plans by inference mechanisms and interfaces with production control systems.

1. Introduction

In 1984 a proposal for a major new project was presented to its sponsors by CAM-I.* The project was referred to as "IMPACT" and it concerned probably the most advanced proposal for Integrated Manufacturing Planning And Control Technology at that time [1]. Budgetwise it may also have been the largest Automated Process Planning project as the required funds were estimated at somewhere between $20-30 million. (Although the project has not yet started). The basic philosophy is still useful for achieving the final goal: automation of Process Planning (Fig. 1). The figure indicates how a computer-based product model has to be made for the Process Planning logic approach, in order to recognise its features and the related dimensional, engineering and technological data. After determination of the operational sequences and the selection of machines the required NC-programs can be generated as well as Process- and Pictorial plans. A connection that has not been indicated is the interface with Supervisory Control Systems governing production in the (partly) automated factory.

At the same time fig. 1 indicates the majority of the problem areas in the development of automated Process Planning Systems. They will result in truly

*CAM-I: "Computer Aided Manufacturing-International" a "not for profit"-organisation, managing industry-sponsored research projects. Arlington, Texas, USA.

NATO ASI Series, Vol. F49
Computer Integrated Manufacturing
Edited by I. Burhan Turksen

Automated Process Planning Systems not being available before 5-10 years from now. Totally integrated systems may take as long as 10-15 years.

The problem areas involve:

- the availability of a CAD-product model that is "digestible" for the Process Planning approach. An important task here is the classification of technological product features.
- the use of a hard- and software system that can handle Process planners experience and logic rules in an economic and user-friendly way.
- the set up of data bases of logic and heuristic rules for Process Planning. They may be company-specific, machineshop-specific (and may be even Plannersmood-specific).
- the interfaces between the various CAD-systems and Process Planning and between Process Planning and Production Control Systems.

These areas will be dealt with in this paper, along with:

- Short overview of Process Planning and the developments until now.
- Some developments taking place in the field of Process Planning.
- Future research topics and encountered problems.

2. Process planning

Process Planning is generally defined as the function within a manufacturing facility that specifies the process and information required to manufacture a part or assembly. Process plans in the machining environment may specify:

- Type of operation to be performed.
- Sequence of operations.
- Machine selection.
- Set-ups and sequence required.
- Tools and fixtures.
- Technological data.
- Calculation of machining times and costs.
- Raw materials required.

as well as other specific information.

Fig. 2 gives an example of a Process Plan. In the information-flow chain, Process Planning can be located between the design and the manufacturing function. So in principle PP is an information transforming activity: Input is an engineering part-drawing and possibly some engineering specifications regarding for instance treatments or inspection procedures. Output is a process-plan that prescribes the manufacturing method as indicated above. (Fig. 3). The Input of data from the drawing into the Process Planning function is performed using a human interface and we will see that by the nature of our drawing-conventions (also adopted by our CAD systems) this interface is highly intelligent compared to any fully automated method. Process planners should have a thourough knowledge of their machining facilities and of the ways in which production is usually performed in their organisation. It is not uncommon that experienced machine-operators are promoted to the function of Process Planner. Their experience forms the major tool for transferring the drawing information into manufacturing specifications. The logic or rather the rules used are generally heuristic which means "aiding to discover" and involves Problemsolving in the absence of precise methods for example mathematical or algorithmic. The Computers widely used in industry at present are good at handling algorithmic or decision type logic rules, but the more complex heuristic rules may require a next generation of hard- and software. The fact that Process Planning has proven to be company-, machineshop- and even planner dependant complicates the requirements for flexibility of automated Process Planning systems. Notwithstanding, Process Planning is of vital importance for the efficiency of the production in manufacturing organisations. It is the basis for machine-shop scheduling, cost-calculations, purchase-information for raw materials, tools and fixtures and generation of shop-floor workorders and required NC programs. Looking at the time manually operating Process Planners spend on various aspects of their work, we might see the following division [2]

* Technical decision 15%.
* Data look-up and calculation 50%.
* Text and document preparation 35%.

Considering the amount of clerical work it is not surprising that computer-assistance has been developed for this part of the work to cut down on the cost of process-planning and to allow the process planner to concentrate on his primary function: Technical decision making.

3. Development of computer assisted Process Planning

The development of Process Planning can be summarized as follows:

- Hand written process plans.
- Clerical support by computers.
- Variant systems. Past experience utilized by means of coding and classification.
- Generative systems, making use of Group technology and/or Expert systems.

Each transition in the development required the availability of new technologies.

3.1 Hand written process plans

This process is obvious. It requires experienced process planners knowing the company way-of-working, available machines, materials, cutting technology etc. In addition a good memory is required to recall which operational sequences have been successfull in the past, along with some endurance, as a lot of simple parts needed almost the same process plan, the writing of which was sometimes very tedious.

3.2 Clerical support by computers

Taking into consideration the large amount of clerical work a process planner had to perform, it is not astounding that use was made of the computer as soon as this was possible. The first systems used were Text-editor type programs. They allowed the process-plans to be edited and stored in an easy "paperless" manner. Generally these systems were offshoots of the systems used for the company's administration and contained little technology. Editor functions enabled the process planner to use standard blocks of text for specific operations with high occurence-frequency. This resulted in prevention of errors as these instructions were more easily recognised by the operator in the machine shop. In addition all kind of lists and look-up tables (Materials, tools, fixtures, machining parameters) could be stored in the computer and could be consulted in an interactive way. Calculations formerly performed with a pocket calculator could also be performed at the terminal in a more user-friendly way. For small and medium size companies employing only a few process planners experience was now stored in an accessible way, and this meant that their experience (generally gained by trial and error) stayed in the company and did not vanish once an experienced process planner retired or left. A group of manufacturers producing a great variety of parts in small lotsizes experienced a specific problem. Their files generally contained thousands of different process plans. It was evident that difficult new plans they had to make had predecessors of similar parts somewhere in their files. These plans had a high practical value since they had been tried out and had been proven to be acceptable. However, retrieving this valuable experience generally would require more time then making a new process plan. Making these new process plans resulted in proliferation of methods and a great lack of standardisation. This lead to the development of systems able to retrieve process plans based on technology: The Variant system.

3.3 Variant systems

Retrieval of relevant technological experience requires corresponding identification. Generally the part-geometry and some technological aspects form a possible key for adequate retrieval.
An ideal situation would exist if a new drawing could be inserted into a CAD system and it would search in its database for the part-drawings which are equal within certain limits. The matching Process-plans, will then give a close representation of what is required for the new part. The closest fitting plan will generally be adapted and this variant plan, based on former experience, has thus been made with little effort. However, CAD systems are not yet able to perform this search within a reasonable time, so another method has been invented: describing the geometry and some technological aspects by a code number. The code number consists of a sequence of numerical digits each describing a certain aspect of a part. For instance mainshape, shape elements, dimensions, tolerance-class etc. In addition lot-size, sequence of manufacturing operations, set up- and machining times could also be represented. The code number gives a general description of a part in such a way that similar parts (each with a different Partnumber) have the same code-numbers. When a new part is designed, the code-number can be generated using a set of definitions and conventions.
Using this newly generated code-number the computer can easily check which parts have the same (or nearly the same) code number. The process plans (but also the drawings) can be used for the new part to be processed.

An important use of code numbers is the application of Group technology. By means of the code numbers parts can be classified into standard groups. New parts will be designed and planned in such a way that they are as near as possible to the standard shape or standard-process plan.

Successful application of the Variant Process planning system depends on the variety of the subject parts spectrum and the possibilities for adapting the Part design to the set standards. Examples of variant process planning systems are: CAPP 3.1 (CAM-I arlington USA)

AUTAP (WZL. Aachen BRD)
CAPSY (IWF Berlin BRD)
MIPREP (TNO, Apeldoorn Netherlands)

Fig. 4 shows the structure of the last mentioned system indicating the coding and retrieval possibilities.

3.4 Advantages and disadvantages of the Variant system

Advantages of the Variant system are:

- There are systems available which have been used succesfully in practice.
- The effort to implement a system in a manufacturing facility compares favourable with the returns on investment.
- By its nature the Variant system tends to group similar parts together and handle them as much as possible in the same way, thus having a positive influence on standardisation.
- Users may extend systems themselves.

Disadvantages are:

- An experienced process-planner will be required to judge if a retrieved plan can be used, and how it should be adapted. There is not always the certainty that the plan bears a logical relation with the plan to be generated, even if the aspects indicated by the code number do bear a resemblance. In fact this depends on the effectiveness of the code number, and thus its complexity.
- The effort that has to be put into coding the piece parts. This effort depends on the complexity of the code and the quantity of digits to be coded. In general it can be stated that the more descriptive (e.g. longer) the code number is, the greater the effort needed in coding will be but also the higher the effectivity of retrieval.. A figure observed in practice roughly gives an average time of 10 minutes required for coding a part. This is the time required for interactively answering questions put forward on the VDU screen of the computer. After all relevant questions have been answered, the software system determines by its decision logic the code number. Practice has shown that a database containing between 2000 and 5000 different parts (depending on the size of the manufacturing facility) is sufficiently large to enable the use of variant process planning and some standardisation. This means, however, that in total 20.000 to 50.000 minutes or 40 to 100 man days have been spent in building up the database (be it very gradually as the Process Planning system can be used from the start). This coding is one of the subjects for automation.

  Two developments may be mentioned here.

  First there is the possibility of reducing the number of questions which have to be answered during coding. This can be achieved by standardising certain so called "Type-parts". For instance shafts, gears, U-brackets etc. Selecting these names from a menu reduces considerably the coding effort. An example is the "Aeroclass" coding system developed by TNO-Netherlands and used by Aerospatiale France.

The second development for coding is automatic coding by the computer. The CAD model of the drawing is compared with various shapes determined by the coding-rules and -definitions. Although it is not certain that a complete code number can be generated in this way, a significant part of it seems possible and the coding effort might be greatly reduced. Research and development in this field however has not yet resulted in a usable system and the question arises if it will be useful before the AI-based generative process planning systems mature to an adult stage.
But even then the role of coding and classification may not be finished. Knowledge bases containing logic rules for determining process plans, may contain such a vast number of different rules referring to various areas that only by dividing them into classes may make the systems to become managable. Selecting the right part of the knowledge base for a certain product may then be performed by coding and classification.

3.5 Generative systems making use of group technology and/or expert systems

3.5.1 General

Generative Process planning systems should be able to generate process plans based on the information provided by part drawings, and relevant engineering specifications and information regarding manufacturing resources available. The data may be provided by human input or automatically from a proper database. The process plan is generated from the above information using the process planning rules.
Contrary to the variant process planning system which tends to standardise by making variants as close as possible to the original, the generative system in extremum produces a process plan which is completely adapted to the subject part by using the rules pertaining to each specific aspect of that part. To prevent this, the rules used to generate the process plans have to be based on machining possibilities available, and machining sequences being determined as standards for the environment. Even for generative systems it is advisable that an experienced process planner checks the generated plan.
Generative process planning need not be the same as automated process planning. The latter involves automatic input of drawing information from a CAD system and the generation of the proper process plan. Automatic interfaces with production control may lead to Integrated Information Processing for design, planning and control of manufacturing; a next level of development.

3.5.2 Levels of process planning

Looking to the type of process planning systems which can be made generative we distinguish between two levels:

- Systems determining the manufacturing (NC) program on one machine.
- Systems determining the operational sequence via various machines and other operations such as heat treatments inspection etc.

Systems of the first category are in fact intelligent NC-program-generators. The problems they solve are of a few orders less complex then those of the systems of the second category. Basically they have to deal with one machine, a known set of tools, and geometrical features which are limited by the possibilities of the one machine. All these factors tremendously reduce the number of possible solutions and hence the selection of a suitable one.

By generative systems we mean, during the rest of this article, the second level systems (determining operational sequences via various machines).

3.5.3 Group technology based generative systems

Before the use of artificial intelligence in expert systems was generally recognised as a possibility for generative process planning, some systems were developed building on the work already done with the variant system.
One example is Genplan [3] developed by Lockheed Georgia USA. Lockheed utilised a special coding and classification system to describe the attributes of parts. By using the experience laid down in some 100.000 process plans it was possible to set-up a data-base of decision rules regarding the manufacturing of parts and the relation with their shape and technological requirements. Since the generated code number of a part described the shape and technological aspects it was possible by computer-analysis to generate a detailed manufacturing plan based on the decision rules mentioned before.
The process plan was presented to the planner who only had to add some minor details. Another example is Autoplan [4] developed by Metcut USA. Part of the system still utilises group-technological retrieval of historical information. However, interactive build-up of new coding- and decisiontrees made it possible to store process planning experience of a company-specific nature. The group technological information was used as input to drive the decision tree. This generated the basic information in the correct order for a new process plan.

3.5.4 Logic-rule driven generative systems

The next step in process planning system development was the use of logic rules to determine the operational sequence and manufacturing details, based on the geometrical and engineering requirements.

The rules could be expressed in decision-tables or decision-trees which could easily be handled by conventional computers.

However, practice showed that generalisation of process-planning rules was very difficult. And the question arises if there realy is process-planning logic. Not only do the rules vary between different companies with similar types of production, even process planners in the same department handle the same part in different ways. This means that adaptability of Process Planning systems has to be very high.

Decision-tables were not flexible enough to support this kind of adaptation, and a different type of rule-handling was required.

A new field in the information technology seemed to give possibilities: Expert systems making use of Artificial intelligence.

Artificial intelligence may be referred to as techniques with which computers simulate the human ability to handle information.

Examples are: automatic interpreters and speach-recognation. Expert systems basically contain a knowledge-base, in which rules and knowledge used by experts in the relevant field is stored, and an "inference engine" manipulating the rules in the knowledge base in such a way that solutions can be generated to various problems. In addition there may be consulting and adapting functions.

An advantage of Expert systems over decision-table systems is, that the knowledge base is functionally separate from the inference engine. This means that changes in the knowledge base do not influence the functioning of the other parts of the system; which is not the case when a decision-table is changed in a system of interlinked decision-tables. And as we have seen this easy adaptation is one of the main requirements for a process planning system.

One of the early process planning systems making use of artificial intelligence is GARI [5]. It is a system that is able to plan prismatic parts like for instance castings which have to be milled and drilled. The knowledge base consists of many differently weighted pieces of advice, represented as production rules. Each consisting of a set of conditions and related pieces of advice: conditions → pieces of advice. For instance (in readable language): If the surface quality of an entity is higher then 32 → then it is advised (with high weight value) to perform grinding as a final operation.

The model of a part to be planned is inserted as a set of features: for instance cylinders, faces, holes, grooves etc.
To each feature geometrical and technological data are attached. Roughing and finishing cuts are considered. GARI generates a machining plan by starting with some general rules describing a certain operational sequence. By an iterative proces assertions are produced which more and more constrain the solution to be found. If a contradiction developes by conflicting rules, one of the rules may be rejected by selecting other assertions that minimize its maximum weight. The resulting plan can be finalized when specific production circumstances like machines to be utilized are known.
Compared to traditional rule-based expert systems the control structure of GARI is rather complex.
In the same period a process planning system TOM was developed [6]. It showed the possibility to generate optimal machining sequences for a given hole geometry.
Another approach is the XPS development (Xperimental Planning System) which is managed by CAM-I Arlington USA.
XPS is a step by step development which should finally lead to a true generative expert system for universal process planning: XPS-E.
XPS-2 is a step forward in developing an automated process planning integrated in a CAD-CAM system [18]. The first step, XPS-1 was a prototype system that provided a basic generative planning capability and a framework for future enhancements. XPS-1 contained logic in the CAPPL language an English-like language specifying logic rules.
Parts data inserted in the system were recognized by means of a dictionary. The rules in the logic rule base can then be applied. Data dictionary and logic rules had to be specified by the user. XPS-2 is an enhanced version of XPS-1 including machine selection and operation grouping. The XPS-2 product definition also made use of the feature specifications, attributes and relationships [10]. The system utilized a work element library containing operation codes, work element type, standard texts for operations and data to enable initial machine selection.

4. Automated Process Planning General, problems, research tasks

4.1 General

Automating process planning in its ideal form should function as follows. The engineering drawing and -specifications of parts to be planned are available in a database having been drawn using a CAD-CAM system. Input of the partnumber and how many are to be manufactured should be sufficient to prompt the system to generate the process plan.

The format in which the plan is generated, is such that it can be presented visually to users by typing, but much more important, it can also be fed into a production control system.

This latter system supplies information for machine-shop scheduling, for tools and fixture preparation, and for materials purchase, and also initiates the issue of work-shop order documents. No human interface is required.

Dream or possibility?

As long as a part can be manufactured using one machine only and then preferably using one set-up, it seems possible to generate a process plan based on the parts features. For instance X-Plane in development at the University of Twente (Netherlands) is a process planning system also capable of generating NC programs [7].

Process planning systems being more or less automated for turned parts or sheet-metal parts are also available or in development, for example proplan [8]. The order of difficulty in these cases is decreased by the fact that the problem is practically two-dimensional.

However, as soon as we have to plan a part that requires operations on different machines and also requires for instance heat treatment cleaning and inspection at specific points during the sequence, or if there is a choice of machines and operation sequences which may be used, it is highly question-able if automated process planning can be realized within a reasonable time-span, with present possibilities.

Not only that the quantity and complexity of rules is too large to handle for our present computers. A much more serious problem is that the rules and decisions process planners are using seem impossible to be generalized. Process planners seem to be able to attack the planning problem at different levels and at each level select the information that is required to make decisions and disregard the unimportant. He starts with an overall view of the outer shape and determines main steps. Still he observes details (like accuracy or surface roughness) which are of importance for the determination of the major operational sequence. We may have to study much more thoroughly the way a human is planning before we can truly start automation of process planning. Moreover the development of computer hard and software should provide the means to handle this way of working, rapidly and in a user-friendly way.

Summarizing, areas for further research are:

- CAD-CAM interface
- knowledge base build-up
- interfaces to production control systems.

4.2 CAD-CAM interface

The CAD/process-planning interface should provide the planning system with the required information in such a form that machining sequences can be deduced by the limited means of the knowledge based sytem. Limited, compared to the ability of a human process planner with his knowledge of drawing conventions. Let us look at a simple example. A drawing like figure 5 would, in most CAD systems, be available in the form of a few vectors indicating points and lines or dash/dot-line. A process planner would recognize in this figure a piece of material with a symmetry-axis and by the default of a second view would probably conclude that the shape indicated was a solid cylinder. Once this conclusion has been reached it is easy to make a choice for machining. This is true for the knowledge based system as well as for the process planner. So a major problem is to make the system understand our drawing conventions. A solution in this direction is already available in the CAD-solid modellers. One difference which has to be dealt with is the fact that a designer generally observes the material that has to be present for his functional requirements, while the process planner will be more interested in the material difference between the blank and the finished product i.e. the material to be removed figure 6 [9].

Proposals have been made for research into the feasibility of a generative process planning system based on the surfaces of the part, specified in a CAD data-base. Relationships have to be established between the surface shapes and surface requirements on one side and the system capabilities that may be used for its manufacture on the other side. For elements of a part with a simple shape this might work. For instance cylinders, conical shapes, threading. However, if we look at figure 7,

7-1 represents a cylinder which can be made with standard tooling on a lathe.

7-2 The same cylinder now requires a different sequence of operations.

7-3 Again the same cylinder but depending on its length L a special tool with a width L may be required, or at least a left- and a righthand-cutting tool in case L is relatively long.
The conclusion may be that a part-surface based system as indicated, may have a problem in deciding how to machine this shape, unless we tell it that shape 7-3 under certain dimensional conditions is a feature named "groove" and grooves will be machined in a specific way. The human process planner appears to be working to a large degree based on manufacturing-features. And it seems that generative process-planning should start from the same basis. This means that the designer making production drawings should work with the capabilities of his specific machine shop represented in a "library" of features. The creativity of the designer is then not used to invent new shapes, but rather within the restrictions of his "machine shop library of features" to create a part that can be planned and manufactured, and at the same time fullfils his functional requirements. This way of working may require a different approach in the development of CAD systems. Two examples may be mentioned here. One is the work on features which has been subcontracted by CAM-I to Deere & Cy, Moline Illinois [10]. The purpose of this work was, to develop a logical means of describing features of an engineering part drawing from a process-planning viewpoint. The work contains hierarchical structures for features along with feature definitions and feature data requirements. An example is given in figure 8. This work is part of the XPS-2 development managed by CAM-I, which should finally lead to the development of an integrated automated process planning system.

The second development is a commercially available CAD system; it is called SMASH (Standard Manufacturing Shape) [11]. It starts with the possibilities in the machineshop, for instance rough and finishing turning or milling, holing, boring, slotting etc. Each manufacturing feature is represented in a CAD-drawing identified by the feature name. It bears all feature data which are required for manufacturing and also the limits set by the machining capabilities. Figure 9 gives an example of such a drawing, the related data and the information for the CAD representation. To conclude we may say that a development in the direction of features integration in CAD systems may help to find a solution for the CAD-process planning interface. Use of solid modelers may benefit this development.

4.3 Expert system approach

Studies into the naturew of process planning have indicated that to a large extent use is made of heuristic rules. The number of rules is considerable.

Consequently expert systems would seem to be a promising approach to automation of process planning [12]. As indicated before expert systems make use of a knowledge-base and an inference engine to produce a process plan. There are various ways in which expert systems can solve the problem of determining an operational sequence given only the part-geometry, the possibilities in the machine shop and the manner of production in that shop. The direction of working in expert systems can be forward, backward or bi-directional. This refers to the possibility of starting with the problem that is posed and ending with the solution or answer, or reverse. It is clear that for instance medical diagnosis or fault finding starts with the "answer" and works backwards to the cause. Process planning, however, generally starts with "forward chaining", using searching rules that fit the problem until the goal (a suitable plan) is reached. Rules are generally of the type if ..... then ...... or condition → action, called "Production rules".

Usually the planning problem is divided into subproblems, facilitating finding the solution. However, subproblems are often interdependent which means that the interaction has to be handled by adding constraints to the sub-part solutions found by introducing new production rules [13]. The system can provide explanations if required by backtracking the rules used to reach the result. In case of decisions not conforming to common practice, it is possible then to change or add rules in order to improve the system. Conflicts between rules can be solved by attacking weights which in turn can be changed by applicable rules. A.I-based systems give many other possibilities, however, for process planning the system described above seems appropriate. An informative reference on A.I. (amongst many others) is [14].

4.4 Knowledgebase build up

Knowledgebase build up refers to extracting, gathering, classifying and storing all information required for process planning. The information will be factual, for example type of machines available, their maximum operational dimensions and power, tools, transport means etc. and rule based, referring to the rules utilized by proces planners when generating process plans.

The rules will help to determine the selection of:

- Processes to produce the required features,
- The sequence of these processes,
- The machines to be used based on dimensions, accuracy required etc.,
- Applicable process, e.g. heat treatment, dimensional inspection, surface treatment etc.

This can be considered the first level of process planning. The second level may detail each of the items specified at the previous level:

- Number and type of set-ups,
- Determination of work elements for each set-up (for instance surfaces to be milled, holes to be drilled),
- Technological data, e.g. feeds, speeds, clamping methods,
- Tools and fixtures required,
- Miscelaneous, e.g. cooling in process measuring etc.

The next level could be the generation of NC programs. This, however, needs not to be part of the Knowledged-base system since automatic NC generating programs are readily available.

The size of a knowledge base, even for one company for all of its parts spectrum is enormous. And the build-up at present must be considered as practically impossible. Logic studies of CRIF Belgium for process planning have indicated this [15] [16]. To solve this problem the knowledge base has to be divided in smaller parts each handling a specific section of the parts spectrum and the production methods. Grouptechnology may be a valuable tool in dividing the knowledge base. To handle these various parts of the knowledge base a superior level has to be introduced: the Meta Knowledge Base handling the utilization and limits of its knowledge. The Meta Knowledge Base may be considered as a generalised system that could be used for all companies in a similar type of industry. However, the specific knowledge base will be highly company specific.

The logic extraction could be performed by asking process planners why they plan the way they do. Recording and statistically evaluating should result in a set of generalized rules useful in practice. There are, however, some serious problems. In the first place the interview. The interviewer should at least be so acquainted with the field, that he understands what the planner is talking about. However, if he is too acquainted it may influence his interviewing. He will not ask the questions which are obvious to him, and he may interpret answers within his own framework of reference. Also his questions may be suggestive in a certain direction. For this reason it has been proposed that the examiner should be a psychologist from origin with knowledge about the process planning field.

Another problem is the inconsistency of planning rules. For some operations the choice is fixed for instance a hole must be drilled before threading. In other cases it is completely dependent on the planners way of working when various sequences are feasible. For instance preference to execute as many operations as possible with the same tool, or rather perform all different operations on one side of a part before the table will be turned and indexed, or make a rough cut and perform another operation before the finishing cut in order to let the subject part cool down. Through experience planners have generally developed their own rules depending on their success in the past.

Some further conclusions of the CRIF study were:

- Logic extraction is feasible within a practical environment in a company with a diversified but stable product spectrum,
- Definition of part families using group technology is a basic requirement enabling the determination of process planning logic,
- Step by step implementation of automated process planning (partsfamily by partsfamily) will prevent enormous practical problems initially.

A next step will be the extraction of planning rules for a well defined set of parts, and thereafter the implementation of the rules in the generative process planning system XPS-2 to investigate its functioning.

4.5 Interfaces to production control systems

It is not only the integration with CAD which makes automated Process Planning feasible and useful, but it is also the link with the Production control systems making use of the information generated in Process Planning. As indicated before this information is used for:

- Machine shop scheduling,
- Cost calculations supporting quotations to be made, make/buy decisions, managements decisions regarding investments,
- Purchase information for raw material, tools, fixtures,
- Generation of shop-floor workorders and other documents,
- Initiation of NC programs to be made,
- Programs for tools to be set-up by the tool departments,
- Workpreparation of stock-room and internal transport.

This information has to be sent to the proper areas of the production control system taking care of the above mentioned functions.

The process plan generated in an automatic system is available as a piece of software. Hence it is possible in principle to transfer it to the production control systems without human intervention. A requirement is then a proper data interface. Both the type of information and the format in which the information has to be supplied, must be specified. At present a process plan Data Interface study is being performed through CAM-I's Factory management

Program and the Process Planning Program [17] of Primary interest was the need for a Factory management system to use process plan data in its decision making process. But beyond that a second goal was to specify an interface in as general a way as possible. The first phase of the study: Development of a preliminary interface has been finished. The second phase will be a testbed implementation using XPS-2 and MADEMA an advanced Factory Management system.

The study concerned the following:

- Definitions of terms important for interfacing, to help establish a terminology that can be shared by both Factory Management and Process Planning,
- Data needs for Factory management and Process Planning were established and laid down in a conceptual datamodel,
- A basic set of routines was specified to store and access data in the conceptual model. The routine will perform the task of translating the physical representation of the FM- or PP system to the conceptual datamodel, and vice-versa. In this way transfer of data becomes independant of the physical structure in which they were generated in the Process Planning System.

5. Conclusions

Within the scope of a tutorial a presentation has been given regarding the development in Process planning leading to automation. It is not yet completely defined along which route these developments will go. Artificial intelligence looks promising but some major problems still have to be solved:

- The CAD-CAM integration,
- Knowledge-base build up,
- Interface to production control systems.

Further it is questionable if one expert system could handle all the aspects of process planning especially at the highest level. These points may deserve major interest for future research.

Literature references

[1] N.N. (1984) IMPACT presentation to ANC/GMP/PPP joint project meeting. CAM-I Arlington USA
[2] Bunce PG (1987) Introduction to computers in process planning. The open University KT20 1215 PT616 Great Britain
[3] Tulkoff I (1981) Lockheeds Genplan, Proceedings of the 18th International technical conference. NC Control Society
[4] Vogel SA, Adlard JE (1981) The Autoplan process planning system. Proceedings of the 18th International technical conference. NC Control Society
[5] Descotte Y, Latombe JC (1981) GARI: a problem solver that plans how to machine mechanical parts. IJCAI 7, Vancouver, Canada
[6] Matsushima K, Okada N, Sata T (1982) The integration of CAD and CAM by application of artificial intelligence techniques. Annals of the CIRP 11/1982
[7] Erve AM, Kals HJJ (1986) X-PLANE a generative CAPP system for parts manufacturing. Annals of the CIRP Vol. 35/1/1986
[8] Philips RH, Mouleeswaran CB (1985) A knowledge based approach to gnerative process planning. Conference proceedings AUTOFACT 1985 Detroit USA
[9] Dommerholt HJ, Kroeze B, Thijssen ME (1987) CAD/CAM-coupling, a multi-client study. Metaalinstituut TNO Apeldoorn The Netherlands
[10] Butterfield WR, Green MK, Scott DC, Stoker WJ (1986) Part features for process planning C-85-PPP-03. CAM-I Arlington USA
[11] N.N. (1987) Examples of DIALOG and MACRO programs. CADMES BV Eindhoven The Netherlands
[12] Amble T, Hystad RA, Haugrud B (1982) Investigation of heuristic approaches to process planning. LA-81-PPP-0S NTH-Sintef report for CAM-I Arlington USA
[13] Stefik MJ (1980) Planning with constraints. Ph. D thesis Stanford University USA
[14] Rich E (1983) Artificial intelligence. McGraw-Hill New York USA
[15] Pinte J, Carlier J, De Vos P (1982) Investigation of process planning methods in industry. R-82-PPP-01 CAM-I Arlington USA
[16] Pinte J, Carlier J (1984) Process planning logic extraction study. R-84-PPP-02 CAM-I Arlington USA
[17] Chryssolouris GM, Sack CF (1986) Process plan data interface study. Phase I report. R-86-FM/PP-01 CAM-I Arlington USA
[18] Sack CF (1986) XPS-2 Final report. M-87-PPP-02 CAM-I Arlington USA

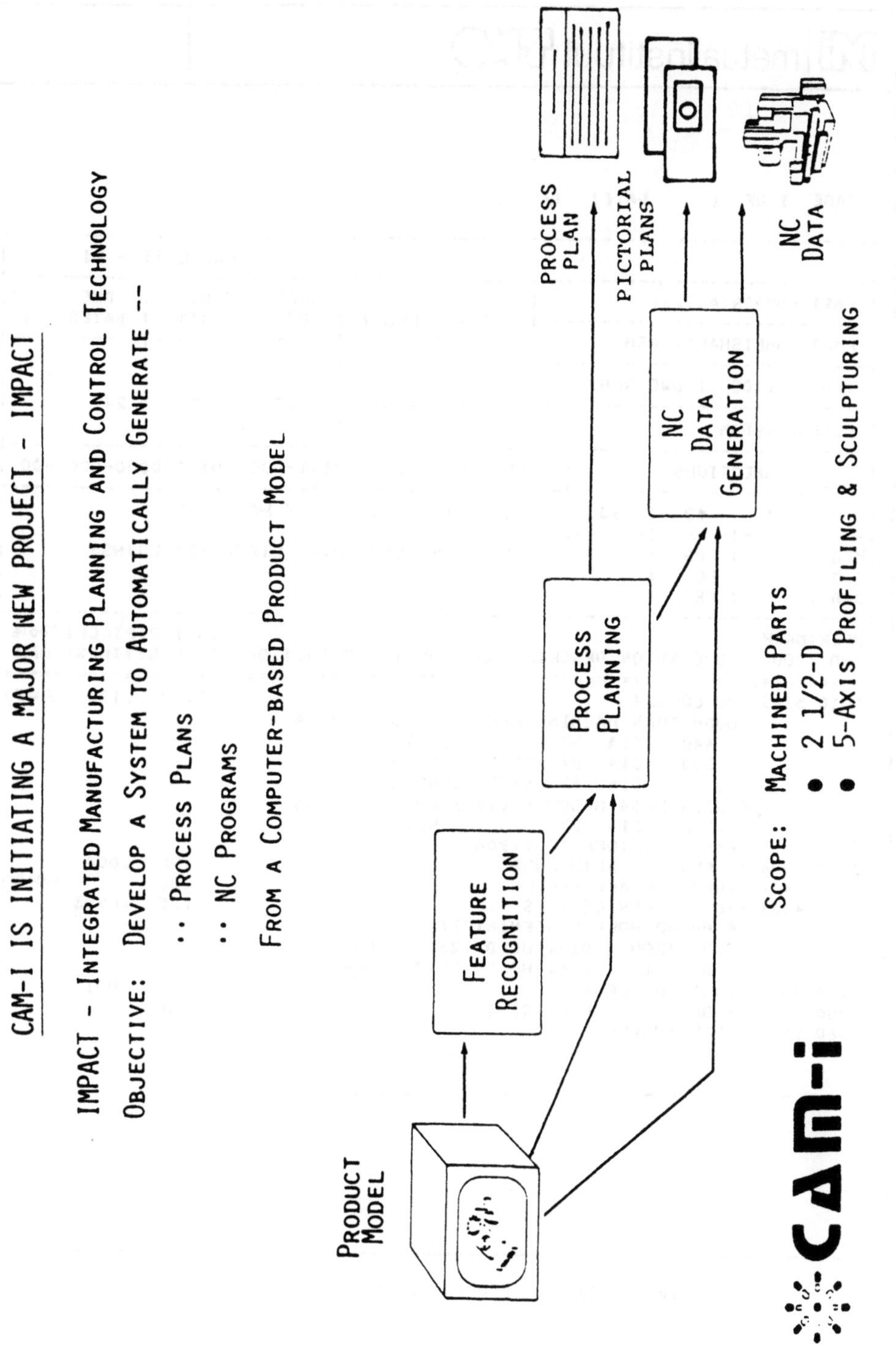

Fig. 1 Impact project

metaalinstituut TNO

```
PAGE  1 OF  1      DATE:  4/ 7/81

-------------------------------------------------------------------------------
:                                                          FACILITY - F1      :
:-----------------------------------------------------------------------------:
: PART NUMBER:A63799          :         :         : ORDER  :MINIMUM:  DUE  :PRI:
:-----------------------------: S/O #   : PRJ #   :  QTY   :  QTY  : DATES : # :
: PART NAME:SHAFT, ARM        :---------:---------:--------:-------:-------:---:
:-----------------------------:         :         :        :       :       :   :
: PLNG REV:02   : DWG REV:0   :         :         :        :       :       :   :
:-----------------------------: A34UB   : 45D3    : 1000   : 935   : 249   : 2 :
: PLANNER:VDAM                :         :         :        :       :       :   :
:-----------------------------------------------------------------------------:
:       INSPECTIONS         : CODE #:1310-1181-2111-0000-0100-0000-0000-00    :
:---------------------------:-------------------------------------------------:
:     :  #1  :  #2   :  #3  : SPECIAL INSTRUCTIONS / HANDLING:                :
:-----:------:-------:------:                                                 :
:MFG  :      : FHB   :      : 1/2" DIA MS-5000 H.R. STEEL (2" LGTHS)          :
:ENG  :      : PC    :      :                                                 :
:Q/A  :      : AH    :      :                                                 :
:-----------------------------------------------------------------------------:
:OPER:MACH:                                             :SETUP:PIECE:Tools.:
: NO :TOOL:  OPERATION DESCRIPTION - ASSY INSTRUCTIONS  :TIMES:TIMES: OPR  :
:====:====:=============================================:=====:=====:======:
0010 5145 S/U COLLET                                      2.00   .173   A457.
          ROUGH TURN MACHINE PER TAPE NO. LS982A
            .440    DIA. BY 1.750  LENGTH
            .300    DIA. BY .8120  LENGTH
            .275    DIA. BY .4375  LENGTH
          FINISH 3/64 GROOVES (TYP) AND CHAMFERS
            .270    DIA. BY .375   LENGTH
          CHAMFER CUTOFF TO 1.906
0015 1026 #2 CENTERS BOTH ENDS                             .25   .004
0020 9401 CARBURIZE AND HARDEN                             .50          WH24
0030 4063 S/U BETWEEN CENTERS                             1.25   .0983
          GRIND OD HOLD CONCENTRICITY
          HOLD .4200    DIM. HOLD .2600    DIM.
          HOLD .2815    DIM. HOLD .2712    DIM.
0040 9501 BLAST TO CLEAN                                         .001
0050 9201 CHROME PLATE PER PRINT                           .38
0060 9805 FINAL INSPECT

-- -------------------------------------------------- --  --  --
```

Fig. 2 Typical process plan

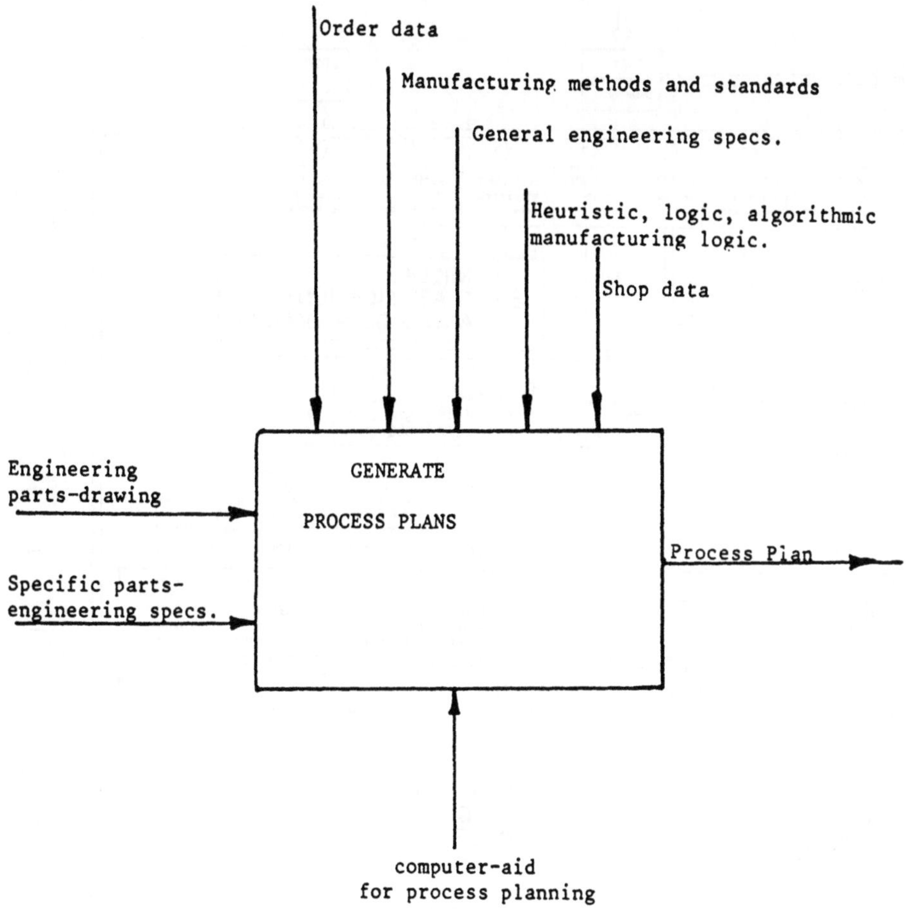
Order data
Manufacturing methods and standards
General engineering specs.
Heuristic, logic, algorithmic manufacturing logic.
Shop data
Engineering parts-drawing
Specific parts-engineering specs.
GENERATE
PROCESS PLANS
Process Plan
computer-aid for process planning

Fig. 3

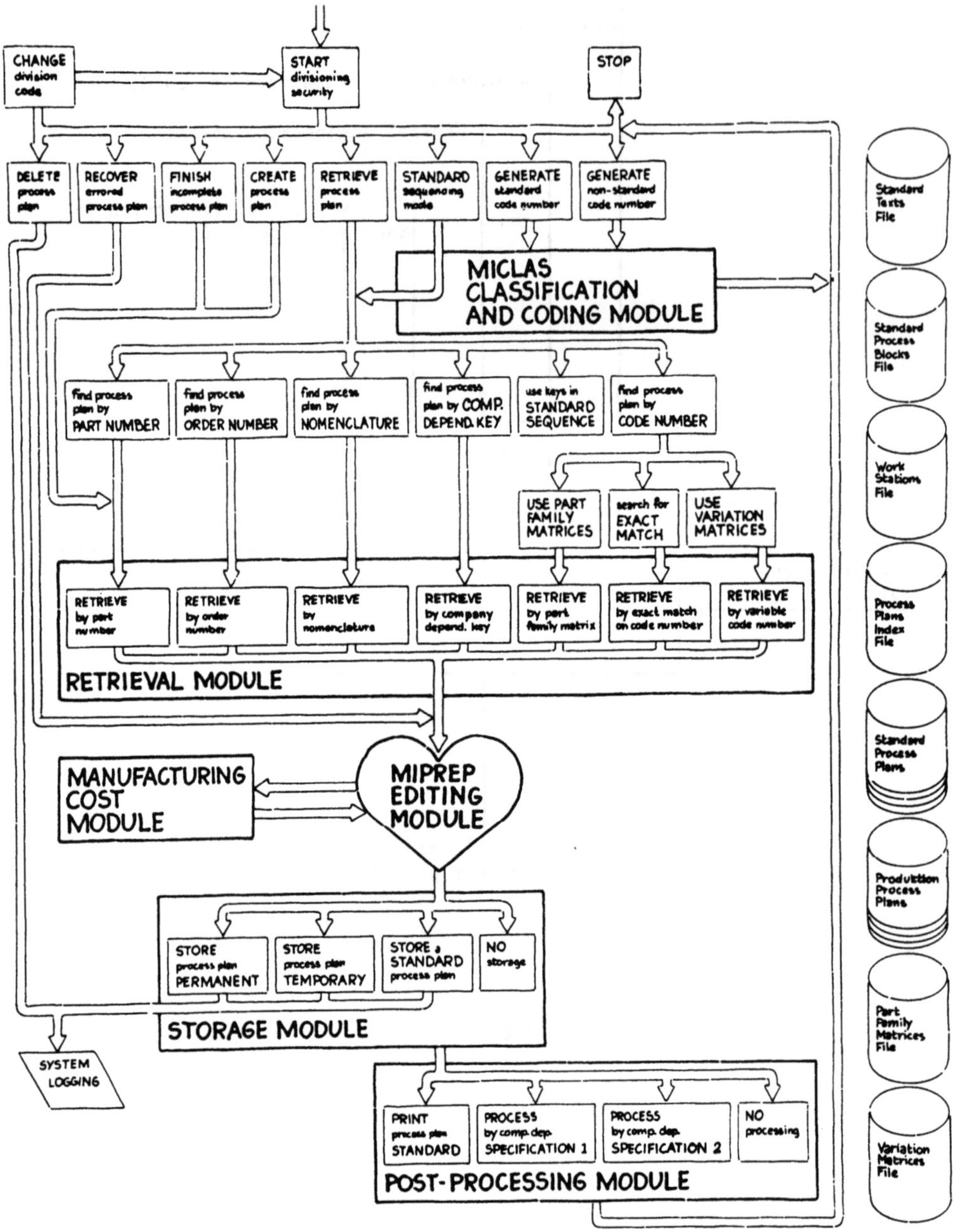

Fig. 4 Process planning system.

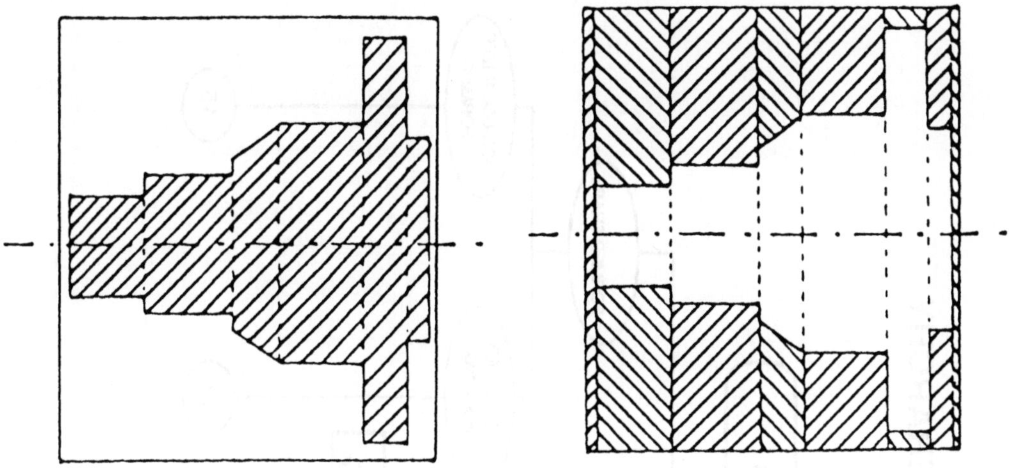

Fig. 6

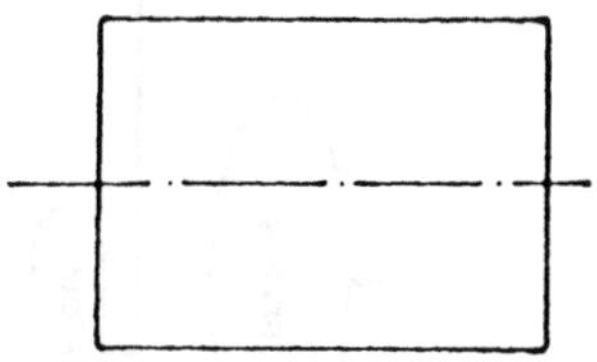

Fig. 5

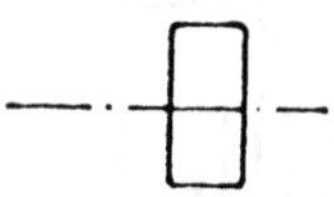

Fig. 7-1

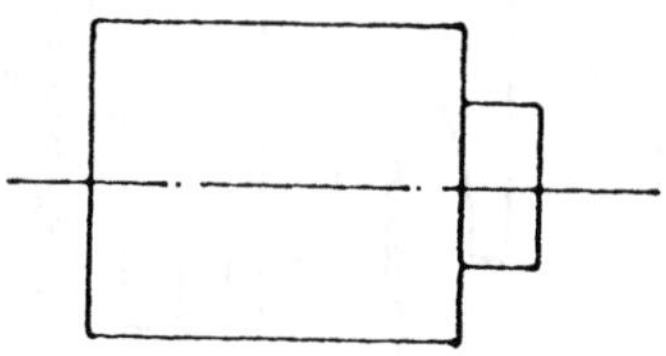

7-2

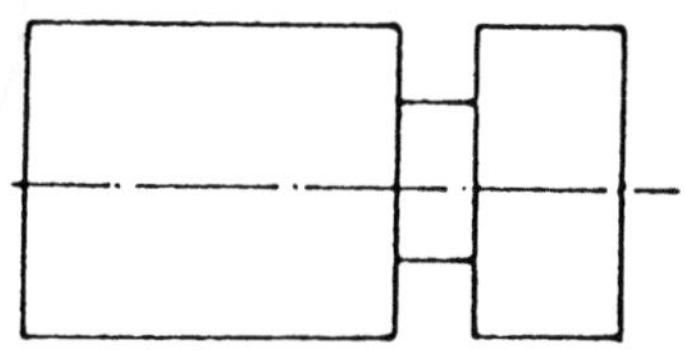

7-3

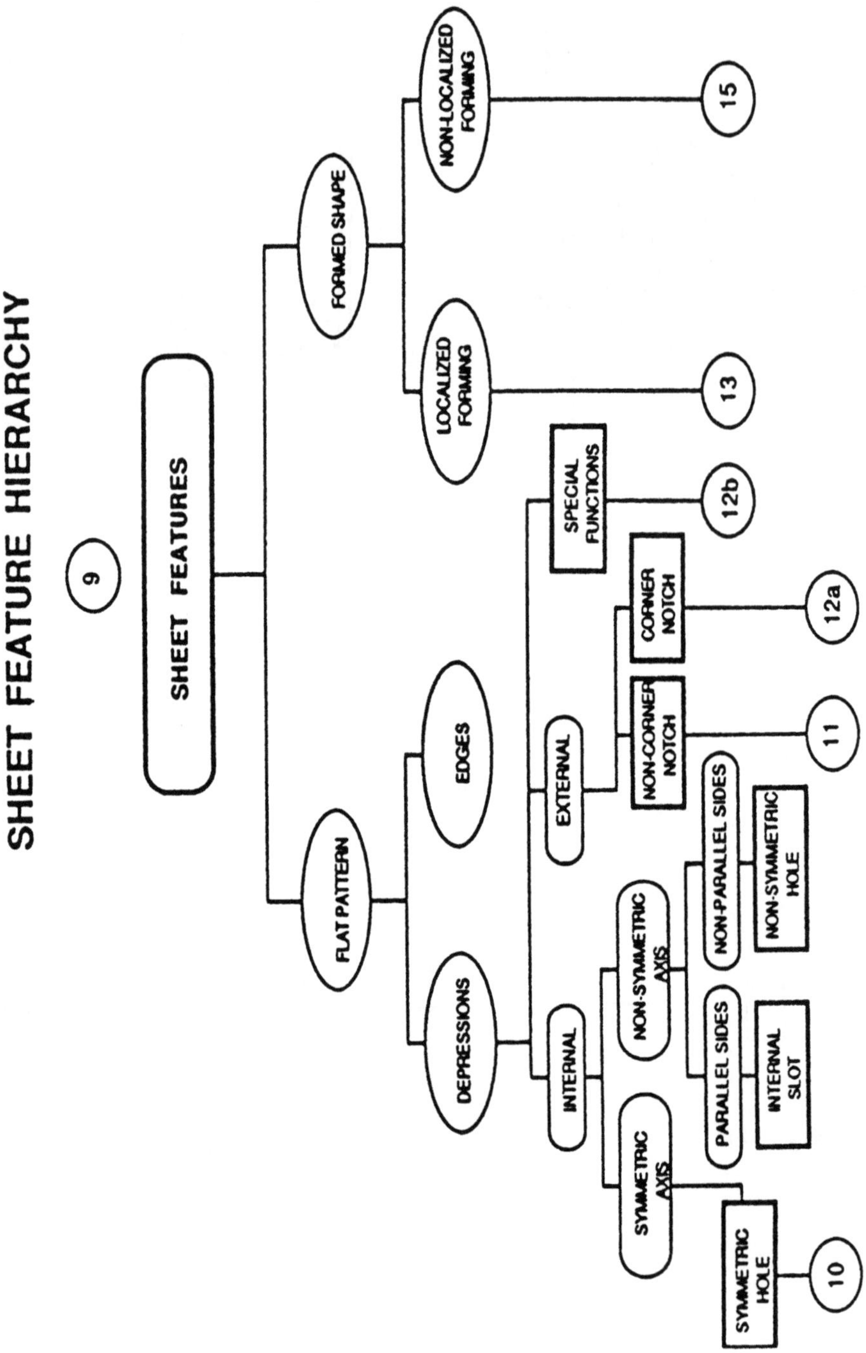

Fig. 8-1 Feature hierarchy

ILLUSTRATION	GROUP S	R	N	DEFINITION
		✓		• FACE: INDICATES SURFACES THAT: 1) MAY OR MAY NOT BE CONCENTRIC WITH THE PART'S A.O.R.. 2) LIE FROM GREATER THAN 45 DEGREES OR LESS THAN 135 DEGREES FROM THE PART'S A.O.R.. 3) CLASSIFY A SINGLE SEGMENT WITH A PORTION QUALIFYING AS A DIAMETER AND THE REMAINDER AS A FACE AS BEING A DIAMETER AND NOT A FACE. 4) A FACE OF A GIVEN DIAMETER, SIZE TOLERANCE, SURFACE FINISH WITH INTERRUPTIONS BY CAVITIES THAT MAY OR MAY NOT BE MACHINED, IS ONE FACE. 5) MAY BE ONE OF THESE TYPES: 1) STRAIGHT, 2) TAPERED, 3) FILLETED, 4) RADIUSED, OR 5) CONTOURED.
KEYWAY SLOT 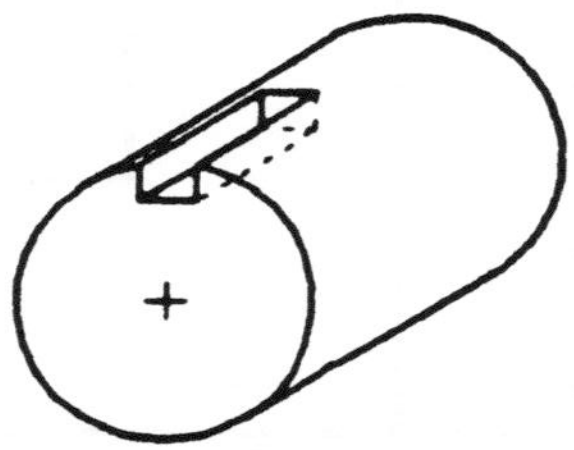		✓		• KEYWAY SLOT: INDICATES A SLOT THAT: 1) IS LOCATED IN AN O.D. OR AN I.D., WITH THE LONGITUDINAL AXIS OF THE SLOT PARALLEL TO THE AXIS OF THE O.D. OR I.D., AND 2) HAS A RECTANGULAR CROSS SECTION. 3) MAY HAVE CORNER RADII OF ANY SIZE 4) A MIDPLANE BETWEEN THE SLOT SIDES WILL INTERSECT THE AXIS OF ROTATION.

Fig. 8-2 Feature definitions

ILLUSTRATION	GROUP S	GROUP R	GROUP N	DATA NEEDED FOR PROCESS PLANNING	
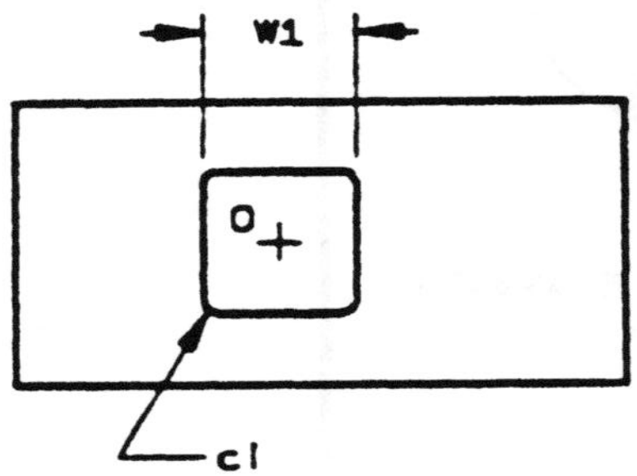	✓			• HOLE SQUARE:	
				PART	CHAR
				LABEL	CHAR
				O = ORIGIN	CHAR
				W1 = WIDTH1	NUM
				WIDTH1TOL	POINTER
				cI = CORNERADI; i=1,4	NUM
				CRADITOL	POINTER
				FACE ENTRY	CHAR
				FACE SIDEI; I=1,N	CHAR
				FACE EXIT	CHAR
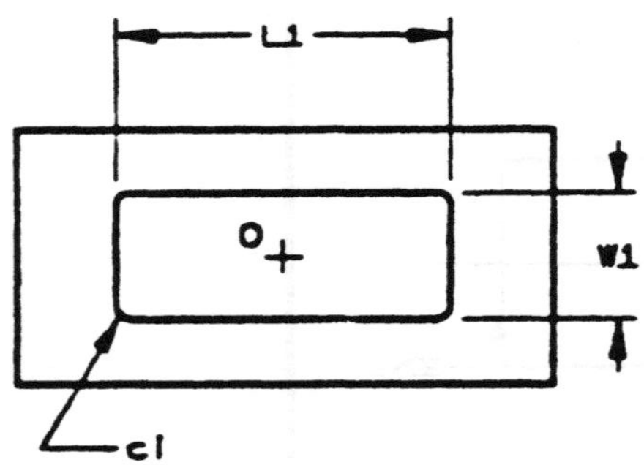	✓			• HOLE RECTANGULAR:	
				PART	CHAR
				LABEL	CHAR
				O = ORIGIN	CHAR
				W1 = WIDTH1	NUM
				WIDTH1TOL	POINTER
				L1 = LENGTH1	NUM
				LENGTH1TOL	POINTER
				cI = CORNERADI; I=1,4	NUM
				CRADITOL	POINTER
				FACE ENTRY	CHAR
				FACE SIDEI; I=1,N	CHAR
				FACE EXIT	CHAR

Fig. 8-3 Feature data

```
Dec 15 12:22 1986  hole2.dat Page 1

$------------------------------------
$
$           DMA2:[4.177]HOLE2.DAT
$
$           1982-05-14  H.RICHTER
$
$       PARTFEATURE DESCRIPTION FOR
$             OBJECT 'HOLE2'
$
$-------------------------------------------------
BEGPRG/ TYPE , HOLE , DIMD1 , D , TOL1 , TOL2 ,  $
                      DIMH1 , H
REAL/ D , TOL1 , TOL2 , H,
STRING/ HOLE , NAME ,  OBJNAM , DIAMOD
DFAULT/ D,12 , TOL1,0 , TOL2,-0.5 , H,16
DFAULT/ DIAMOD    , '$$UNDEFINED$$'
DFAULT/ HOLE , 'HOLE2'
$
$    GEOMETRIC DESCRIPTION
$
R = D / 2
M = R * 1.3
IF ( LOGIC/R,GT,5) LAB10
M = R * 1.5
LAB10)  V = R * 0.57735
$
$      POINTS
$
P1 = P/0,0
P2 = P/R,0
P3 = P/R,-H+V
P4 = P/0,-H
P5 = P/-R,-H+V
P6 = P/-R,0
P7 = P/0,3
P8 = P/M,0
P9 = P/0,-M
P10 = P/-M,0
P11 = P/0,M
P12 = P/0,-H-3
$
$    CIRCLE
$
C1 = C/CENTER,P1,RADIUS,R
$
NAME=STRING/ HOLE ',' D ',' H
TOPOCC/P1,P2,P3,P4,P5,P6,P7,P8,P9,P10,P11,P12,C1,NAME
$
$     GENERATE         OBJECT STATEMENT
$
STM11 = ST/ OBJNAM '=OBJECT/TYPE ,' HOLE ' ,DIMD1, ' D ',' TOL1 ',' TOL2 ',$'
STM12 = ST/ '                           DIMH1 , ' H
GENERATE STM11,STM12
```

Fig. 9 Part feature

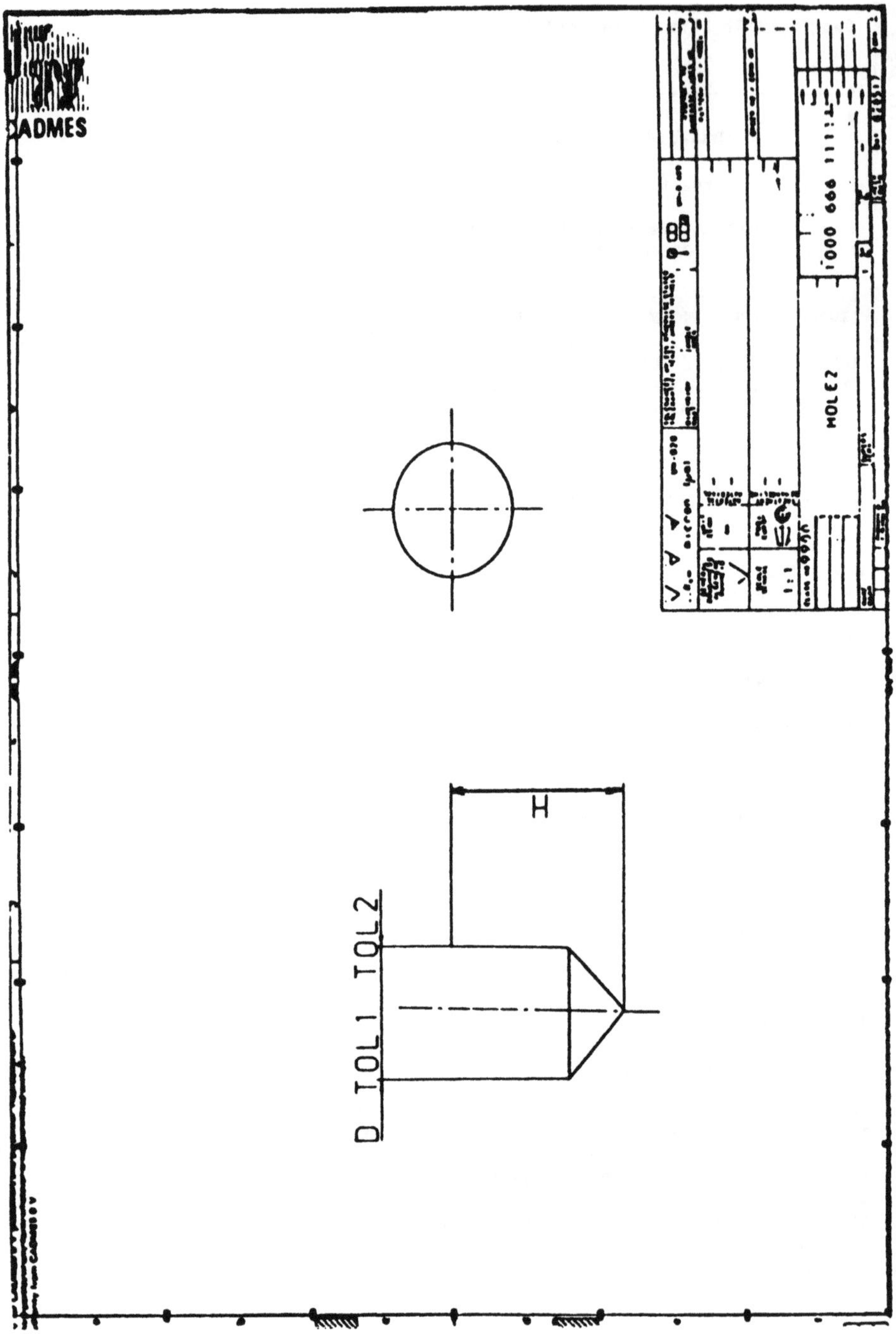

Fig.9 Part featuredrawing.

ON SYSTEMS ENGINEERING FOR COMPUTER INTEGRATED MANUFACTURING

Kiyoji Asai

Department of Industrial Management
Osaka Institute of Technology, Osaka,
535, Japan

ABSTRACT

Factory Automation (FA) serves to promote the diversity and flexibility of production through the use of hardware and software, such as NC, MC, industrial robot, automatic warehouse, AGV, CAD-CAM, and production control information system. The recent rend of FA is toward systematic, rather than isolated, equipment and toward more software than hardware.

This is a report on the actual condition of FA. Data from two years of operation in Japanese business are up for analysis. Technical and labor problems are also taken up for discussion, along with solutions for them. Observations are then made on the future development of FA in Japan and other countries. In view of the growing weight of software in FA, some comment is made on ideas and trends of rolesharing among software, hardware and man.

Next, the systems engineering for the computer integrated manufacturing (CIM) descrived above is shown. It will be important for such a large scale system as CIM.

1. INDUSTRIAL ROBOTS AND SYSTEMS – OUTLINE

The following is an outline of the types and characteristics of the various industrial robots currently used in Japan.

An industrial robot is defined as an automatic machine capable of performing diverse industrial operations. Many serve much the same use as human arms or hands and are operated by computer (microprocessor). In Japan, the term "indu-

NATO ASI Series, Vol. F49
Computer Integrated Manufacturing
Edited by I. Burhan Türksen

strial robot" also includes simpler devices, defined as devoid of any computer equipment by JIS (Japanese Industrial Standards) criteria and handless or armless NC machine tools (NC robots). Manual manipulators and sequence robots are so-called low-grade robots not enquipped with auxiliary microprocessors. The rest, with the exception of NC robots, are microprocessor equipped. There are two types of NC robot: those operated by sequential commands recorded on tapes and cards, and those which employ a micro-processor. A manipulator refers to a manually-operated robot, such as the Magic Hand. A sequence robot faithfully repeats a series of actions in a set sequence, under set conditions and at set positions. While prescribed settings in a fixed sequence robot do not permit simple modification, the set program of a variable sequence robot can be easily changed. The microprocessor in a playback robot can programmed with sequence, positions and other information on an action series which must be remembered and repeated by the robot. Lastly, an NC robot refers to a so-called NC machine tool which is capable of repeating a set of actions numerically specified on tapes or cards. Intelligent robots are presently under development to meet the important new requirements of visual sensing and recognition. While technologically feasible, the development of high-performance intelligent robots remains limited by their high cost.

Furthermore, FMS (Flexible Manufacturing System) components, such as conveyor robots and machining centers, are also considered as kinds of robots. A conveyor robot is one which carries work pieces (i.e. materials and parts for processing or assembly) to machine tools or assembly robots, and later retrieves the piece when the procedure is completed. A machining center, on the other hand, is a machine tool equipped with computerized automatic processing and tool exchange functions.

Next, one must consider the operational characteristics of industrial robots. The types of robots described above are controlled and operated either individually or in a group by an operator using a control panel with a display (Fig. 1).

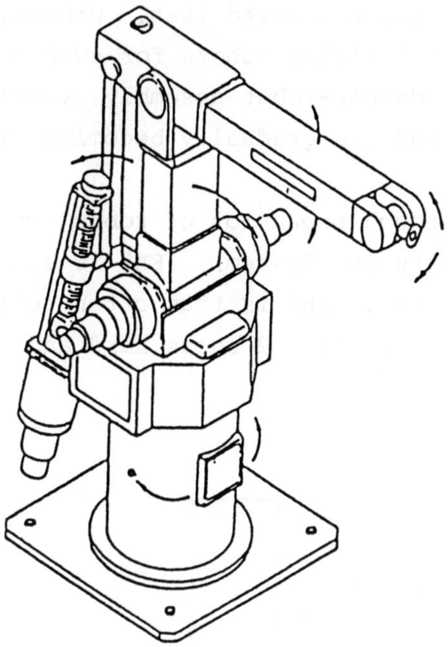

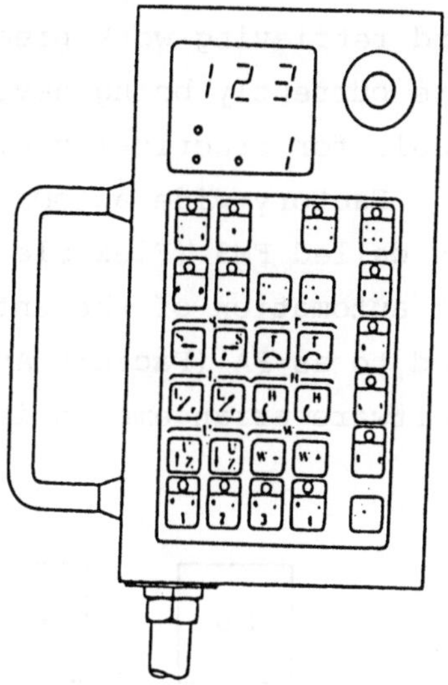

Control Panel

Fig.1 An example of an industrial robot and control panel.

Presently, control panel and display are combined in a single unit installed on each robot. The unit enables instruction (called "teaching" in playback robots) of work procedures and programs and gives commands to robots. However for some robots, such as painting robots, instructions are provided by manual or manipulator-aided movement of the robot arm. In such a case the robot is directly operated to produce arm movement.

Development of a single control panel and display unit capable of operating several types of robots will probably come in the near future. When realized, it will eliminate the inconvenience of using different control panels and display units for different purposes, thus significantly reducing operational mistakes by operators.

The limited functions of individual industrial robots have been increasingly compensated for through the establishment of systems combining robots with other machines. Earlier versions of these systems, called processing cells or FMC (Flexible Manufacturing Cells), combined an armed robot with an NC machine tool. The robot would supply the work piece to the NC and then retrieve and transfer the processed item. Others, including assembly machines which employ robots for supplying and retrieving work pieces and double-robot assembly systems, are currently being developed and are gradually becoming available for practical use.

Factory-wide automation through expansion of robot systems is called FMS (Flexible Manufacturing System). FMS is aimed at automation of the entire factory, and is thus often referred to as FA (Factory Automation). Fig. 2 shows hard and software arrangements in FA.

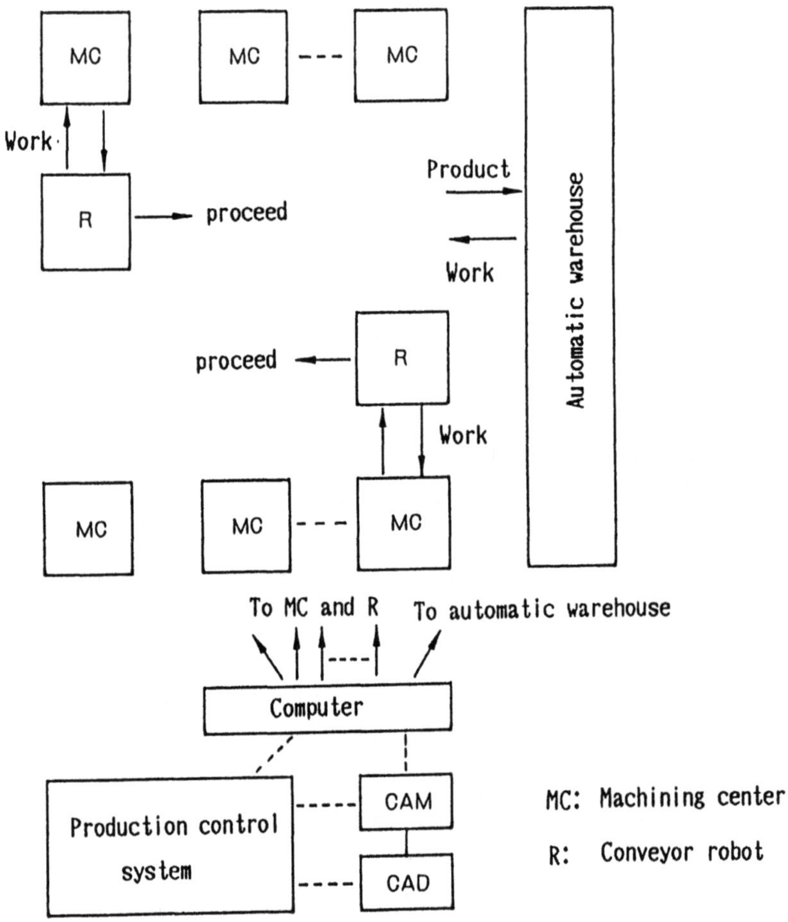

Fig.2 An example of an FA system.

In the diagram, each Machining Center MC (multi-functional machine tools capable of automatic tool exchange) performs a number of different machining processes. Conveyor robots R receive the workpiece from the automatic warehouse, approach the appropriate MC, supply the piece for processing, retrieve it when completed and proceed to the next MC. The entire procedure is repeated automatically. When all MC processing is completed, the robots replace the pieces in the automatic warehouse. The above process is completely computerized. CAD (Computer-aided Design) and CAM (Compute-aided Manufacturing) systems, or a factory-wide production control system, are added to MC's and robots, for software support. The computer-aided production control system determines types and numbers of products according to management strategies based on market research, and plans production schedules. As this hardware and software are not yet completed, future systemization is anticipated. Although this FA is being designed for manufacturing, it is expected to be applied to various other operations as well.

In looking back the history of the development of Factory Automation, we can see the process of gradual changes of the concept initiated with the applications to the discrete work stations as NC-machines. Machining centers and industrial robots, etc., which were then integrated into broader systems. They have progressed further into more information-intensive ones, added with CAD and CAM, and production control information systems, and are now developing into Computer-Integrated Manufacturing (CIM), with both the hard system and soft system functioning just like two wheels of a cart. Mode of the progress in FA, however, has not followed the same pattern but appeared differently from manufacturer to manufacturer; CAD or CAM has emerged first in some companies, while priority have been placed on the introduction of production control information system in others, with this section having reached the level of the full-fledged utilization earlier than other facets of the intended factory automatization. Generally speaking, it appears that business concerned both at home and abroad are concentrating their efforts to the refinement of their production control information system at the moment. In other words, they are trying to materialize further streamlining of the attempts of factory automatization that were initiated on "ad hoc" basis in more systematic form, by integrating with the competent production control information systems.

In order to see the tendency of FA described above, a research was conducted on part of the enterprises who had implemented FA during the past two years, focusing our attention on the determination of what were the subjects of their major concern. Data obtained through the research are shown in Fig. 3. For your reference, items used for evaluation of the effects of the FA undertaken by these corporations were also shown.

As can be seen from these data, it is obvious that as a recent trend, subjects of their major concern are shifting from such applications to the discrete installations as NC machines, MCs, industrial robots, unmanned warehouses and conveyance systems, etc., to the more software-oriented production control information systems integrated in the related networks.

Although the trend of the times seems to aim at such network configuration as mentioned above, we observe many cases where soft systems employed are not adequately coordinated with the related hard systems. More specifically, there still can be seen inadequate matching between CAD and CAM applied in a particular manufacturing system, between the production control information system and the related CAD/CAM, between the different pieces of equipment used in various FA systems, or between the production control information system made available under such a system and individual workers and/or the equipment controlled under the system. This problem of poor matching of these fac-ors would be the subjects yet to be solved in future.

Along with such a more systematization-oriented concept on FA as mentioned above, there is another trend that has come into fore lately, advocating the need of reviewing FA from the very beginning, which reflects the stagnant economic situation as we are faced today.

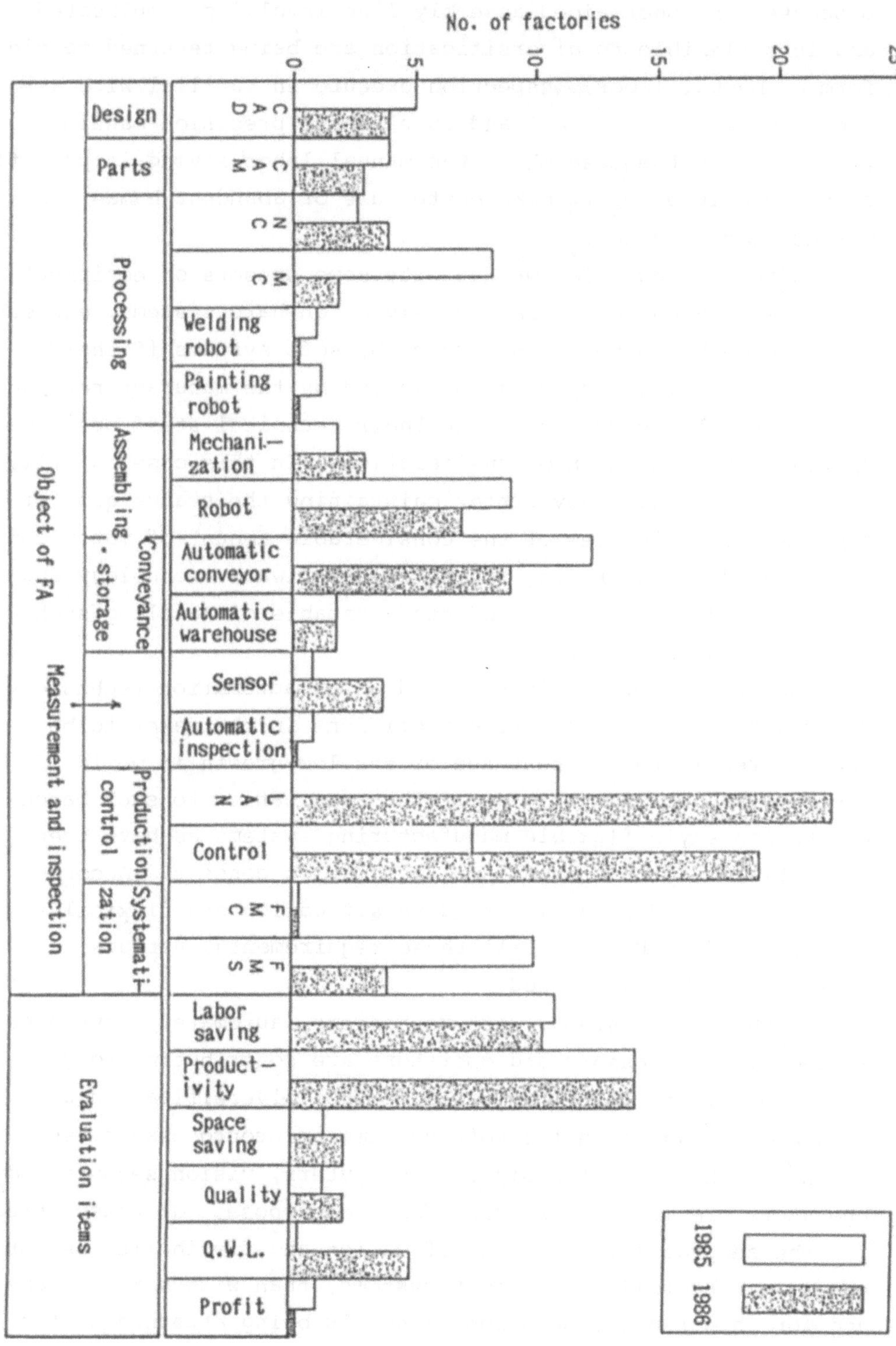

No. of investigated factories: 35(1985), 33(1986)

Fig.3 Major concern and evaluation items in FA in Japan.

The latter trend is typically seen in such cases where the processes in robot-aided assembly line involving complicated and less flexible to diversification are being returned to old form of manual labor, inspection execute in parallel with assembly work done on small and complicated precision machine tools, and in the case where the manual labor-saving is put off on purpose in order to make better use of abandunt human resources under its control.

As another example, we can cite some sensors or equipment or systems intended for FA are relying on hard systems, whereas such are ought to be taken care of by soft systems if their original technical expertise possessed by the manufacturers and the technical fields into which their technical staff are committed to are taken into consideration. In this case as well, it is assumed that they prefer maintaining the status quo for the time being in view of the considerable amount of money and time needed for upgrading the related software technology and for securing enough number of staff capable of handling such software required.

Factory Automation is utilized as an automation technology devised to cope with severe competitions in the maket to be encountered in the present age of the low-growth in world economy, and as such it aims at the application to such batch-production style flexible manufacturing systems as have been impracticable with conventional automation concept. Accordingly, higher adaptability to diversification and flexibility are required of FA. To fill these requirements, various measures are being adopted.

In the hard systems, soft mechanisms, automated conveyance systems, visual sensors and operators are contributing to giveing adaptability to the required need of diversification and flexibility, while in the soft systems, microprocessors used in the "mechatronics" equipment, computers, vision sensors and operators are likewise serving for the purpose. Of these, need for the adaptability to diversification and flexibility is more easily can be filled with soft systems, than with hard systems, and for this reason, much importance is being attached to the computer programs, processing of sensor-originated information, or to the thinking and judgements made by humans. Part of the direct labor that cannot be dealt with software is taken care of with competent hard system, with various measures being devised to realize more flexible mechanisms. Articulated robots, mechanisms that can change their heights and widths, magazines and automated conveyance systems, etc. are typical examples. A radio-controlled automated conveyance system using no tracks is introduced, which can travel to and fro between another plant building, going upstairs or downstairs as desired.

What play an important role in filling the need for the diversification and flexibility are sensors used for making judgements in the preceding stage to the above-mentioned hard and soft systems as to what sort of work is then being executed. In a batch production with rather frequent changes in models of products, it is necessary to constantly check and have accurate information as to the models of the products and to what types of products are the work-in-process or parts are intended for. This is particularly important in case of a mixed assembly line. Sensors used for the purpose come in various types; from the highly mechanical contact type (for recognizing shapes of objects) to such advanced types as mark card reading type and vision sensors. In other words, they cover a wide range of types from the mere hardware type to those with higher weights of software employed in them.

Such advanced technology that can cope with the need for the diversification and flexibility constitutes the very core of the FA. In case of batch-type production involving fairly many models of products and frequent changes in setup, the complicated manufacturing processes involved make it difficult to solve the problems by merely from the technical point of view, and moreover it becomes hard to pay off from the economic point of view as well because of the increased capital investment cost involved. This is especially evident in the assembly lines having more complicated processes, thus leading to the tendency preferring reliance on manual labor to the FA.

If the requirements of further diversification and flexibility become mandatory, reliance on manual labor or introduction of "expert system" may have to be considered. The "expert system" is intended primarily to utilize the knowledge and experience of skilled workers and experts skilled in the art saved in the computers, in applications where the properties of objects handled are not well known with no standard processing, operation and control techniques have not yet been established. As such, it is considered to be able to substitute humans who excel in their adaptability to diversification and flexibility.

2. SYSTEMS ENGINEERING FOR ROBOT INTRODUCTION

The development of technology required for the practical introduction of industrial robots to industry (i.e. Systems Engineering, SE) lags far behind the recent remarkable improvement in robot performance. For maximum efficiency, SE must consider robots as parts of the entire FA, not as independent components.

Many Japanese industries now recognize the importance of SE in obtaining sufficient return on the large investiments in plant and equipment associated with robot introduction. An increasing number of industries are hiring systems engineers for their expertise in robot introduction and utilization.

SE consists of a 3-stage process: 1) preparation before introduction, 2) introduction procedures and 3) post-introduction engineering. Preparations before robot introduction include examination of the current production system, identification and organization of its problems, and consideration of possible solutions (e.g. automation method selection, production system improvement, or design modification). Introduction procedures include selection of operations, prediction of advantages and dis-advantages, in-house PR, examination and design of manufacturing processes, robot model selection, operator training and final execution of introduction plan.

The following two items must be considered before introducing an industrial robot.

(1) IE improvement

Optimum production efficiency of existing equipment must be carefully examined prior to introducing any new automation or hardware. Next, IE improvements such as reexamination of equipment layout and worker assignments, production plan improvement of process and quality control, development of worker abilities, examination of their performance and improvement of their morale must be executed.

This kind of production improvement requires no investment in plant or equipment, and constitutes an important preparatory step for future automation, and serves as a base for future plans.

(2) Consideration of various automation methods

Any mechanization, automation, information communications or instrumentation, other than industrial robots, which is capable of productivity improvement must be considered. Optimum equipment introduction must be determined according to production targets and economic feasibility. If introduction of industrial robots and related systems (NC, MC, automatic warehouse, FMS, etc.) still proves most appropriate for a particular factory, SE should be executed.

Industrial robots and related systems are most effective when employed in the flexible production of multiple product types in varying small or medium quantities. Conventional continuous automation, conversely, is more appropriate for mass production of a small number of fixed product types.

Below, the specific procedures necessary to ensure successful SE are described in detail (Fig. 4).

(a) Operation selection

Operations ensuring effective utilization of industrial robots are selected by considering robot performance, the company's technological ability to develop peripheral equipment, and personnel requirements.

If FA goes still further, functions expected of the workers assigned to the workfloors will become further limited. In the factory automatization, on the other hand, proportion of soft systems will become higher than that of the related hard systems as can be seen as the general tendency shown in Fig. 5.

As the soft systems here, we can cite the afore-mentioned production control information system, as well as CAD/CAM, computers and various "mechatronics" equipment. What propor-

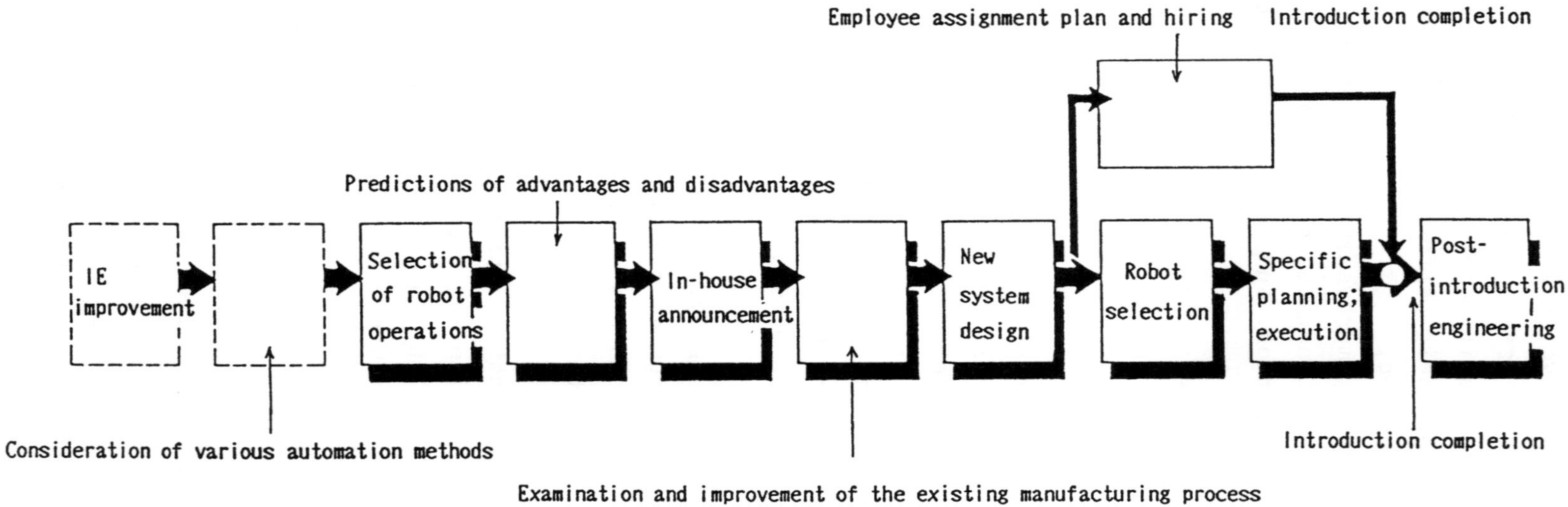

Fig.4 Procedures for industrial robot introduction.

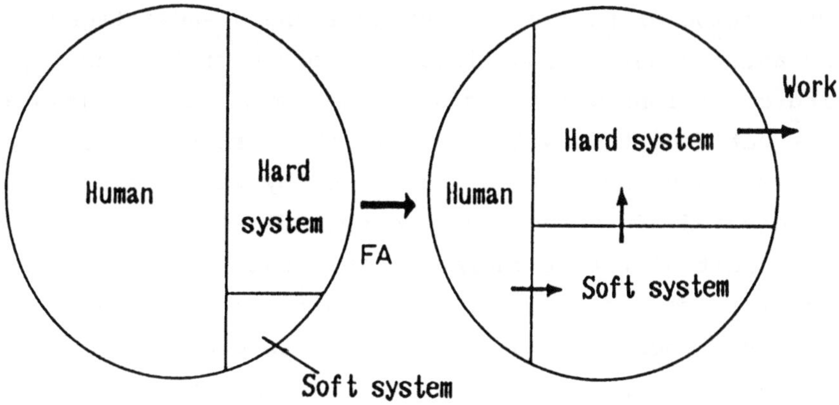

Fig.5 Partial charge of work by human, hard system and soft system.

tions are to be shared by humans, and the hard and soft systems in such future FA systems is much dependent on the quality and volume of the human resources possessed by individual corporations and on their individual technical standards. It would be a natural tendency that the business enterprises holding many technical staff members will rack their brains to how make best use of them, with those committed so far to machinery intending transforming themselves into hard system-oriented business, and those committed to electric (electronics) and information-intensive businesses aiming at software-oriented sectors of industry. Standing from the more broader perspective, however, we should aim at the goal of optimum sharing among human resources and hard and soft systems, so that each of them can be best use of exhibiting their intrinsic features and advantages. Also, from the economic point of view, it would be wise to enhance the weight of soft system over hard systems, to complement accuracy of machines by the use of feedback function, to convert part of the information processing into software and to adopt the method of controlling with information, etc. Although it takes time and cost for initial preparation and for subsequent improvements and/or modifications, soft systems, if they are not subject to not much changes in use, have advantages in no wear in use and virtually -permanent life.

(b) Prediction of possible advantages and disadvantages

Advantages may include increased productivity and improved working conditions while disadvantages may involve adaptation of work patterns to new technology, the worker's psychological alienation, and physical dangers posed by the robots.

Predicated advantages must be specifically evaluated in terms of initial target increases in productivity and the degree of improvement in working conditions, while adequate, comprehensive measures must be devised to counter any foreseeable disadvantages accompanying robot introduction. Details will be discussed in section 3.

(c) In-house announcement

After determining operations and clearly identifying and evaluating the predictable advantages and disadvantages of robot introduction, make an announcements to those involved in order to obtain their cooperation in examining the current conditions of selected operations, education and employee transfer. Personnel involved should be those workers engaged in the selected operation and other related procedures, supervisors and labor union executives.

(d) Examination and improvement of current manufacturing processes

With the cooperation of supervisors and workers involved in selected and related operations, examine current production flow, methods and equipment and devise possible improvements as a basis for the subsequent design of manufacturing processes.

(e) New system design

A new overall production system must be designed by integrating possible improvements and process modifications required with robot introduction, as determined by examination of the current manufacturing situation. Carefully reconsider the total production system and robot and accordingly plan such items as layout, worker assignments and production line balance (i.e. balanced distribution of work load throughout the production line).

Such planning should eliminate stock or any other operative

problems in the plants before and behind a robotized plant. Furthermore, necessary equipment and safety devices around the robot must be designed, a production plan for the entire production system devised, and individual worker responsibilities determined.

(f) Robot selection; staff planning and hiring

The new system design described above is based on general assumptions concerning robot types and necessary staff. At this point, the robot type appropriate to the new system and the number of and requirements for staff members should be determined.

First, select a robot from among the standard models offered by the manufacturer. Due to low investment return and technical problems, only a limited range of standard robots is available, thus a model selected for a particular operation may have either excessive or deficient performance. In such a case, one may request partial adjustment of the model, or a more expensive solution, place a special order for a model primarily designed by one's own company. Auxiliary devices and peripheral equipment for robot reinforcement can thus be designed and produced within the company.

Planning and hiring of robot-related staff must begin as early as possible to prepare for initial operation.

Required staff includes the operator and, if possible, maintenance and systems engineers. The systems engineer is in charge of solving various system problems regarding robot introduction, production system design and software development for robot application.

An addition, further significant productivity increases can be expected if an R & D engineer is included in robot improvement, new automation development, and the design and manufacturing of auxiliary devices and peripheral equipment. Staff members recruited from other sections or through new hiring must receive education and training before beginning their new assignments.

A lack of the desired robot model or staff members will necessitate modification of the above-mentioned new system design.

(g) Concrete planning and execution

New system planning and design, robot purchase, auxiliary devices and peripheral equipment preparation, planned layout execution, employee organization and assignment, test operation and adjustment will be proceeded. Detailed planning between these two steps will ensure a smooth transition from the previous system to new system.

(h) Post-introduction engineering

The new system, when finally put into operation at a factory, will initially require further modifications and adjustments to meet target performance.

Such aftercare is particularly important when adopting new technology. In addition, data and experience gained during robot introduction must be recorded and organized for future reference, especially for subsequent robot introduction in other production systems. These on-going procedures involved in new system operation, are included in what is generally called "post-introduction engineering", a critically important task in its own right.

Steps (a) - (h) represent the general procedures involved in industrial robot intorduction at a factory.

3. IMPACTS OF ROBOT APPLICATION - OUTLINE

There are two problems that are main impacts of robot applications. These are employment problems and safety problems.

3.1 Employment Problems

(1) Development of automation and its influence on employment

Influences on employment by past automation, industrial robots and Flexible Manufacturing Systems are shown in Table 1.

(2) Technology required for robot engineers

Robot engineers will be classified as operators, maintenance engineers, systems engineers and development engineers. The technologies required for these engineers are shown in Table 2.

For reference, some of the results of the employment problems questionnaire for corporations in the Osaka area in 1983 are shown in Fig. 6 - Fig. 9.

Outline of the introductions of industrial robots and related apparatus for Factory Automation are shown in Fig. 6 and Fig. 7 respectively. The data about changes in work are shown in Fig. 8 and Fig. 9. It is obvious that watching work and set-up for work are increasing.

3.2 Safety Problems

Industrial robots are a kind of Mechatronics apparatus. A block diagram of a common Mechatronics apparatus is shown in Fig. 10 to explain safety problems of industrial robots. The causes of accidents or critical conditions are shown in Fig. 11. The results of the questionnaire on critical conditions in production lines of corporations in the Osaka area in 1983 are shown in Fig. 12 and Table 3.

Table 1. Comparison of influence on employment by three kinds of automation.

i) Past automation a)Very high productivity in mass production line b)Improvement of work environment c)Job assignment after education and trainning d)Decrease of many workers e)Increase of watching work
ii) Industrial robots a)High productivity in job shop type production line b)Release of workers from dirty work (welding, painting etc.) c)Decrease of workers d)Change to VDT work e)Arising of safty problem
iii) FMS a)Very high productivity in job shop type production line b)Improvement of work environment c)Job assignment after education and trainning d)Decrease of many workers e)Increase of work which is operation and programming of computer, watching for line, set-up of work and cleaning of machines etc. f)To the work in night and day shifts

Table 2. Subjects of technology required for robot engineers.

Technology	Subjects
Mechanical Engineering	Mechanics(M,D) Machining(O,S,D) Precision engineering(M,D) Oil hydraulic and pneumatic engineering(M,D)
Common Engineering	Systems engineering(M,S,D) Control engineering: Servomechanism(M,S,D); Sequential control, Computer control(S,D) Cybernetics(D) Sensor technology(M,S,D)
Electrical Engineering	Fundametal mathematics of logic (M,S,D) Outline of electro-magnetics(M,S,D) Personal Computer: Hard(M,D); Operation(O,M,S,D); Soft(O,M,S,D) Outline of IC(M,D) Outline of electronic circuit (M,S,D) Motor and electro-magnetic device (M,S,D)

O:Operator, M:Maintenance engineer, S:Systems engineer, D:Development engineer

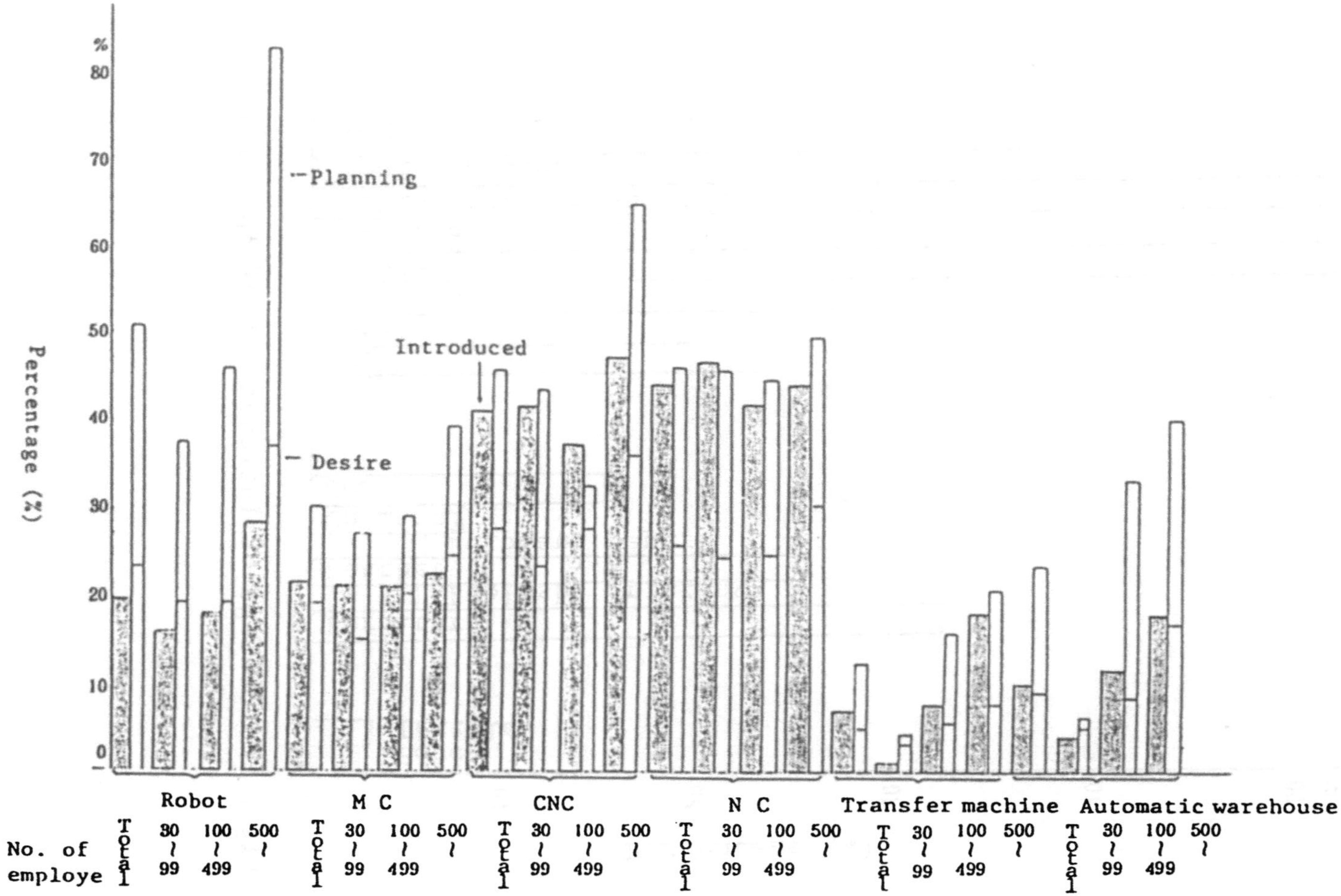

Fig.6 Results of questionnaire on introduction of apparatus for FA (Osaka Area, 1983).

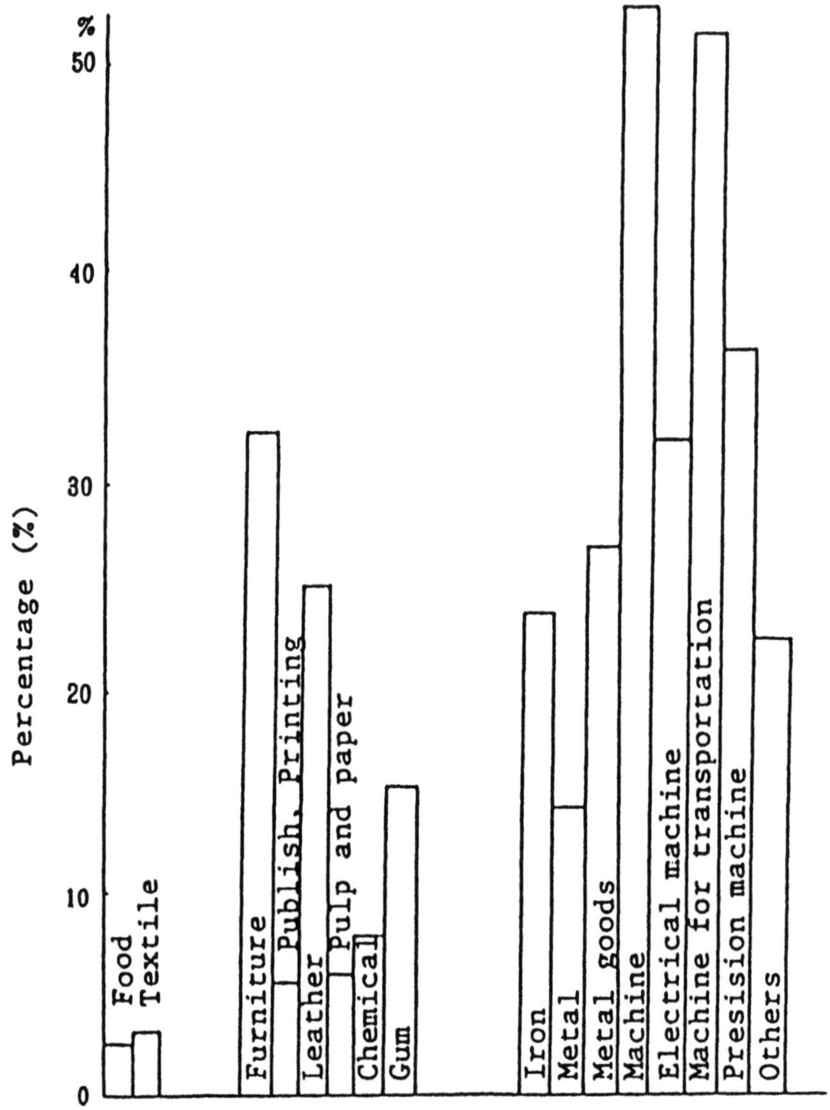

Fig.7 Percentage of Introduction of Apparatus for FA.

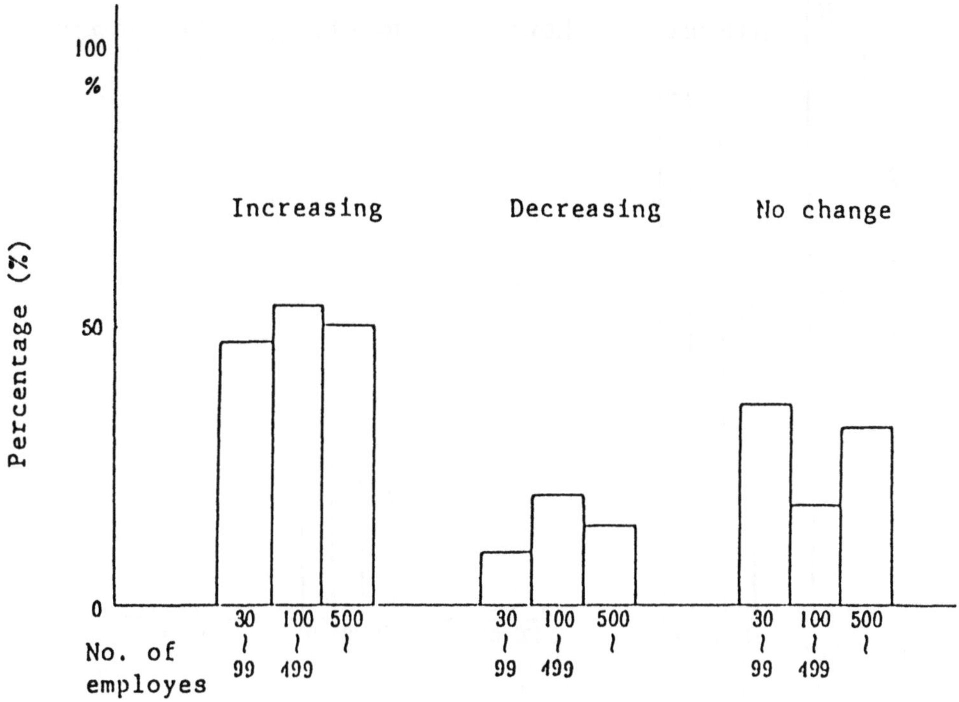

Fig.8 Data of increasing of watching work.

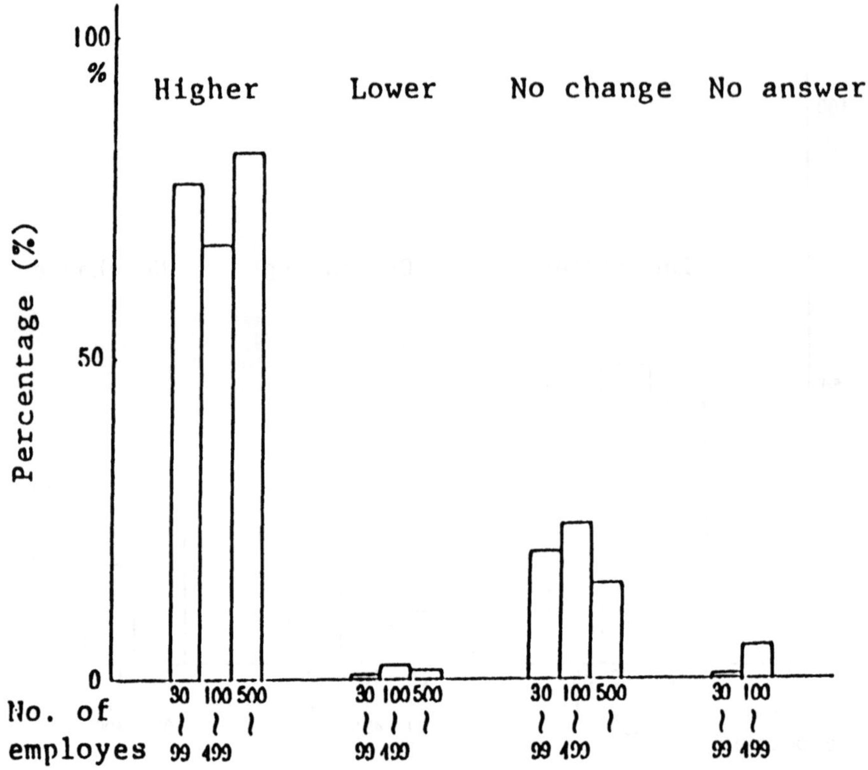

Fig.9 Importance of set-up for work.

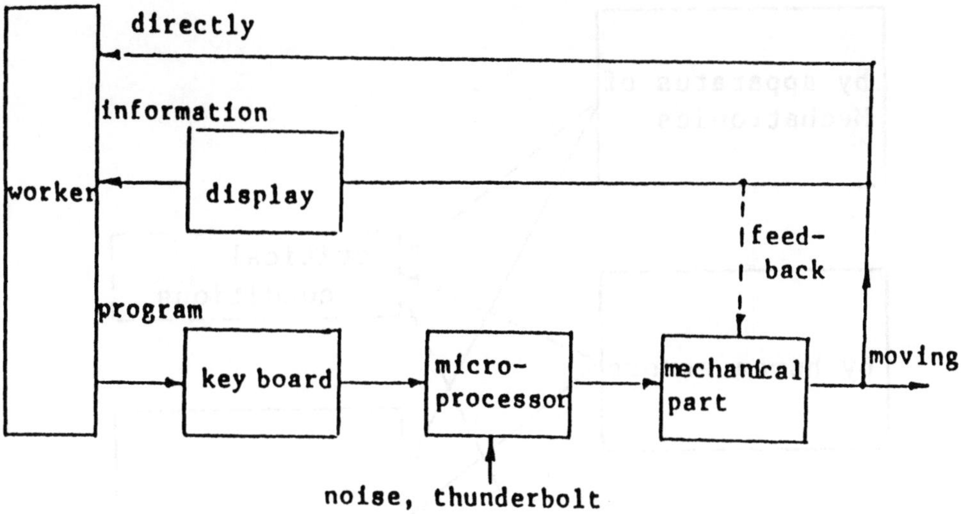

Fig. 10 Block diagram of apparatus of Mechatronics.

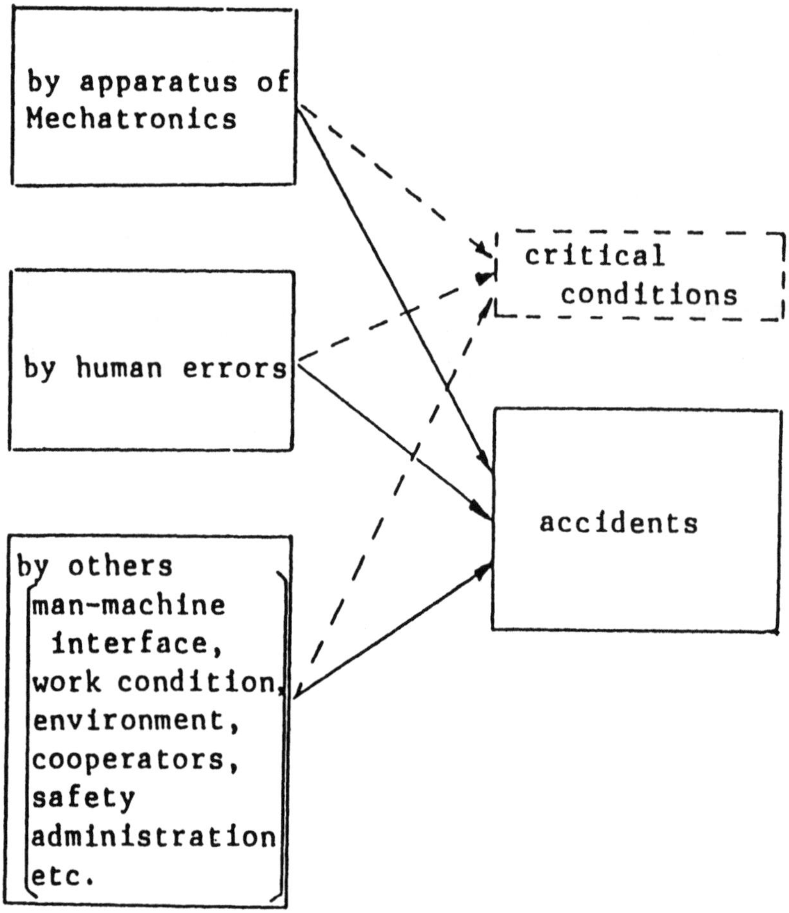

Fig.11 Causes of accidents or critical conditions.

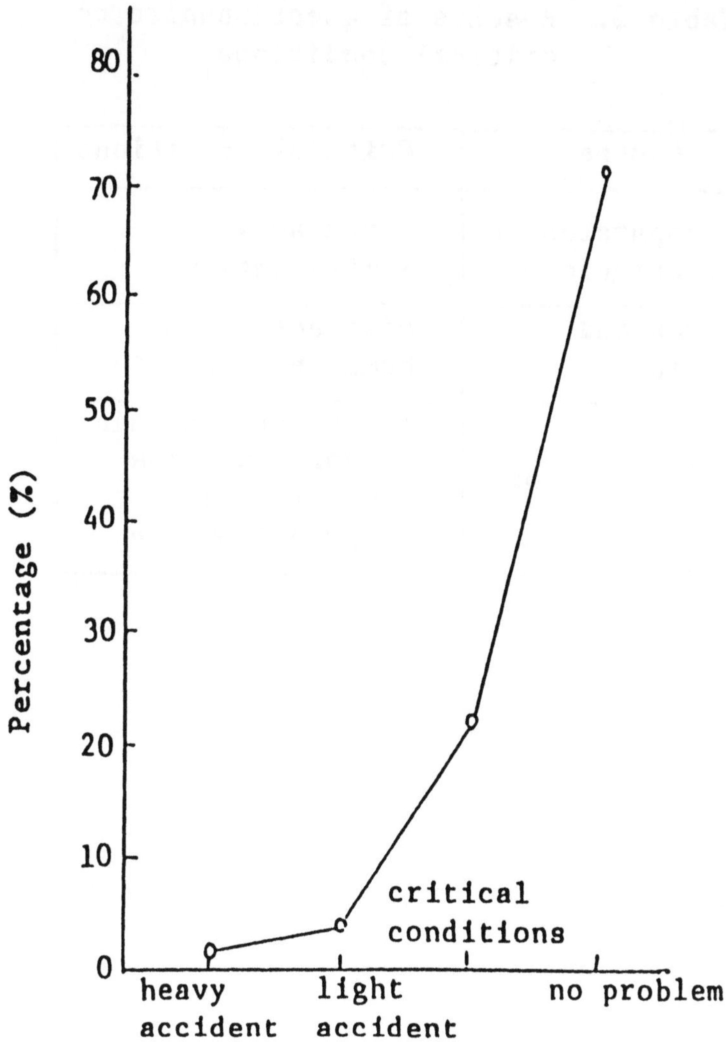

Fig.12 Results of questionnaire on accidents and critical conditions.

Table 3. Results of questionnaire on critical conditions.

Causes	Critical conditions
by apparatus of Mechatronics	· run away · miss action
by man and machine	miss action and human errors
by human errors	· men enter within dangerous area · program miss · operation miss

Factory Automation and Robot Engineerings
(Current Technologies on Robotics and Factories in Japan)

Satoru Takashima
Technical Institute,
Japan Society for the Promotion of Machine Industry, Japan

ABSTRACT

Practical Robots and Factory Automation (FA) have been remarkable in Japan. This paper presents robot engineerings and applications in Japan mainly, moreover introduces the research and developments (R&D) of robots and robot systems in the industries: (classification, language, intelligence, and standardization) and FA.

1. Introduction

Machines were created by Human being to keep themselves since they constructed a tool for use in ancient times. Sciences and technologies which began by emerging gradually, later remarkable progress.
Industries have always been in needs of higher technologies, because new machine systems which engineerings of control and information constructure in demand on automation and MECHATRONICS.
Nowadays, we are researching some intelligent machines. It is found that the existing intelligent machine is only a robot, which consists of mechanisms, computers, sensors and others.
But two kinds of robot systems have been developed: practical robots and advanced robots. Practical robots generally work in the following application fields: transfer, handling, spot-welding, spot-welding painting, assembly, palletizing, inspection and deburring etc. Advanced robots are still in experiment stages or development stages. Three types of robots have been developed as the actual robots: manufacturing robots, service robots and future robots.
In Japan, robotics studies have focused communication, intelligence, power, architecture and standardization.
The first report is Japanese technical concept of industrial robots on classification, and outline of Japanese robot history. The second report is applications of industrial robots in Japanese manufacturing, the third report is robotics R&D tendency of industrial robots production and utilizations in Japan. Finally, the conclusion describes future industrial robots in Japan.[1)]

2. Japanese industrial robots

Industrial robots classified by the Japanese Industrial

NATO ASI Series, Vol. F49
Computer Integrated Manufacturing
Edited by I. Burhan Turksen

standard (JIS) [cooperating with the Japan Industrial robots Association (JIRA)].

(1)General classification
(a)operating robot
(b)sequence control robot
(c)playback robot
(d)numerical controlled robot
(e)intelligent robot
(f)sensory controlled robot
(g)adaptive controlled robot
(h)learning controlled robot

(2)Classification by control method
(a)servo controlled robot
(b)nonservo controlled robot
(c)continuous path (CP) controlled robot
(d)point to point (PTP) controlled robot

(3)Classification by motion
(a)Cartesian coordinate robot
(b)cylindrical coordinate robot
(c)polar coordinate robot : spherical coordinate robot
(d)articulated robot

The Japanese robot's history begins with KARAKURI-NINGYOH (puppets created in EDO), and R&D of Japanese industrial robots began in 1967 after when Unimate robot and Versatran robot were introduced to Japan in 1962. Since then, robots were almost exclusively used in factories, robots have generally been meant of manufacturing robots in Japan. The national robotics projects have aimed at the applications for automation and improvement of productive and working environments. Moreover, it has recently been tried to use robots in space fields, in ocean fields and in nuclear power fields as well as in manufacturing fields (FA, FMS, CAD/CAM, CAE, CIM etc.). Therefore, new technologies, new materials, new architecture, control system and communication systems have become necessary to construct the advanced robotics with higher level engineerings with sensing technologies, computer science, artificial intelligence, multi-technologies and integrated systems.

3. Robotics in manufacturing

Here reports only manufacturing robot systems to say indoor system. In Japan, it has attempted to utilize industrial robots instead of human workers in order to increase productivity in advanced FA. Many robots have been used in factories ; Automobiles, Machinery and so on. And then it has already been steadied to use robots in loading/unloading and spot-welding. Moreover new industrial robots have become of dexterity and flexibility, consequently robot's working fields extend to arc-welding, assembling to need of software: programming language and intelligence for communication systems.

3.1. Application system

In Japan, existing applications of robots are as follows.

(1) Arc-welding (Photo. 1)

Arc-welding requires more skill and dexterity than spot-weldings. Automatic arc-welding entails the following problems:

(a) welding characteristics of materials and methods are varied,

(b) velocity control of a torch should be stable,

(c) skillful motion is required.

Many practical arc-welding robots have rapidly been developed and utilized in Japan, depending on development of automatic welders.

(2) Assembling

Assembly robots have been utilized in electric and automobile industries. Conditions to succeed in robot applications for various industries are shown as follows:

(a) assemblings are composite workings of handling, transfers and locating parts and tools, because assembly robots are used of grasping, grapping, and locating characteristics.

(b) assembly robots are increased of productivity, flexibility and dexterity, in manufacturing systems, with high speed motion, high degree precision and skillfulness.

(c) it is important to use robots without unnecessary function and ability(ie. all too high flexibility).

(d) parts and products designs have been changed for assembly robots.

Most assembly robots are SCARA (Selective Compliance Assembly Robot Arm) robots in Japan. R&D of sensors (vision, tactile, proximity, etc.), sensory controls (high speed a and quality processing), and mechanisms (high performance becoming lightweight, etc.) are necessary.

(3) Deburring (Photo. 2)

Nowadays, robots have been working in finishing processes. Practical finishing robots will step by step be used in deburring of plastics and aluminums, and in polishing of cast metals. Automatic and flexible finishings using robots entail many problems. These problems are adaptive control of motion and force, and cooperative control of locations and workings between parts and tools. It is necessary in machinery robot systems to develop of high level sensory systems with sensors of vision, tactile and force, and to develop of cutting processing and algorithm.

3.2. Architecture

The industrial robot's architecture consists of mechanisms and control systems (Figure 3). Machine coordinate systems are referenced to robot's mechanism, and task coordinate systems are mainly referenced to control systems. The robot is constructed of part mechanisms which are generally body, legs, feet, arms, hands, and fingers, but it is defined by the International Organization for Standardization (ISO) Technical committee (TC)184 / Subcommittee(SC)2 that industrial robots are Manipulating Industrial Robot (MIR).

The most important function of industrial robots is end effector motion. It is included of control systems which include working control and communication control (Figure 4).

Working control includes motion control and acting control, and communication control includes servo control, teach/program control and adaptive control.

Motion control also includes joint control and effective control which commands robot each motion : mechanical joint motion and end effector motion. End effector motion has basically three types motion : moving to a point, moving from a point to a point, and moving on a continuous path of mechanical joint motion with interpolations, control of a velocity and an acceleration. Acting control includes process control and object control. Process control has to make the working sequence, and object control has to make robot's operation with motion speed and postures, of moving robot's foot, leg, manipulator and tools.

According to the view of robot's basic mechanism, the mechanism includes drive mechanism (see 3.3. Drive mechanism) and stand mechanism which support robot's base and builds. Robot's mechanical joint motion is referenced to machine coordinate systems (base coordinate system*), and other hand one's effector motion is referenced to task coordinate systems (world coordinate system / mechanical interface coordinate system*). And then one of mechanical joint motion is only mobile motion but for locomotive motion in future.

Accordingly, it will be unnecessary for us to study robot motion except manipulating motion, because the practical robot's mechanical joint motion is as follows.

(a) up-down turning of arm
(b) up-down traverse of arm
(c) out-in traverse of arm
(d) right-left turning of arm; rotation of arm
(e) right-left traverse of arm
(f) revolution of arm
(g) bending of arm
(h) bending of hand
(i) revolution of hand; swivel of hand

3.3. Drive Mechanism

Electrical drives, hydraulic drives and pneumatic drives are utilized in robot articulation. Many robots have suitable actuators for applications in each industrial fields. In general robot which requires powerful articulation will have mighty actuators, one which requires high speed or velocity will have lightweight actuators, and one which requires high accuracy will have special compound actuators.

The following applications have been made ;

(1) Electrical drives : assembly, arc-welding, measuring,

painting, palletizing, spreading, inspection,(with flexibility, skill and dexterity);

(2) Hydraulic drives : transfer, handling, spot-welding, assembling, palletizing, (with powerful and productivity);

(3) Pneumatic drives : transfer, loading/unloading, handling ,(with high speed and productivity);

The existing robot drives have electric-hydraulic and electric pneumatic methods compounded with plural drives.
Use of electrical drives is increasing because high performance servo motors were developed with lower inertia moment and high torque. Moreover, DC/AC servo motors have become compact and lightweight. New converters can change rotary motion to straight motion, and reversibly can also easily change straight motion to rotary motion. Also, newly developed mechanical power transmissions have replaced chainsprockets, ballscrew-reducers, etc.
One of the new power transmissions is called Harmonic Drive. It has unique operation, compact and light construction, and offers advantages of high reduction ratio and very low backlash. It has three elements : Circular Spline, Flexspline, Wave Generator (Figure 5).
New actuators have utilized brushless motors, stepping motors and liner motors. The drive mechanism becomes more compact and simple, for example, which utilized direct drive with transmissible function.

3.4. Materials

Since the 1950's research related to industrial robots have focused on development of materials, in order to produce light weight, cost robots. Several types materials have been developed; however, these materials are currently too expensive for use in industrial robot fabrication.
An aluminum alloy is currently being tested for fabrication of lightweight robot with high productivity. It is also possible at the research stage to make use of a shape memory alloy which has a quality named of Shape Memory Effect (SME).[2] A university and a research institute have conducted experiments on the SME control system (Figure 6).
Other interesting developments are a rubber actuator[3] with which a robot's arms are constructed, resembling a human arm and muscular system (Figure 7), it has a new actuator which uses metal hydrides with a thermal-mechanical energy conversion.
In Japan, R&D of robot materials is important problems especially for robot mechanisms. The development projects of new materials will be increased in the industries, because robot materials are soon replaced without exception.

3.5. Communication system

Communication is very important for robots in manufacturing

system. Development of manufacturing systems has brought a recent increase in robot usage, consequently a complex communication system is need of refined hardware and software. In a robot communication system, hardware includes teaching device, a robot controller, a computer, an interface ; and software includes a teaching/showing methodology, a control system, a language system, a simulation system. Here the important and necessary robot languages, control systems, interfaces, simulations and modelings are confirmed.
Robot languages have only been developed for assembly systems to date. Both control systems and interfaces depend on each application systems. Accordingly, one system is easily unable to connect with another system under existing economical and technological conditions.
Moreover, a simulation and a modeling are individually researched or developed at universities, research institutes and government laboratories only.
Communication systems may be classified according to the following phases, styles(operations) and categories(CAD/CAM).

(1) real communication systems
  (a) teach and control robots [on site style]
  (b) programmed control robots [off site style]
  (c) communicate with robot system [on site style]
  (d) communicate with computer systems to control robots [off site style]

(2) virtual communication systems
  (a) design a virtual robot and a robot system[CAD]
  (b) input a model of robot and a model of robot system [CAD/CAM]
  (c) simulate a robot and a robot system [CAD/CAM]
  (d) analysis a robot and a robot system [CAD/CAM]
  (e) fabricate a robot and a robot system [CAD/CAM]
  (f) program to operate a robot and a robot system [CAM]
  (g) produce a robot motions [CAM]

In real communication systems, an operator directly controls the robot on site, and controls through data command off site. There is difference between a real communication system [off site style] and virtual communication systems. Operator can take care of robots on real time in real communication system, but can't do in virtual communication system. However, real communication system is need of higher quality in processing for robots than other systems. Also it is difficult too much to realize virtual communication system which requires robots higher accuracy and resolution, because model of virtual communication system only in software is not real. It can be not always to keep high performance, reliability, adaptability, between virtual communication system and robots.
Since the community of data structures and the interfaces between each systems individually make any differences, the communication systems need to standardize functions and data structures. The communication systems mentioned above are able to be shown in Figure 8.
With regard to practical robotics/graphics systems around the world, McAuto's Robotic system PLACE is utilized in USA

(Figure 9). In Japan, the following systems have gradually been developed.

(1) Cooperative robot motion system; Hokkaido university
(2) Distributed robot system; Yokogawa hokushin
(3) Robot programming system using Interactive Graphics; Hitachi
(4) The interactive graphics programming and simulations system; IBM Japan

3.6. Robot Language

Japanese technologies have resulted in so many robot application systems in industries, but few advanced robot software systems have utilized except teaching/playback systems and scalar robot systems in Japan. The majority of the above mentioned practical robot applications utilized teaching/playback control, but assembly robot languages have been developed to control only SCARA type robots. It has generally based on horizontal articulated motions on three degrees of freedom motion with a compliance control function. Assembly SCARA type robots have more than one or two degrees of freedom motion such as up-down traverse of hand on basic articulations of themselves. In SCARA type robot diffusion, robot language systems have been able to process sensory informations for actual assembling. In Japan, the existing robot languages are of primitive levels as follows:

(a) ARCLE(Assembly Robot Control Language)[6]; Nissan
(b) ARL(Assembly Robot Language)[7]; Hitachi
(c) AR-Basic; Daikin
(d) LUNA(Language for Users Needs and Aims); Sony
(e) MELFA(MEL for Factory robot language); Mitsubishi electric
(f) PASLA(Programmable Assembly robot Language); NEC
(g) PLAW(Programming Language for Welding robot); Komatsu
(h) SCOL(Symbolic Code Language for robot); Toshiba
(i) SERF(Sankyo Easy Robotics Formula); Sankyoseiki
(j) SPEL(Suwa seiko Production Equipment Language);[8] Suwa seiko**
(k) Yaskawa C. Robot language Yaskawa

3.7. Standardization

It is hopeful that the standard for industrial robot is proposed by robot users. Also, regarding problems concerning the industrial robot, JIRA has made the following standardization to the present in Japan as follows :

(a) Glossary of Terms for Industrial Robots
(b) Graphical Symbols for Industrial Robot
(c) Methods of Indicating Characteristics of Industrial Robots and Measuring Their Functions
(d) Safety Standard on Industrial Robots
(e) Standardization of Wiring and Piping on Industrial Robots
(f) Standardization of Indicators Appearing on the

Operation Panel of Industrial Robot
(g) Standardization of Modularization of Industrial Robot
(h) Standardization of Electric-Signals of Industrial Robot
(i) Standardization of Joints Arms and Hands of Industrial Robot
(j) Standardization of Robot Language

4. R&D in Japan

In the past, main R&D of robots in FA was on high power, high speed and high accuracy. However, at present, the focus of R&D has shifted to adaptability, operationality and reliability. It is need of human interface between operators and robots, such as the relationship between operators and machines in man-machine system, when system engineer designs robot systems in CAD/CAM, FA, FMS, CIM.[10]
Now, there are many problems for using robots and robots system. One of these is a few practical application supplied from the each universities, research institutes, government laboratories, and other associations which conduct R&D for robotics individually.
Because, it is currently done in Japan that "National R&D Project on Advanced Robotics" has made progress of robotics, and "International Robotics & Factory Automation Center(IROFA)" has been established ; Japanese robotics will be very hopeful. The technical status of robots, and their Application's Reference.[11]
Here describes communication and intelligence of many problems : Robot language and sensory controlled operation.

4.1. Robot languages

Above mentioned assembly robot languages have been developed, so many software have been presented. Especially it is remarkable to research and develop robot languages and simulators.
R&D of robot languages are active forward a gloval level, several classification methods for robot languages have proposed, for instance Bonner et al. proposed classification by description (programming) level,[4] and Inoue has suggested considering, executing language functions.[5]

Bonner's classification	Inoue's classification
micro-computer level	command level
point to point level	primitive motion level
primitive motion level	structured motion level
structured programming level	object situation level
task oriented level	task working level

The following describes of
DECISION MAKING OF CONTROL METHOD WITH LANGUAGES.
The adaptive control method is selected for the robots, in

general, conversation is completed by an ability of a listener rather than a talker, so that communication in a robot system is not completed by an operator, but by a robot. Accordingly, the necessary and sufficient condition which decided the level of a robot language is depending to the set of robot functions. This condition J can be expressed in the set form by robot function K.

$$J = \text{solve } (K = K_1 \cap K_2 \cap \cdots\cdots\cdots \cap K_n)$$

where K : functional elements, control solution S result in union of a prime condition factor J.

$$S = J_1 \cup J_2 \cup J_3 \cup \cdots\cdots \cup J_{m-1} \cup J_m$$

A control execution E becomes subsets of control solution S.

$$E \in S$$

A robot language systems is designed with control solution S namely a control method. Language systems depend on communication tools : computer system. In the design of a robot Language system, it is necessary to consider both language expression methods and robot control methods.
Therefore, a robot language Lr is producted by a robot's Or and a robot procedure function Pfr.

$$Lr \ni (Or \cap Pfr)$$

Robot language instructions Lr should reach the total sum of possibilities of practical function expression, on a control solution S, because Lr is expressed by the effective cooperatings with acting function Af and control execution E .

$$Lr_i \fallingdotseq \int_O \int_P E_O \cap Af_{OP} \qquad : \text{vagueness!}$$

4.2. Intelligence

Artificial intelligence (AI) is too strange to be explained, since the existence of AI is the dream of human beings.
However, the R&D of the basic AI have been proceeded, and the few intelligent machines were fabricated.An AI should be integrated all of functions of knowledge, inference, cognition; thinking, learning, understanding, recognition, communication, decision making, and representation. Prof. M. Minsky describes the AI R&D which proceeded "Intelligent Machinery ", in his "Steps toward Artificial Intelligence" [MIT].
Japan began to investigate of new computer technologies in 1979, with the Fifth Generation Computer System (FGCS) project resulted in the foundation of the Institute for New Generation Computer Technology (ICOT) in 1982 .
Japanese R&D has resulted in the Personal Sequential Inference Machine (PSI) and the Sequential Inference Machine Programming and Operating System (SIMPOS). However, R&D of new prototype

robot systems have just begun. Recent examples of this development using only sensors include ;

(1) Seam Tracking Robot System by Molten pool Vision sensor (Yaskawa Electric)
This arc-welding robot has a hand assembled Sensor and Torch[Motoman-L10W](Photo. 3).
A fiber scope always is possible to recognize a molten pool during arc-welding. While a laser source works to find the seam by leading of welding torch. A fiber scope and a laser source are possible to rotate independently, and manipulator rolls by means of a motor and a gears. Both axes of the optical sensors are coincident with a molten pool.

(2) Deburring Robot System by force sensor tool (Meidensha Electric)
An advanced robot has been developed to meet the problems for economy quality and functions in the industry [MeiRobo GWR60]. This robot controls a hand itself by the force control system which is sensing of load variations (Photo. 4).

(3) Assembly robot system by cooperative two arms (Fujitsu)
An articulated robot with six degrees of freedom as an arm [FAROT-M6 250II] has two arms. One arm can supply, transfer, and pose a part which robot handled and grasped. Another arm can insert the part into another part. And this robot is possible to complete products using two arms of itself (Photo. 5).

4.3. Investigation for the standardization

At ISO, new TC184 "Industrial Automation Systems" / SC2 "Industrial Robots" were established in 1984. The five titles in SC2 are as follows :

(1) WG-1 : Terminology and Graphic Representation
(2) WG-2 : Performance Criteria and Related Testing Method
(3) WG-3 : Safety
(4) WG-4 : Requirements for Programming Methods and Data Communications
(5) WG-5 : Mechanical Interfaces

TC184 has five SC1-5s, and more standards have been established by the International Electrotechnical Commission (IEC), therefore ISO TC184/SC2 should make liaisons with other SC's and IEC as much as possible. Investigation for ISO/TC184 started from 1986.

5. Conclusion

In the past, main focus of R&D in robotics was on high power, high speed and high accuracy. However, at present, the focus has shifted to adaptability, operationality and reliability. It is need of human interfaces between operators and robots, such as the relationship between operators and machines in man machine system, when system engineering designs robot systems

in CAD/CAM, FA, FMS, CIM.[10]
Problems of robotics, in future, are to adjust advanced robots in the integrated systems which are need of the high level existed technology and manufacturing including of new technology and manufacturing researched and developed in all fields. Moreover, manufacturing will have a big problem that the higher level advanced robot is coming, the more impossible robots mass production isn't.
The prosperity of robot's industry in Japan has been based on the national character: diligence and adaptability. In the industry, Japanese workers have coexistence and coprosperity with tools and machines, themselves.
However, in Japan, there are many problems for using robots and robot systems. One of these is a few practical application supplied from the each universities, research institute, government laboratories, and other associations which conduct R&D for robotics individually. However, it is currently done in Japan that "National R&D Project on Advanced Robotics" has made progress of robotics, and "International Robotics & Factory Automation Center" has been established; Thus, Japanese robotics will be very hopeful. The technical status of robots, and the R&D activity in Japan, was described in "Intelligent Robots and their Applications" Reference.
Incidentally, I hope future robots which can know location and situation by themselves in working space of the world. Therefore, We should develop looking eyes and feeling skin for robots. It will do over again to R&D of Intelligent robot as well as life. It is necessary for robots have possess the following characteristics:

(a) adaptability
(b) self-recognition
(c) recognition of environment

1) Japanese Industrial Standard "Glossary of Terms for Industrial a Robots" JIS B-0134-1979, Japanese Standard Association
2) Y.Yamamoto, "Development of Arm with 5 fingers, an application of SMA to Robotics", ROBOT no.46 pp81-86, January 1985
3) S.Kobayashi, et al., "Artist Robots", ROBOT no.47 pp.43-49 April 1985
4) A.Bonnor, G.Shin K. "A Comparative Study of Robot Languages" Computer IEEE vol.14 no.12 pp82-96, December 1982
5) H.Inoue, "Towards an Advanced Robot Programming", Journal of Robotics Society of Japan, vol.2 no.2 pp3-6, April 1984
6) H.Suda, "ARCLE(Assembly Robot Control Language)", Journal of Robotics Society of Japan, vol.2 no.6, December 1984
7) S.Mohri, et al, "ARL(Assembly Robot Language)", Journal of Robotics Society of Japan, vol.2 no.6, December 1984
8)K.Saito, "SPEL(Suwa Seiko Production Equipment Language with CAD)", Journal of Robotics Society of Japan, vol.2 no.6, December 1984

9) FGCS'84, ICOT, November 1984
10) S.Takashima, "CAD system for CAM assisted with Robots", Operations Research, no.7 pp511-525, September 1982
11) T.Kitagawa, "Computer Science & Technologies", Japan Annual Reviews in Electronics, "Intelligent Robots and their Applications", Computer & Telecommunications, pp189-275.

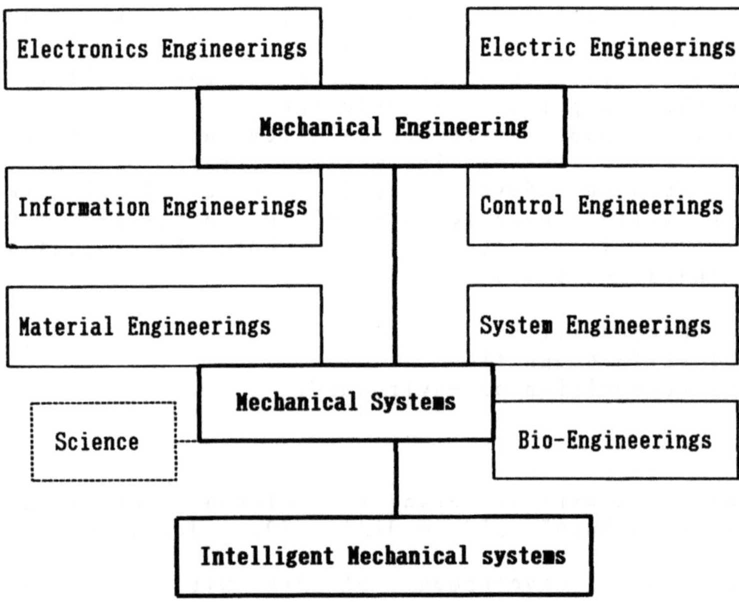

figure 1 Intelligent Machines / System Engineerings

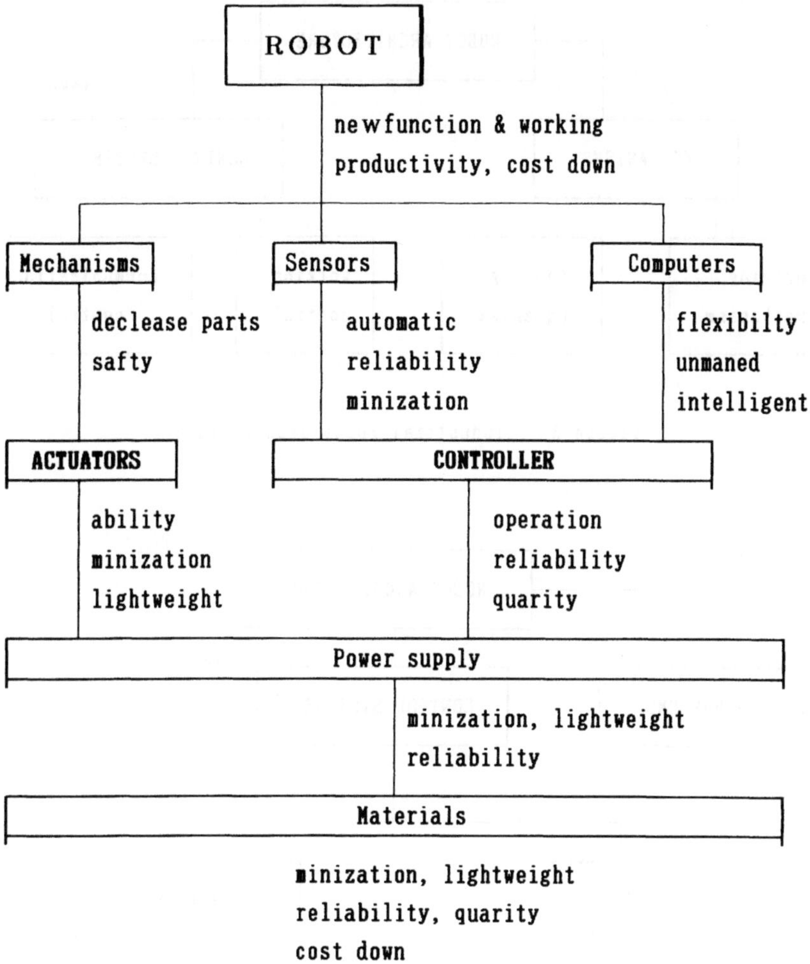

figure 2 The functions & characteristics of a robot
(improvement points to be hoped on each parts)

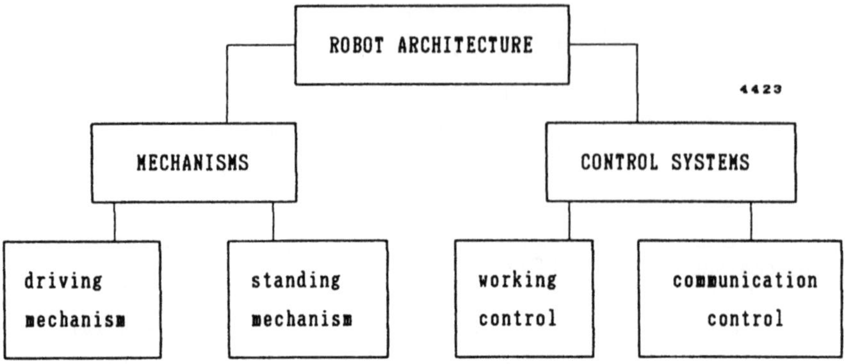

figure 3 Industrial robot architecture

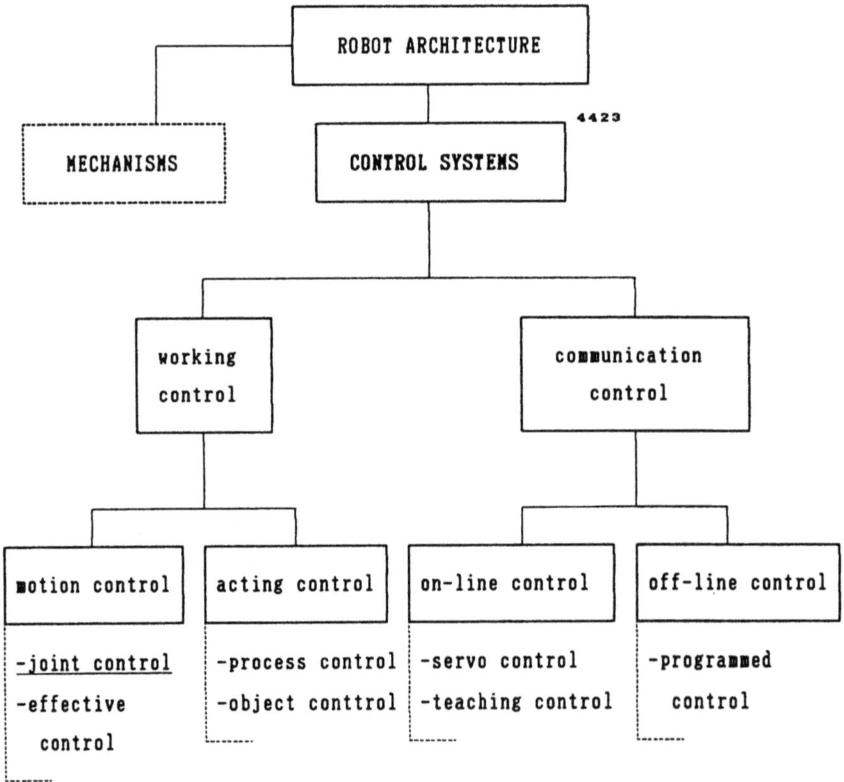

figure 4 Industrial robot control architecture

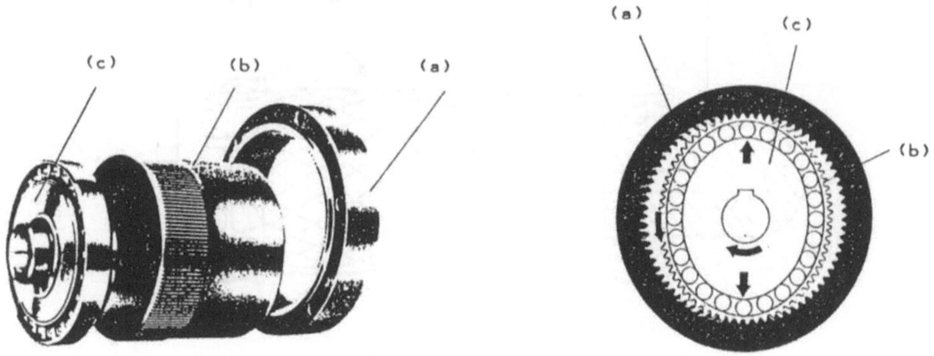

(a) circular spline (b) flex spline (c) wave generator

figure 5 A harmonic drive and basic operations

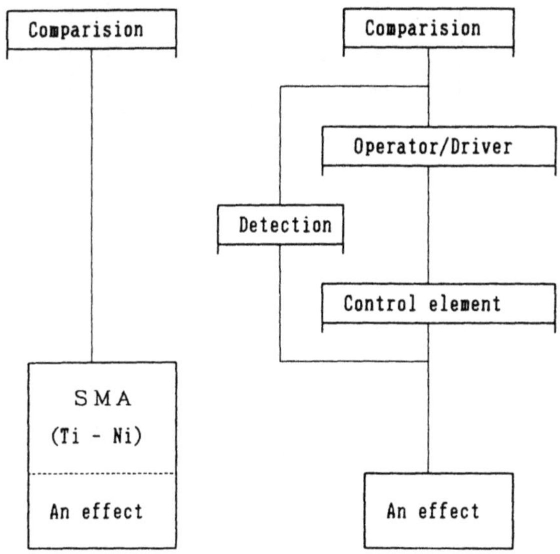

(a) SME control system (b) Servo control system

figure 6 Simple SME control system compare with existing control system

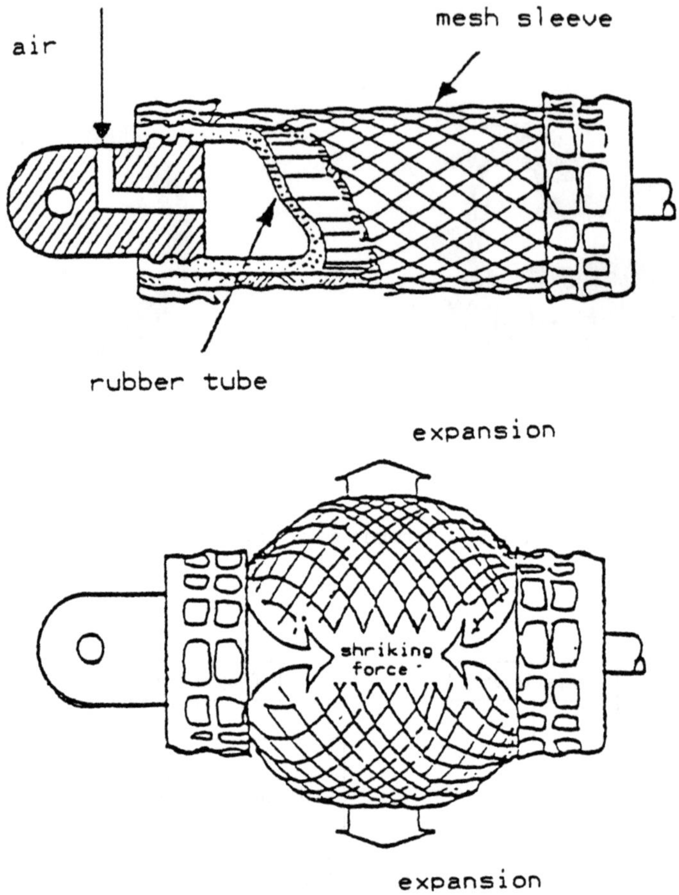

figure 7 A construction of rubber actuator

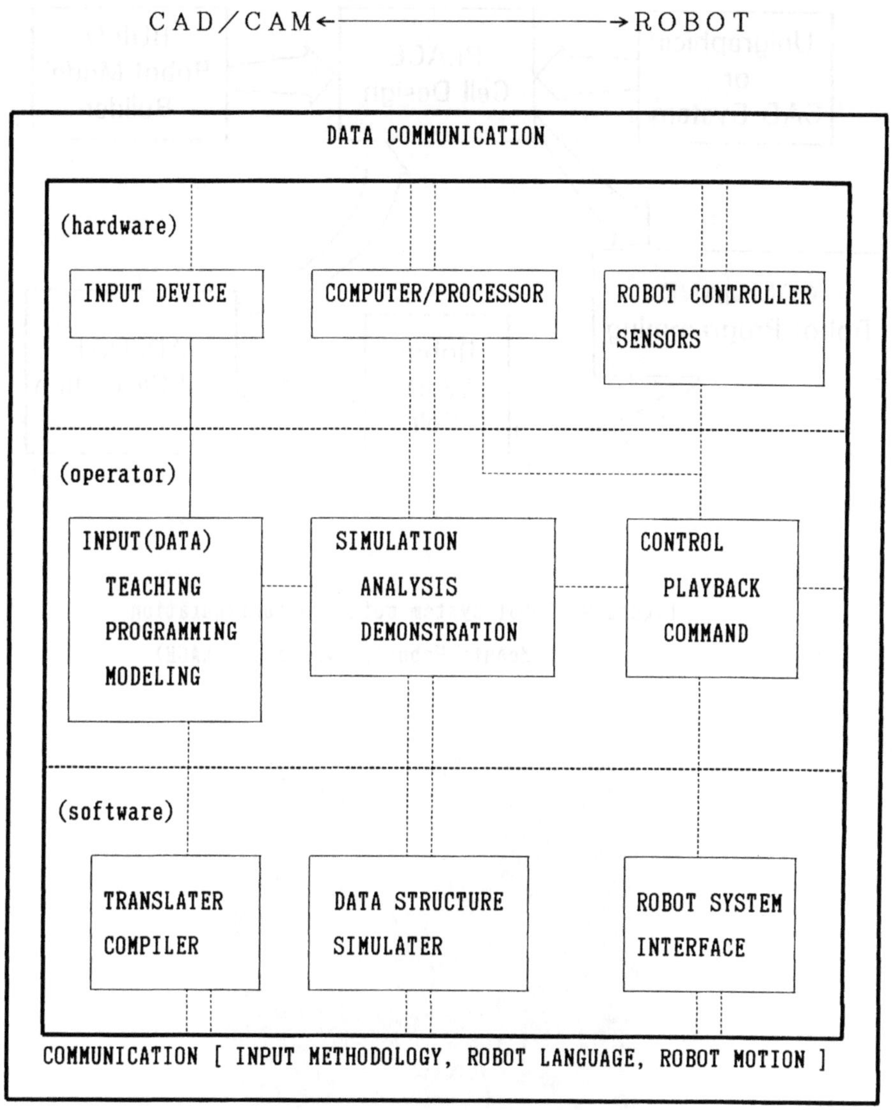

figure 8 Communication System on CAD/CAM

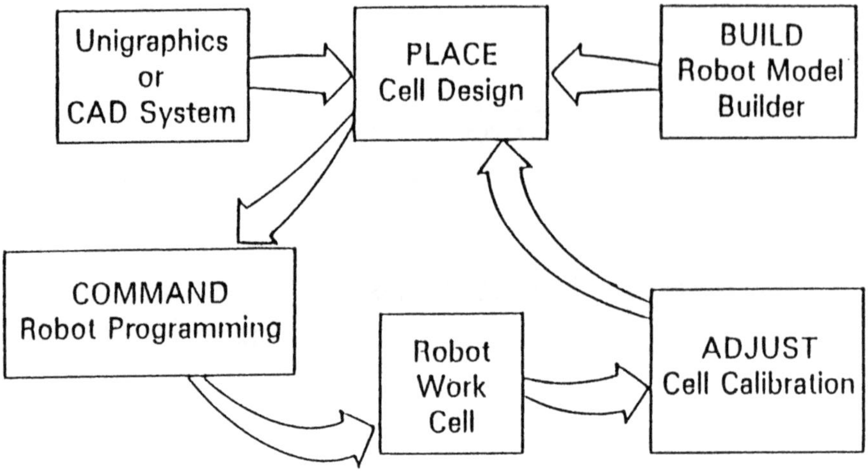

figure 9 Robot system software configuration
(McAuto Robotic Systems : PLACE)

photo.1 An arc-welding robot

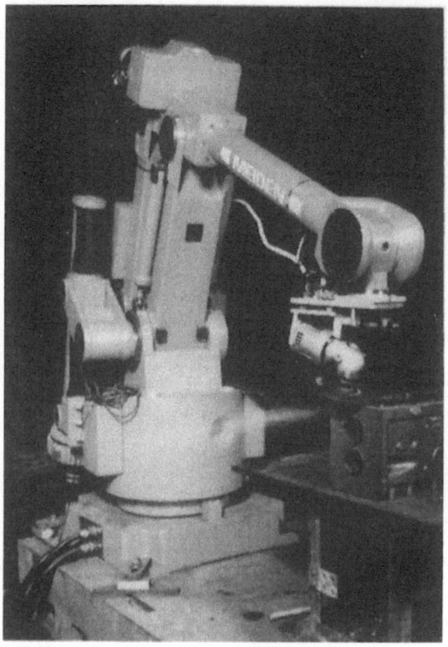

photo.2 A deburring robot

photo.3 A robot hand with a vision sensor

photo.4 A deburring tool with a force sensor

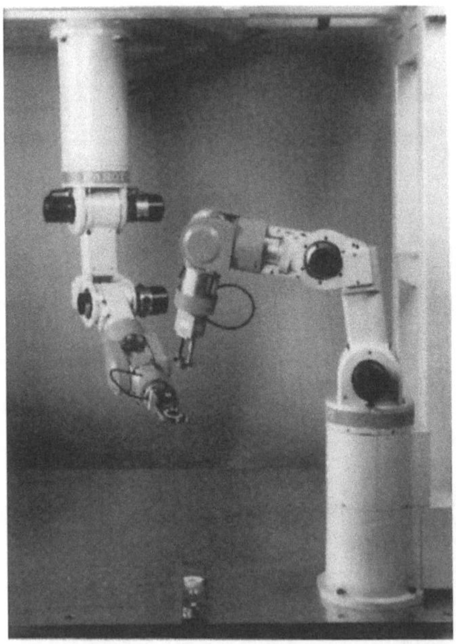

photo.5 An assembly with cooperative two arms

PART 2: NEW DIRECTIONS

NEW DIRECTIONS IN DECISION SUPPORT FOR MANUFACTURING

Abhijit Chaudhury, Shimon Y. Nof and Andrew B. Whinston*

Purdue University

West Lafayette, IN 47907

ABSTRACT

Accepting the view of a decision support system (DSS) as being composed of three generic components (the language system, the knowledge system, and the problem processing system), we investigate several issues relating to a flexible manufacturing system (FMS) that have an impact on the design of the three DSS components and their interrelationships.

It is observed that an FMS is a distributed decision making structure composed of autonomous cells. Concepts and methods from economics and distributed computation together with a language and formalism supplied by modal and temporal logic can constitute an effective set of representation and problem processing means for building decision support systems for FMS.

*A. Chaudhury and A.B. Whinston are with the Krannert Graduate School of Management. S.Y. Nof is with the Department of Industrial Engineering.

NATO ASI Series, Vol. F49
Computer Integrated Manufacturing
Edited by I. Burhan Turksen

1. INTRODUCTION

With the advent of computers on the production floor, there is a major revolution impacting the world of manufacturing. Computers are being introduced to control, monitor and supervise processes in a more cost-effective manner. In this paper, we explore developments in areas of distributed computation, modern logic and economic theory and study how they can aid us in building a decision support system (DSS) in an automated manufacturing environment.

We will view a DSS as being composed of three generic components: the language system, the knowledge system and the problem processing system. Using the above framework for a DSS, we investigate several issues relating to an FMS system that have an impact on the design of the three DSS components and their interrelationships. The issues we study are:

(1) Flexible manufacturing system: An overview
(2) Framework for a decision support system for an FMS
(3) Modeling of the distributed planning and control structure
(4) Intermodule communication and message protocol
(5) Issues relating to the concurrency and temporal relationship of processes
(6) Knowledge acquisition and reasoning with knowledge

2. FLEXIBLE MANUFACTURING SYSTEM (FMS): AN OVERVIEW

There are two kinds of machine configurations used in flexible manufacturing systems: functional layout and cellular layout.

In functional layout, functionally similar machines are grouped together. In cellular layout, which is also known as a group technology layout, various machines pertaining to one group of operations are grouped together. Large scale FMSs are generally implemented using the cellular layout. Jobs which enter the system are initially assigned to a cell for a family of operations and then move from one cell to another for different classes of operations before making their exit from the system.

A cell consists of several machine tools served usually by robots or other automated material handling devices. The cell supervisory system controls the machine tools and the robots which constitute the cell. This is the level in the control hierarchy where a human operator usually intervenes to modify and preempt the commands issued by the automated control systems in the FMS.

For material handling and material movement among cells, automated guided vehicles (AGV) or powered conveyers are commonly used. This material handling system is under the control of a shop level supervisory system. Many shop level control functions are still carried out by human operators in interaction with the shop supervisory system.

The cells are connected with each other and the shop control system through a communication network. The shop supervisory system is in turn connected with the CAD/CAM system and the office information system. A typical configuration of an FMS with component cells is shown in Figure 1.

An FMS can therefore be described as a collection of islands of machine-tools connected together by communication links and material handling means. An FMS is a dynamic system, where a wide variety of jobs are continuously introduced into the system in a random fashion. These jobs are analyzed into operations, the operations are scheduled on various machine-cells and the completed jobs are removed from the system, all with or without minimal human intervention.

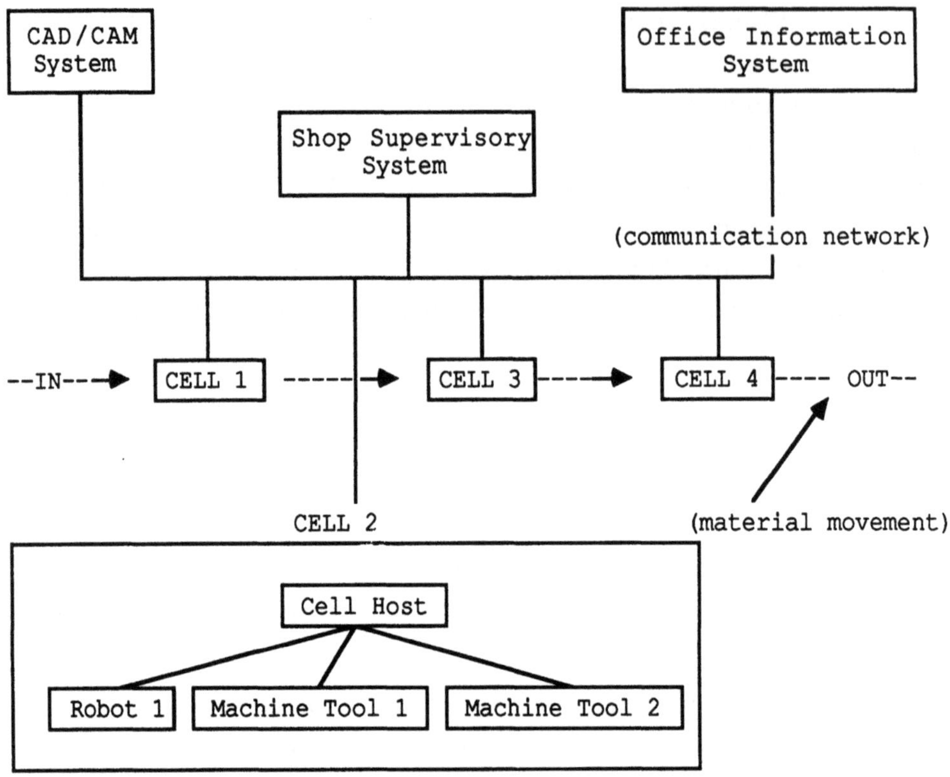

Figure 1. An FMS with Component Cells

2.1. Hierarchical Planning System in An FMS

Like any manufacturing hierarchy, the decision making process in an FMS spans its multi-level structure, from that of an individual robot to a machine-cell and right up to the facility level. Usually the highest level decisions and longest planning horizon exist at the topmost level, and the planning horizon decreases progressively, as we go down the levels of hierarchy.

De, Nof and Whinston describe the planning and control tasks at different levels of a manufacturing hierarchy in [12], a summary of which is shown in Table 1.

The planning and control function exercised at each level of an FMS can be described as follows:

Table 1. Manufacturing Levels and Planning & Control Tasks

LEVEL	PLANNING & CONTROL TASKS	TIME-HORIZON
Facility Level	Plant-wide database, Master Schedule, MRP	months/years
Shop Level	Process planning, scheduling of jobs to cells, feedback to higher levels	weeks/months
Cell Level	Scheduling of jobs to machines, dispatching, failure recovery, feedback to higher levels	hours/days
Single Machine-tool Level	Scheduling of individual operations like load, move, grip, etc.	seconds/hours

1. At the lowest level in an FMS, the single machine-tool level, we have the individual machines or the robot controllers. Cutowsky et al. [10] list the functions of a machine controller as follows: (a) send and receive messages from the cell-controller, (b) allow the cell controller to access the machine and controller state, (c) receive input from the sensors, and (d) decompose a task into elementary operations and send output signals to motors, solenoids, valves, etc.

   A machine controller should be able to cope with contingencies like occasional failure which may result from various causes like faulty parts, tool breakage, robot drift, etc. For a controller to be classified as intelligent and autonomous, it must exhibit rudimentary abilities towards error detection and error recovery.

2. The next level is that of a cell. The cellular layout is a common configuration used to arrange an FMS system. A manufacturing cell, as described earlier, is a group of machine tools and associated material handling equipment that is managed by a supervisory computer called the cell-host. Cells behave as independent units, which are tied together to form an FMS. The equipment that moves material through the cell and the cell-host that supervises

all cell activities provide the means of autonomy. The cell-host has to serve several functions, some of which are: (a) decompose a cell-level task into machine-level tasks and schedule these tasks within the cell, (b) maintain a database of part details and process details, (c) execute the software for gracefully degrading the system in the event of a failure, and (d) instruct the individual elements under its control, such as the machine tools and robots, as to the tasks each must perform.

3. The level above that of a cell is the shop level. The decision and control functions exercised by the shop-level supervisory system are very similar to that of a cell-host. Also, the system at this level has to control the dispatch of material handling equipment, monitor the condition of various cells, collect data, etc. The shop supervisory system interacts with the engineering CAD/CAM system and the company-wide office information system.

4. The highest level is that of a facility. The premanufacturing phase of planning is accomplished here, such as: MRP, master production schedule, batching, and detailed release date schedule for introducing the jobs into the shop.

As can be seen from the above, there are three generic planning and control functions applicable to all levels of decision making in an FMS: (1) task decomposition, (2) allocation of tasks and resources to lower level elements, and (3) feedback to a higher level and updating of records.

The exercise of these functions at any level requires exchange of information and decision making. These decision issues may be structured or unstructured problems like scheduling and error recovery. Also the systems in an FMS have to exchange information with systems beyond the FMS boundaries. The CAD/CAM system and the Office Information system interface with the FMS system so that design data, process data, material data, and order and billing data systems can be integrated where possible with the manufacturing control system. To assist the control systems and the human operators in an FMS to make these decisions and to provide them with the necessary data we need a decision support system. In the next section we discuss the generic structure of a DSS and also investigate how the requirements of an FMS influence the design of its decision support system.

3. DECISION SUPPORT SYSTEM FOR AN FMS

A Decision Support System (DSS) has been defined as a computer based system that assists decision makers, whether human or computers, to make appropriate decisions. As described in [5], a DSS can be viewed as having the following three distinct components (refer to Figure 2):

(1) A language system (LS)

(2) A knowledge system (KS)

(3) A problem processing system (PPS)

3.1. Language System

A language system (LS), is the total of all linguistic facilities made available to the decision makers by a decision support system. A language system encompasses languages for information storage into and retrieval from the knowledge base and also languages required for computation. It is characterized by the syntax and semantics that it furnishes to decision makers. It is therefore a vehicle that allows decision makers to express themselves and at the same time it limits the permissible expressions allowed in the decision support system. The language system adopted has a strong influence on the kind of user interface one can have in the decision support system.

Language systems vary in terms of the level of procedurality that they require for expressing a problem. At one extreme are

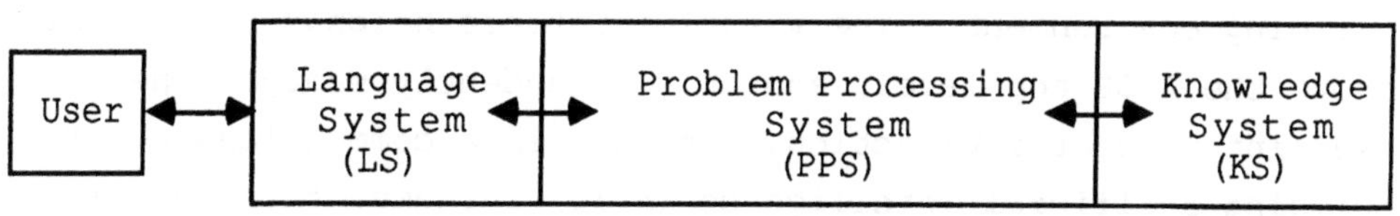

Figure 2. Structure of a Decision Support System

procedural languages that allow the user to state a problem by specifying the procedural steps involved in solving that problem. At the other extreme are non-procedural languages that allow a problem to be specified by merely stating the characteristics of the solution.

For an FMS, the LS must have the necessary syntax to allow the decision maker at any level to represent the problem adequately. Decision makers at different levels in an FMS use different levels of abstraction to describe a problem. The LS must ensure that its communication to a particular decision maker is appropriate to the decision maker's level of abstraction. For instance at a facility level, a decision maker will be dealing with aggregate issues like 'manufacturing', 'shipping', 'quality-control', etc. At the shop-level the concern will be with detailed activities like 'machining', 'assembly', etc.; and at the machine cell level, with a more granular level of activities like 'drilling', 'milling', etc. The LS for an FMS can be viewed as being composed of several language systems, each appropriate for a particular hierarchical level of decision making.

The level of abstraction is one of the factors that determines the syntax and symbols of the language at any level of hierarchy. The kind of user also determines the type of language. In an FMS environment the user could be either a computer or a human operator. The human operator on the shop floor is not likely to be an expert programmer. If the user is a human operator, the language used by the DSS for that user should have a strong visual component. It should be able to represent in a natural fashion such concepts as flow chart, process network and job-orientation. In [26], Tortora et al. present an icon based visual extension of the LISP language called ICONLISP which is used for implementing a visual programming environment. The visual interface belongs to a level above the LISP top level, and it independently manages the user/interpreter interaction. As LISP is a common language for building artificial intelligence (A.I.) systems, this system permits a common environment to implement not only a suitable user interface but also the other two components of a DSS -- the knowledge system and the problem processing system.

The FMS as we have seen is a multi-computer system, where a single instruction from a higher level has to travel through several layers of computer control systems before it gets physically executed. The language system in such an environment should not only take care of the human/DSS interaction as discussed in the previous paragraph, but also machine/DSS interaction. These computer systems at various levels in the DSS are part of the hierarchical control system of the FMS. In a hierarchical control system, each node receives its instruction from the next higher level. The instruction generally consists of a goal, a stop function (to be executed in the event of system failure), and a specification of a return message. The interaction between the DSS and the machine-controllers will generally concern these instruction messages. The language system adopted for a particular level for machine/DSS interaction should have the features of a robot control language and should have the capacity to express the following:

- Goal primitives as assigned by the higher level
- State primitives to describe a given state
- Action primitives which are used to execute the instruction and also the stop instruction
- Language syntax that permits appropriate command messages, report messages, error messages and error recovery instructions. Lokshin et al. [17] describe similar language facilities that have been incorporated into hierarchical robot control languages.

3.2. <u>Knowledge System</u>

A knowledge system (KS) is characterized by the facilities it furnishes for the representation and organization of knowledge. A KS in a decision support system contains the large volume of facts relating to the problem domain. It may consist of data-files and databases. The knowledge representation method utilized by a particular knowledge system may be thought of as a set of rules according to which knowledge is expressed for the purpose of retention within a DSS.

Knowledge represented in a KS must be retained in an organized and systematic manner. There are a variety of knowledge representation methods. KSs vary not only in terms of knowledge they contain, but also in differing approaches that they utilize to represent and organize knowledge. If a database is employed, it may offer relational, hierarchical or network constructs. Other knowledge representation methods include those used in the A.I. area, such as the predicate calculus method, the production system approach, and the semantic net approach. Conventional mathematical representation techniques such as the matrix method, algebraic equations and calculus are also used, particularly in systems which are employing a systems-theoretic model for planning and control. For a manufacturing environment, currently the frame representation method for representing knowledge is attracting considerable interests from researchers. In a programming environment, the frame representation is implemented using an object-oriented paradigm, which will be discussed, in brief, in Section 4.

For an FMS the knowledge system has some typical requirements to meet, which are unique to an FMS environment. An FMS is a collection of decision makers, spread over several levels of a hierarchy, who infrequently exchange information to update their databases. Exchange of information consumes both time and resources and therefore only the most essential information is exchanged between decision makers. An FMS is a dynamic system which is continuously undergoing state transformation. As the information exchange takes place only infrequently, the knowledge available to the various decision makers may not be consistent. Such a situation rules out the use of usual A.I. methods of knowledge representation, which are based on propositional logic, because any method based on logical deduction would soon grind to a halt due to contradiction among databases. We will, in Section 8, study other methods of A.I. type knowledge representation, that are more suited to such a distributed knowledge environment of an FMS.

3.3. Problem Processing System

The main function of a DSS is to take strings of symbols organized according to an LS syntax and to take strings of symbols according to KS representation rules and to produce information that supports decisions. To fulfill this function, there needs to be an interfacing mechanism between expressions of knowledge in the KS and an expression of a problem in an LS. This interfacing mechanism that results in information to support decision making is referred to as a PPS.

A PPS, as the software which actually works on a problem, provides the formal specification of a DSS's behavior. A PPS must possess the ability to explicitly recognize a problem by transferring the problem statements into an appropriate executable plan of action. For an LS that requires only a procedural problem statement, the PPS problem recognition ability is at most rudimentary. Non-procedural problem statements may necessitate a more sophisticated problem recognition ability. If the user directly specifies or selects the model through the LS, then there is no need for the PPS to recognize the modeling problem. A highly sophisticated problem recognition ability is required if a PPS itself is to select or formulate the model. In general, given a problem, a PPS should be able to recognize the problem, formulate a model or choose an appropriate model from a repertory of models, analyze the solution, and communicate the results to the user through the language system.

In the context of an FMS, an essential ability that a PPS must possess is that of model management. The model management function is required to organize and implement various decision models. In an FMS, as we have seen earlier, there is a wide variety of decision issues at various levels which span the entire spectrum of structuredness from the most to the least structured decisions.

A model management capability facilitates users to access various models. The task of model management involves the representation, storage, retrieval and execution of algorithms at decision points in the system. There are many kinds of models which are relevant to the needs of decision makers in an

FMS. These are analytical or operations research types of models, simulation models, game-theoretic models, and A.I. models based on a deductive logic formalism. To manage the control mechanism in a dynamic system like an FMS, the PPS should have access to models that represent static manufacturing elements and also dynamic elements such as system goals, machine status, etc. The temporal representation of such knowledge is useful for control systems that are intended for real-time control and are using some A.I. paradigm of knowledge representation and problem processing.

Traditionally, the DSS has been viewed as a single user or a homogeneous multiple user system. However, this is far from the case in flexible manufacturing systems. Such systems are characterized by multiple users, several hierarchies, and users with limited and inconsistent knowledge and different perceptions. For the purpose of study and design, we can classify DSSs for such distributed environments along several dimensions. Although there is a continuum of possibilities along each dimension, some of them could be represented as dichotomous variables for the sake of simplicity. Some of these dimensions are:

- Number of users
- Number of hierarchies and consequently the requirement of number of languages
- Control philosophy of the system, whether the control among the user group is exercised in a hierarchical fashion or by means of a democratic mechanism
- Interrelationship of local and global goals, whether the local goals of a user are compatible with the goals of the system

In the next section we will investigate various group and distributed decision making systems and models which are relevant to flexible manufacturing systems. These may be incorporated into the problem processing system of a DSS. Also the architecture of a DSS, in terms of distribution of models and databases among various nodes, is affected by the kind of group problem-solving process that the DSS is designed to support. We will first study the general characteristics of decentralized systems and then study how subject areas like distributed computation and economic theory can help us in designing distributed decision making structures, which the DSS in turn has to support.

4. MODELING THE DISTRIBUTED DECISION MAKING STRUCTURE

4.1. Decentralized System

A decentralized system like an FMS can be perceived as a collection of autonomous modules which communicate with one another for a common goal. There are many examples of decentralized systems in this world, a ubiquitous example is the world telephone system. It consists of millions of individual telephone units all across the world; and different parts of the system are operated by cooperating organizations with different standards, hardware and languages. Other examples of decentralized systems can be found in large distributed computation systems, electronic financial systems, etc. Gray [13] lists the common characteristics shared by these distributed systems, as follows:

- Diverse organizations and organizational procedures
- Diverse hardware and software architecture
- Diverse site environments with local goals and a common set of global goals, etc.

Flexible manufacturing systems can be centralized or decentralized to different degrees. A decentralized FMS would be one where the cells enjoy a large measure of autonomy. Instead of a hierarchical planning system deciding on the allocation of tasks and resources to cells, such decisions are taken by the cells acting in concert. The 'best' form of decentralized organization depends on the balance of advantages and disadvantages of decentralization for the particular application at hand. The main benefits of decentralization in the context of an FMS, are:

- Integration of diverse types of systems
- Organizational autonomy -- Each module can be organized in a way best suited to the module and its purpose without much concern for its neighbors

- Modularity -- A decentralized system is necessarily modular. In such a system it is easy to add capacity and the change is non-disruptive in nature.
- Ease of control -- When systems become very large, centralized control is no longer technically feasible. One has to adopt a 'cooperative' type of control architecture to decompose the control problem into manageable pieces.

4.2. Distributed Computation

There is considerable similarity among the issues of concern in a distributed computer system and an FMS. The core decision issues in a distributed computation system are that of allocation of resources and tasks to a team of autonomous processors. This is similar to the generic decision issues described for FMSs earlier in this paper. Like an FMS, a distributed computational system is also dynamic and decisions have to be made in a real-time environment. There are however differences too. In an FMS, the individual cells are rarely identical, whereas in computing it is common to have processors that are identical. Similarly there is a much greater variety of tasks and resources to be allocated in an environment of an FMS, than in the world of computing.

A distributed system can be represented in a knowledge system of a DSS using the powerful and elegant concept of an 'object'. The 'object' concept forms the basis of the control architecture for many distributed operating systems. A system can be viewed as a collection of objects which exchange messages. An object is basically a data abstraction with a set of operations defined on it. Individual objects communicate by sending messages to each other and higher level objects can be constructed from primitive objects.

The object model is better able to express the distributed nature of systems like an FMS, the parallelism inherent in such systems, and the inter-module interactions. This makes the model ideal for describing problems in a knowledge system of a DSS, where it is necessary to represent a collection of objects that interact. The object model has the semantic power to represent a live system of interacting and independent modules and their features and can be used to build a real-time control system for an FMS.

In terms of a distributed computation system, an FMS can be described as a team of heterogeneous, intelligent processors that are logically integrated. A multi-user and distributed DSS for such an environment must cope with the following system characteristics of an FMS:

- A multiplicity of processors distributed over several levels of hierarchy
- A physical distribution of processors
- An autonomous and intelligent nature of processors which can take decisions over a wide range of issues that affects its performance without human intervention
- An intermodule communication system for sending messages for instruction, control and record updating

For a distributed DSS, the design of the communication system that ties together the various independent nodes into an integrated whole plays a crucial role in the success of the system. The communication component is often viewed as the fourth generic component of a distributed DSS structure, which is also the approach we adopt in this paper.

4.3. Economic Theory Approach

The basic decisions which need to be made at all levels of an FMS concern allocation of tasks and resources among competing modules. Both of these can be viewed as allocation of some commodity among agents. Economic theory has long concerned itself with the central problem of optimal allocation of resources and commodities among consumers in order to maximize their utilities. Depending on the nature of interaction between agents, economic theory attempts to outline optimal behavior in several different frameworks such as: market structure, game theory and team theory.

A market is an efficient resource allocation mechanism in economics. The market serves as a location where agents can exchange information about price and supply, and it helps to coordinate the action of numerous agents in a highly decentralized manner.

A variant of market mechanism is the auction method. Shaw and Whinston [23,24] propose the so-called contract-net approach. This approach uses an auction-bid-award paradigm for distributing tasks among competing cells in an FMS. Whenever a cell wants to farm out a task to another cell, it announces an auction. Other cell qualified to execute the task, submit their bids. The value of a bid depends on the present load in the cell and the efficiency with which the cell can execute the task. The 'announcer' then awards the task to the lowest bidder. A negotiation protocol helps to arrange the bidding and award process.

In problems that deal with allocation of resources or tasks among competing agents, the action of other agents invariably have an effect on the outcome of a given agent's action. Game theory has been used for modelling interaction among agents where the outcome or the payoff to one agent depends not only on his action, but also on the action of other agents.

Game theory, under the title of 'cooperative games' has studied the interaction of n-agents who with cooperation can improve the amount of realized payoffs compared with that obtained without cooperation. A cooperative game is one where the agents can enter into binding agreements. The study of cooperative games begins with the specification of possible agreements available to two or more players where their interests conflict and focuses on the selection of agreements. In contrast, the study of non-cooperative games focuses directly on problems of individual choice with the principal objective of obtaining equilibrium points. Formal game theoretic analyses of cooperative situations were first provided by Nash [21]. Nash noted that many agreements might be stable, that is, if an agreement were to be reached, all parties would voluntarily carry out the agreement. In other situations only the presence of a 'mediator' is able to bring about an agreement. In some cases an 'arbitrator' is necessary for an agreement to be reached, and later enforced. In the context of an FMS, the cells could be viewed as the players and the FMS control level as the mediator or arbitrator. The system architecture of a distributed DSS may be designed to reflect this arrangement.

Game theory does not concern itself with the process of acquisition of necessary information for the players to play. On the other hand, in team theory the central concern is the efficient way of providing and allocating information to a team of decision makers. This characterizes the situation in an FMS. Cells in an FMS share common global objectives, have only partial information about the total state and have to incur additional costs for collecting more information.

The economic theory of teams studies the case in which several persons perform various tasks including those of collecting information and making decisions. In this theory, all the members strive to optimize a common set of goals, unlike in game theory where each member has its own individual payoff which it seeks to maximize.

Team theory as developed by Marschak and his collaborators, and as described in [18], is basically models that derive a solution or an optimal point. It does not specify the process that needs to occur for the solution to be obtained. In [19], an extensive form of the solution is described by Moore and Whinston. In this paper the process of information acquisition is as much a subject of concern as actually obtaining the solution.

4.4. Analytical Models

Considerable work is also under way in developing analytical models for an FMS and its various sub-systems. For a survey of such models readers may consult [8]. These models form part of the repertoire of models to which the PPS in a decision support system needs to have access. As pointed out in [4], up to now the dominant tool for modelling FMSs has been simulation. Most analytical models have been based on queueing theory. These analytical models help to evaluate the performance of a proposed design and help determine the system parameters during the design phase. For instance, in [25], a queueing network of multi-server queues is analyzed for determination of machine group sizes.

For real-time control, hierarchical control systems have been suggested in [16] and [1]. These proposed systems have programming models that seek to optimize a job sequencing and job release problem. Every time a new state results due to completion of a job or a machine breakdown or due to other causes, a new optimal solution is obtained. In [9] a software system under development for real-time planning is described. The system addresses the issue of assigning resources by formulating a multiple attribute decision making model which is executed on a real-time basis every time there is a need for making an assignment decision.

For the wide variety of decision issues that arise in an FMS, we can construct a typology of decision models based on classification along several dimensions. Such a typology can be used by a KS for organizing the collection of models. Some of the relevant dimensions for classification are as follows:

- Single actor or multiple actors
- Single goal or multiple goals
- Real-time or off-line decision making
- Single step decision or a sequence of decisions
- Cooperative or non-cooperative framework for group decision making

The various decision models referred to in this paper can be classified as follows (refer to Figure 3).

Except for the system and control theory approach, the above approaches have not been generally used for real-time decision making. With the growth in speed of computer hardware, there is an increased interest among researchers for constructing real-time control schemes that incorporate such models. It should also be noted that as yet there is no single decision model that incorporates all relevant dimensions of decision making in an FMS such as: multiple actors, real-time, cooperative framework, and sequential nature of decision making. For that matter, there is yet to be a satisfactory formulation of the concept of optimality in a setting like that of an FMS. The purpose of a multi-user DSS, as proposed here, is to provide a conceptual framework for implementing a structure for distributed and democratic decision making that can deal with unstructured issues.

In the next section we study the distributed information structure of an FMS and the communication component of a multi-user DSS for a flexible manufacturing system. As noted earlier we treat the communication component of a multi-user DSS as a distinct fourth component of a distributed DSS. We will next study key issues relevent to the design of the communication component such as the communication architecture and the message protocol.

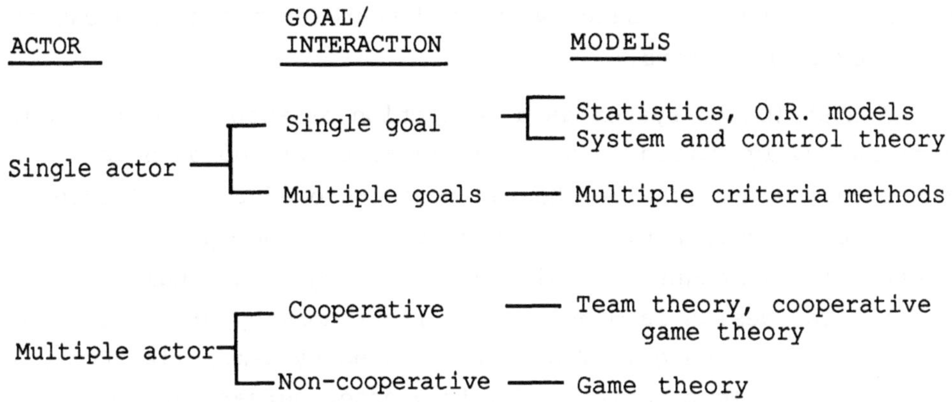

Figure 3. Classification of Decision Models

5. INTERMODULE COMMUNICATION

A key characteristic of any distributed information structure is the lack of global and centralized knowledge. The total knowledge base of an FMS is partitioned among individual cells and the higher level FMS control system. It is difficult for each cell to know everything about other cells, even though to make an optimal scheduling decision the cells need to have current data about the whole system. Generally transfer of information among processors in a distributed system is slower and therefore more expensive than local computation.

In a distributed system with autonomous nodes it is the communication system and the message protocol that hold the system together. The parts of such a system are not tightly integrated, they are heterogeneous and the integration of functionalities is brought about by having all separate parts conform to a common message protocol.

Gray [13] notes that a message protocol is an ideal way to structure the interaction between nodes in a distributed computation system. It has the benefits of: (1) encapsulation; (2) abstraction (i.e., providing a simple conceptual interface to the next level of abstraction). A processor sending an instruction does not need to know the internal rules or nature of the receiver node; and (3) autonomy. Internal methods and procedures of a node can change without disturbing the communication from other processors.

A distributed system as a collection of autonomous nodes is more naturally suited to a 'requester-server' approach for inter-cell communication than a 'remote-procedure-call' type input/output approach. A 'requester-server' design has many technical advantages in addition to its organizational advantages. The design is modular and offers more oppurtunities for non-disruptive change. Truly distributed systems, like FMSs, have to adopt some variant of a 'requester-server' approach to fit naturally with the idea of individual node autonomy.

The communication architecture and the message protocol of a multi-user DSS is dependent on the setting in which the multi-user interaction takes place. The communication component of a multi-user DSS plays a critical role in the success of the DSS. The communication system should possess the necessary mechanism to arrange data-sharing among users, democratic problem solving among users if that is the prevalent philosophy of control, and integration of individual goals and perceptions of users. In [6], Bui and Jarke, propose a separate protocol layer over and above the ISO-7 layer reference scheme for communication, to automate the group solving process in a multi-user DSS.

In general, the design of the communication component of a multi-user DSS has to address the following issues:

- Distribution of data and decision models among the nodes in the network
- Problem-solving protocol to implement the chosen group-solving philosophy
- Message protocol to permit exchange of messages to support the problem-solving protocol
- Network architecture to be used (i.e., a choice has to be made among possible configurations such as star, bus, loop, etc.)
- Communication rates and reliability
- Synchronous vs. asynchronous communication
- Choice of transmission medium and communication hardware

In the next two sections, we present some of the logical tools available with non-standard logic to model the dynamic knowledge and reasoning characteristics of the machine-cells. By the term 'non-standard' we mean any logic system other than 'first-order-logic' (FOL). Non-standard logical tools can be used for representing and processing knowledge in a multi-user DSS for an FMS.

6. CONCURRENCY AND TEMPORAL LOGIC

Concurrency is an essential and pervasive issue in the study of FMS systems. The FMS is a dynamic system having several cells, each cell consisting of several machine-tools, and all the cells and the machine tools are in operation most of the time. To control such a dynamic system and to coordinate activities on a real-time basis, the control system should be able to address the issue of concurrency. In the case of a multi-user DSS, the knowledge system and the problem processing system should be able to cope adequately with the issue of concurrency.

A Petri net representation has been the common means to describe the degree of dependence or concurrency between processes. Bullers, Nof and Whinston [7] have used a Petri net to model the relationship of processes in the domain of manufacturing. Processes can be in various relation to each other. Two processes can be strictly concurrent like the two arms of a robot lifting a heavy object, or they can be in sequence, in an arbitrary order or totally unrelated. Petri net defines 'synchronic distances' as a means of obtaining quantitative information about two related processes avoiding the notion of explicit time. For details on Petri net theory readers may refer to [22].

However, there are many occasions where an explicit representation of time is required. The problem processing system in DSSs for an FMS should have the capacity to relate two different processes and to coordinate and reason about events which become true at different points. For this, the capacity for representation of time is unavoidable. Existing means of inferencing do not support the concept of temporal truth of propositions. Propositional temporal logic can furnish us with the means to represent causal relationships among processes and events spread over time.

Temporal logic as described by Turner in [27] builds on FOL and has the following additional operators:

1. FA : A is true at some time in the future
2. PA : A was true at some time in the past
3. GA : A will be true at all future time
4. HA : A always has been true in the past

Allen in [2] has introduced temporal reasoning in A.I. applications. He employs a many sorted logic with variables ranging over 'properties', 'time-intervals', and 'events'. De, Pan and Whinston in [11] have described a temporal logic for database applications. They define primitive relations such as 'during', 'before', 'meets', etc., which are represented by a predicate in the logic and governed by a set of axioms. Features of temporal logic need to be incorporated in the knowledge system and the problem processing system of a DSS for it to prove effective in an FMS environment.

7. REASONING WITH KNOWLEDGE

As described earlier, an FMS is a real-time dynamic system characterized by a distributed information structure. Each cell in the FMS is continuously undergoing state transformation with a resultant change in its database. As information sharing is an expensive process the information is exchanged by cells only at periodic intervals and only the most essential information is exchanged. The information the cells have about other cells is partial and dated which results in various cells having inconsistent databases. We can view each cell as having its own 'world-view' and the same proposition may have different truth-values in different 'world-views'.

As intelligent systems grow in size they need to acquire the ability for meta-level knowledge that characterizes their state of knowledge. An intelligent robot of today is not equipped with this capacity of meta-knowledge, consequently for any query it has to search its entire database to formulate a response. This cannot be a practical solution for a machine cell that has a large database and seeks to exercise intelligent control. It must have knowledge about what it has in its database, what it can possibly know or not know and therefore obviate the need to scan the entire database at every query. Also when two or more cells cooperate they will need to reason about each other's goals, databases or world-views.

This brings us to the realm of reasoning about knowledge and beliefs. A persistent theme is the issue of 'common knowledge'. This issue has been studied by Aumann [3] in economic theory and by Halpern [14] and Moses [20] in the field of distributed computation. Achieving the state of common knowledge is useful to any coordinated action by a group of independent agents. It is therefore a relevant subject of concern in studying the knowledge structure of a group of autonomous and cooperating cells in an FMS.

In the context of an FMS a group of cells can be said to have a common knowledge of a proposition 'X', if every member of the group knows that 'X' is true, and also every cell knows, that every cell knows, that every cell knows.........{ an infinite sequence} that 'X' is true. In [20] it is argued that while common knowledge is desirable it is unattainable in many realistic settings. To have common knowledge it is necessary to have guaranteed communication among agents and the means for simultaneous change of their knowledge states. A hierarchy of weakened versions of common knowledge has been suggested which is suitable for settings where these conditions are not met.

FOL, which is commonly used in A.I., does not allow us to represent and model structures of distributed knowledge. In order to formalize such a situation we need a language. Halpern has shown in [14] that modal logic provides us with such a language and that 'Kripke structure' provides a formal tool for giving semantics to this language.

A 'Kripke structure' is defined as a {W, R, V}, where W stands for the set of worlds, R represents the accessibility relation among these worlds and V is the value assignment function that determines the truth value of a proposition in a particular world. Its syntax is based on the S-5 modal logic system, which has an additional operator to FOL: 'K', which stands for 'knows' in the semantics associated with Kripke structures. A distributed structure can be characterized by axioms due to Hintikka [15]:

(A1)	All propositions of FOL	
(A2)	$KX \wedge K(X \rightarrow Y) \rightarrow KY$	
(A3)	$KX \rightarrow X$	(knowledge)
(A4)	$KX \rightarrow KKX$	(positive introspection)
(A5)	$\sim KX \rightarrow K(\sim KX)$	(negative introspection)

The rules of deduction in such a system are as follows: (here 'A|B' implies that given proposition 'A' one can infer proposition 'B'):

(R1)	$X, X \rightarrow Y \mid Y$	(modus ponens: given X and $X \rightarrow Y$, one can infer Y)
(R2)	$X \mid KX$	(generalization: given X one can infer KX)

With a structure of knowledge as above, we can define a hierarchy of knowledge among agents which is usually present in the FMS among its various cells, as follows:

- "Someone knows X " : $\exists i\ K_i X$
- "Everyone knows X" : $\forall i\ K_i X$
- "Everyone knows, everyone knows X" : E^2X. Similarly E^kX

- "Common knowledge X" : $X \wedge EX \wedge E^2X \wedge E^3X.........E^{\infty}X$

Summarizing, we see that modal logic and semantic systems like Kripke structures supply us with the formal tools to describe and model the distributed knowledge structure of an FMS. These tools can be useful in building the knowledge system and the problem processing system of a multi-user decision support system.

8. CONCLUSION

In conclusion, we have observed that an FMS is a distributed decision making structure composed of autonomous cells. Such distributed structures of decision making are pervasive in the world we live in. Researchers in economics have long studied decision making methods in a variety of settings such as markets, game theory or team theory. Distributed computation also concerns itself with a similar core set of problems, of distributing tasks and resources among a team of autonomous processors. Concepts and methods from economics and distributed computation supplemented with a language and formalism supplied by modal and temporal logic, can constitute an effective set of representation and problem processing means for building decision support systems for FMSs.

9. REFERENCES

1. Akella, R., Choong, Y. and Gerhwin, S. "Performances of Hierarchical Production Scheduling Policy," *IEEE Transaction*, CHMT Vol. 3, 1984, pp. 225-240.
2. Allen, J. "Maintaining Knowledge about Temporal Intervals," *Communications of ACM*, Vol. 26, No 11, 1983, pp. 832-843.
3. Aumann, R. "Agreeing to Disagree," *Annals of Statistics*, Vol. 4, No. 6, 1977, pp. 1236-1239.
4. Barash, M.M., et al. "Optimal Planning of Computerized Manufacturing System," *Proceedings of the Eighth NSF Grantees Conference on Production Research and Technology*, 1981.
5. Bonczek, R., Holsapple, C. and Whinston, A. *Foundations of Decision Support Systems*, Academic Press, 1981.
6. Bui, T. and Jarke, M. "Communication Requirement of a Group Decision Support System," *Proceedings of the Nineteenth Hawaii Conference on System Sciences*, 1986, pp. 524-533.

7. Bullers, W., Nof, S.Y. and Whinston, A.B. "Control and Decision Support in Automatic Manufacturing Systems," *AIIE Transaction*, Vol. 12, No. 2, 1980, pp. 156-169.
8. Buzacott, J.A. and Yao, D. "Flexible Manufacturing Systems: A Review of Analytical Models," *Management Science*, Vol. 32, No. 7, 1986, pp. 890-905.
9. Chryssolouris, G. "Madema: A Decision Making Concept for CIM," CAM-1 Fourteenth Annual Meeting and Technical Conference, San-Diego, 1985.
10. Cutowsky, M.R., Russel, P.S. and Milligan, R. "The Design of Flexible Machining Cells for Small Batch Production", *Journal of Manufacturing Systems*, Vol. 3, No. 1, 1984, pp. 39-59.
11. De, S., Pan, S. and Whinston, A. "Natural Language Query Processing in a Temporal Database," *Data and Knowledge Engineering*, Vol. 1, No. 1, 1985, pp. 3-15.
12. De, S., Nof, S.Y. and Whinston, A. "Decision Support in Computer Integrated Manufacturing", *Decision Support Systems*, Vol. 1, No. 1, 1985, pp. 37-56.
13. Gray, J.N. "An Approach to Decentralized Computer Systems," *IEEE Transaction on Software Engineering*, Vol. SE 12, No. 6, 1986, pp. 684-692.
14. Halpern, J.Y. "Reasoning About Knowledge: An Overview," *Proceedings on Theoretical Aspects of Reasoning about Knowledge*, Monterey, California, 1986, pp. 1-17.
15. Hintikka, J. "Knowledge and Belief," Cornell University Press, 1962.
16. Kimemia, J.G. and Gerswin, S.B. "Flow Ooptimization in Flexible Manufacturing Systems," *IJPR*, Vol. 23, No. 1, 1985, pp. 81-96.
17. Lokshin, A., et al. "Towards a Hierarchical Robot Control Language," *IEEE Workshop on Languages for Automation*, Singapore, 1986, pp. 3-8.
18. Marschak, J. and Radner, R. *Economic Theory of Teams*, Yale University Press, 1972.
19. Moore, J. and Whinston, A. "A Model of Decision Making with Sequential Information Acquisition : Part 1," *Decision Support Systems*, Vol. 2, No. 4, 1986, pp. 285-307.
20. Moses, Y.O. "Knowledge in a Distributed Environment," Ph.D. Thesis, Dept. of Computer Science, Stanford University, 1986.
21. Nash, J.F. "Two-Person Cooperative Games," *Econometrica*, Vol. 21, No. 7, 1953, pp. 128-140.
22. Reisig, W. *Petrinets: An Introduction*, Springer-Verlag, New-York, 1982.
23. Shaw, M.J. "The Design of a Distributed Knowledge-based System for the Intelligent Manufacturing System," Ph.D. Thesis, Krannert Graduate School of Management, Purdue University, 1984.
24. Shaw, M.J. and Whinston, A. "A Distributed Knowledge-based Approach to Flexible Automation: The Contract-net Framework," forthcoming in *International Journal of Flexible Manufacturing Systems*, Vol. 1, No. 1, 1987.
25. Stecke, K.E. and Solberg, J.J. "The Optimality of Unbalancing both Work-loads and Machine Group Sizes in Closed Queueing Networks of Multi-server Queues", *O.R.*, Vol. 33, No. 7, 1985, pp. 882-910.

26. Tortora, G., et al. "Iconlisp: An Example of a Visual Programming Language," IEEE Computer Society Workshop in Visual Languages, Dallas, 1986, pp. 22-25.
27. Turner, R. *Logic for Artificial Intelligence*, Ellis Horwood, U.K., 1984.

INDUSTRIAL APPLICATIONS OF ARTIFICIAL INTELLIGENCE[1]

Mark S. Fox

Intelligent Systems Laboratory
Robotics Institute
Carnegie Mellon University
Pittsburgh, Pennsylvania 15213

Carnegie Group Inc
Five PPG Place
Pittsburgh, Pennsylvania 15222

Abstract

This paper reviews current and future applications of Artificial Intelligence (AI) and Knowledge-Based systems to manufacturing. This is not a review of robotics technology, but focuses, instead, on manufacturing decision problems. Manufacturing, in this case, refers to the entire product life cycle: product design, production planning, production, distribution, and field service and reclamation. The review focuses on where, at each point in the product life cycle, there are problems to be solved, where AI is currently being applied, and where it may be applied in the future. The results of a recent survey note that research and development in this area has increased significantly in the 1980s. Most focus on spot applications of the technology. More recent work is taking a systemic view of manufacturing.

1. Introduction

The purpose of this paper is to review current and future applications of Artificial Intelligence (AI) and Knowledge-Based systems to manufacturing. This is not a a review of robotics technology, but focuses, instead, on manufacturing decision problems. Manufacturing, in this case, refers to the entire product life cycle:

- product design,
- production planning,
- production,
- distribution, and
- field service and reclamation.

The review focuses on where, at each point in the product life cycle, there are problems to be solved, where AI is currently being applied, and where it may be applied in the future.

[1]Reprint of paper which appeared in *Robotics 2 (1986)*, Elsevier Science Publishers B.V. (North Holland)

NATO ASI Series, Vol. F49
Computer Integrated Manufacturing
Edited by I. Burhan Turksen

2. What is Artificial Intelligence

Artificial Intelligence is the science concerned with the creation of machine intelligence which is able to perform tasks heretofore only performed by people. Much of this machine intelligence is symbolic and heuristic. Artificial Intelligence deviated from the juggernaut of computer research in the early 50s by exploring how computers can be used for more than just numeric processing. Back during the days when languages like COBOL and FORTRAN were being defined, people at Carnegie-Mellon University and MIT were investigating the simulation of human problem- solving on a computer. Some of the first programs at that time were being applied to solving logic problems as found in the book "Principle Mathematic" by Whitehead and Russell (Newell & Simon, 1956). Problems such as chess, checkers, image understanding, etc. began to be investigated. As a matter of fact, one interesting sets of problems that people chose to demonstrate the AI techniques could solve problems at the same level that humans could, were chosen from the intelligence test that we normally give to students to measure their IQ (Evans, 1968). It so happens that computers are very good at solving them. The difficulty in the development of machine intelligence lies in programming computers to perform common sense reasoning, i.e. reasoning about every-day occurrences which people find easy to do.

The types of problems which AI attempts to solve are nonlinear and combinatorially complex (e.g., planning/scheduling, image understanding, etc.). The impact of their being non-linear is that there do not exist algorithms which will provide optimal solutions in polynomial time. Hence, the use of symbolic, heuristic knowledge, or "rules of thumb", play a major role in AI systems.

AI research can be divided into two basic categories. Knowledge representation is concerned with *how* to represent knowledge in a computer understandable form, so that systems can act in an intelligent manner. Consider a description of an activity that occurs on a factory floor.

> The milling operation precedes inspection. It is composed of two steps: setup and run. Setup takes an hour and piece time is 30 minutes. Two resources are required by the operation. A wrench and an operator. The wrench can be found in the tool crib and is only required during setup. The operation is performed in cost center 84.

The question is: How would one represent the knowledge embedded in that paragraph in a computer? Typically, a record in a data base which contains fields in some language is created. Describing the basics of an operation, there may be a field in a record which describes the tooling required, type of operator, who the operator is, setup time, run time, and the next operation in a sequence of operations to produce a product. The problem is: How does a computer understand the meaning of the record. The answer is that the program which uses it projects its own interpretation. The goal of knowledge representation research is to move away from ad hoc representations of knowledge to a semantics-based representation which identifies the levels of representation and their machine interpretation (Brachman, 1979). One level of representation, the conceptual level, provides a standard semantics which can be used across organizations and tasks, such as simulation, scheduling, accounting, etc. Figure 2-1 is an example of the representation of the knowledge in that paragraph. It is relational in form. Knowledge representation research focuses on the identification standard types of nodes and relations. Figure 2-1 divides the knowledge into two types: activity and state. A state description describes a snapshot of the world before an activity is performed. For example "cost center 48 possess a wrench" is a state description. It must be true in order to enable the milling activity to occur. States and activities are linked via causal relations. A state describes what must be true of the world to enable an activity to occur. In addition, activities may be defined at multiple levels of abstraction. Milling is refined into two sub-activities: the setup and running of the machine. Lastly, we must represent time. Setup, in time, occurs before the milling run. Time is not absolute, it is relative. When describing the factory floor, it is atypical to use absolute time periods, instead, activities are described as preceding each other; and hence, once the time of one activity is determined the time of all the other activities related to it can be determined.

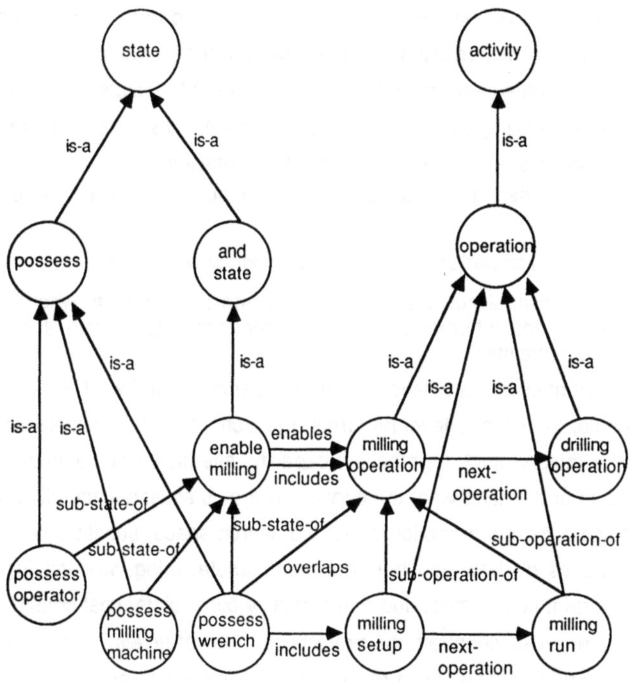

Figure 2-1: Activity Semantic Network

This is a small glimpse of what people in AI have been doing in the area of knowledge representation. For more details on conceptual level representations of activities see (Sathi, Fox & Greenberg, 1985).

The second category of AI is search. Problems are solved by performing search. Search in scheduling may be described as a search which incrementally builds alternative schedules. First, starting with a single order, it generates alternative first operations. Then, for each operation, it generates alternative machines on which to perform that operation, and for each machine it generates alternative queue positions, that is, times to perform that operation. There may be other alternatives such as alternative shifts and substitute tooling and fixtures, which expand the tree into a larger network. Once the first operation is fully defined, the search proceeds with the next operation that follows it. Search can be performed in a forward manner or in a backward manner, starting from the due date.

This is a very simple example of search. This type of search is similar to playing chess by considering all the alternatives. In a single factory which has 85 orders, 10 operations and only one substitutable machine, there are over 10^{800} alternative schedules that one could create. So, while this kind of search an is interesting, theoretical technique, practically, it is unusable because there are too many alternatives to consider. So, search has to be much smarter; in fact, search has to be structured in a manner that the search space is reduced from 10^{800} to something much smaller and manageable. The requirement to reduce the combinatorics of search has led to more complex and sophisticated search architectures. Figure 2-2 depicts the search architecture for the ISIS job shop scheduling system (Fox & Smith, 1984). ISIS has a hierarchical search architecture where it divides search at each level into three phases:

1. pre-analysis of the search space to choose the operators which generate the search space,
2. the search phase which performs the actual problem-solving, and
3. post-analysis which analyzes the results of the search to identify whether it was successful or not and whether it should be performed again with a change in operators and constraints.

These phases are performed at each level within the search hierarchy. The first level chooses what order to schedule, the second level performs a search where it considers only a subset of the information in the factory, such as machine capacity, alternative operations, due dates and start dates, and tries to understand what the capacity is that is available, identify bottlenecks etc. The third level considers more information about the search space, substitutability of machines, tooling, shifts, constraints on how you perform your manufacturing, working in process, time constraints, how important is it to meet due dates, quality constraints, cost constraints, etc. The search results of higher levels constrain the processing of lower levels. Finally, the fourth level determines the time periods during which resources are allocated to operation. So, the system successively refines its schedules by starting from the aggregate and refining it via search to a detailed complete schedule by using a little bit of information to make more global decisions and then refining those decisions by including more and more detail. Search-based systems like ISIS use sophisticated search techniques to reduce the size of the search space while still generating a good schedule.

The combination of representation and search are, in essence, the two legs of AI. It is upon these two legs that knowledge-based systems are constructed.

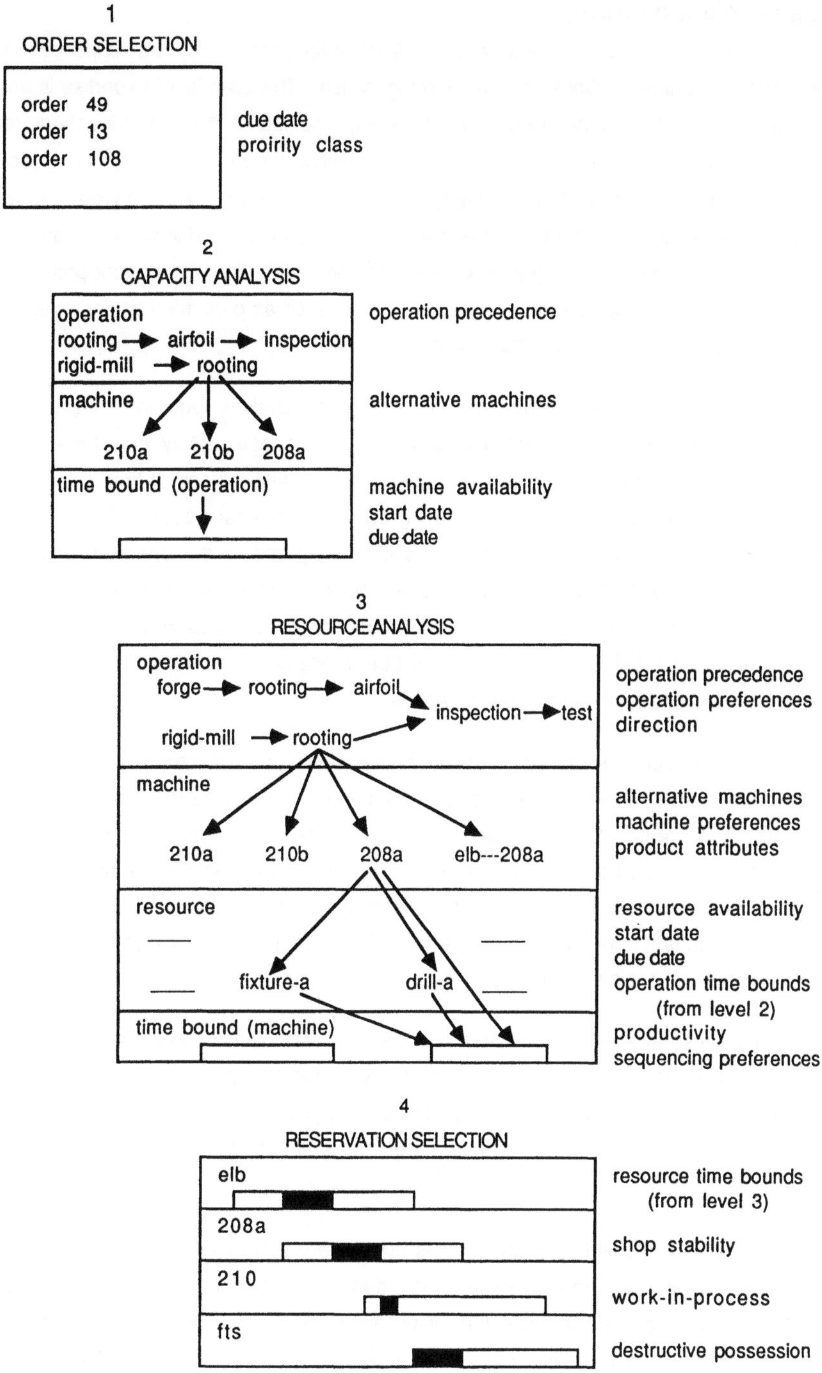

Figure 2-2: ISIS Search Hierarchy

3. Issues in Manufacturing

There are a number of reasons why Artificial Intelligence may be important in the manufacturing environment. Consider the following issues. The scarcity of expertise is endemic to many corporations. Each time I visit a manufacturing organization there is at least one request of the following type: "Mark, we have an expert in a particular area and he/she is going to retire in two years. We want to save that knowledge for the organization." Not only do they want to capture that knowledge before the person retires, but they also want to distribute it throughout the organization. The issue of capturing scarce expertise and distributing it is a major problem faced by manufacturing organizations today. It occurs in tasks such as process planning, diagnosis of machinery, scheduling, and engineering design.

Another issue being faced in manufacturing is decision complexity. Decision complexity arises when there is a large number of choices from which to choose. They may be engineering choices of how to design a product, or they may be scheduling choices of how to produce the product. It is the case that flexible manufacturing systems, are exacerbating the problem. The added flexibility on the factory floor provides the scheduling person with more alternatives, hence more choices of how to produce product. Today, making scheduling decisions in very rigid manufacturing environments is too difficult. As these environments become more flexible, the complexity of decisions increases. The same is true in engineering: the product complexity is increasing and, hence, design complexity increases.

Information is also becoming more complex. Many people use the term 'paperless factory' or 'paperless office'. The problem of how to get information from the factory floor on-line is now being solved. This creates another problem: how to reduce on-line information to only what is necessary for an individual to make a decision? Putting information on-line does not make it accessible to the individual decision maker. Reducing 1000 pages to a single screen full of pertinent information is an unsolved problem today. The intelligent reduction of information on an individual-by-individual basis is a long way off.

Decision timeliness is another issue which is coupled with decision complexity. Not only is decision complexity increasing, but the time to make a decision is decreasing. For example, programmable automation reduces the setup time of machinery, hence the scheduling decisions have to consider more alternatives in less time: how can faster, smarter decision-making systems be built?

Lastly, there is an issue of coordination. It is now known that design is intimately connected with production, distribution and field service. If a design is not optimized for manufacturability, assembly, distribution, or field service, then it will increase its cost of manufacturing with the possibility of reduced quality. The question is: how can designs be coordinated with all the down-stream activities?

These are some of the issues that affect the ability to increase the quality and the productivity of manufacturing operations. The question arises again as to whether Artificial Intelligence is a useful technology.

4. AI in Manufacturing Survey

The results of a recent survey show the extent to which AI is being applied to manufacturing problems (Fox, 1986). The survey shows that in the 60s and 70s there was very little work being done (figure 4-1). But in the 1980s there are at least 68 systems in research, 38 in development, 9 in field test and 14 in production use. With a response of about 125 systems this represents about a quarter of the number of real systems being investigated today. There are closer to 500 systems that are under construction around the world today using Artificial Intelligence techniques.

The survey demonstrates that a number of people who believe AI will have an impact. The question is: where? In the following, I review the areas of impact by examining each phase of manufacturing.

5. AI in Design

The first phase of the product life cycle is design (figure 5-1). Design is composed of a product specification followed by the actual design and then its validation. In parallel, a number of management activities are performed. AI is being applied to each of these phases. Today, specifications are natural language descriptions written, in the case of commercial products, on a few pages, or in the case of military products, on thousands of pages. It is necessary to analyze these specifications to determine their completeness and consistency.

Artificial Intelligence has been applied to the acquisition and analysis of specifications. XSEL (McDermott, 1982) is an example of a system which works with the computer sales person to acquire customer product requirements. The XSEL system has a natural language interface for the specification of customer goals; it analyzes customer needs and queries the customer in order to complete the model of the customer's needs. This system is in production use. XSEL deals with a portion of the specification problem. The general ability to analyze complete specifications, identifying consistencies and incompleteness is beyond the state of the art but is something to which AI techniques can be applied.

Once the specification is received and its completeness and consistency verified, design may proceed. Design is one of the most difficult and creative of the manufacturing tasks. It requires not only education but much experience. Consequently it is a knowledge-intensive task which utilizes both analytic and heuristic knowledge, and therefore makes AI techniques appropriate.

There are at least four categories of design. The first is *selection*, which maps functional requirements onto the attributes of an existing product line, hence selecting a product from an existing set. For example, XSEL maps user functional requirements onto product attributes and selects the appropriate products.

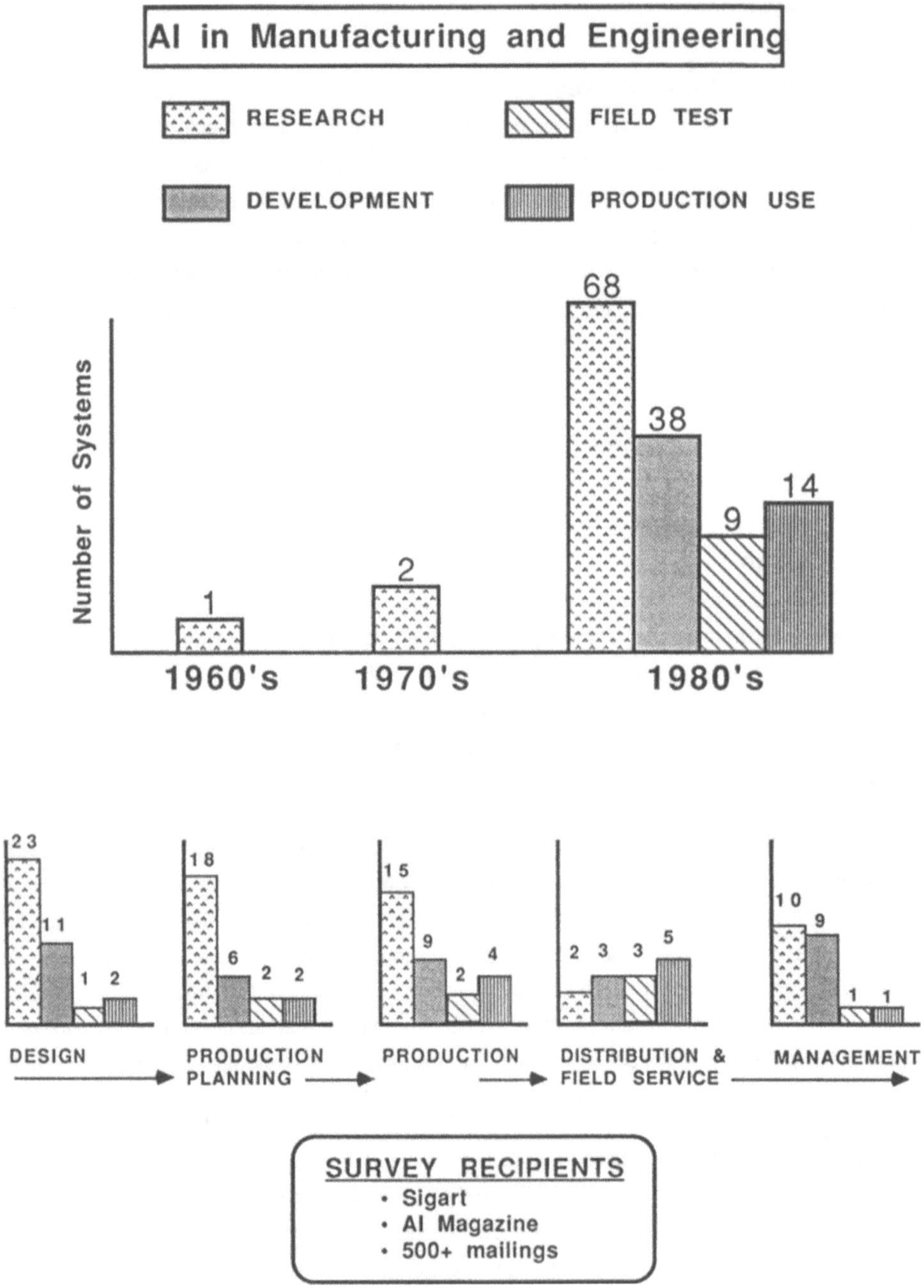

Figure 4-1: AI in Manufacturing Survey

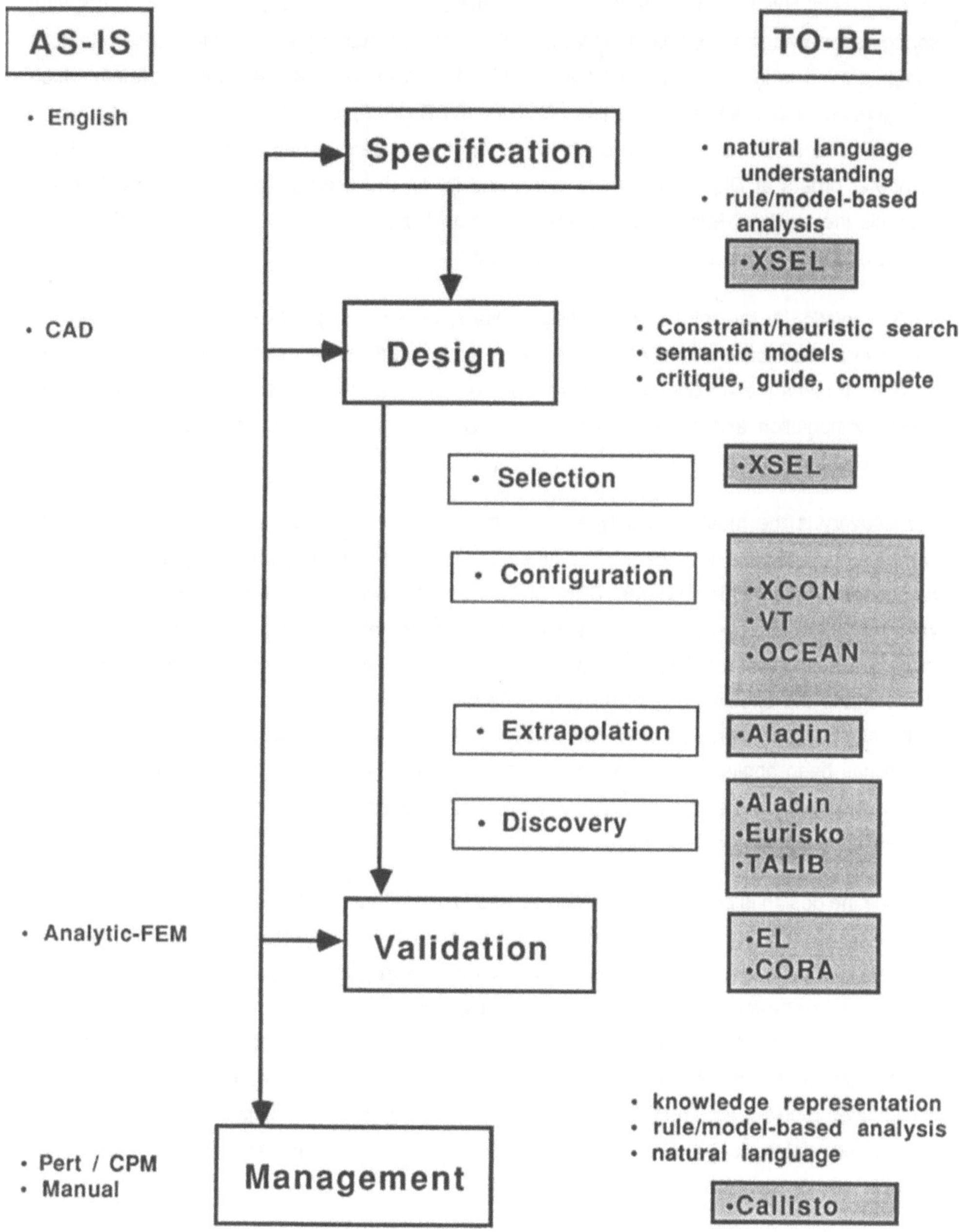

Figure 5-1: AI in Design

Configuration is a second type of design and is a little more difficult. The problem is to combine semi-finished components to form a final product. There are a number of successful configuration systems. XCON/R1 (McDermott, 1982) configures the VAX and PDP/11 lines of computers. It is in production use today. VT (Marcus et al., 1986) is another system which configures elevators and is in field test. OCEAN (NCR, 1985) is another computer configuration system. The XCON system has been in use for three years. It has configured over 100,000 computer orders at Digital Equipment Corp. and is, on the average, over 99% correct, which exceeds the ability of human configurers. It is also faster, configuring systems in 3-5 minutes versus one hour to three hours for human configurers.

Extrapolation is the third type of design. Here, an existing product is altered to meet the customer's specifications. ALADIN (Farinacci et al., 1986) is an example of an extrapolation system. It takes the specification of the properties for an aluminum alloy, and alters an existing alloy's composition and thermal-mechanical processing, so that a new alloy which meets the customer's needs is defined.

Discovery is the most difficult type of design. It is similar to configuration in that it takes components and combines them. The difference is that configuration uses more semi-finished components, i.e., memory boards, CPU, busses, etc. for computers, whereas discovery begins with transistors and resistors. The distance in the functionality of the components from the end product determines whether it is discovery or configuration. There are systems which demonstrate the effectiveness of AI techniques on discovery problems. ALADIN can start just with aluminum to design an aluminum alloy. EURISCO (Lenat, 1983) is a discovery system which has been applied to the design of VLSI circuitry. TALIB (Kim & McDermott, 1983) is a system which designs electronic circuitry. All of these systems have been able to design/discover interesting functional systems out of basic primitives.

Once the design is complete, *validation* is used to verify that the form of the design behaves as defined by the functional specification. Simulation is used most often; for example, finite element analysis is used to test stresses in mechanical parts, and in electronic parts simulation is used to verify circuit behavior. The use of AI for validation differs in that it attempts to replicate what humans do when they validate a design: people have the ability to look at the structure of a design and infer functionality with minimal use of simulation. Two systems have been investigated in this area: The CONSTRAINTS system (Steele, 1980) infers functionality from the form of an electronic circuitry, and the CORA system (Westinghouse, 1985), verifies relay protection systems which protect power lines.

Lastly, AI is also being applied to the management of engineering projects. In particular, the Callisto system (Sathi et al., 1986) focuses on the management of product definition, and the activities performed to design the product. It provides a knowledge representation for the representation of dynamic objects and activities, and captures expertise for the management of activities.

6. AI in Production Planning

Production planning (figure 6-1) is the second stage of the manufacturing product life cycle. It takes a product design and definition of the production facilities as input. It forecasts customer demand, plans the process, lays out the facilities, specifies maintenance and trains the workers.

AI is being applied to each of the tasks. In the case of forecasting there are researchers in marketing groups who are looking at rule-based techniques for simulating market behavior. These systems go beyond the typical or techniques which tend to predict the future based on the past by having a greater understanding of the individual's needs. The ROME system (Kosy & Dhar, 1983) focuses on another part of the forecasting problem, the analysis of resource plans. It is an intelligent VISICALC tm where the input is natural language and the system automatically analyzes the data to see if production goals are being met. If not, it figures out why.

A number of process planning systems are being investigated; ESP (Lindsay, 1985) has been developed for planning sheet metal production. XPSE (CAMI, 1984), a continuation of the GARI system (Descotte & Latombe, 1981), is looking at the process planning of three-dimensional mechanical parts. Both are ruled-based systems.

Process programming, for electronic applications programming, is straightforward and can be automated using conventional techniques. But for mechanical processes, such as the assembly of a carburetor, we do not know how to automate the programming of the assembly process. The application of AI to this problem is being investigated for the assembly of a power supply (Chang & Wee, 1985).

Another aspect of the process programming task is the selection of supporting resources such as cutting fluids for machining operations. GREASE (Mogush et al., 1986) is a system which selects cutting fluids for machining operations. It is now in field test.

Facility layout is concerned with how to organize production facilities based on process plans and product forecasts. The FADES system (Fisher, 1984) has been used to select the parameters of an Operations Research technique, which then determines the facility layout.

In maintenance design, AI is being used to develop maintenance procedures. RACE (Honeywell, 1985) is one such system now in field test.

Lastly, training is an important aspect of manufacturing. The impact of poorly trained workers can be devastating on production. Typically, training is performed manually, though simulations have been used for aircraft and nuclear power stations. What AI systems brings to the table is the ability to capture human expertise and using it to train new workers. It is also the case that

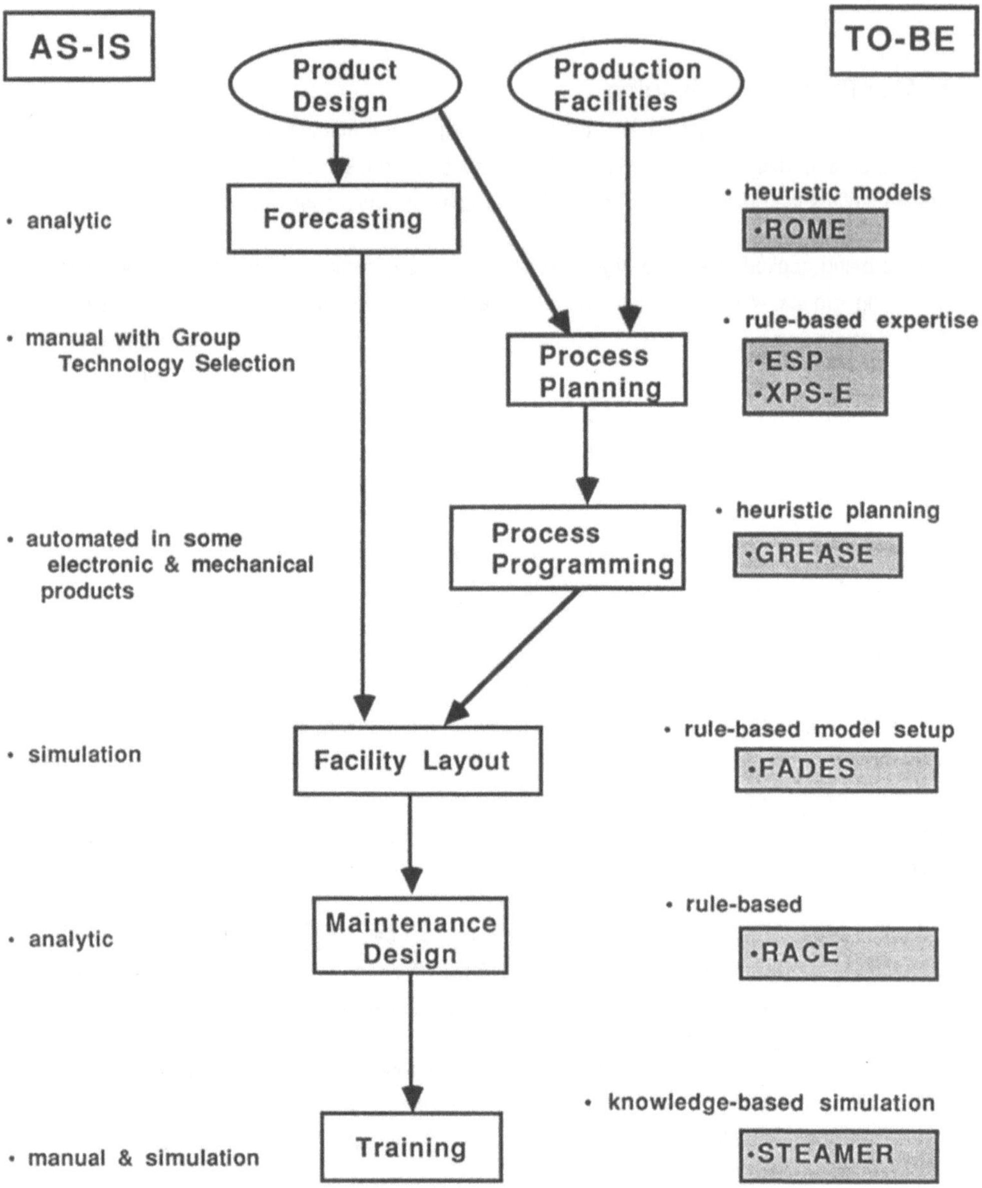

Figure 6-1: AI in Production Planning

expert simulations using Artificial Intelligence are being used. For example, Simulation Craft[tm] (CGI, 1986) is an expert manufacturing simulator. It captures and provides scarce simulation, statistics and manufacturing expertise in support of the simulation life cycle: model specification, experiment design, and result analysis.

7. AI in Production

Production (figure 7-1) is concerned with the planning, scheduling, management of the shop floor, control of cells, inspection of product and maintenance of processes.

Scheduling is a combinatorial problem whose complexity, in many cases, exceeds the ability of human schedulers. AI has been applied to scheduling. ISIS (Fox & Smith, 1984) is a system which uses constraint-directed search techniques for the scheduling of job-shops. Constraint-directed search uses constraints to reduce the combinatorics of the search resulting in schedules which satisfy some constraints while relaxing others.

IMACS (Haley et al., 1983) focuses on flow shop scheduling. The important concept in IMACS is its approach to managing a schedule. IMACS believes in Murphy's Law. Murphy's Law states that whatever can go wrong will go wrong. Rather than design the shop floor control systems to expect the schedule will be followed, it expects that deviations will occur and focuses on their detection and repair.

The same concept of looking for deviations is also used at the cell level by the TRANSCELL system (Bourne & Fox, 1984) and the SCD system (Tashiro et al., 1985) developed at Hitachi. These are rule-based systems which react to change and identify the next activity to perform.

For inspection, both rule-based and model-based techniques are used to diagnose computers and printed wire boards. IDT (Shubin & Ulrich, 1982) is a system for testing computers that are assembled, and IPWBIS (Seto et al., 1982) is a system for inspecting inner layers of printed wire boards and identifying where in the production process the error was introduced.

8. AI in Distribution and Field Service

Distribution and field service (figure 8-1) are the final phases of the manufacturing product life cycle. Distribution begins with the design of the organization: where to manufacture, how much to manufacture, how much to inventory, what transportation routes to use, what type of transportation to use, etc. Other distribution tasks are order entry, product installation planning, and diagnosis and repair of the product at the customer's site. AI again is being applied to each of these tasks. The INET system uses knowledge-based simulation techniques to model organization structures, simulate them and analyze automatically the results. XSEL is used for sales order entry. The output of XSEL is then provided to a distribution system, ILOG, which decides where to source the order, where to assemble, and how to ship it. Each of these systems use rules to capture and apply expertise, and heuristic search to improve upon the solution.

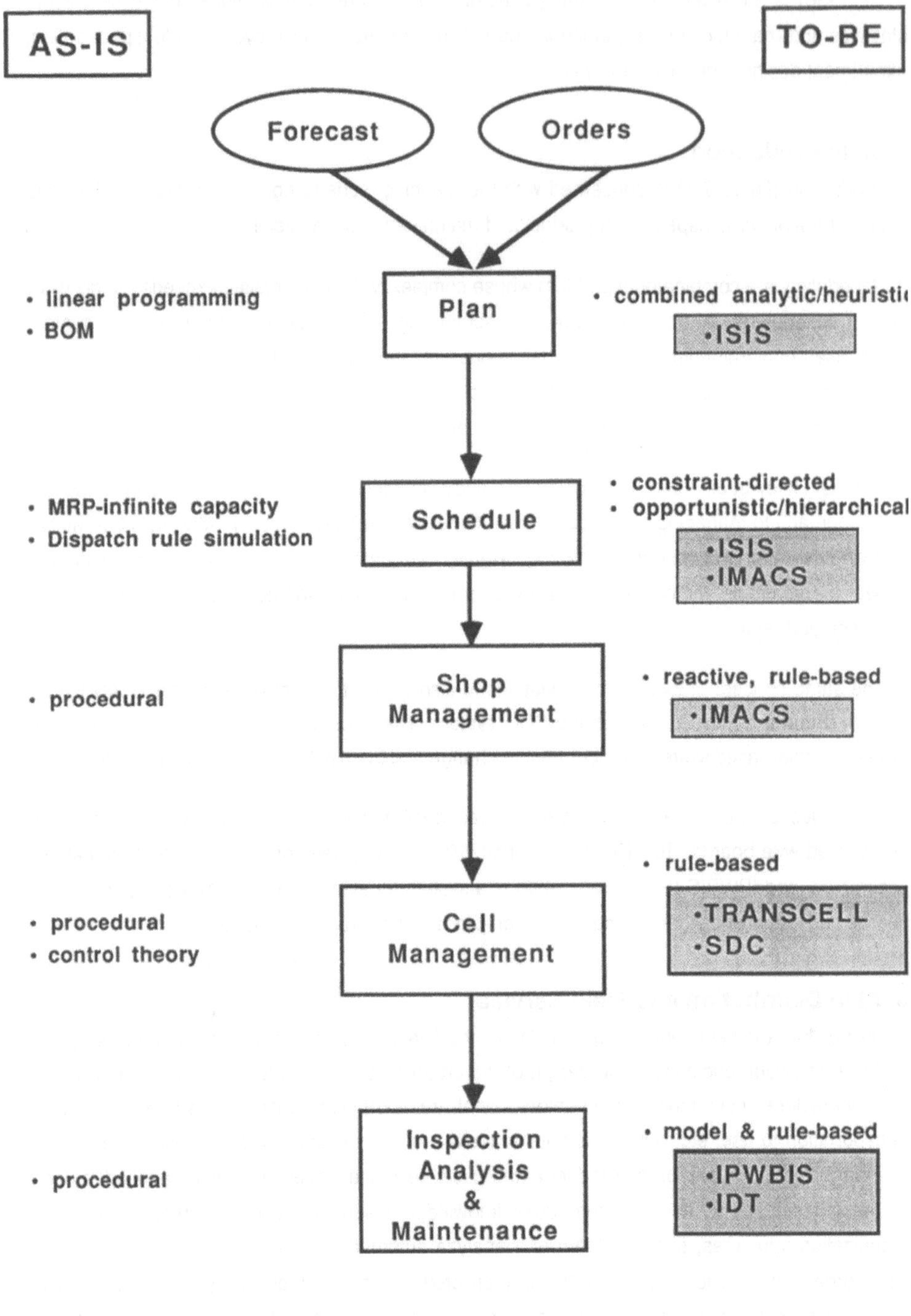

Figure 7-1: AI in Production

Installation planning has also been attacked using AI. The XSITE system (McDermott & Steele, 1981) decides how to lay out the computer at the customer's site.

Lastly, diagnosis and repair is an area that has received a great deal of attention from AI. PDS (aka GENAID) (Fox et al., 1983) is a system which, using sensor input, performs diagnosis of steam turbines and generators, using sensor input. It has been in production use for more than a year. PDS is interesting because it has to worry about sensor inputs. The problem with sensor inputs is that the sensors degrade in many situations more quickly than the process itself. Diagnosis in this environment must determine whether the sensors are incorrect or whether the process itself is incorrect. That makes the problem a lot more difficult.

ACE (Stolfo & Vesonder, 1982) is a system which performs diagnosis of cable problems for telephone cabling. DOC is a system which performs diagnosis of computers and CATS is a system for the diagnosis of diesel locomotives. These are all rule-based systems which use heuristics, knowledge of situation and action. Current research in AI focuses on model-based reasoning, representing the physical structure and operating theory of the product, i.e. chemistry, physics, thermo dynamics, etc. For example, the DART system (Genesereth, 1982) diagnoses computers based upon their structure.

9. Future Trends

In the preceding, I have examined each phase of manufacturing: design, production planning, production, distribution, field service; and I have identified point applications of Artificial Intelligence. The question is: What is the future of applying Artificial Intelligence to manufacturing? What needs to be done is to take a system's view of the manufacturing process. One view is strategic: How can a product be designed so that it optimizes all the down-stream activities? I call this "Design Fusion" (figure 9-1). The problem is to represent and utilize all the knowledge about planning, production, distribution and field service, during the design of the product, so that down-stream activities are optimized.

The second system's view is called "Parts on Demand": How is an existing product pulled through the organization? Parts on demand decisions are concerned with tactical issues of: Where to get the product? How to get it more quickly? If the product does not exist, how can it be produced quickly? Or in the case where the design of the product no longer exists, how can it be reverse engineered? By taking a more global strategic view and tactical view of pushing new designs and pulling existing products within the organization, it becomes possible to see how spot applications of AI can be further optimized.

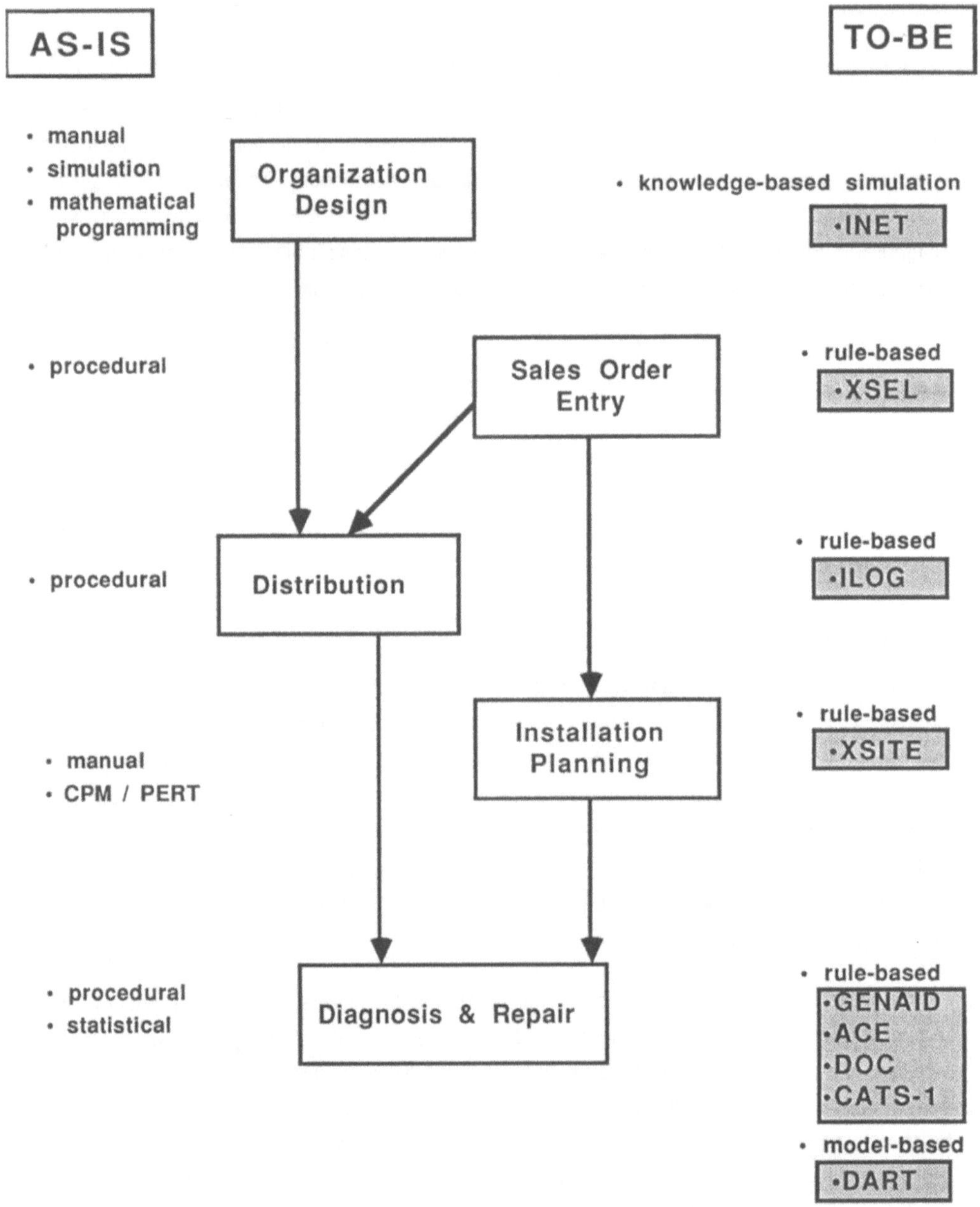

Figure 8-1: AI in Distribution and Field Service

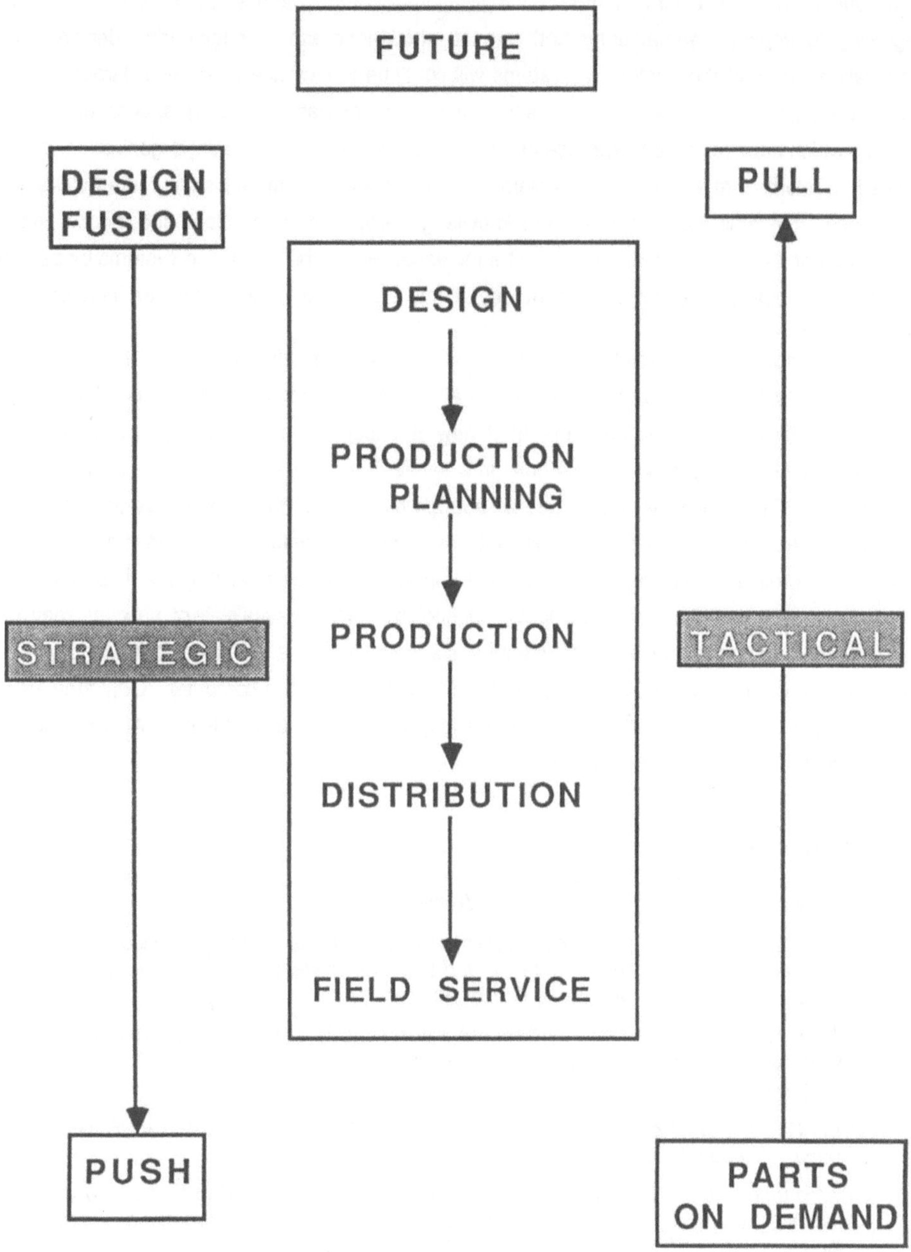

Figure 9-1: Future Trends

10. Conclusion

In conclusion, there exist a number of applications of AI in manufacturing today. They are beginning to impact manufacturing both on the shop floor and in engineering design. My expectations are that the number of systems will continue to increase at an even larger rate as corporations acquire more AI expertise and feel more comfortable about its application. More systems which capture scarce expertise and make it available throughout the organization will be created. Systems that enhance our problem solving by making better decisions more quickly will be created. Systems that integrate more knowledge about the factory floor and, hence, make better decisions will be created. There will be increased accessibility to these systems by people who are not computer-oriented through the use of natural language and explanation facilities.

The future appears to be rosy, but there are barriers looming along the road. The problem is that there are few people who really understand AI. There are people who have book knowledge of AI but do not understand AI from a point of view of actually building real, large systems. It so happens that AI is like a guild. It is a guild in a sense that there are masters, journeymen and apprentices. The problem is, there are not enough masters for the apprentices to work with. Because of this, many organizations will not have enough masters or any masters at all to guarantee a project's success. There will be a number of failures over the next five years, as corporations stub their toes in applying the technology. The possibilities of success may be enhanced if the corporation works with masters, starts small, lowers its expectations, and makes sure that the problem can be done by people before solving it with a computer. Only then will it be possible to successfully apply Artificial Intelligence to industrial problems. AI has a lot of promise but it must be applied cautiously.

11. References

1. Bourne D., and M.S. Fox, (1984), "Autonomous Manufacturing: Automating the Job-Shop", *Computer*, Vol 17 No 9, pp. 76-88.
2. Brachman R.J., (1979), "On the Epistemological Status of Semantic Networks" in *Associative Networks: Representation and Use of Knowledge by Computers*, N.V. Findler (Ed.), pp. 3-50, New York: Academic Press.
3. Chang K.H., and W.G. Wee, (1985), "A Knowledge Based Planning System for Mechanical Assembly Using Robots", *Proceedings of the 22nd IEEE Design Automation Conference*, pp. 330-336.
4. CGI, (1986), "Simulation Craft", Carnegie Group Inc., Commerce Court at Station Square, Pittsburgh, PA 15219.
5. Descotte Y. and J-C Latombe, (1981), "GARI: A Problem Solver That Plans How to Machine Mechanical Parts", *Proceedings of the Seventh International Joint Conference on Artificial Intelligence*, pp. 766-772, Vancouver B.C.
6. Evans T.G., (1968), "A Program for the Solution of Geometric-Analogy Intelligence Test Questions", in *Semantic Information Processing*, M. Minsky (Ed.), Cambridge MA: The MIT Press.
7. Farinacci M.L., M.S. Fox, I. Hulthage, M.D. Rychener, (1986), "The Development of ALADIN, An Expert System for Aluminum Alloy Design", *Artificial Intelligence in*

Manufacturing, Thomas Bernold, (Ed.), Springer-Verlag, to appear.

8. Fisher E.L., (1984), "Knowledge-Based Facilities Design", Ph.D. Thesis, School of Industrial Engineering, Purdue University, W. Lafayette, Indiana.

9. Fox M.S., (1986), "AI in Manufacturing: A Survey", *AI Magazine*, to appear.

10. Fox M.S., S. Lowenfeld, and P. Kleinosky, (1983), "Techniques for Sensor-Based Diagnosis", *Proceedings of the International Joint Conference on Artificial Intelligence*, August 1983, Los Altos, CA: William Kaufman Inc.

11. Fox M., and S. Smith, (1984), "ISIS: A Knowledge-Based System for Factory Scheduling", *International Journal of Expert Systems*, Vol. 1, No. 1.

12. Genesereth M.R., (1982), "Diagnosis Using Hierarchical Design Models", *Proceedings of the National Conference on Artificial Intelligence*, pp. 278-283, Los Altos CA: William Kaufman Inc.

13. Haley P., J. Kowalski, J. McDermott, and R. McWhorter, (1983), "PTRANS: A Rule-Based Management Assistant", Technical Report, Computer Science Dept., Carnegie-Mellon University.

14. Honeywell (1985), Personal communication with Honeywell Corporation.

15. Kim J. and J. McDermott, (1983), "TALIB: An IC Layout Design Assistant", *Proceedings of the National Conference on Artificial Intelligence*, pp. 197-201, Los Altos, CA: William Kaufmann Inc.

16. Kosy D.and V. S. Dhar, (1983), "Knowledge-Based Support System for Long Range Planning", Technical Report, Robotics Institute, Carnegie-Mellon University, Pittsburgh, Pennsylvania, December, 1983.

17. Lenat D.B., (1983), "Theory Formation by Heuristic Search. The Nature of Heuristics II: Background and Examples", *Artificial Intelligence*, Vol 21 No 1, pp. 31-60.

18. Lindsay K.J., (1985), "Frame-Based Knowledge Representation for Processing Planning", *Proceedings of the First Annual Aerospace Applications of Artificial Intelligence*, Dayton OH: Wright-Patterson Air Force Base Materials Lab, Sept. 1985.

19. Marcus S., J. Stout, and J. McDermott, (1986), "VT: An Expert Elevator Configurer", Technical Report, Computer Science Dept., Carnegie-Mellon University, Pittsburgh PA, to appear.

20. McDermott J., (1982a), "R1: A Rule-Based Configurer of Computer Systems", *Artificial Intelligence*, Vol 19 No. 1, pp. 39-88.

21. McDermott J., (1982b), "XSEL: A Computer Sales Person's Assistant", In *Machine Intelligence 10*, J.E Hayes, D. Michie, and Y.H Pao (Eds.), John Wiley & Sons.

22. McDermott J., and B. Steele, (1981), "Extending a Knowledge-Based System to Deal with Ad Hoc Constraints", *Proceedings of the Seventh International Joint Conference on Artificial Intelligence*, Vancouver B.C., August 1981.

23. Mogush J.E., D. Carrega, P. Spirtes, and M.S. Fox, (1986), "Issues in Diagnosis and Treatment: A Case Study in Cutting Fluid Selection", Technical Report, Robotics Institute, Carnegie-Mellon University, Pittsburgh PA, to appear.

24. Newell A., and H.A. Simon, (1956), "The Logic Theory Machine: A Complex Information Processing System", *IRE Transactions on Information Theory*, Vol. IT-2, No. 3, pp. 61-79.

25. NCR (1985), Personal communication with National Cash Register Corporation.

26. Reddy Y.V., M.S. Fox, K. Doyle, and J. Arnold, (1983), "INET: A Knowledge-Based Simulation Approach to Distribution Analysis", *Proceedings of the IEEE Computer Society Trends and Applications*, National Bureau of Standards, Washington DC.

27. Sathi A., M.S. Fox, M. Greenberg, (1985), "Representation of Activity Knowledge for Project Management", *IEEE Transactions on Pattern Analysis and Machine Intelligence*, Vol. PAMI-7, No. 5, September, 1985, pp. 531- 552. Also appeared as Technical Report CMU-RI-TR-85-17, Robotics Institute, Carnegie-Mellon University.

28. Sathi A., T. Morton, and S. Roth, (1986), "Callisto: An Intelligent Project Management Systems", *AI Magazine*, to appear.

29. Seto J., R. Thibadeau, and M. Friedman, (1982), "Automatic Inspection for Printed Wiring", Technical Report CMU-RI-TR-82-16, Robotics Institute, Carnegie-Mellon University, Pittsburgh PA.

30. Shubin H., and J.W. Ulrich, (1982), "IDT: An Intelligent Diagnostic Tool", *Proceedings of the National Conference on Artificial Intelligence*, Aug. 1983, Los Altos, CA: William Kaufman Inc.

31. Steele G.L., (1980), The Definition and Implementation of a Computer Programming Language Based on Constraints", (PhD Thesis), MIT tech. rep. AI-TR-595, Cambridge MA.

32. Stolfo A. and G.T. Vesonder, (1982), "ACE: An Expert System Supporting Analysis and Management Decision Making", Technical Report, Computer Science Dept., Columbia University, New York N.Y.

33. Tashiro T., N. Komoda, I. Tsushima, and K. Matsumoto, (1985), "Rule Based Control of Factory Automation Systems - An Approach with a Production System" Technical Report, Systems Development Lab., Hitachi Ltd.

34. Westinghouse (1985), Personal communication with Westinghouse Electric Corporation's Productivity and Quality Center.

Intelligent Knowledge Bases of CAD Environments: The Hybrid System "KANON"†

Hans-J. Schneider and Dimitris Karagiannis

Technical University of Berlin
Franklinstr. 28/29; FR. 5-8
D-1000 Berlin 10

Abstract

The CAD-/CAM-area can be seen as an interdisciplinary field between the mechanical engineers and the computer scientists. Regarding this development the software-engineering aspect has been dominant up to now on the part of computer science. The reason why is that in the meantime already good and refined systems exist on the market (CATIA,CADAM, CAEDS, CADIS, ICEM). But the integration capability of such systems has hardly been taken into consideration. Only since CIM‡ and the combining of CAD-/CAM-systems have been discussed one tries to integrate these systems. Two fundamental integration approaches can be differentiated:

A) Extension of the existing sytems for universal interfaces which enable their integration;

B) Development of new systems in which the integration aspect is part of the system.

In this connection the integration of AI-methods and existing database system-, software-engineering- up to information system approaches offers - in our opinion - suitable ways of solutions which shall be explained afterwards.

Starting out from chapter 1, which describes the problems of conventional CAD-/CAM-systems, the hybrid system KANON with its application possibilities in a CIM-environment is introduced in chapter 2, so that useful application possibilities of a KBS regarding the development and realization of CAD-/CAM-systems can be explained in chapter 3.

† The name KANON originates from the greek word "κανων" which means "rule"

‡ This article deliberately leaves out those CIM-aspects which deal with the planning and control, like PPC or CA(P)P, as there the so-called planning paradigm of AI is well to the fore, which plays a subordinated part in this case.

NATO ASI Series, Vol. F49
Computer Integrated Manufacturing
Edited by I. Burhan Turksen

1. CIM: INTEGRATION OF CAD-/CAM-SYSTEMS

Data management

The main problem concerning the integration of CAD-/CAM-systems arises when a common database is being developed, as the systems have specially cut and therefore efficient data and file administration. There exist e.g. fast storage algorithms for storing line- or volume- constructions which have been developed with CAD-systems. But these algorithms are not suited for the optimal storage and management of NC-programs.
In order to solve storage problems of CAD-systems DBMS have been used for storing and managing structures which are presented in geometrical form. In this process three forms of coupling between a CAD-system and a DBMS arose:

1. Parallel operation between CAD-system and database
2. Coupling of CAD-system and database
3. Integration of a database system in a CAD-system.

Concerning the *parallel operation* both systems work on separate stored data. After changing a CAD-design only the processed changes or the complete design will be recorded by a separate program in a database. The only advantage of this method is that in generally no changes of the existing CAD-system and database system must be carried out. It is disadvantageous that an extraction of the changes from the stored CAD-data is very costly or even totally impossible, and that considerable changes of data in the database take place during the acceptance of each complete drawing with NGI (Non-Graphical-Information) even when only a small part of the data has changed. This configuration is almost exclusively suited for the batch-operation.
Concerning *coupled systems* a CAD-system and a database system are bound jointly through a user program. The data exchange takes place by means of the user program (program interface). The flexibility is very high regarding this form of coupling, but the programming needed for realizing the coupling is very costly.
The last possibility of coupling, the *integration*, represents the most comfortable solution. The CAD-system can store its data directly in a database. This form however is so special that it can not be used by other CIM-parts (components, aspects) anymore. In addition, the efficiency of existing systems on the CAD-side as well as on the database-side is very limited. The flexibility and efficiency of a separate database system is often lacking on the part of the database, concerning

both the available data structures and the possible retrieval- and evaluation procedures. But the increasing distribution of data in all the cases where the NGI is conciderably more extensive than the CAD-data itself is even more crucial. Then the CAD-data is scattered so widely that very large amounts of data must be processed in order to generate the CAD-data itself or to change a graphic. Here also, the problem occurs that it is hardly possible to work interactively.

Therefore the second form of coupling is the obvious method for binding a CAD-system in CIM, as the processing speed regarding the first form is too low, and periods of time, in which the database is inconsistent resp. areas of the database must be locked to prevent these inconsistencies, arise which are too long. The third coupling does not provide a good solution because the processing speed is too low and one database is only used especially for the CAD-part. A separate database will also be needed e.g. for the CAM-part, which would contradict the idea of CIM (Spur and Krause, 1984).

There are two problem areas concerning the second form. The first one is the processing speed, for even the use of large DBMS for storing and managing graphical objects does not solve this problem. A study in (Grill et. al., 1987) e.g. shows that the response times for the users of graphical interactive systems range in realms which clearly exceed the limit of acceptance. The second problem area concerns the programming needed for realizing a coupling.

Both problem areas could be reduced considerably by using KBS. On the one hand such a KBS must make suitable interfaces between the DBMS and the CAD-system available, but on the other hand the KBS must also provide an environment in which the DBMS as well as the CAD-system are embeded. In the following there will be a short explanation of how a system integration under CIM can be carried out.

In order to achieve the integration of CAD-/CAM-systems one needs a construction model which is able to represent all the important physical, geometrical and technological features of technical objects.

The following points about it have to be noticed in particular:

- Inclusion of a design- and construction methodology;
- Adequate representation and management of CIM-objects in data structures;
- Design of an integrated user interface;
- Inclusion of a production- and assembly-methodology.

Up to now a suitable construction methodology for all parts of CIM exists only in theory. For converting these methods the support on the computer-side is often

lacking. It is characteristic of the construction processes that the methods stand out due to their variety and multitude. A construction model shows high dynamics and just a small share of statics. This means in regard to the computer assistance that it must be very flexible and that it has to react to changes with low expense.
Something similar goes for the representation, too. The construction model consists of a lot of different data structures. Data structures are required e.g. to classify a part according to its features, like standard part, norm part, separate part, construction group, etc.. Apart from this classification one also needs hierarchies to establish relations. The semantic networks would be a suitable representation form. Figure 1 shows as an example the general representation of a part as a semantic network

User interaction

The design and functionality of a user interface determines decisively the acceptance and the effectivety of a CIM-system. Some requirements about comfortable user interfaces are described in the following:
Suitable Visualization: Graphical information as e.g. the form of parts and tools, tool paths and that sort of thing must be visualized suitably. Non-Graphical Information as e.g. details about surface treatment, heat resistence of materials, weight and price can be represented e.g. through icons. The representation within windows comes to the fore more and more. One possibility would be to divide the screen into two areas. One area could represent the graphical information, the other the non-graphical ones within windows.

Graphical and alphanumeric oriented query- and input functions: The user must be able to query the information of the construction model comfortably. This includes the query functions of a common database query language as well as the possibilities of direct "leafing through" catalogs which contain mixed graphical and non-graphical information.
The "navigation" in a "data room" of icons represented database information (Eberlein, 1984) would be an especially simple and clear query mode.

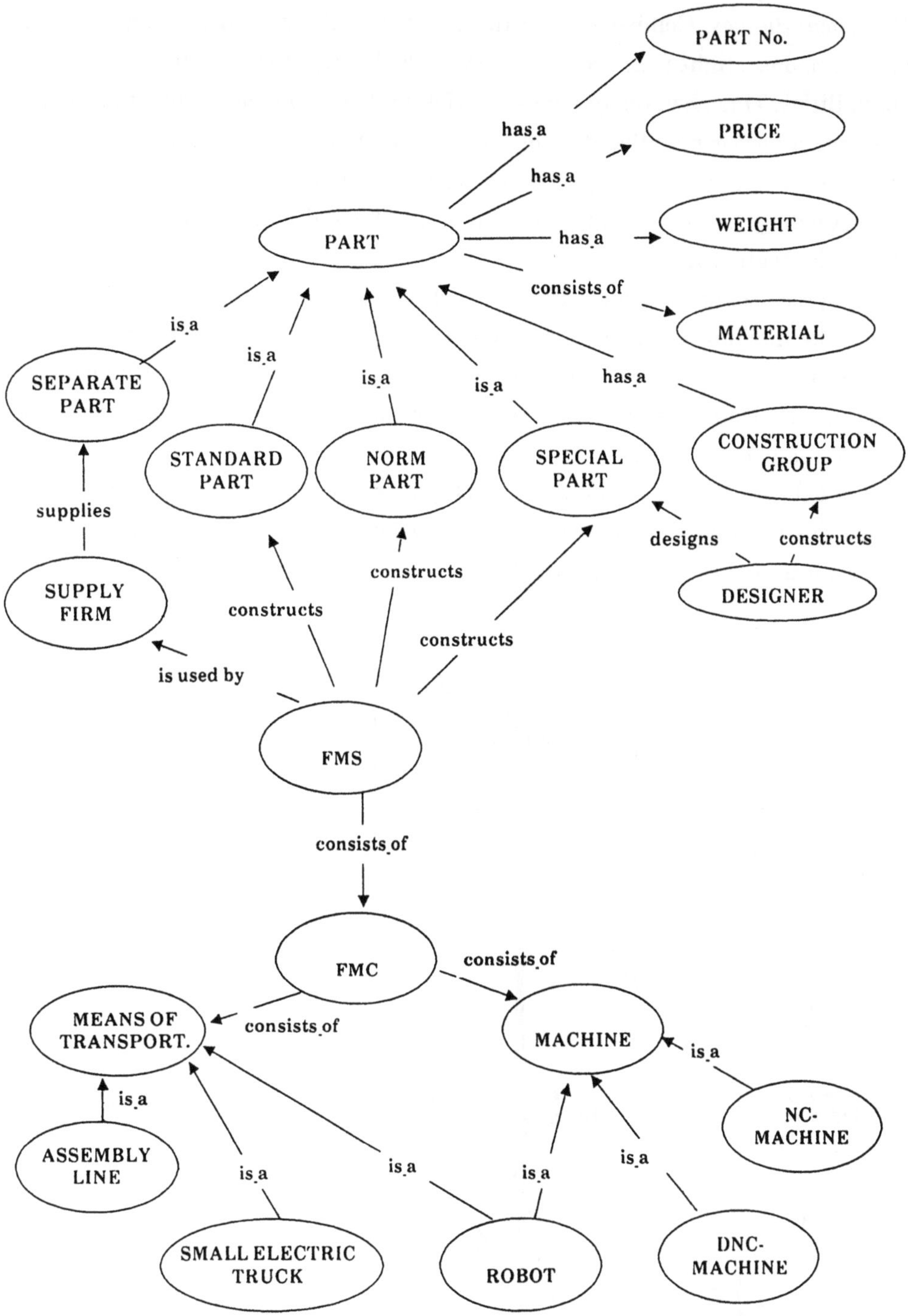

Figure 1. Representation of a part as a semantic network

Consistent dialogs: Compared with the conventional dialog the graphical input devices (mouse, light pen) provide a considerably larger spectrum of interaction possibilities. This however involves the risk that the application modules do not use any uniform convention during the input syntax. In the process considerable problems during the dialog run can arise. The application of formal methods, as e.g. state diagrams for the specification of graphical dialogs in the CIM-area, is still at the beginning.

Design processes often include iterative actions. The user interface of the system must take this into consideration by e.g. making it possible to store, modify and re-execute supervisory sequences or to cancel command jobs again. By means of suitable definition possiblities, e.g. through command-macros, the user should be able to create his own work environment.

Figure 2 shows the construction of a CAD-system with the individual components which are necessary to integrate such a system (Eberlein, 1984). The special components are discribed in Figure 2.

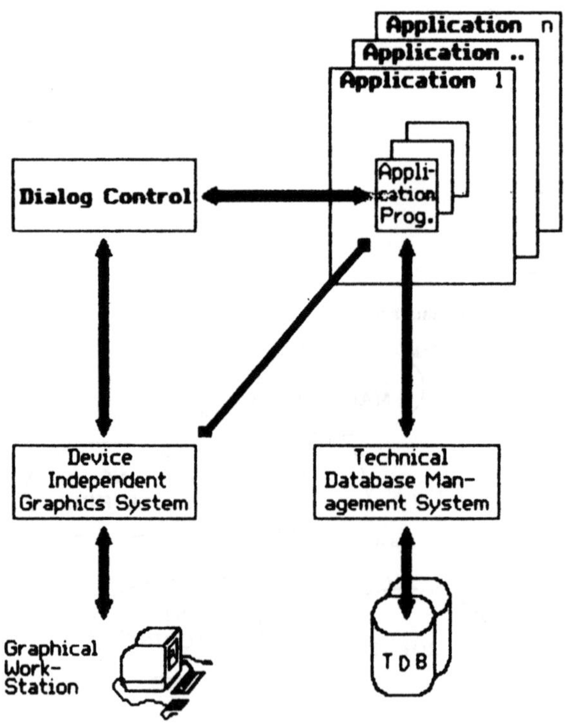

Figure 2. CAD-system with integrated Database (DB)

The CAD-system consists of the following components:

A) *Device independent graphics system*

B) *Dialog control component*

C) *Technical Database Management System*

D) *Application programs*

In the process several application programs form an application which is able to run. The dialog control component and the device independent graphics system are bound and added to an application program while the data transfer with the Technical DBMS runs through local-net-communication.

A) Device independent graphics system

In contrast to non-graphical systems the device independency of graphical systems is much harder to realize. And yet, the device independency regarding alphanumeric display devices is relatively easy to solve through the application of mask generators or by using tables in which capabilities and special control codes of single terminals are classified.
The situation here is much more complicated because of the big difference concerning the kind of graphical systems. E.g. the various quantizations of screens or the different functionalities caused by differing screen technologies like Vector-Refresh, Raster-Scan or storage screen have to be noted.
Therefore interfaces were designed especially for graphical devices which make it possible for the application program to address these machines device-independently. The Graphics Kernal System (GKS) or the CORE-proposal, should here be pointed out as an internationally standardized interface.
Such device-independent graphical systems abstract from the specifics of the single graphical devices and provide abstract graphical functions on their interface. These systems are subprogram packets which have a possibility to be integrated into a CIM-system.
As graphical representations and operations are very time-consuming one has to accept losses of performance - compared with the direct device control - which arise mainly from coordinate transformations and clipping-processes (Encarnacao and Schlechtendahl, 1983).
Therefore it should be possible for certain applications to also do without a certain program comfort, but instead achieve a faster execution time. This would be useful when "leafing through" catalogs which requires a fast picture transfer from the database to the graphical output device.

B) Dialog control component

The dialog control component provides a uniform interaction control with the following functions:

- Managing of windows on the screen of the graphical device;
- Managing of dynamic menus;
- Output of error messages and auxiliary texts, dialog protocols;
- Syntax examination of the input concerning single dialog steps;
- Control of the dialog run (sequence).

The dialog control component should be highly parametricable from software technical points of view. The techniques would be e.g. the storing of control data like menu descriptions or dialog sequences in the database, the generation of this component or the application of preprocessors.
Here it depends on the special system design whether the application programs only select the graphical device through the dialog control component or whether they may also use direct functions of the device-independent graphics system.

C) Technical Database Management System (TDBMS)

The data of the construction model is managed by the TDBMS. The TDBMS has to know the scheme of the construction model, and must contain the following functions:

- Definition and Modification the scheme of the construction model;
- Access of application modules to data of the construction model;
- Data control (data protection, precautions for the multi user operation, recovery in the case of an error).

The technical database is generally distributed among several computers. While the data transfer between the server workstation concept, which ranks on top of such a hierarchy, is mostly not contemporary, and could not be realized with the methods developed in connection with distributed databases (Schneider, 1982) or computer networks, the user at the graphical workstation needs a direct and fast access to the design-, analysis- and test data which are active just at that moment.
A two-stage hierarchy is always being used in which the division computer contains the technical database. The application modules run on computers which are coupled with the graphical workstations and the modules have access to the

construction model of the TDB through a local network. These special graphical computers contain a running time system which transfers the data of the division computer, which were required by the application programs, into the virtual address space of the computer and manages them. After the modification of the data they are being rewritten into the TDB.

D) Application programs

The application programs support single phases of the construction process through certain functions. Geometrical operations are required e.g. during the character design for representing and modifying outlines, or during the NC-programing for ascertaining the tool paths. Up to now such functions are not integrated components of systems, but additional performances of programs. Application programs which support a whole process are called application program chains. In the following the question is asked of how such systems can be made more efficient, and not the question of what the new system can do what the old one is not capable of. In spite of different construction types there are recurrent kernal application programs:

Geometrical modelling: From an abstract data type point of view this program realizes different operations, like scrolling, rotation, section or combining on objects of a geometrical type.

Documentation component (longtime storage component): The documentation component stores parts of the construction model in archives in the long term. In the process it uses the TDBMS for storing references on data media which are kept externally.

Data transfer in an external data format: Conversion programs enable the transfer of information of the construction model into another format. This data exchange contains the complete data structure of the part.

There exist certain data formats for the transfer of geometrical features within the construction model. The source-, resp. goal system needs one converter each, which changes the description of the construction model from the TDB into a neutral exchange format, resp. converts it the other way around.

Release management: An internal testing process releases a construction. The norms for such a release depend considerably on the firm. This program realizes the necessary data and the control of the running of the release process.

Management of CIM-objects

Two solution approaches, the Knowledge Based Approach and the extended DB Approach, which support the storing, managing and representation of complex CIM-objcets, are being dealt within this article.

Figure 3 shows the application areas of the two solution approaches. The application possibilities of the Knowledge Based Approach will be explained with the KANON-system in the following chapter. The current research activities and results in the area of the extended DB approach would be handled in a forthcoming paper.
Works in the area of data models for Non-Standard-Database-Systems as well as the application of the traditional E/R-model (Chen, 1976) for complex-object-modelling, semantic data models and especially the extensions of the relational model for direct representation of hierarchic relations are part of this (Wedekind, 1981). The introduction of the generic type COMPONENT_OF, the NF^2-relations model and the Extended-NF^2-relations model are counted among this group. Furthermore proposals for new transaction mechanisms in connection with a version management as well as solution approaches for the data storage during the task data processing are shown in this report.

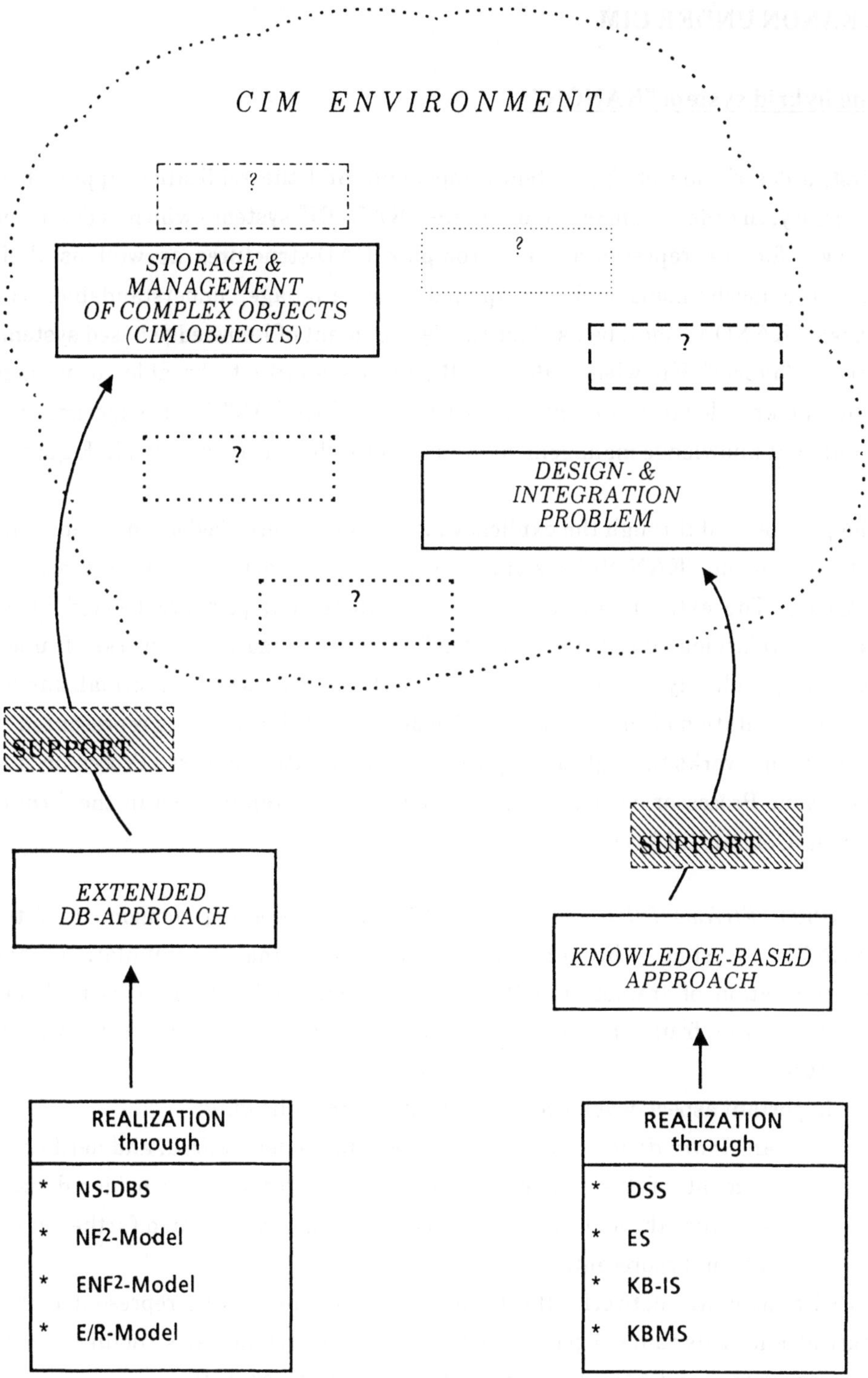

Figure 3. Solution approaches in a CIM-Environment

2. KANON UNDER CIM

The hybrid system "KANON"

First, a description of the system components and the realization approach is necessary, in order to make the use of the "KANON" system - which is described below - for the representation of complex CAD-structures as well as their inference mechanisms and the applied rule language, understandable. The system "KANON" shall be used for the development of knowledge based systems with a "hybrid" Knowledge Base (KB), i.e. its ability to be able to manage different knowledge representation structures. "KANON" is an experimental, intelligent knowledge management system. Its architecture is shown in Figure 4.

The part defined through the explicitly given facts is described as an extensional database in the "KANON" system, and the derived facts form the intensional database. The extensional as well as the intensional part are located in an external relational database. "KANON" provides semantic networks, frames, rules and explicitly stored relations as knowledge representaion formalisms for the representation of the extensions. The derivable information is represented in semantic networks through the support of the Semantic Networks Processor, and the Frame Processor ensures that the information is represented in the form of frames (Gebhardt, 1987).

FRAPRO *(FRAme PROcessor)*: is the "frames" processing component of the "KANON" system through which the recording, manipulation and transformation of frames into the relational knowledge base is carried out. Therefore, the frame processor models the object "frame" as a knowledge structure.
The important aspect developed here is based on the following:

> an *"individual"* of the object frame is not to be realized, and neither an element of the frame-structure, but a *"frame"* is considered as a structure through the frame processor and is represented further in the internal representation.

As with semantic networks the frame structure can also be represented in a statical and a dynamical part structure, and the dynamical structure can be found in frames under the description "Procedural Attachment".

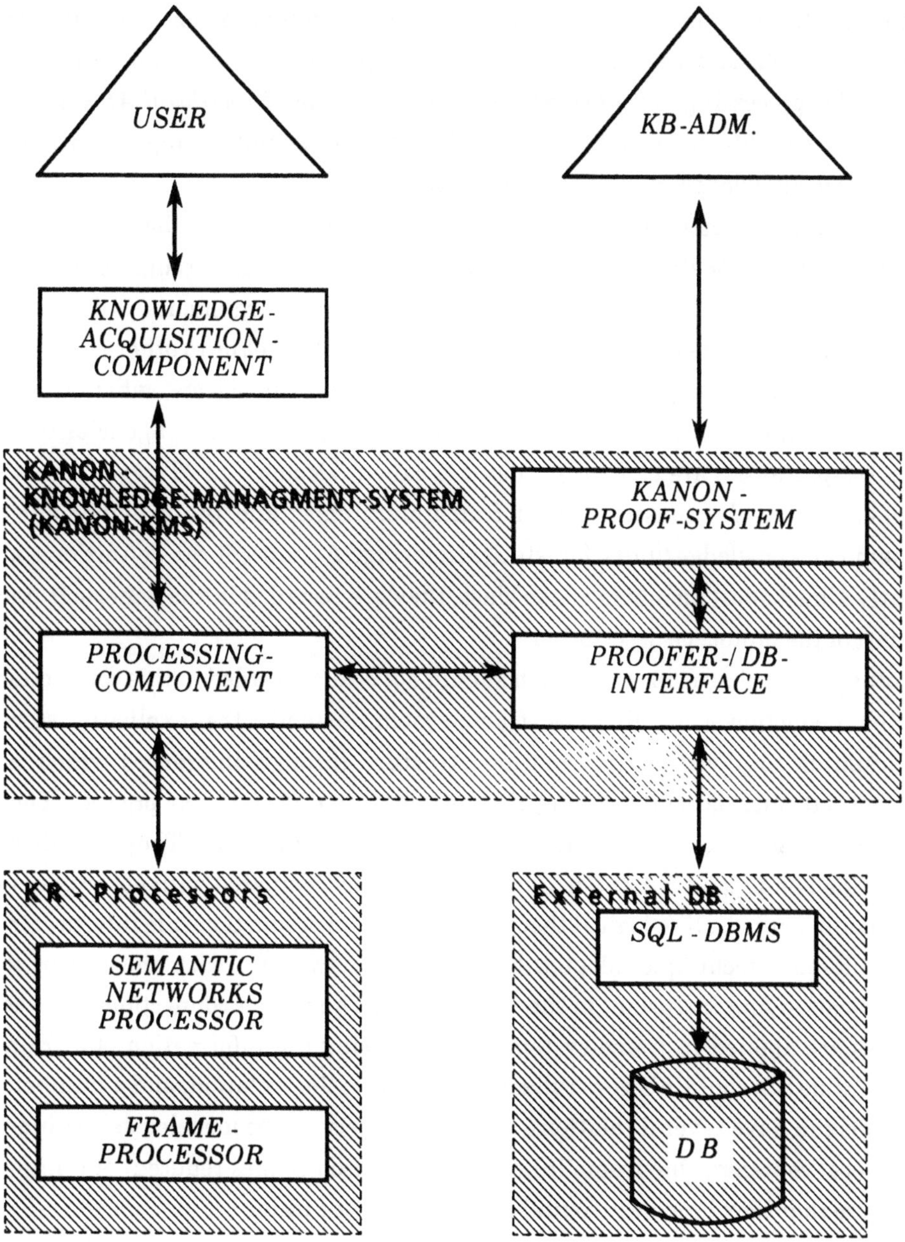

Figure 4. Architecture of the hybrid system "KANON"

Although frame-languages have no declarative procedure for modelling the behavior of objects, they have a number of possibilities to bind procedural knowledge processing to frames by means of rules. This *"Procedural Attachment"* makes it possible to model the behavior of objects and to represent better specialized knowledge in certain application areas.

One technique of using the Procedural Attachment consists in the definition of functions (called *"methods"* or *"servants"*) which is activated through an indirect call (*"message sending"*). The function, together with its arguments, is defined in a slot and provides result values to another facet of the slot or to other slots. This technique offers the possibility to generate dynamical objects, whereas it is another technique to define procedures which is activated when agencies of the slots (resp. frames) are changing; they are called *"Demons"* or *"Trigger"*. This is applied during the construction of frame-hierarchies and the development of dynamical knowledge units for statical storing. As procedural definitions are mostly aimed at all members of one class, they are recorded in the generic frames as part of the prototype description.

Another frame characteristic, which should not be disregarded in this connection, is the deduction capability. A frame-system extends the explicitly stored information to a virtual information structure through inferences which are carried out automatically during the retrieval- and description process. These inferences are based on the frame-structural characteristics. They are firmly predefined and stand out due to a higher efficiency compared with other general deduction methods, like logical theorem-proofer and production rule interpreters.

The inference technique adopted by semantic networks , with the use of the classification and inclusion, is the inheritance. Thus, such connections between frames are applied in frame-systems for increasing the information of a frame. Each frame can contain a generalization-/classification-relation, and a number of systems can have several relations, so that an object can be modeled as a subterm of several contents. One frame inherits all slots (sometimes reduced) of a higher frame so that they become a virtual component of the frame.

A multiple inheritance is mostly supported if there are several generalizations/-classifications. E.g. if there is a frame *"ball-bearing"* with a slot *"colour"* and the value assignment *"black"* and an individual of the class of ball-bearings as a frame with the name *"KLOA1"*, and a classification relation between the individual and the generic frame, the slot *"colour"* with the assignment *"black"* becomes virtually a slot of the frame *"KLOA1"*, so that it can be infered that *"KLOA1"* is black. For reasons of clarification it should be pointed out that this

relation is not only valid between individuals and generic objects (prototype description), but also between generic concepts (frames).
Another application of frames in knowledge based systems is described with the system KEE (Knowledge Engineering Environment) in (Fikes and Keller, 1985).

SENPRO *(SEmantic Networks PROcessor):* is the semantic networks processing component of the "KANON" system, through which the recording, manipulation and transformation of semantic networks in the relational knowledge base is processed. In this connection two aspects of the semantic networks and especially of the SNePS are examined,

a) the possibility of reproducing the **static** and
b) the **dynamic structure.**

Especially the SNePS (Semantic Network Processing System) of Stuart C. Shapiro are dealt with in "KANON" SENPRO. They serve the presentation and deduction of knowledge on the basis of semantic networks. The articles (Shapiro, 1977) and (Shapiro, 1979) are recommended as fundamentals for the SNePS, whereby the article in (McKay and Shapiro, 1982), in which SNePS is introduced for the recognition of speech, and in (Maida and Shapiro, 1981), where a component for recursive rule definition and derivation is defined, describe further applications.
On the one hand the semantic network contains a static structure, which is represented through nodes and edges, by which explicit knowledge is represented. On the other hand, the possibility is created - through the introduction of rule nodes - to deduct knowledge under the application of rules.
The types of nodes which can be used for object modelling can be represented within SNePS and especially by "KANON" as follows:

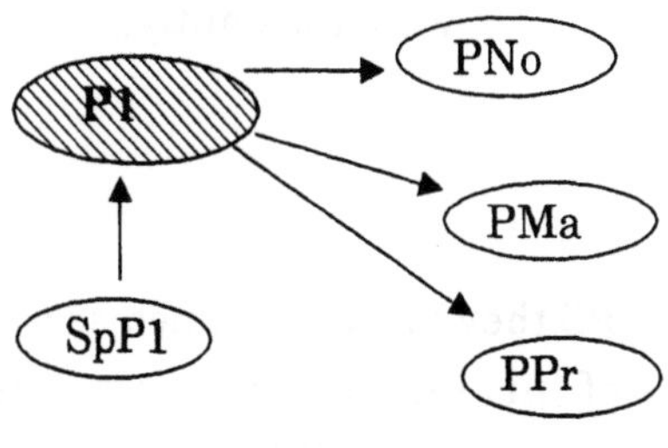

1. Through *base nodes* a concept is represented as a constant node in the semantic network.

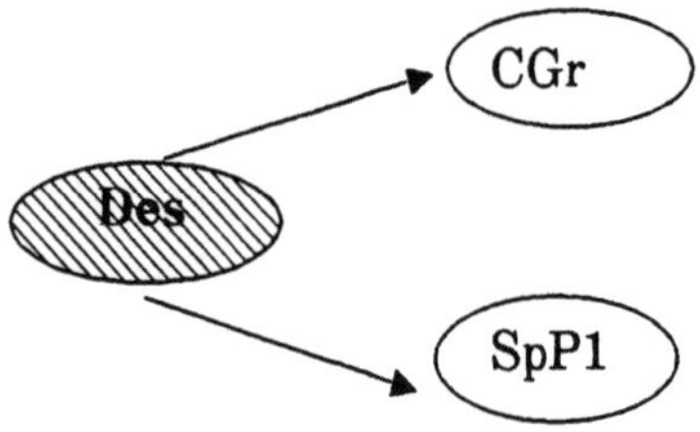

2. Through *assertion nodes* relations with base nodes are established. In this connection descending edges are pointed from the assertion node towards the base nodes.

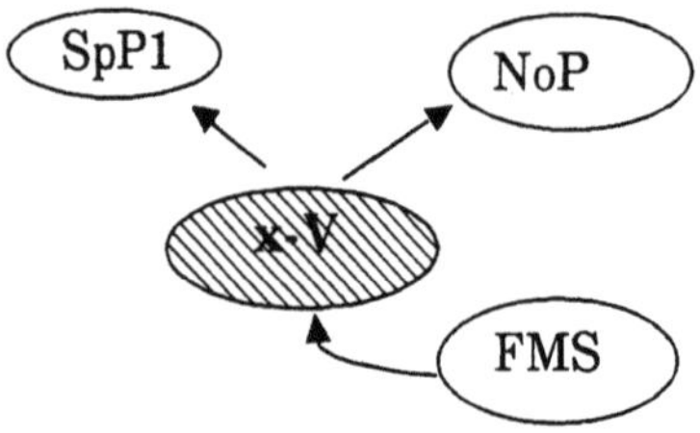

3. *Variable nodes* can adopt different values from the area of base nodes.

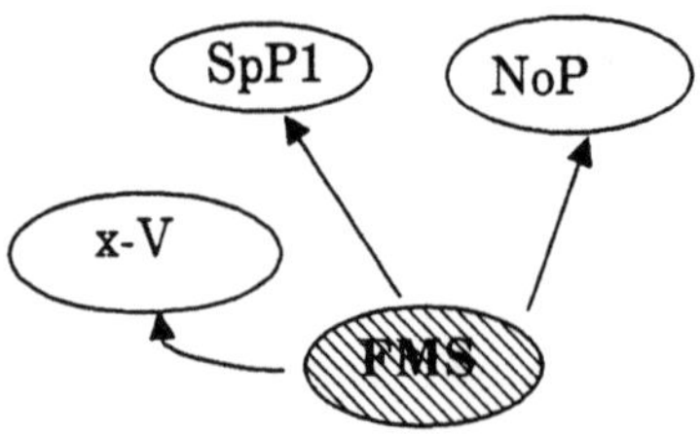

4. The only difference between *pattern nodes* and assertion nodes is that - where pattern nodes are concerned - at least one descending edge points towards a variable node.

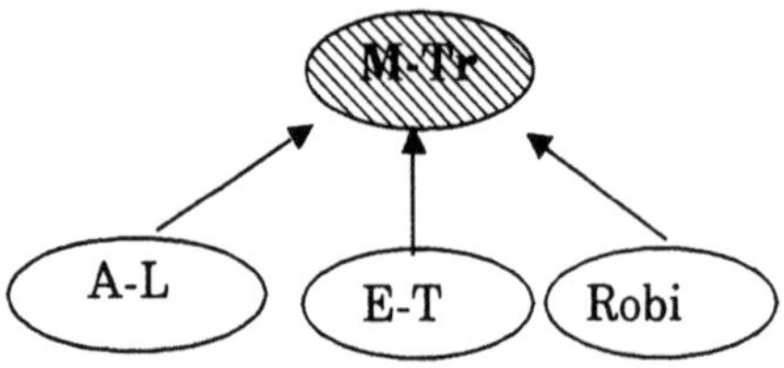

5. *Temporary variables* are nodes which do not exist in the network constantly.

Before applying corresponding relations in SNePS they must be defined as a set, as an assertion node is always representative of this set, and has - from every element of the set - one edge each towards the concept node. When applying an

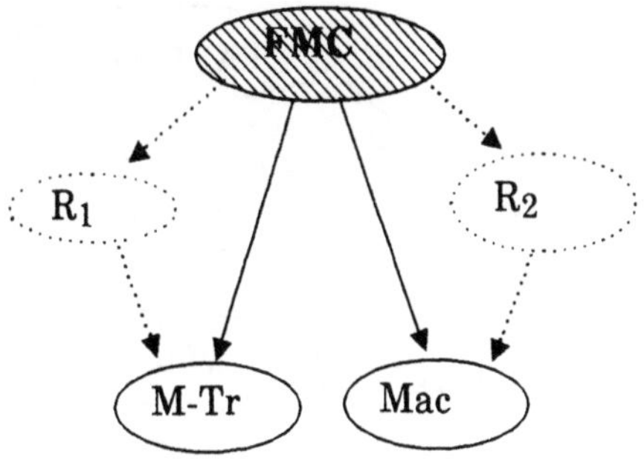

6. *Temporary pattern nodes* have the function of pattern nodes regarding temporary variables.

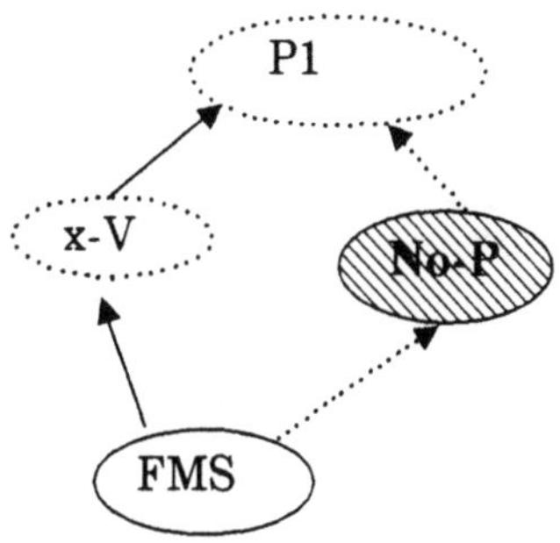

7. Through *auxiliary nodes*, nodes can be provided with a marker or a value can be assigned to a variable node. Auxiliary nodes can only be connected with other nodes through auxiliary edges.

edge set for the representation of relations between nodes, one node has to be assigned to each element of this set.

The dynamic structure of the network has been realized in the "KANON" system through the allocation of rule-systems to *rule nodes* which vary depending on the interpretation of the network.

A dynamic structure results from the introduction of rule nodes. Through rule nodes implicitly existing knowledge can be put explicitly in a semantic network by means of an inference mechanism.

The representation of the semantic network which presents a part (see Figure 1) can also be realized through the FMC-Rule System (FMC-RS). It is obvious that it's possible to create different FMC by using different FMC-RS (see Figure 5).

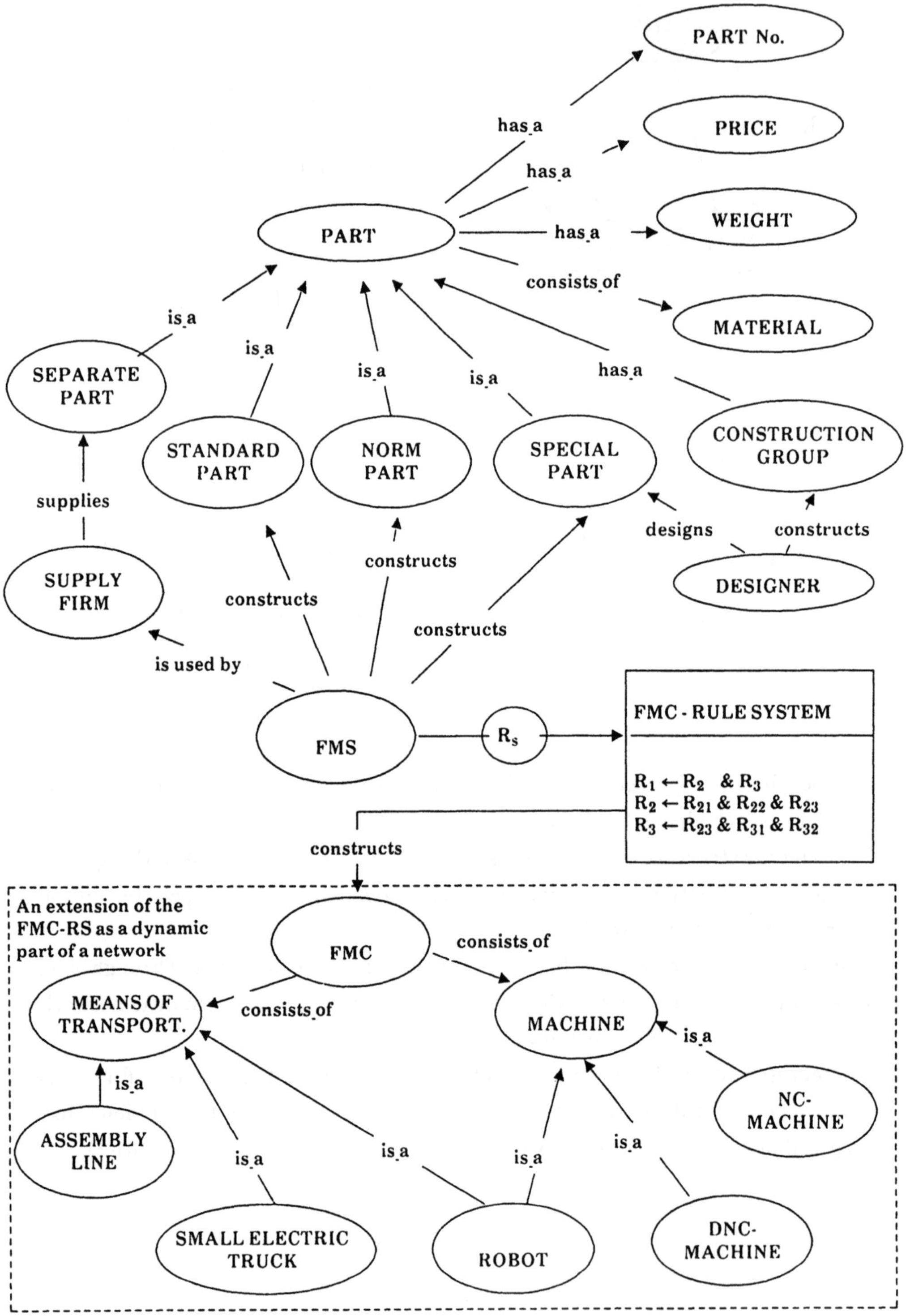

Figure 5. A FMC extension by using the FMC-RS R_S

Conclusions can be derived from a number of requirements through logical *"or"*-links or logical *"and"*-links. The number of possible arguments is composed of further rule nodes or assertion nodes, pattern nodes or temporary pattern nodes. There is the possibility to bind variable nodes to a rule node with the help of quantifiers. It concerns the All-quantifier and the Existency-quantifier. Several variables may be bound to one rule node by applying only one quantifier. Variables can be bound either through a directly higher rule node or through a pattern node to which a rule node is directly superordinated.
Both components serve the re-establishment of the RKB (Relational Knowledge Base) on the corresponding KRS, and their management and manipulation are subject to the "KANON" knowledge management system.

KANON-KMS *(Knowledge Management System):* is the deductive component of the "KANON" system. It consists of the "KANON" proof system and the interfaces belonging to the external data base. The main task of the proofer is the interpretation resp. the compilation of the rule language. The shaping of the model form and the communication of the entire system with the outside world is carried out through the rule language. The processing component supports the running of the user interaction and manages the knowledge base with the help of the database-/user-interface.
A detailed description of the architecture and realization of the "KANON" Knowledge Management System (KANON-KMS), which is regarded as an important system part, is shown in Figure 6 and will be explained below.

The stored data in databases of External Database Management Systems is located on the lowest level, whereas - in this connection - it has to be differentiated between:

- Extensional databases explicitly contain the (base-) relations, the relational presentation of rules as well as the belonging scheme information
- Intensional databases contain the internal representation of rules and meta rules.
- Builtin-databases contain all Builtin-relations which are located in the system in the form of PROLOG predicates as a component of the database.

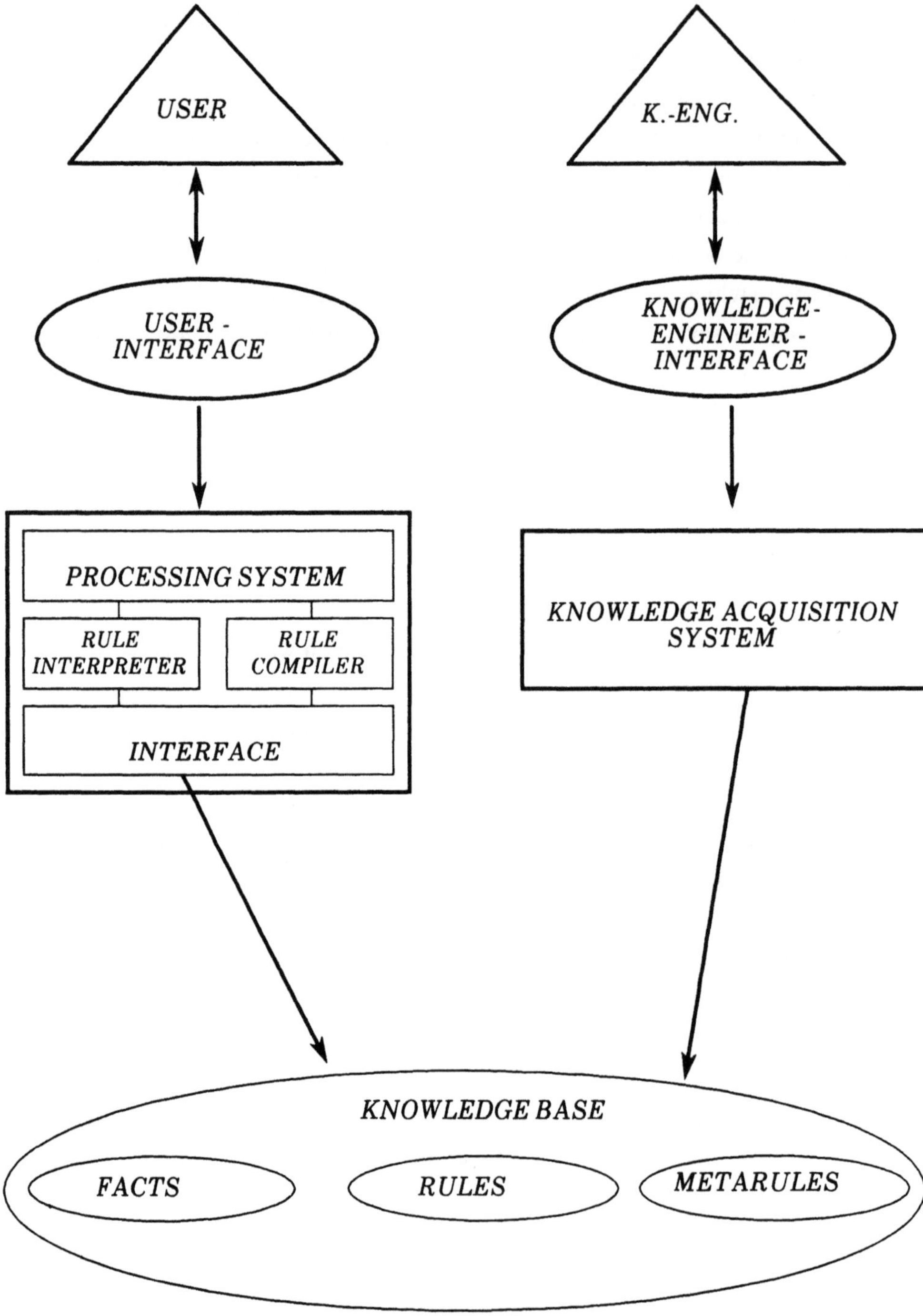

Figure 6. KANON-KMS overall view of the system

The selection of the corresponding database for retrieval- or data manipulation operations is carried out through the access system by means of the deduction processor. Together with the rule compiler and rule interpreter it forms the deduction component. Above this component lies the processing system which controls the entire run of the query with the control component. The access to this level is carried out through the user interface with a variety of subcomponents, like rule editor, meta rule editor, SQL, deductive data management commands etc.. A complete description of the "KANON" system is found in (Karagiannis, 1987).
The KANON-processing system offers a variety of capabilities, starting with an extended SQL-Language up to a knowledge acquisition possibility.

The following functions are available:

- Inserting tuples into the database (INSERT)
- Deleting tuples (DELETE)
- Dropping entire relations (DROP)
- Selecting tuples (SELECT)
- Counting tuples (COUNT)
- Sorting relations (SORT)
- Displaying tuples (DISPLAY)

In addition, there are functions which are substantiated through the deductive character. These are:

- for defining/creating rules (CREATE_RULE)
- for deleting rules (DELETE_RULE)
- for logical querying (QUERY)
- for defining/creating meta rules (CREATE_METARULE)
- for deleting meta rules (DELETE_METARULE)

In order to be able to carry out several manipulation operations without consistency checks between operations the KANON-KMS has a transaction concept. Operations can be joined together with this concept so that all of the mentioned consistency checks are not carried out until the end of the transaction.

The Supervisor Component (SC) (Figure 7) is a system implemented in PROLOG with which the decision about the type of derivation process is made, i.e. that the corresponding components are activated from simple queries about explicitly

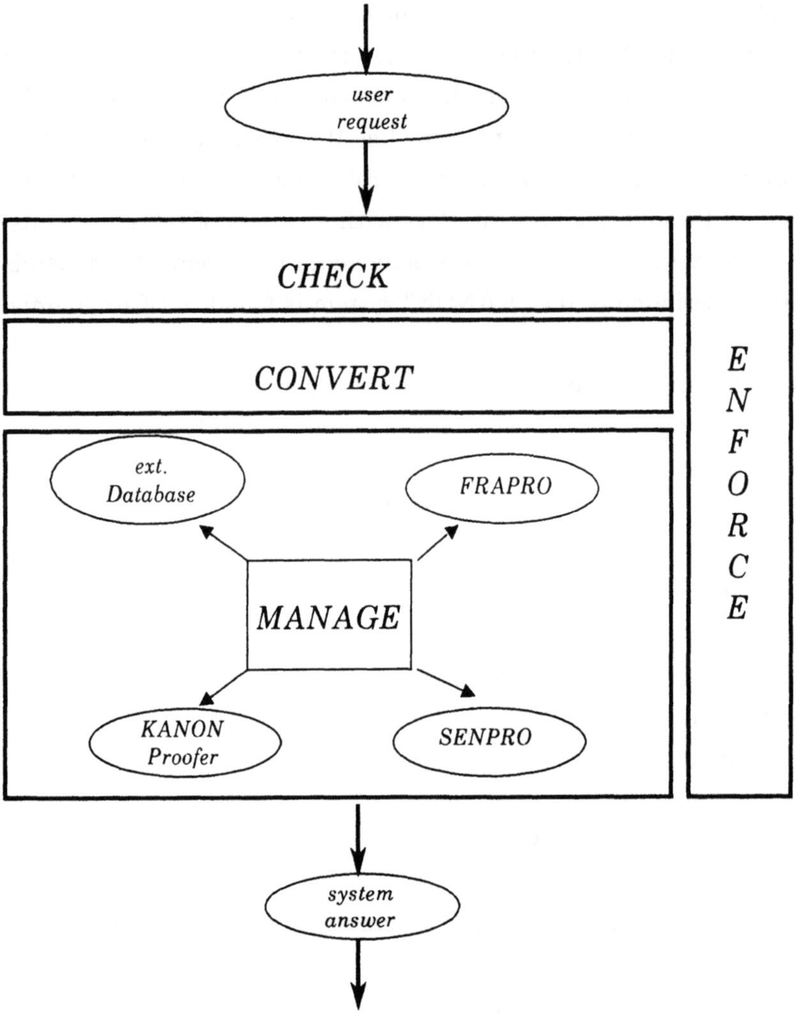

Figure 7. Structure of "KANON" Supervisor Component

stored knowledge to queries about implicit knowledge of the database in different knowledge representation structures (frames, semantic networks and logical rules).

Queries for the SC can be asked through the user interface. During a query through derived relations the SC calls the rule interpreter which deduces expressions of the virtual relations from extensional relations, which are kept in the extensional database (EDB), and from rules, which are kept in the intensional database (IDB). During queries regarding extensional relations the SC has direct access to the relations which are kept in the EDB. During queries regarding frames resp. semantic networks the corresponding processors are activated.

The knowledge management system realized in "KANON" shall support the user with his work concerning various types of KRS (Knowledge Representation Structure) which he must use for adequately handling his problem.
A constant/general concept is applied for an exemplary implementation of the four components KNOWLEDGE-AQUISITION SYSTEM, FRAPRO, SENPRO and KANON-PROOF-mechanism. A KMS is required as the basis for the conception of the FRAPRO-, SENPRO-processors.
The performance capacity of such systems can mainly be characterized by the following points:

a) The user describes his model by means of a rule system without con-sidering how the system stores the internal representation.
b) The user has a uniform interface to the entire system which is realized with the help of a *"rule language"*.
c) Inconsistencies or repeats are identified and announced to the user. New entries are compared with the knowledge available in the system.
d) The processing of large amounts of data is possible through the coupling to external database systems.
e) Adequate forms ascertaining knowledge (rule-editor, metarule-editor as well as explaination- and integrity/constraints- editor).
f) The possibility of the transformation of knowledge between different re-presentation structures and
g) their reproduction on a uniform structure is by no means trivial.

If, however, one confines oneself to special KRS, especially if a claim for completeness of the representation of the semantic level is made, the requirements mentioned above can definitely be fulfilled. When manipulation or provision of KRS is mentioned in the folllowing, a transformation between semantic networks, frames and relations shall be understood. This shall promote a better understanding and serve concrete statements within the realized system. Dynamical extensions of the system with new structures are planned as far as the existing system architecture is concerned. In our case the *"rule language"* takes on this task.
The essential prerequisite for fulfilling these requirements was the development of a representation form for the manipulation of the KMS. In our case the *"rule language"* takes on this task.
In addition, the chosen KRS has to be transformed into an adequate form. It should enable the provision-/recording in the knowledge base and the representation of the existing knowledge in the wanted representation form.

The component for the coupling to the external database has been taken from the actual "KANON" system. It forms a generally applicable interface which is independent of the underlying database. The selection of the external database is carried out through PARAMETRIZATION of the corresponding *coupling*-Predicates. The coupling with an external database is subject of the articles (Dadam, 1986), (Vassiliou, 1985) and (Ceri et. al., 1986).
The relational presentation of rules in the "KANON" system is realized through the rule editor. The automatically issued relations *"conceptual_constructs"*, *"representation"* and *"base-scheme"* form the interface to the KANON-proof mechanism. The tuples described through relations have to be derived in different ways depending of the attributes *"type_of_inference"* of the relation *"conceptual_constructs"*. These must not only be deduced with the help of the KANON-proof system, but they can also be derived by activating autonomous systems or they can be queried directly. This is especially valid for the extensional facts. All of the KRS are represented on the same internal format in the interface, in order to provide the information which is part of the proof of a rule. The following structure (Figure 8) for the proof mechanism of the "KANON" system results from that:

Application possibilities

The KBS "KANON" can be used under CIM as an **interface, knowledge management system** and as a **knowledge based environment.** In the following, these application possibilities are explained in detail from a conceptional point of view.

"KANON" *as an interface*

Up to now the area of interfaces under KBS has not been particularly taken into consideration. But as conventional interfaces can only transform more or less data from one format into another one, the application of KBS presents itself in this case as KBS do not only comply with the syntax during a transformation, but also with the semantics of the data. With the help of "KANON" one can not only convert the proportions of parts, which were designed and which have to be produced, from the CAD-format into a CAM-format, one can also derive CAM-knowledge from CAD-knowledge.

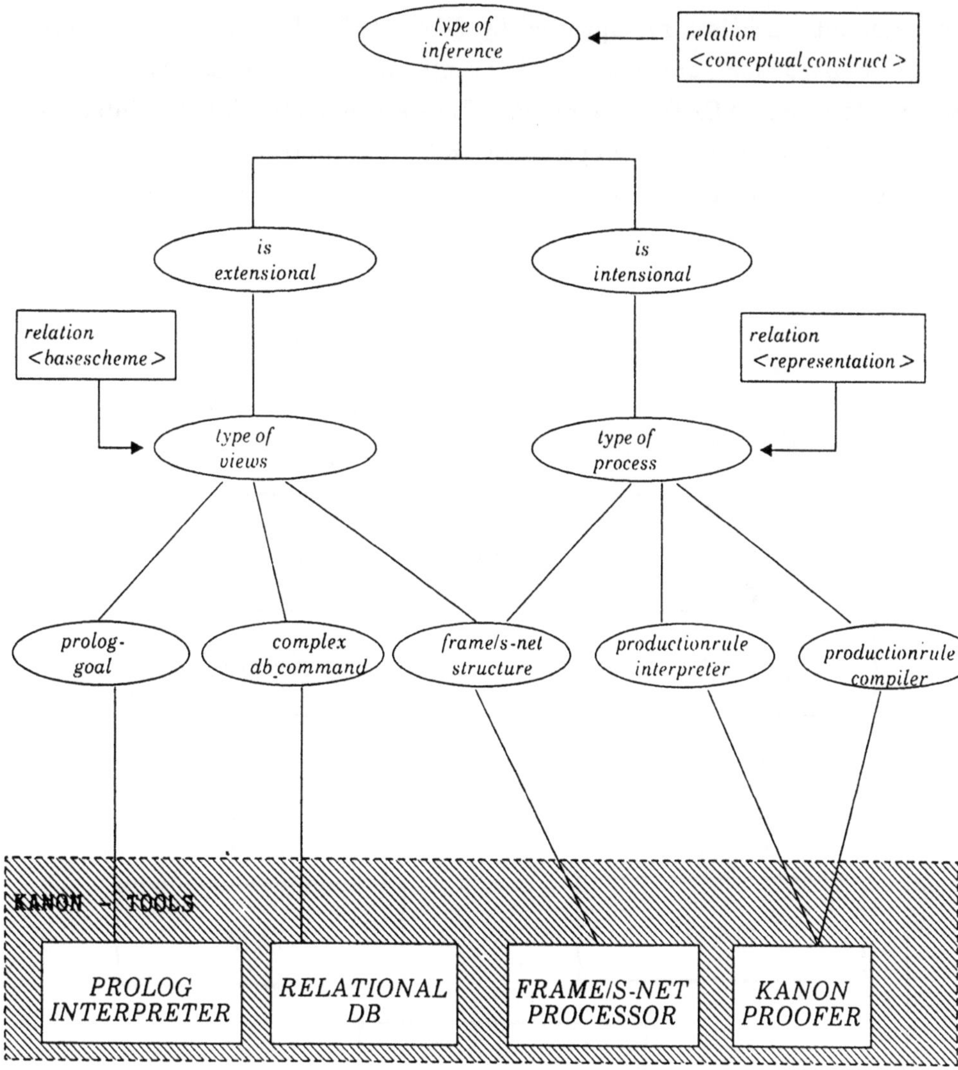

Figure 8. Decision tree for the proof of a user query in the "KANON" system

The user interface is also supported by "KANON". This concerns especially design-, construction-, production- and assembly inputs of the user. These inputs are transformed into CAD-/CAM-formats. Figure 9 shows the different interfaces to the extent of how they are supported by "KANON". In this connection it has to be taken into account that each time the figures a), b) and c) concern the same "KANON" system.

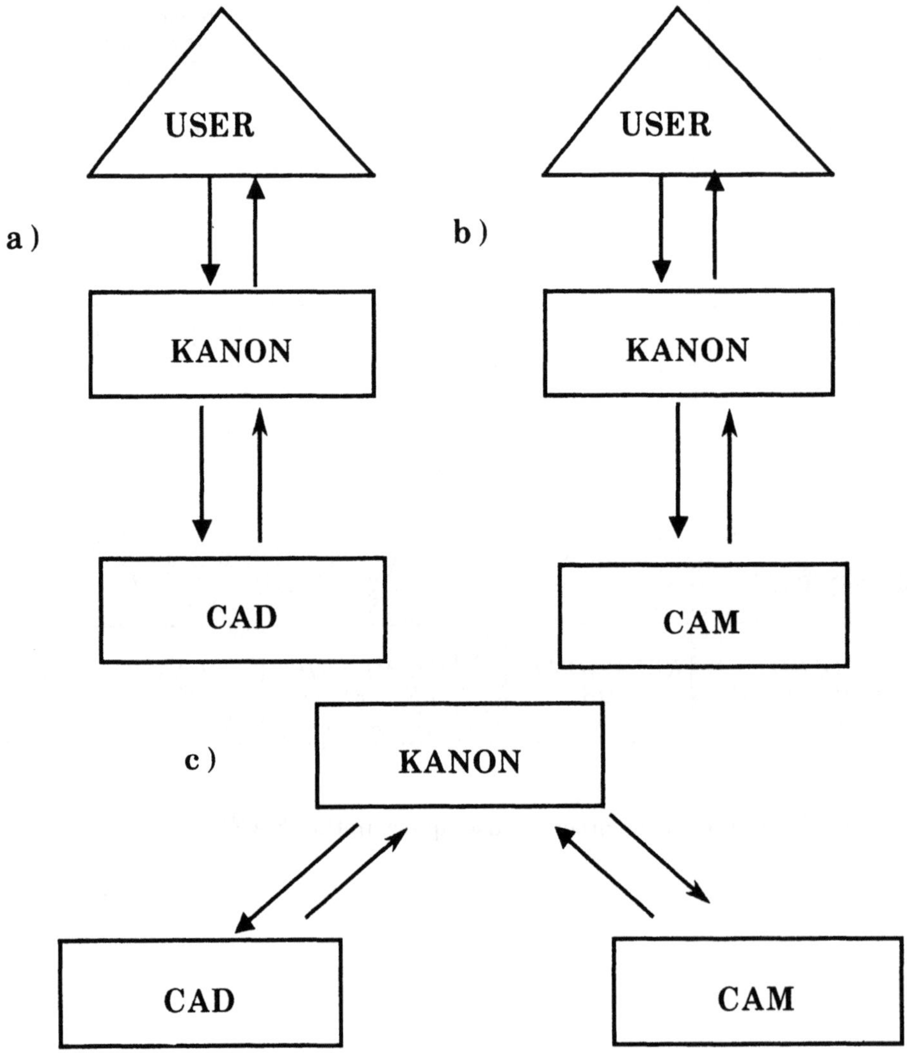

Figure 9. "KANON" as an interface

"KANON" *as Knowledge-based Management System (KBMS)*

As already mentioned, large amounts of data accumulate in the CAD- and also in the CAM-areal. In addition, it makes it more difficult that these data are complicated to manage. As "KANON" offers a coupling to a relational DB the processing of these amounts of data is much easier and more efficient. Especially in the area where conventional DBMS have great difficulties, namely the checking of Non-Standard Data Types, lies the strength of "KANON". But the DBMS also have problems concerning the efficient management of CIM-objects. Figure 10 shows "KANON" as a knowledge management system. Such management systems are also called KBMS (Knowledge Based Management System)

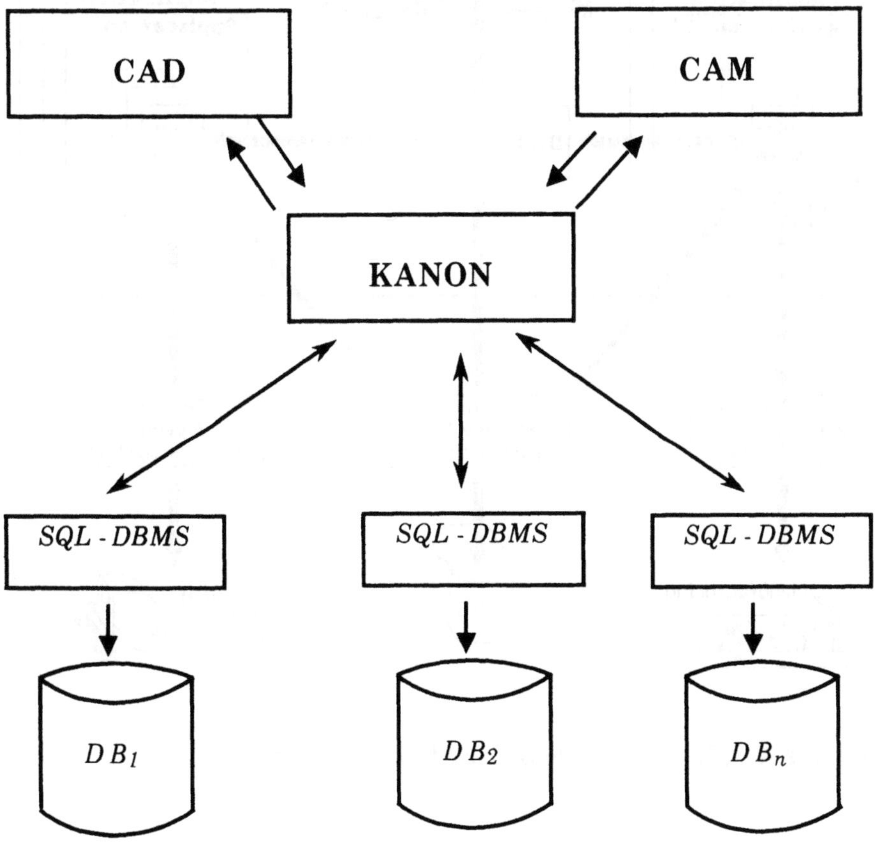

Figure 10. "KANON" as a knowledge based management system

"KANON" *as a knowledge based environment*

In order to integrate CAD-/CAM-systems it is not sufficient to integrate the data management and processing, but the dialog components and the user interface must also be integrated. Figure 11 shows "KANON" as an environment adding to the structure of a CAD-system as shown in Figure 2. In addition, the CAM-area has also been taken into consideration during the integration.

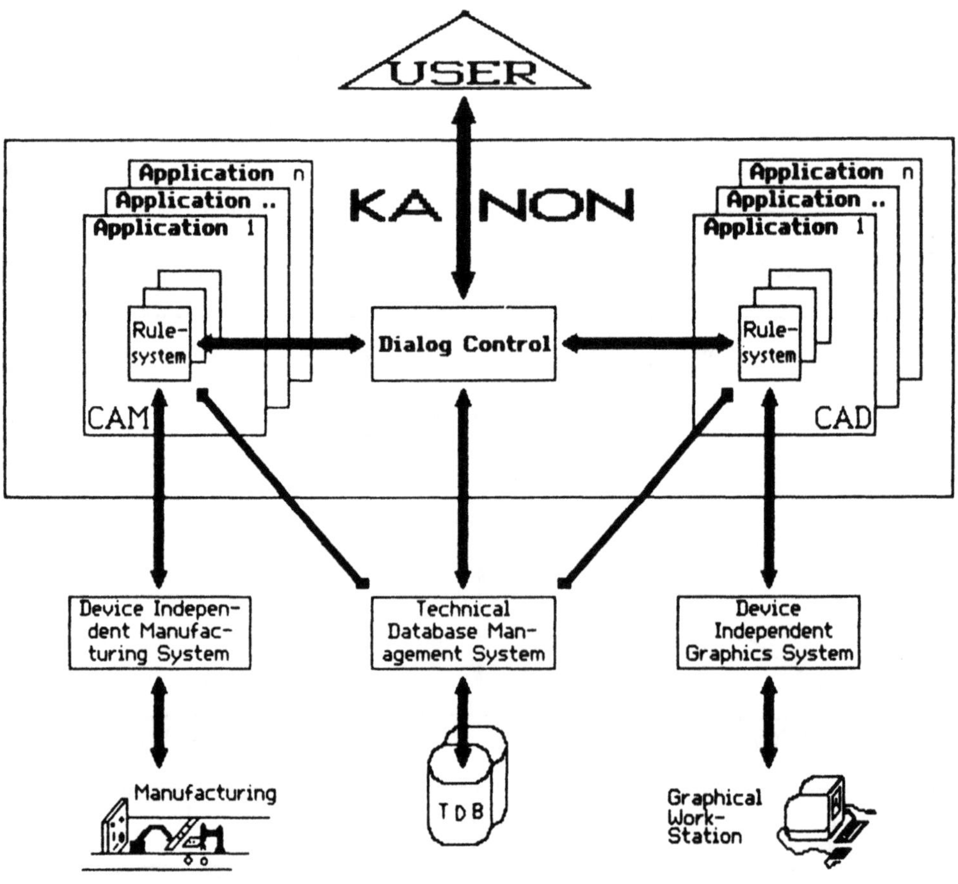

Figure 11. "KANON" as a knowledge based environment

3. REPRESENTATION OF CIM-STRUCTURES THROUGH "KANON"

The hybrid KBMS "KANON" provides - as already mentioned - the three representation types: logical rules, S-networks and frames. With the help of these representation types parts from the CAD-/CAM-areas can be modeled. A ball-bearing serves as an example for a construction- and production part.
Figure 12 shows an S-*net*work which represents such a bearing. A ball-bearing is a construction group which has a number and which was put together by a designer. It consists of an outer ring, balls and an inner ring. The outer ring was designed by the designer of the ball-bearing. It has a number, consists of steel and has-a depth of 10 mm. The outer ring is produced by the flexible production system 1. 20 pieces from the balls are needed. They have the number 1234 and a diameter of 4 mm. The balls are standard parts and they are stored in stockroom 5. The inner ring has a number, costs 0,50 DM and weights 100 g. It is an extra special buy part and is supplied by the TU-Berlin. All three parts (outer ring, balls, inner ring) are assembled by the flexible assembly system 11.

For reasons of clarity only a small part was represented here. It would be beyond the scope of a single drawing if e.g. one wanted to represent an entire car. Therefore, "KANON" provides coarsening - and refinement possibilities for S-networks during graphical representation.

Logical Rules are applied for supporting the S-networks. The following paragrahs show three rules which represent the link of the construction and of the production.

Logical Rule 1 indicates whether a part has to be ordered at a supply firm. The procedure is that first it is checked whether there actually is a firm that supplies such a part with the given attributes. Here it may not concern a norm part or a special part. If, in addition, an equivalent standard part exists, then an order is only placed if the price of the extra special buy part is lower than the one of the standard part.

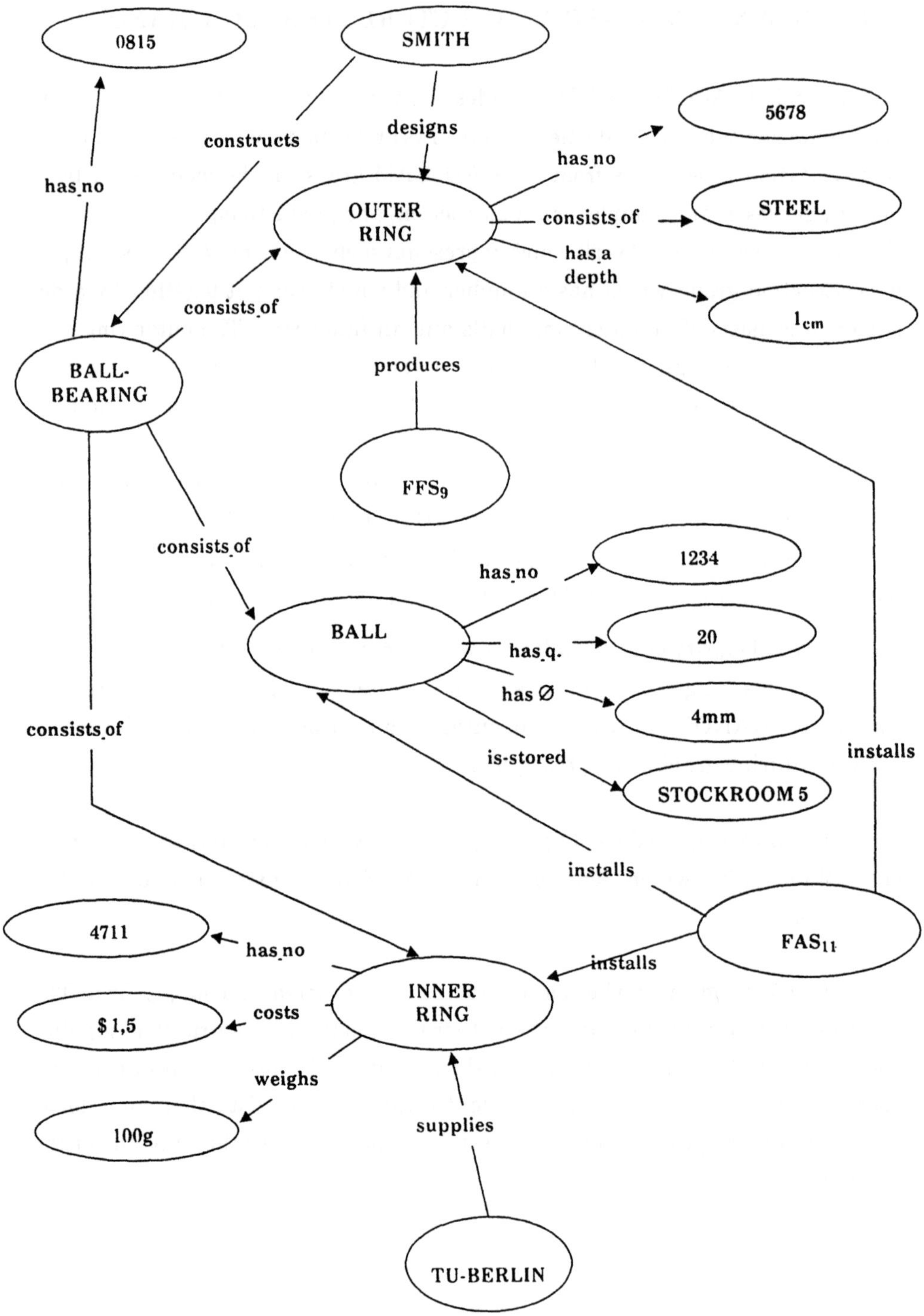

Figure 12. Representation of a ball-bearing through an S-network

```
Logical-Rule 1:

create rule buy extra separate parts
              (part no : *t, part desc. : *d, firm : *f
               price : *p, radius : *r, material : *m)
      if      extra separate part
              (part no : *t, part desc. : *d, firm : *f
               price : *p, radius : *r, material : *m)
      and     not norm part
              (part desc. : *d, radius : *r, material : *m)
      and     not special part
              (part desc. : *d, radius : *r, material : *m)
      and     standard part
              (part desc. : *d, price : *pp, radius : *r, material : *m)
      and     smaller
              (price1 : *p, price2: *pp).
```

Logical Rule 2 checks whether there still is a sufficient quantity of a standard part in stock. In order to support the S-network in Logical Rule 2 it would e.g. be checked whether 20 balls are still in stock.

```
Logical-Rule 2:

create rule in stock
              (stock no : *l, part no : *t, part desc. : *b,
               radius : *r, material : *m, quantity : *a)
      if      standard part
              (part no : *t, part desc. : *b
               radius : *r, material : *m)
      and     stock room
              (stock no : *l, part no : *t,
               part desc. : *b, quantity : *a).
```

Logical Rule 3 defines whether a ring has to be manufactured . First it is checked whether the requested part is a special part (constructed by a designer). There may not exist an equivalent part neither in stock or at a supply firm. This rule also forms the link to the third representation form.

```
Logical-Rule 3:

create rule ring manufacturing
              (part no : *t, part desc. : *d, i-radius : *i,
               a-radius : *a, weight : *g, material : *m, depth : *d)
      if      special part
              (part no : *t, part desc. : *d, i-radius : *i,
               a-radius : *a, weight : *g, material : *m, depth : *d)
      and     not stock room
              (part no : *t, part desc. : *d, i-radius : *i)
      and     not extra separate part
              (part no : *t, part desc. : *d, i-radius : *i).
```

The following section show a part of the production with the help of *frames*.
The *first frame* has the name FMS_1 and is a 'child' of FMAS. The agencies of FMS_1 are the transportation system, FMC_2 and FMC_3. The task of this flexible manufacturing system is the manufacturing of the outer ring. It consists of the flexible production cells FMC_2 and FMC_3. The transportation system consists of the assembly line AL_7 and the small electric trucks ST_6 and ST_7.

```
FRAME FMS1:

( FMS1
   ( AKO              ( $VALUE      (FMAS)))
   ( INSTANCE         ( $VALUE      (TRANSPORTATION SYSTEM)
                                    (FMC2) (FMC3)))
   ( TASK             ( $VALUE      (OUTER RING PRODUCTION)))
   ( CELL             ( $IF-NEEDED (FMC2)
                                    (FMC3)))
   ( TRANSPORTATION  ( $IF-NEEDED ((AL7) (TYPE: ASSEMBLY
                                                     LINE))
   (                                ((ST6) (TYPE: SMALLELECTRIC
                                                     TRUCK))
   (                                ((ST7) (TYPE: SMALLELECTRIC
                                                  TRUCK)))))
```

The *second frame* (FMC_2) and the *third frame* (FMC_3) have the frame which has just been described for a 'father'. Their structure is similar to the one of the first frame. FMC_2 has additional slots for the sawing distance (10 mm) and the type of saw blade (SB_3). FMC_2 represents the sawing process during the production of the outer ring, FMC_3 the milling process. The structure of these three frames has been kept very simple. Normally such a frame would also have to represent error messages and exceptions of a real production world in a suitable way. Thus, e.g. an error procedure should start regarding a frame with the task 'drilling', when a drill breaks off.

4. SUMMARY

According to today's research level the two integration approaches presented here have gained the same acceptance. If, however, one follows the developments further it becomes apparent that the KBS become more and more successful in the long term, because both disadvantages of the systems, the low processing speed and the enormous need for storage capacity, push more and more into the background. Thus, special hardware-components are developed for the FGCS-program (Fifth-Generation-Computer-System) which shall take the high performance requirements into account (Moto-Oka, 1982). The requirements for the

```
FRAME FMC3:

( FMC3
   ( AKO                 ( $VALUE      (FMS1)))
   ( INSTANCE            ( $VALUE      (RING PRODUCTION)
                                       (BALL PRODUCTION)))
   ( TASK                ( $VALUE      (MILLING)))
   ( MACHINES            ( $IF-NEEDED ((F2) (TYPE: MILLING
                                                   CUTTER))
                                       ((R8) (TYPE: ROBOT))
                                       ((R9) (TYPE: ROBOT))))
   ( MILL RADIUS         ( $VALUE      (20 mm)))
   ( MILL HEAD           ( $VALUE      (MH7)))
   ( MILL DEPTH          ( $VALUE      (10 mm))))
```

```
FRAME FMC2:

( FMC2
   ( AKO                 ( $VALUE      (FMS1)))
   ( INSTANCE            ( $VALUE      (RING PRODUCTION)
                                       (BALL PRODUCTION)))
   ( TASK                ( $VALUE      (SAWING)))
   ( MACHINES            ( $IF-NEEDED ((S5) (TYPE: SAW))
                                       ((R5) (TYPE: ROBOT))
                                       ((R6) (TYPE: ROBOT))))
   ( SAWING DISTANCE ( $VALUE      (10 mm)))
   ( SAW BLADE           ( $VALUE      (SB3))))
```

processing speed of the FGCS are around 10 to the 12th up to 10 to the 14th instructions per second. The database machines shall be able to store 100 - 1.000 billion bytes. The intelligent interface systems are designed for 10.000 different words of the natural language, for 2.000 grammatical rules and for 10.000 graphical elements (Schneider, 1986).

5. ACKNOWLEDGEMENTS

The authors would like to thank the members of the research group DBS/KI at the faculty of computer science of the TU-Berlin. Special thanks is meant for Mr. Dipl.-Inform. G. Eilers for his valuable criticism and helpful discussion. We want to thank Mr. Dipl.-Inform. A. Elsholtz for his assistance and his useful suggestions in the area of NDBS for CIM applications.

6. Glossary

CAD	Computer Aided Design
PPC	Production Planning & Control
CAM	Computer Aided Manufacturing
CA(P)P	Computer Aided (Process) Planning
FAS	Flexible Assembly System
FM&AS	Flexible Manufacturing & Assembly System
FMS	Flexible Manufacturing System
FMC	Flexible Manufacturing Cell
CIM	Computer Intregrated Manufacturing
GKS	Graphical Kernal Systems
TDBS	Technical Database Management Systems
TDB	Technical Database
KRS	Knowledge Representation Structure
DSS	Decision Support System
KB-IS	Knowledge Based Information System
CADAM	CAD System
CATIA	" " " "
CAEDS	" " " "
CADIS	" " " "
ICEM	" " " "
KBMS	Knowledge Based Management System
KBS	Knowledge Based System
NDBS	Non- Standard Database
DBMS	Database Management System
NGI	Non-Graphical Information
NC-M	Numeric Control Machine
DNC-P/M	Direct Numeric Control Program/Machine
(E)NF2-Model	(Extended) (Normal Form)2-Model
E-/ R -Model	Entity Relationship Model
AL	Assembly Line
ST	Small electric Truck

7. References

CAD-Handbuch (1984):
Auswahl und Einsatz von CAD-Systemen, Springer-Verlag, 1984

Ceri, S.; Gottlob, G.; Wiederhold, G. (1986):
Interfacing Relational Databases and Prolog Efficiency, in: Kerschberg, L. (Ed.), Proc. of the first International Conference on Expert Database Systems, Charleston, South Carolina, April 1986

Chen, P.P. (1976):
The Entity-Relationship Model,
ACM Transactions on Database Systems, 1976

Dadam, P. (1986):
Forschung und Entwicklung im Datenbank-Management-System-Bereich,
in Informatik-Fachberichte, Bd. 126, Springer-Verlag, 1986

Eberlein, W. (1984):
CAD-Datenbanksysteme, Springer-Verlag, 1984

Encarnacao, J. (1980):
CAD - Modelling, Systems Engineering, CAD - Systems
Lecture Notes in Computer Science, Vol. 89, Springer-Verlag, 1980

Encarnacao, J.; Schlechtendahl, E.G. (1983):
CAD, Fundamentals and System Architectures, Springer-Verlag, 1983

Encarnacao, J.; Schuster, R.; Vöge, E. (1986):
Product Data Interfaces in CAD/CAM Applications - Design, Implementation, Experiences, Springer-Verlag, 1986

Findler, N. V. (1979):
Representation and Use of Knowledge for Computers,
in Associative Networks, New York, Academic Press, 1979

Fikes, R.; Kehler, T. (1985):
The role of frame-based representation in reasoning,
Communications of the ACM, Vol. 28 No 9, 1985

Gebhardt, F. (1987):
Semantisches Wissen in Datenbanken - Ein Literaturbericht,
in Informatik Spektrum, Springer-Verlag, 10/1987, pp. 79-98

Grill, E.; Flittner, J.; Rauch, W. (1987):
Integration von CAD, CAE, CAM über relationale DB,
in: Information Management, CW-Communications, 1/87

Karagiannis, D. (1987):
Bereitstellung und Manipulation von WRS: Das hybride System KANON,
Dissertation., FB Informatik, TU-Berlin, 1987

Maida, A. S.: Shapiro, S.C. (1981):
Intensional Concepts in Propositional Semantic Networks,
in Cognitive Science 65, 1981, pp. 291-330

McKay,D. P.; Shapiro, S. C. (1981):
Using Active Connection Graphs for Reasoning with Recursive Rules, in Proceedings of the 7th IJCAI, 1981

Moto-Oka, T. (1982):
Fifth Generation Computer Systems, North-Holland, Amsterdam, 1982

Renn, W. (1986):
Struktur und Aufbau prozessnaher Steuergeräte zur Verkettung, in: flexiblen Fertigungssytemen 58, Forschung und Praxis, Springer-Verlag, 1986

Schneider, H.-J. (1982):
distributed data bases, Proc. of the Second International Symposium on Distributed Data Bases, Berlin, FRG, Sept. 1.-3. North-Holland, 1982

Schneider, H.-J. (1986):
Systems of the 5th Generation and their Impact on the Qualification, Training, and Retraining of Management, in: Robotics, Vol. 2, No. 4, North-Holland, 1986

Shapiro, S.C. (1979):
The SNePS Semantic Network Processing System, in Associative Networks, New York: Academic Press, 1979, pp. 51-92

Shapiro, S. C. (1977):
Representating and Locating Deduction Rules, in a Semantic Network, in Sigart Newsletter, Nr. 63, 1977, pp. 14-18

Spur, G.; Krause, F.-L. (1984):
CAD-Technik, Hanser-Verlag, 1984

Vassiliou, Y. (1985):
Integrating Database Management and Expert Systems, in Informatik-Fachberichte, Bd. 94, Springer-Verlag, 1985

Wedekind, H. (1981):
Datenbanksysteme I, Wissenschaftsverlag, 1981

An Information System Architecture for Robot Cell Programming

L.M. Camarinha-Matos; A. Steiger-Garção
Universidade Nova de Lisboa - Robotics Group
Quinta da Torre - 2825 Monte Caparica - Portugal

Abstract

Programming a robot cell is a complex problem implying contributions from several scientific disciplines with a major need of integration. More than a functional integration, the problem is mainly one of information integration. The Information System is, therefore, the key element for integration of all components of the programming system.
A conceptual structure for the Information System is proposed and some preliminary implementation results discussed with special emphasis on the interface between cell programming and upper level aspects of the CIM system. Some guidelines for future work are presented.

1. INTRODUCTION

CIM is frequently presented as the integration of "automated islands", however some of these islands are not completely automated yet. This is the case of robot cells.
The problem of robot cell programming is a complex one implying contributions from several scientific disciplines with a major need of integration.
The desire of more flexibility and reduction of programming time calls for a progressive level of automatization of the process.

An architecture for an assembly robotic cell programming system was proposed [1] [8] conceiving a development in two phases:

-Off-line: where a generic plan / high level program is produced based on the fact that a reasonable information is available a-priori in a station of average structure. A special emphasis is put on off-line simulation.

-On-line: detailing (specialized planning) and readjustment of the solution as a function of sensorial information acquisition.
Fig. 1 shows a general sketch of this conceptual architecture.

NATO ASI Series, Vol. F49
Computer Integrated Manufacturing
Edited by I. Burhan Turksen

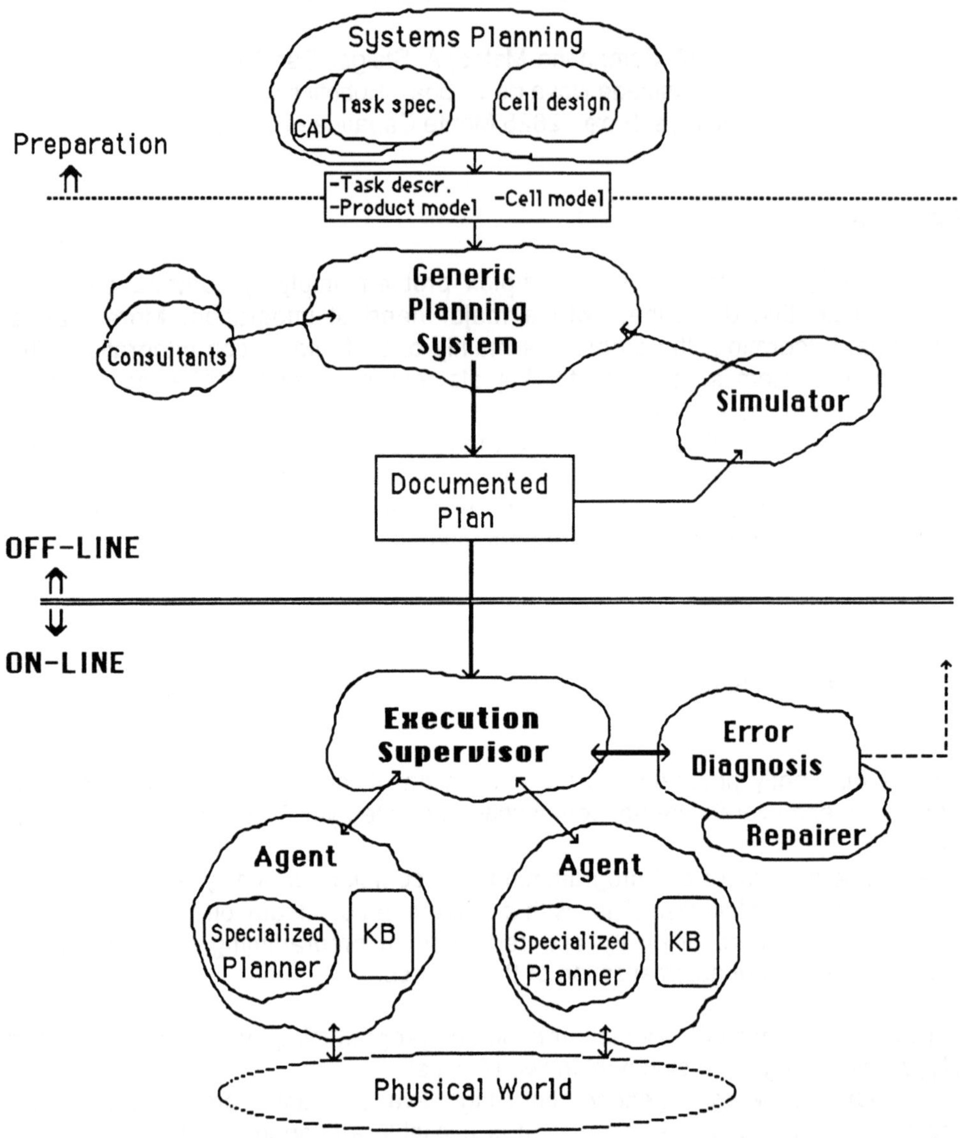

Fig. 1 - Cell programming system architecture

Brief description of components:

-*Systems planning:* this component is responsible for the task specification (and preparation) - product modeling, assembly process specification - and cell design.

-*Generic planner:* module that will produce a generic plan / program for the intended task. This process can be performed in an automatic (implicit programming) or in an interactive way (with human help).

-*Consultants:* specialized modules to assist the planner. For instance, a trajectory planner, or the human operator (interactive approach).

-*Simulator:* module that will allow testing the generic plan thoroughness and suggest eventual changes. In order to have a realistic testing, special emphasis is put on simulation of the sensorial subsystem, emulating the on-line phase [1][9].

-*Supervisor:* this is a high level interpreter that executes the generic plan, assigning each segment of it to the agents (actuation, perception).
The agents have a local planning capability to adapt to the dynamic information acquired during execution (for instance, grasping planner).

-*Diagnosis and error recovery:* module that analyses possible errors based on sensorial information and proposes corrective steps or returns control to the general planner in case of severe problems.

This architecture establishes the framework for the research work being done by the Robotics Group of UNL.
The aspects related to the Information System are partially carried out under the framework of an EEC Esprit Project (Nr. 623 - Operational Control for Robot System Integration into CIM) and a project on CAD/CAM and Robotics (Portuguese-American Foundation).

As it becomes apparent, the problem complexity requires contributions from various scientific areas - where most work has been done in a somewhat isolated way or with partial degrees of interaction - and, therefore, a further important step is the attempt to integrate all these results.

Our statement is that the Information System is the key element for integration of all components of the programming system. More than.a functional integration, the problem is mainly one of information integration.
Therefore, it is necessary to understand which are the information (data, knowledge) requirements for each functional block and which are the contributions of each block. Finally, an architecture integrating all the information "chunks" and information flows is needed.
An integration of tools is also a natural need, since the information serving different purposes is resident on distinct representation environments (CAD, KBS, RDB, etc).

2. CONCEPTUAL ARCHITECTURE

2.1 Requirements

Design (product and cell) and process planning sub-systems are not truly components of the programming system. We would rather say they "perform" a preparatory phase that is of vital importance for the programming activity. In this sense they represent the interface between the programming system and the upper levels of the CIM system. Then, they must be taken in consideration in terms of their contributions to the IS.

As output, this phase will produce a complete specification of the task, including
-product model: list of components and model of each part
-a description of the assembly process
-a description/model of the assembly cell.

In order to select / propose a cell, it is necessary to have a library of models of components (technical, economical information). A simplified ("qualitative") model of product parts is also used.

Interactive programming can be seen as a subset of the automatic programming in terms of information needs. This phase will receive the results of the systems planning. A different model of cell components and product parts is necessary. For instance, an execution model of the robot, or an abstraction of parts' geometric model (grasping points, relative position in the assembly, etc.).

For simulation (programming level) we need the same information as for planning, with good geometric models (for graphic simulation).

Execution monitoring requires:
-executive models of cell components
-derived models of product components to support the perception process
-a rule base for error diagnosis and recovery.

This information will have a support partly distributed.

As can be seen, there is a need for information with a quasi-static character (libraries of components, parts, ...) and other dynamic (for instance, the current position of the robot).

By other hand, some objects can be seen with different levels of abstraction in different "areas" of the programming system.

2.2 IS General Model

An approach for information integration is depicted in fig 2.

Brief description of components:

-*Product model:* description of the product to be assembled and its composing parts.

-*Cell model:* description of all components of the assembly cell (robot, conveyor, sensors, etc.) and the inter-relations among them.

-*Assembly graph:* a partially ordered graph representing a high level specification of the assembly task to be performed. This description is done using task level operations and not robot level operations. The graph nodes will include additional information like the approach direction, kind of tool, technological restrictions, etc.

-*Technological KB:* contains technological knowledge to be used during the systems planning process.

-*Documented plan / program:* represents a specialization of the task specification, now at cell operations level - again a partially ordered graph of operations and task knowledge necessary for execution monitoring.

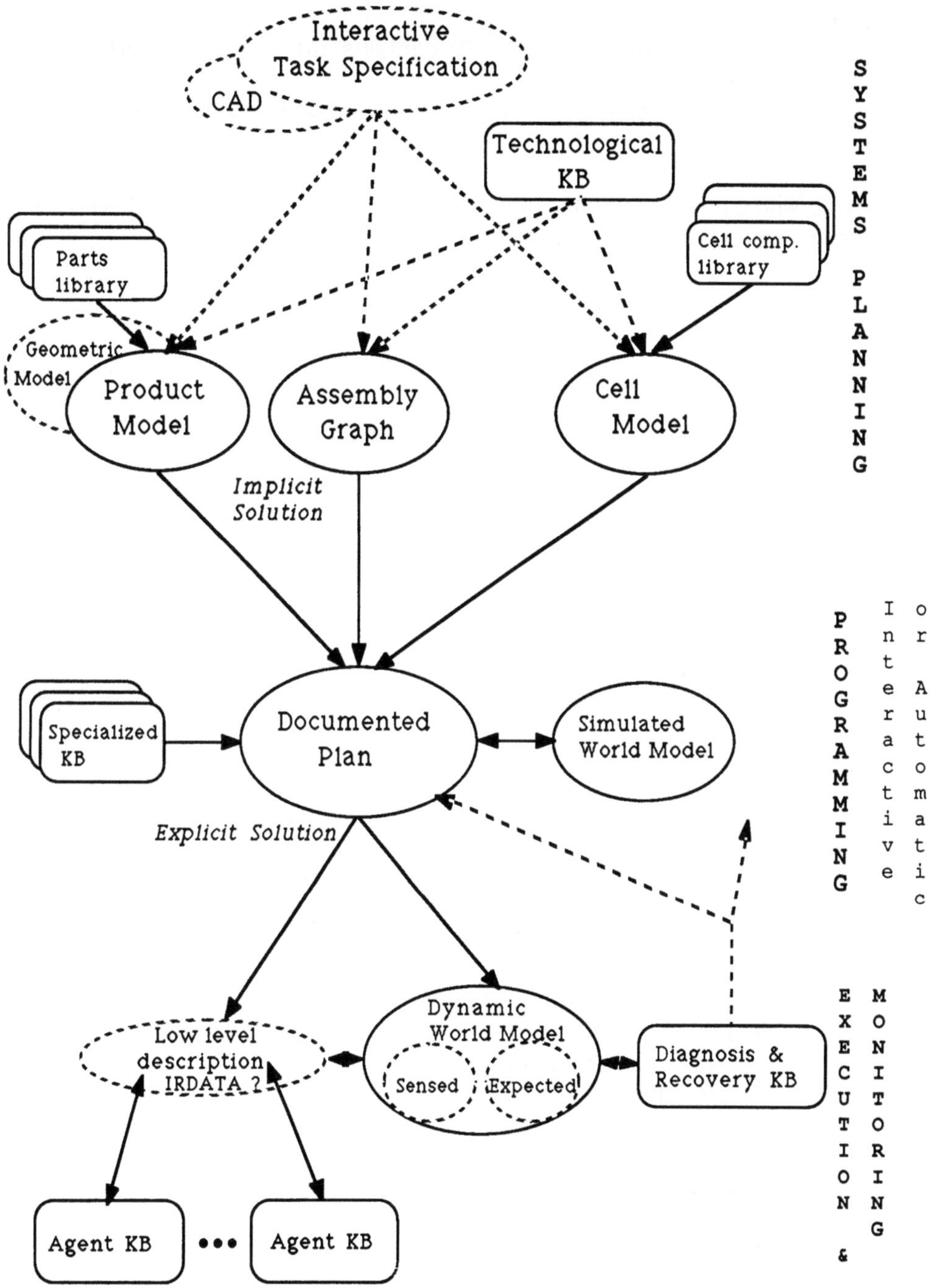

Fig. 2 - Information System - main components

-*World model* - dynamic and simulated: represent a model of the dynamic aspects of plan execution (expected and sensed).

-*Specialized KB:* represents the knowledge that supports the specialized consultants.

-*Diagnosis & error recovery KB:* contains the rules that direct error diagnosis and recovery.

-*Agents' KB:* specialized knowledge that support agents' local planning and perception.

Associated with each information "node" there are two aspects: the concepts and concrete instances (for a specific task, cell ...).

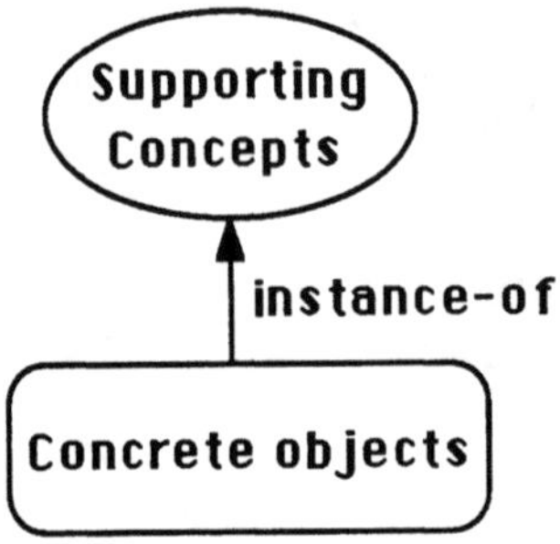

Fig.3 Concepts and instances

For example, associated with "Assembly Graph" the concepts of operation, node, link, etc., can be found. A concrete instance will be the description of a specific task.

Concerning implementation aspects, in order to evaluate / select a representation tool, a list of requirements was established [2] under the scope of the following aspects:

.Potentialities for the representation of several kinds of objects.
.Static and dynamic aspects.
.Communication / interfacing with other systems.
.User interface
  -Graphical visualization of the information structure
  -Active images
  -Development of end-user interfaces
  -Sub-models.
.Explicit and implicit information.
.Multi-user access.
.Facilities for information acquisition
  -Automatic validation
  -Dynamic modification of the information model.
.Tool capacity.
.Availability and portability.
.Imprecise knowledge representation.

A market survey on data and knowledge management systems was made [3], and, as a result, it was concluded that no single tool fullfills our needs. Therefore, we need an integration of tools (CAD, RDB, KBS). The frame / object oriented representation paradigm was pointed as a very promising technique to be used as the kernel for integration.

3. DEVELOPMENT APPROACH

This paper doesn't present a finished work, but intends to reflect the ideas on the IS concept on the current stage of our research.
Various topics are being worked out in our group under the framework of the general model discussed. Covering all the different subjects isn't our present goal but we are benefiting from the cooperation with other partners of the Esprit project.
We think this kind of work must be done in an interactive way - trying some cases and coming back to refine the model accordingly to the experience acquired.
Under this perspective, some partial results are described in section 3.1 and next steps towards an integration are presented in section 3.2.

3.1 Some results

i. Product modeling

Besides the direct geometric information there is a need for some derived information and also complementary (non-geometric) information.
Commercialy available CAD systems don't allow an easy extension of their information representation capabilities.
Connecting a CAD system to a frame representation system and extract (from CAD) an abstraction of the geometric model is our current approach [4]. This abstraction will be complemented with the information necessary to the assembly planning/programming process.
Mainly for evaluation purposes, a small experimental frame engine was implemented in Prolog-2 on a Sperry PC/IT and a connection to PADL-2 was made (Fig.4).

Information extracted from CAD:

.The assembly structure
.Absolute and relative position of parts
.Attributes (material, color, ...)
.Grasping points
.Enclosing box.

In order to represent products and parts the concepts of *assembly, solid, family, part_type* and *part* have been defined.
The abstractions extracted from CAD are represented as instances of these concepts.

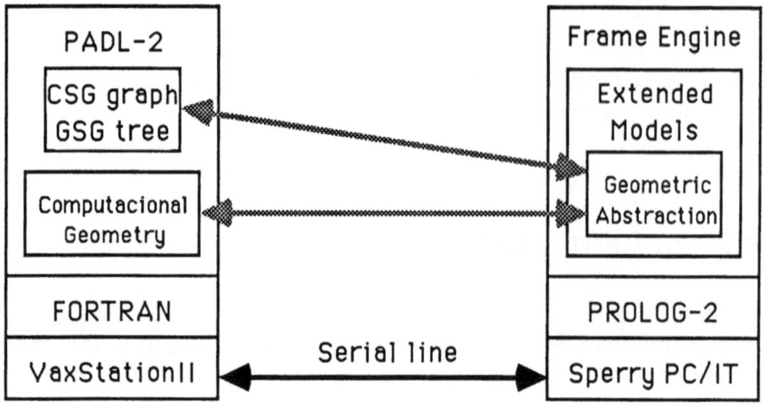

Fig. 4 Connection PADL-2 <-> Frame Engine

A partial example for the Cranfield Benchmark:

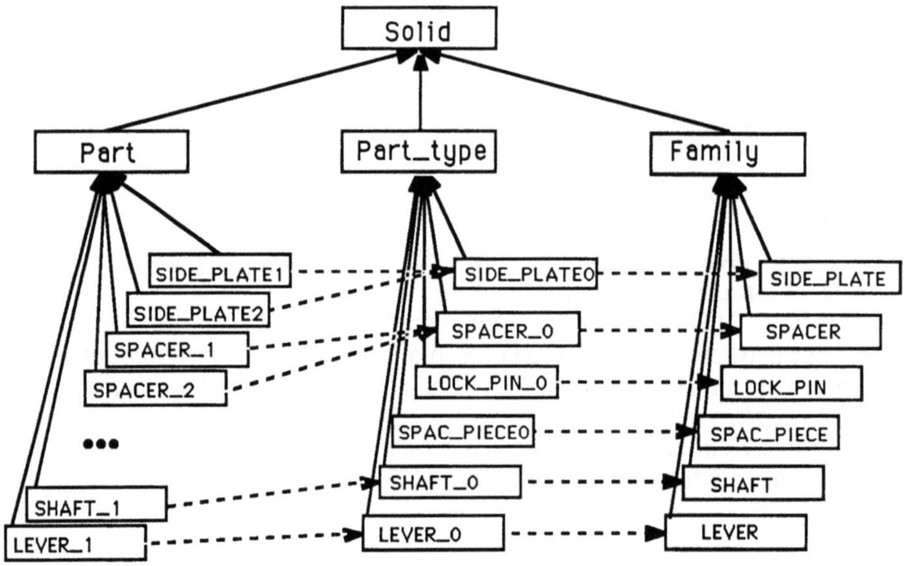

Fig. 5 Semantic network for parts of Cranfield Benchmark

frame 'SIDE_PLATE_1'
slot ako *value* part
slot instance_of *value* 'SIDE_PLATE_0'
slot ref *value* ref(0.0, 0.0, -1.0, 0.0, 1.0, 0.0,
1.0, 0.0, 0.0, 0.0, 0.0, 10.0)
slot box *value* box(-1.0, 0.0, 0.0, 111.0, 125.0, 10.0)
...

Some attributes can be extracted from CAD during the initial loading of the frame structure or on demand via the if_read demon. In the case under consideration both examples are present: *position* (ref) and *enclosing box* are extracted during the loading, and *grasp points* on demand.
Object oriented programming allows a bi-directional interface, i.e., methods are used to access PADL-2 from the frame engine.

```
frame geom_object
        slot ako            value  thing
        slot status         value  ready
        slot error          value  []
        slot context        method context
        slot padl2open      method padl2open
        slot padl2close     method padl2close
        slot padl2run       method padle2run
        slot padl2stop      method padl2stop
        slot prostruct      method prostruct
        slot proshow        method proshow
        slot erase          method call_padl1
        slot ugb            method call_padl1
        slot center         method call_padl3
        slot disp           method call_padl3
        slot draw           method call_padl3
        slot shade          method call_padl3
        ...
```

All geometric objects (assemblies, solids, ...) inherit these methods. The solid modeling system can be seen as a geometric reasoning server.

One product is represented as a list of parts / subassemblies.

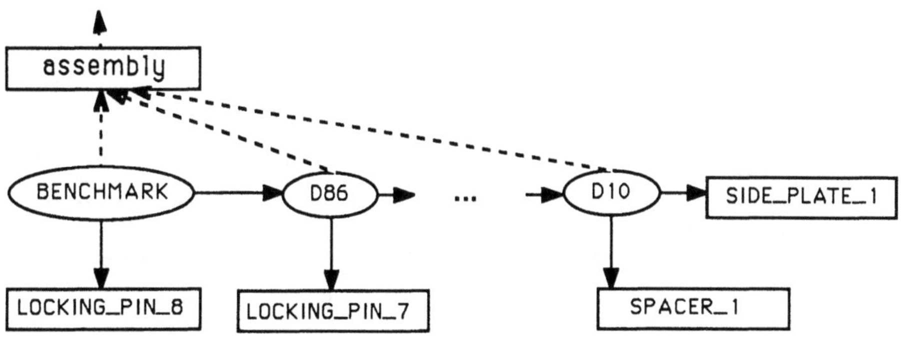

Fig. 6 Product composition

```
frame 'BENCHMARK'
        slot ako value assembly
        slot ref value ref(1.0, 0.0, 0.0, 0.0, 1.0, 0.0,
                           0.0, 0.0, 1.0, 0.0, 0.0, 0.0)
        slot box value box(-1.0, 0.0, 0.0, 111.0, 159.0, 30.0)
        slot has_component(1) value 'LOCKING_PIN_8'
        slot has_component(2) value 'D86'
        ...
```

This representation reflects only the structure of the product and not the assembly graph.
This information will be complemented by systems planning: new attributes, relation to the workcell, relation to the assembly graph, ...

As CAD models internal representation has to be accessed , this work depends on the CAD system we are working with. "Shifting" to another CAD system may imply a considerable effort. Therefore, a standardization on solid models representation would be very useful. Some efforts have been put on this problem.
IGES is one result, however in the preliminary versions it is relatively poor in terms of solid modeling.
An Esprit project have tackled this problem and produced a more powerful specification - CAD*I [7].
The basic idea behind these systems is the interchange of information by means of neutral files. But, in the present case, an interactive exchange based on a direct connection facility is necessary.
By other hand, these standardization proposals are basically concerned with geometric models and not with modeling of the assembly process. Therefore, additional information that can be derived from CAD data is not standardized. This is an open question with implications in the system architecture.

ii. Assembly graph

Associated with the Assembly Graph "node" there are some supporting concepts. For instance the concept of (task level) operation, used to describe the task to be done.

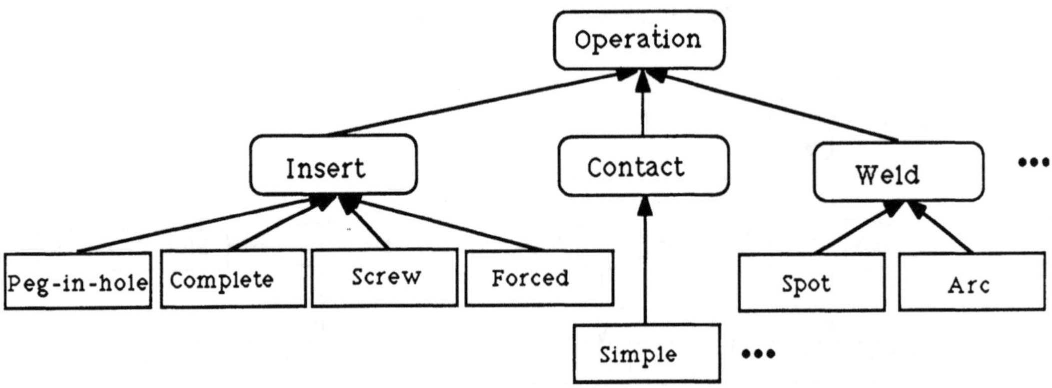

Fig. 7 Example of task level operations

frame INSERT
slot ako *value* Assembly-operation
slot component1 *value* <part or assembly>
slot component2 *value* <part or assembly>
slot approach_dir *value* <edge>
slot default_tool *value* <tool>
...

frame PEG-IN-HOLE
slot ako *value* INSERT
slot tolerance *value* <real>
...

We have been reasoning in terms of an assembly station. However the IS model can easily be generalized to support other kinds of cells.
For instance, instead of having one set of operations (task level) a library of sets could be implemented.

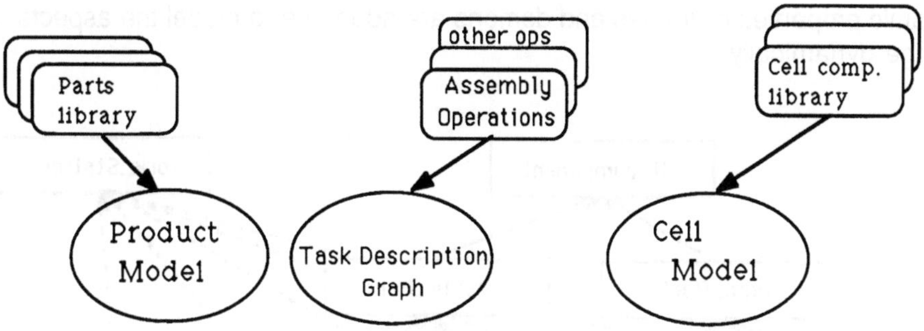

Fig. 8 Generalization of the Assembly Graph

In this way, the concepts inherent to the "Assembly Graph" correspond to an instance / specialization of a generic "Task description" concept.

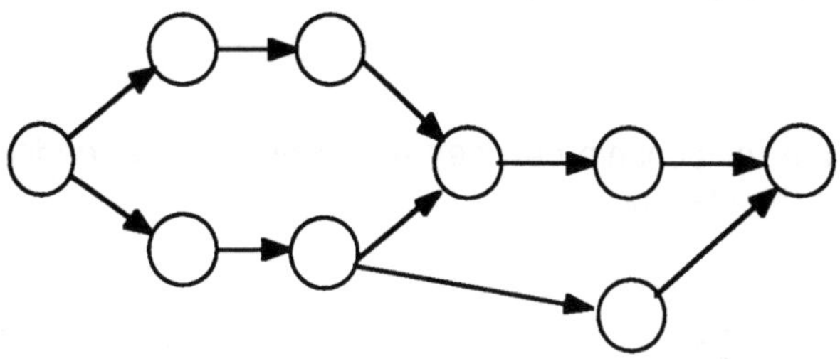

Fig. 9 Example of Assembly Graph

An example of node:

frame OP_n
slot ako *value* Peg-in-Hole
slot belongs_to *value* Cranfield_Benchmark
slot component1 *value* lock_pin1
slot component2 *value* sideplate1
slot approach_dir *value* z
slot tool *value* gripper2
slot next
slot prev
...

We are assuming that much of the preparing work for the programming process have been done at this stage. For instance, the grasping points, the approaching direction for the assembly operation, etc. have been established (probably using an interactive graphical facility for the specification of working points, what is in accordance with new developments in graphical systems).

iii. Aspects of cell model

The frame/oop paradigm is also adequate for cell modeling.
The concept specialization notion can be used to represent different views of each object.
Dynamic properties of frames and demons are adequate to model the aspects that change dynamically.

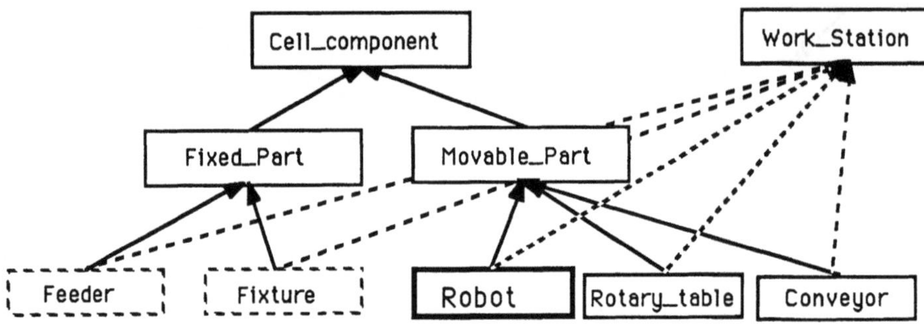

Fig. 10 Example of cell model

Fig 11 illustrates the relationship between a part type and a feeder - the origin of that kind of parts in the station.

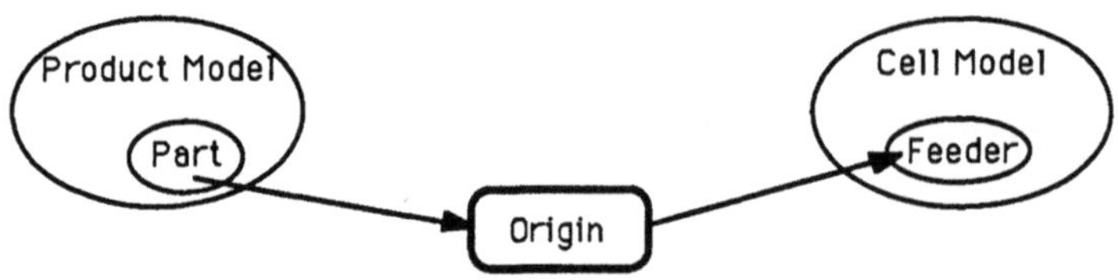

Fig. 11 Relationship between parts and feeders

Relationships between parts and operations of the assembly graph can also be represented.

Besides the generic relation ako, it is useful to have some domain specific relations with special inheritance properties. For instance, a relation *current-tool* could be used to connect a robot with a tool. The robot operations set is enriched with the operations of the attached tool inherited through this relation.
We can think also of the inverse relation *current-user* to connect a tool to its holder.
For instance, with Knowledge Craft it is very easy to create new relations and specify their inheritance behaviour.

iv. Plan / program

This is a refinement of the task description expressed in terms of cell level operators. It can be represented by a partially ordered graph easily implemented in frames.

v.World Model - dynamic and simulated

The triple (Product Model, Assembly Graph, Cell Model) - that represents effectively the task description - progresses in successive detail levels across the programming system architecture from the systems planning to the agents level. This information is passed from the programming phase to the execution phase as plan documentation.
Some parcels of information (some slots and links) are instantiated only at execution (or simulation) time when their values are known, representing the dynamic evolution of the task execution. Therefore, the world model is distributed by the objects already modeled in the IS.
The notion of context (or world) present in some frame engines is useful to represent this dynamic evolution. A root context holds the static structure / concepts and a child context (simulation context, for instance) represents the changes in the model.

vi. Agents

KB for perception

The models used for perception can be derived - by an off-line training process [9] - from the geometric models of the parts to be recognized.

OOP in robot control

Existing manipulator level languages are not adequate for an upper level extension if we want a more abstract programming system or concept generalization towards cell programming. However, these languages incorporate some useful low-level features: robot models (geometric, kinematic), trajectory interpolation, interfaces to the physical systems, etc, that we would like to use - as a supporting layer - in more advanced programming systems.
On the other hand, different robots (with different languages) can be used in the same cell. Some other components have their own controllers (Programmable Logic Controllers, for instance) with their own command (or rudimentary language) facilities. The need is therefore for a kind of "cell level language" able to integrate in an uniforme way this diversity of "low" level (isolated) controllers.
Our approach was to connect an LM-controlled robot to a frame engine [5] in order to integrate the desired functionality in the system.
On top of LM, an interpreter of a subset of the language - the robot related commands - was written in LM as a mean to recuperate the desired functionality.
On the frame engine side, a robot (and other cell elements) executive model was

developed. The commands to the robot controller (LM interpreter) were implemented as methods and demons.

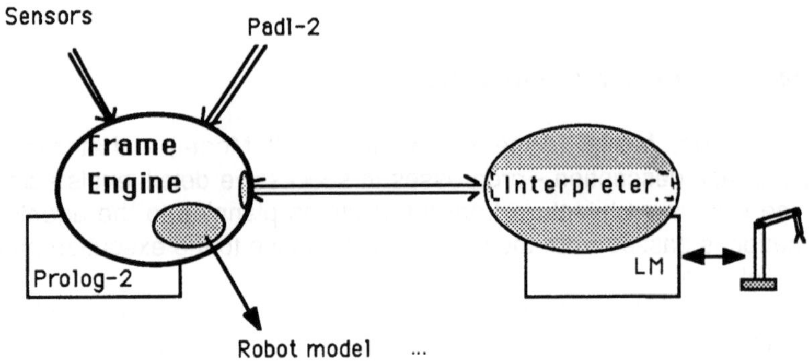

Fig. 12 Integration of robot controllers

frame robot
slot ako *value* movable_component
slot part_of *value* work_station
slot in_work_station_position *value* [1.,0.,0.,0.,1.,0.,0.,0.,1.,100.,100.,0.]
slot robot_move *method* move_fn
slot robot_moving *if_read* get_robot_moving
slot gripper_width *if_read* get_gripper_width
slot robot_moving_time *if_read* get_robot_moving_time
slot soft_home *method* move_to_soft
slot open_gripper *method* open_gripper_fn
slot close_gripper *method* close_gripper_fn
...

This approach is also valid for the integration of other components (conveyors, rotary tables, vision, ...).

Some other works for the recuperation of existing controllers / languages adopt a "compile-download" approach.
In our view, the interactive connection provided by the interpretative approach - where feedback is possible - is a mandatory condition to implement an execution supervision with error diagnosis and automatic plan repair.
Consideration of parallelism at lower level as well as at the upper level is a major need. An extension of the frame/OOP concept to incorporate concurrency could provide a valuable tool.

3.2 Future work

In a preliminar phase a small experimental frame engine was used mainly to evaluate the adequacy of the frame/oop paradigm.

As a next step the use of a more powerful system like Knowledge Craft [6] is planned.
Knowledge Craft is a programming environment for the development of knowledge based systems including:

-a more generic and powerful frame representation system - allowing semantic network representation with pre-defined and user-defined inheritance mechanisms

-object oriented programming and demons

-a forward inference system (OPS 5 like)

-a backward inference system (Prolog like)

-metaknowledge.

The initial Prolog based frame engine being a small, light system, running on a PC, will continue being useful as a delivery system (shop floor level) and, therefore should be improved.
Another important step will be the integration of other works being developed on our group on knowledge based perception.

An extension to other types of cells (e.g., DNC machining cells) and to relations between cells is also planned.

Acknowledgments

This work is partly funded by the European Community under the ESPRIT Programme. However the views expressed in this paper are primarily those of the authors, and are not necessarily shared by all members of the Consortium involved in the project.
Other funding was provided by Instituto Nacional de Investigação, Junta Nacional de Investigação Científica e Tecnológica, and Fundação Luso-Americana para o Desenvolvimento.

References

[1] Robotic Cell Programming: A Knowledge Based Approach
L.M. Camarinha-Matos; A. Steiger-Garção
Proceedings of the Iasted/Afcet Int. Symposium on Robotics and Artificial Intelligence, University Paul Sabatier, Toulouse 18-20 Jun 86.

[2] Information System: Rules for tool selection
L. Camarinha; L. Correia; A. Barbosa
Esprit 623 working paper EP-UNL-01.87/1, Jan 87.

[3] Market Survey on Data and Knowledge Management Tools - Comparative Analysis
L. Camarinha; D. Ferreira; J. Moura-Pires; L. Correia; A. Barbosa
Esprit 623 working paper EP-UNL-02.87/1, Feb 87.

[4] Integrating CAD in the CIM Information System
L.Camarinha; A. Steiger; J. Batista
Esprit 623 working paper EP-UNL-06.87/1, Jun 87.
[5] An experiment with object oriented programming for robot control
L. Camarinha; A. Barbosa
Esprit 623 working paper EP-UNL-07.87/1, Jul 87.
[6] Knowledge Craft User's Manual
Carnegie Group, Inc., Apr 1987.
[7] Specification of a CAD*I Neutral File for Solids
E.G. Schlechtendahl (Ed.)
Springer-Verlag, 1986
[8] A Conceptual Structure for a Robot Station Programming System
A. Steiger-Garção; L.M. Camarinha-Matos
To be published in Robotics (North-Holland), vol.3, 1987.
[9] A Knowledge Based Approach for Multisensorial Integration
A. Steiger-Garção; L.M. Camarinha-Matos
in "Languages for Sensor Based Control in Robotics", Springer-Verlag, NATO ASI Series F, Vol. 29, 1987.

PART 3: MANAGEMENT OF UNCERTAINTY

A COMPUTATIONAL THEORY OF DISPOSITIONS

Lotfi A. Zadeh*
Computer Science Division
University of California at Berkeley
Berkeley, CA 94720

ABSTRACT

A ***disposition*** may be interpreted is a proposition which is preponderantly, but not necessarily always, true. In this sense, ***birds can fly*** is a disposition, as are the propositions ***Swedes are blond***, ***snow is white*** and ***slimness is attractive***.

An idea which underlies the theory described in this paper is that a disposition may be viewed as a proposition with implicit fuzzy quantifiers which are approximations to *all* and *always*, e.g., *almost all, almost always, most, frequently, usually*, etc. For example, ***birds can fly*** may be interpreted as the result of suppressing the fuzzy quantifier *most* in the proposition ***most birds can fly***. Similarly, ***young men like young women*** may be read as ***most young men like mostly young women***. The process of transforming a disposition into a proposition with explicit fuzzy quantifiers is referred to as *explicitation* or *restoration*.

Explicitation sets the stage for representing the meaning of a disposition through the use of test-score semantics (Zadeh, 1978, 1982). In this approach to semantics, a proposition, p, is viewed as a collection of interrelated elastic constraints, and the meaning of p is represented as a procedure which tests, scores and aggregates the constraints which are induced by p.

The paper closes with a description of an approach to reasoning with dispositions which is based on the concept of a fuzzy syllogism. Syllogistic reasoning with dispositions has an important bearing on commonsense reasoning as well as on the management of uncertainty in expert systems. As a simple application of the techniques described in this paper, we formulate a definition of *typicality* and establish a connection between the typical and usual values of a variable.

1. Introduction

Informally, a disposition is a proposition which is preponderantly, but not necessarily always, true. Simple examples of dispositions are: *Snow is white*, *exercise is*

* To Didier Dubois and Henri Prade. Research supported in part by NASA Grant NCC2-275, and NSF Grants IST-8320416 and DCR-8513139.

NATO ASI Series, Vol. F49
Computer Integrated Manufacturing
Edited by I. Burhan Türksen

good for your health, long sentences are more difficult to parse than short sentences, a cup of coffee costs about fifty cents, slimness is attractive, a bachelor is an unmarried man, etc. Dispositions play a central role in human reasoning, since much of human knowledge and, especially, commonsense knowledge, may be viewed as a collection of dispositions.

The concept of a disposition gives rise to a number of related concepts among which is the concept of a *dispositional predicate.* Familiar examples of unary predicates of this type are: *healthy, honest, optimist, safe*, etc., with binary dispositional predicates exemplified by: *taller than* in *Swedes are taller than Frenchmen, like* in *Italians are like Spaniards, like* in *young men like young women, smokes* in *Ron smokes cigarettes*, and *relevant* in *age is relevant to health.* Another related concept is that of a *dispositional definition* which is exemplified by: *a spinster is an unmarried woman past the common age for marrying, an island is a tract of land surrounded by water*, etc. It is of interest to note that most of the definitions found in a dictionary are dispositional definitions in the sense that it is almost always possible to find (a) examples of objects that satisfy the definition and yet do not fit the concept, or (b) fit the concept but do not satisfy the definition.

The basic idea underlying the approach described in this paper is that a disposition may be viewed as a proposition with suppressed, or, more generally, implicit fuzzy quantifiers such as *most, almost all, almost always, usually, rarely, much of the time,* etc[1].

To illustrate, the disposition *a cup of coffee costs about fifty cents* may be viewed as the result of suppression of the fuzzy quantifier *usually* in the proposition *usually a cup of coffee costs about fifty cents.* Similarly, the disposition *young men like young women* may be interpreted as *most young men like mostly young women.* It should be stressed, however, that *restoration* (or *explicitation*)—viewed as the inverse of suppression—is an interpretation-dependent process in the sense that, in general, a disposition may be interpreted in different ways depending on the manner in which the fuzzy quantifiers are restored and defined. For example, *slimness is attractive* may be interpreted in a number of ways, among them: (a) *most slim people are attractive*; (b) *most attractive people are slim*; and (c) *the proportion of attractive people in the population of slim people is substantially larger than in the general population.*

The implicit presence of fuzzy quantifiers stands in the way of representing the meaning of dispositional concepts through the use of conventional methods based on truth-conditional, possible-world or model-theoretic semantics (Cresswell, 1973; McCawley, 1981; Miller and Johnson-Laird, 1976). In the computational approach which is described in this paper, a fuzzy quantifier is manipulated as a fuzzy number. This idea serves two purposes. First, it provides a basis for representing the meaning of dispositions; and second, it opens a way of reasoning with dispositions through the

---

[1] In the literature of linguistics, logic and philosophy of languages, fuzzy quantifiers are usually referred to as *vague* or *generalized* quantifiers (Barwise and Cooper, 1981; Peterson, 1979). In the approach described in this paper, a fuzzy quantifier is interpreted as a fuzzy number which provides an approximate characterization of absolute or relative cardinality. Fuzzy numbers may be manipulated through the use of fuzzy arithmetic (Kaufmann and Gupta, 1985), which is a generalization of interval arithmetic (Moore, 1966).

use of a collection of syllogisms. This aspect of the concept of a disposition is of relevance to default reasoning and non-monotonic logic (McCarthy, 1980; McDermott and Doyle, 1980; McDermott, 1982; Reiter, 1983).

To illustrate the manner in which fuzzy quantifiers may be manipulated as fuzzy numbers, assume that, after restoration, two dispositions d_1 and d_2 may be expressed as propositions of the form

$$p_1 = Q_1\, A's \text{ are } B's \tag{1.1}$$

$$p_2 = Q_2\, B's \text{ are } C's\ , \tag{1.2}$$

in which Q_1 and Q_2 are fuzzy quantifiers, and A, B and C are fuzzy predicates. For example,

$$p_1 = \textit{most students are undergraduates} \tag{1.3}$$

$$p_2 = \textit{most undergraduates are young}\ .$$

By treating p_1 and p_2 as the minor and major premises in a syllogism, the following *chaining* syllogism may be established if $B \subset A$ (Zadeh, 1983):

$$\begin{array}{l} Q_1\, A's \text{ are } B's \\ \underline{Q_2\, B's \text{ are } C's} \\ \geq (Q_1 \otimes Q_2)\, A's \text{ are } C's\ , \end{array} \tag{1.4}$$

in which $Q_1 \otimes Q_2$ represents the product of the fuzzy numbers Q_1 and Q_2 (Figure 1), and $\geq (Q_1 \otimes Q_2)$ should be read as "at least $Q_1 \otimes Q_2$." As shown in Figure 1, Q_1 and Q_2 are defined by their respective possibility distributions, which means that if the value of Q_1 at the point u is α, then α represents the possibility that the proportion of $A's$ in $B's$ is u.

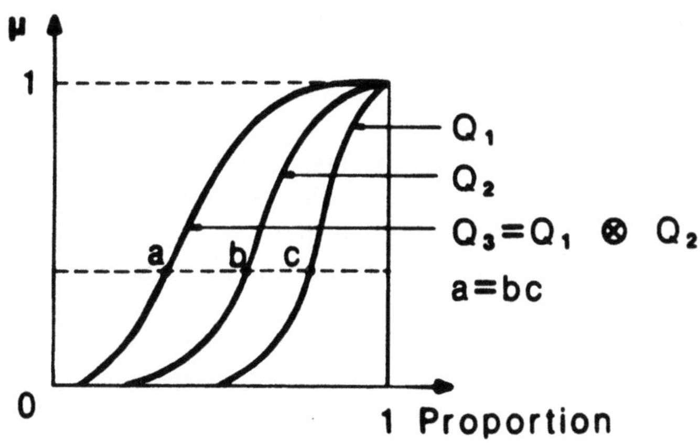

Figure 1. Multiplication of fuzzy quantifiers.

In the special case where p_1 and p_2 are expressed by (1.3), the chaining syllogism yields

$$\frac{\begin{array}{l}\textit{most students are undergraduates}\\ \textit{most undergraduates are young}\end{array}}{\textit{most}^2\ \textit{students are young}}$$

where $most^2$ represents the product of the fuzzy number *most* with itself (Figure 2).

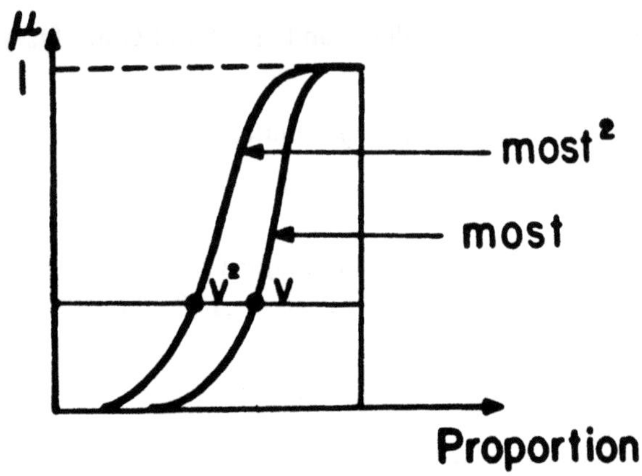

Figure 2. Representation of *most* and $most^2$.

2. Meaning Representation and Test-Score Semantics

To represent the meaning of a disposition, d, we employ a two-stage process. First, the suppressed fuzzy quantifiers in d are restored, resulting in a fuzzily quantified proposition p. Then, the meaning of p is represented—through the use of test-score semantics (Zadeh, 1978, 1982)—as a procedure which acts on a collection of relations in an explanatory database and returns a test score which represents the degree of compatibility of p with the database. In effect, this implies that p may be viewed as a collection of elastic constraints which are tested, scored and aggregated by the meaning-representation procedure. In test-score semantics, these elastic constraints play a role which is analogous to that of truth-conditions in truth-conditional semantics (Cresswell, 1973).

In more specific terms, the process of meaning representation in test-score semantics involves three distinct phases. In Phase 1, an *explanatory database frame* or *EDF*, for short, is constructed. *EDF* consists of a collection of relational frames, i.e., names of relations, names of attributes and attribute domains whose meaning is assumed to be known. For example in the case of the proposition

$$p = \textit{Carol lives in a small city near San Francisco}\ , \qquad (2.1)$$

the *EDF* may consist of the following relations (in which "+" should be read as ***and***):

$$EDF = RESIDENCE\ [Name;\ City] + POPULATION\ [City;\ Population] + SMALL\ [Population;\ \mu] + DISTANCE\ [City1;\ City2;\ Distance] + NEAR\ [Distance;\ \mu]\ . \qquad (2.2)$$

In *RESIDENCE*, the attribute ***City*** is the name of the city in which ***Name*** resides; in *POPULATION*, ***Population*** is the population of ***City***; in *SMALL*, μ is the degree to which a city whose population is the value of the attribute ***Population***, is small; in *DISTANCE*, ***Distance*** is the distance between ***City1*** and ***City2***; and in *NEAR*, μ is the degree to which two cities whose distance from one another is the value of ***Distance***, are near each other. The relations in question may be interpreted as constraints on the attribute variables which are associated with them. In the case of the relations *SMALL* and *NEAR*, the constraints are elastic; in the case of *RESIDENCE*, *POPULATION* and *DISTANCE*, the constraints are inelastic. In essence, the role played by the concept of an explanatory database frame in test-score semantics is similar to that of a collection of possible worlds in truth-conditional and possible-world semantics. In this sense, an explanatory database *ED* which is an instantiation of *EDF*, may be interpreted as a possible world, i.e., an instantiation of a collection of possible worlds. The main difference between the two concepts is that the elasticity of constraints plays a central role in test-score semantics and is ignored in truth-conditional and possible-world semantics.

The testing of elastic constraints constitutes Phase 2 of the meaning representation process. More specifically, in Phase 2 a test procedure is constructed which acts on the relations in *EDF* and yields the test scores which represent the degrees to which the elastic constraints are satisfied. For example, in the case of p, the test procedure would yield the test scores for the constraints induced by *RESIDENCE*, *POPULATION*, *SMALL*, *DISTANCE* and *NEAR*. Normally, the test scores are numbers in the unit interval [0,1]. More generally, the test scores may be elements of a partially ordered set or possibility/probability distributions.

In Phase 3, the partial test scores are aggregated into an overall test score, τ, which, in general, is a vector which serves to represent the degree of compatibility of p with *ED*, that is, an explanatory database which is an instantiation of *EDF*. As in the case of partial test scores, the components of τ are numbers in the unit interval or, more generally, possibility/probability distributions over the interval. When τ is a single number, which is usually the case, it may be interpreted as the truth value of p with respect to *ED* or, equivalently, as the degree of possibility of *ED* given p.

As a simple illustration of the application of test score semantics, consider the proposition expressed by (2.1). In this case, for the *EDF* defined by (2.2), the test procedure which computes the overall test score may be described as follows:

1. Determine the city in which Carol resides:

$$City(Carol) = {}_{City}RESIDENCE\ [Name = Carol]$$

In this expression , we use the notation ${}_{Y}R[X = a]$ to signify that X is bound to a in R and the resulting relation is projected on Y, yielding the values of Y in the tuples in which $X = a$. If there is only one such tuple, then ${}_{Y}R[X = a]$ is simply the value of Y corresponding to $X = a$.

2. Determine the population of the city in which Carol resides:

$$P = {}_{Population}POPULATION\ [City = City(Carol)]\ .$$

3. Determine the distance between San Francisco and the city in which Carol resides:

$$D = {}_{Distance}DISTANCE\ [City1 = SF;\ City2 = City(Carol)]$$

4. Test the elastic constraint induced by the fuzzy predicate ***small***:

$$\tau_1 = {}_{\mu}SMALL\ [Population = P].$$

5. Test the elastic constraint induced by the fuzzy predicate ***near***:

$$\tau_2 = {}_{\mu}NEAR\ [Distance = D].$$

6 Compute the overall test score by aggregating the partial test scores τ_1 and τ_2. For this purpose, we shall use the min operator $\wedge$ as the aggregation operator, yielding

$$\tau = \tau_1 \wedge \tau_1\ , \tag{2.3}$$

which signifies that the overall test score is taken to be the smaller of the operands of $\wedge$. The overall test score, as expressed by (2.3), represents the compatibility of $p =$ *Carol lives in a small town near San Francisco* with the data resident in the explanatory database.

In testing the constituent relations in *EDF*, it is helpful to have a collection of standardized aggregation rules for computing the test score of a combination of elastic constraints $C_1, \ldots, C_k$ from the knowledge of the test score of each constraint considered in isolation. For the most part, such rules are default rules in the sense that they are intended to be used in the absence of alternative rules supplied by the user. This applies, in particular, to the rules pertaining to such modifiers as *very* and *more or less*.

In test-score semantics, the basic aggregation rules are the following.

Rules pertaining to modification

If the test score for an elastic constraint C in a specified context is τ, then in the same context the test score for

(*a*) *not* C is $1 - \tau$ (negation)

(*b*) *very* C is τ^2 (concentration)

(*c*) *more or less* C is τ . (diffusion)

Rules pertaining to composition

If the test scores for elastic constraints C_1 and C_2 in a specified context are τ_1 and τ_2, respectively, then in the same context the test score for

(*a*) C_1 and C_2 is $\tau_1 \wedge \tau_2$ (conjunction)

(*b*) C_1 *or* C_2 is $\tau_1 \vee \tau_2$ (disjunction)

(*c*) *If* C_1 *then* C_2 is $1 \wedge (1 - \tau_1 + \tau_2)$. (implication)

where $\wedge$ = min and $\vee$ = max.

Rules pertaining to quantification

The rules in question apply to propositions of the general form *Q A's are B's*, where Q is a fuzzy quantifier, e.g., *most*, *many*, *several*, *few*, etc.

To make the concept of a fuzzy quantifier meaningful, it is necessary to define a way of counting the number of elements in a fuzzy set or, equivalently, to determine its cardinality. There are several ways in which this can be done (Zadeh, 1983a). For our purposes, it will suffice to employ the concept of a *sigma-count*, which is defined as follows.

Let F be a fuzzy subset $U = \{u_1, \ldots, u_n\}$

expressed symbolically as

$$F = \mu_1/u_1 + \ldots + \mu_n/u_n = \Sigma_i \mu_i/u_i ,$$

in which the term μ_i/u_i, $i = 1, \ldots, n$, signifies that μ_i is the grade of membership of u_i in F, and the plus sign represents the union.

The *sigma-count* of F is defined as the arithmetic sum of the μ_i, i.e.,

$$\Sigma Count(F) = \Sigma_i \mu_i, \quad i = 1, \ldots, n , \tag{2.4}$$

with the understanding that the sum may be rounded if necessary to the nearest integer. Furthermore, one may stipulate that terms with grades of membership below a specified threshold be excluded from the summation in order to keep a large number of terms with low grades of membership from becoming count-equivalent to a small number of terms with high membership.

The *relative sigma-count*, denoted by $\Sigma Count(F/G)$, may be interpreted as the proportion of elements of F which are in G. More explicitly,

$$\Sigma Count(F/G) = \frac{\Sigma Count(F \cap G)}{\Sigma Count(G)} , \tag{2.5}$$

where $F \cap G$, the intersection of F and G, is defined by

$$\mu_{F \cap G}(u) = \mu_F(u) \wedge \mu_G(u) , \quad u \in U .$$

Thus, in terms of the membership functions of F and G, the relative sigma-count of F and G is given by

$$\Sigma Count(F/G) = \frac{\Sigma_i \mu_F(u_i) \wedge \mu_G(u_i)}{\Sigma_i \mu_G(u_i)} . \tag{2.6}$$

The concept of a relative sigma-count provides a basis for representing the meaning of a proposition of the general form

$$p = Q \ A's \ are \ B's, \tag{2.7}$$

where Q is a fuzzy quantifier, and A and B are fuzzy predicates. More specifically, assume that the *EDF* associated with p consists of three relations:

$$EDF = Q[v; \mu] + A[u; \mu] + B[w; \mu] , \tag{2.8}$$

in which v is a point in the unit interval, and μ in Q is the grade of membership of v in the fuzzy number which represents Q; μ in A is the grade of membership of u in A; and μ in B is the grade of membership of w in B, with u and w representing points in the domains of A and B, respectively.

With the *EDF* defined by (2.8), the overall test score for p may be expressed as the translation rule

$$p \rightarrow \tau = {}_{\mu}Q \ [v = \Sigma Count(B/A)] , \tag{2.9}$$

in which the right-hand member is the test procedure which computes the value of τ and thus represents the meaning of p.

As an illustration of translation rule (2.9), consider the proposition

$$p = \textit{most physicians are more affluent than all but a few musicians}, \quad (2.10)$$

with which we associate the *EDF*:

$$\begin{aligned} EDF = {} & PHYSICIAN\ [Name;\ Income] + \\ & MUSICIAN\ [Name;\ Income] + \\ & MUCH.MORE\ [Income1;\ Income2;\ \mu] + \\ & MOST\ [Proportion;\ \mu] + \\ & FEW\ [Proportion;\ \mu]\ . \end{aligned} \quad (2.11)$$

The steps in the test procedure which lead to the overall test score may be described as follows.

1. Let $Name_i$ be the name of the *ith* individual in *PHYSICIAN*. For each $Name_i$, $i = 1, \ldots, m$, find his/her income:

$$IP_i = {}_{Income}PHYSICIAN\ [Name = Name_i]\ .$$

2. For each $Name_j$, $j = 1, \ldots, n$, in *MUSICIAN*, find his/her income:

$$IM_j = {}_{Income}MUSICIAN\ [Name = Name_j]\ .$$

3. For each pair $(Name_i, Name_j)$ find the degree to which $Name_i$ is much more affluent than $Name_j$ (assuming that affluence is measured by income):

$$A_{ij} = {}_{\mu}MUCH.MORE\ [Income1 = IP_i;\ Income2 = IM_j]\ .$$

4. For each $Name_i$, compute the relative sigma-count of musicians in relation to whom $Name_i$ is much more affluent:

$$P_i = \frac{1}{n} \Sigma_j A_{ij}$$

5. Test the constraint on P_i induced by the fuzzy quantifier *all but a few*:

$$\tau_i = 1 - {}_{\mu}FEW\ [Proportion = P_i]\ .$$

6. Compute the relative sigma-count of physicians who satisfy the constraint *much more affluent than all but a few musicians*:

$$\sigma = \frac{1}{m} \Sigma_i \tau_i\ .$$

7. Compute the test score for the constraint on σ induced by the fuzzy quantifier *most*:

$$\tau = {}_{\mu}MOST\,[Proportion = \sigma] \ . \tag{2.12}$$

The value of τ expressed by (2.12) represents the overall test score which defines the compatibility of p (2.10) with the explanatory database. The meaning of p is defined by the test procedure which computes τ.

3. The Scope of a Fuzzy Quantifier

In dealing with the conventional quantifiers *all* and *some* in first-order logic, the scope of a quantifier plays an essential role in defining its meaning. In the case of a fuzzy quantifier which is characterized by a relative sigma-count, what matters is the identity of the sets which enter into the relative sigma-count. Thus, if the sigma-count is of the form $\Sigma Count(B/A)$, which should be read as the proportion of $B's$ in $A's$, then B and A will be referred to as the *n-set* (with n standing for numerator) and *b-set* (with b standing for base), respectively. The ordered pair {n-set, b-set}, then, may be viewed as a generalization of the concept of the scope of a quantifier. Note, however, that, in this sense, the scope of a fuzzy quantifier is a semantic rather than syntactic concept.

As a simple illustration, consider the proposition

$$p = most\ students\ are\ undergraduates \ .$$

In this case, the n-set of *most* is *undergraduates*, the b-set is *students*, and the scope of *most* is the pair {*undergraduates*, *students*}.

As an additional illustration of the interaction between scope and meaning, consider the disposition

$$d = young\ men\ like\ young\ women \ . \tag{3.1}$$

Among the possible interpretations of this disposition, we shall focus our attention on the following (the symbol *rd* denotes a restoration of a disposition):

$$rd_1 = most\ young\ men\ like\ most\ young\ women$$

$$rd_2 = most\ young\ men\ like\ mostly\ young\ women \ .$$

To place in evidence the difference between rd_1 and rd_2, it is expedient to express them in the form

$$rd_1 = most\ young\ men\ P_1$$

$$rd_2 = most\ young\ men\ P_2 \ ,$$

where P_1 and P_2 are the fuzzy predicates

$$P_1 = likes\ most\ young\ women$$

and

$$P_2 = likes\ mostly\ young\ women\ ,$$

with the understanding that, for grammatical correctness, *likes* in P_1 and P_2 should be replaced by *like* when P_1 and P_2 act as constituents of rd_1 and rd_2. In more explicit terms, P_1 and P_2 may be expressed as

$$P_1 = P_1[Name;\mu] \tag{3.2}$$

$$P_2 = P_2[Name;\mu]\ ,$$

in which *Name* is the name of a male person and μ is the degree to which the person in question satisfies the predicate. (Equivalently, μ is the grade of membership of the person in the fuzzy set which represents the extension of the predicate.)

To represent the meaning of P_1 and P_2 through the use of test-score semantics, we assume that the *EDF* consists of the following relations (Zadeh, 1983b):

$$\begin{aligned} EDF = {} & POPULATION[Name;\ Age;\ Sex] + \\ & LIKE[Name1;\ Name2;\ \mu] + \\ & YOUNG[Age;\ \mu] + \\ & MOST[Proportion;\ \mu]\ . \end{aligned}$$

In *LIKE*, μ is the degree to which *Name*1 likes *Name*2 ; and in *YOUNG*, μ is the degree to which a person whose age is *Age* is young.

First, we shall represent the meaning of P_1 by the following test procedure.

1. Divide *POPULATION* into the population of males, *M.POPULATION*, and the population of females, *F.POPULATION*:

$$M.POPULATION = {}_{Name,Age}POPULATION[Sex{=}Male]$$

$$F.POPULATON = {}_{Name,Age}POPULATION[Sex{=}Female]\ ,$$

where ${}_{Name,Age}POPULATION$ denotes the projection of *POPULATION* on the attributes *Name* and *Age*.

2. For each $Name_j$, $j = 1, \ldots, L$, in *F.POPULATION*, find the age of $Name_j$:

$$A_j = {}_{Age}F.POPULATION[Name{=}Name_j]\ .$$

3. For each $Name_j$, find the degree to which $Name_j$ is young:

$$\alpha_j = {}_{\mu}YOUNG[Age{=}A_j]\ ,$$

where α_j may be interpreted as the grade of membership of $Name_j$ in the fuzzy set, *YW*, of young women.

4. For each $Name_i$, $i=1, \ldots, K$, in *M.POPULATION*, find the age of $Name_i$:

$$B_i = {}_{Age}M.POPULATION[Name=Name_i] \ .$$

5. For each $Name_j$, find the degree to which $Name_i$ likes $Name_j$:

$$\beta_{ij} = {}_{\mu}LIKE[Name1 = Name_i; Name2 = Name_j] \ ,$$

with the understanding that β_{ij} may be interpreted as the grade of membership of $Name_j$ in the fuzzy set, WL_i, of women whom $Name_i$ likes.

6. For each $Name_j$ find the degree to which $Name_i$ likes $Name_j$ and $Name_j$ is young:

$$\gamma_{ij} = \alpha_j \wedge \beta_{ij} \ .$$

Note: As in previous examples, we employ the aggregation operator min ($\wedge$) to represent the meaning of conjunction. In effect, γ_{ij} is the grade of membership of $Name_j$ in the intersection of the fuzzy sets WL_i and YW.

7. Compute the relative sigma-count of women whom $Name_i$ likes among young women:

$$\begin{aligned} \rho_i &= \Sigma Count(WL_i/YW) \qquad (3.4) \\ &= \frac{\Sigma Count(WL_i \cap YW)}{\Sigma Count(YW)} \\ &= \frac{\Sigma_j \gamma_{ij}}{\Sigma \alpha_j} \\ &= \frac{\Sigma_j \alpha_j \wedge \beta_{ij}}{\Sigma_j \alpha_j} \ . \end{aligned}$$

8. Compute the test score for the constraint induced by *MOST*:

$$\tau_i = {}_{\mu}MOST[Proportion = \rho_i] \ . \qquad (3.5)$$

This test-score way be interpreted as the degree to which $Name_i$ satisfies P_1, i.e.,

$$\tau_i = {}_{\mu}P_1[Name = Name_i] \ .$$

The test procedure described above represents the meaning of P_1. In effect, it tests the constraint expressed by the proposition

$$\Sigma Count(WL_i/YW) \text{ is } MOST$$

and implies that the n-set and the b-set for the quantifier *most* in P_1 are given by:

$$n\text{-}set \;=\; WL_i \;=\; {}_{Name2}LIKE[Name1 = Name_i] \cap F.POPULATION$$

and

$$b\text{-}set = YW = YOUNG \cap F.POPULATION \;.$$

By contrast, in the case of P_2, the identities of the n-set and the b-set are interchanged, i.e.,

$$n\text{-}set \;=\; YW$$

and

$$b\text{-}set \;=\; WL_i \;,$$

which implies that the constraint which defines P_2 is expressed by

$$\Sigma Count(YW/WL_i) \text{ is } MOST \;.$$

Thus, whereas the scope of the quantifier *most* in P_1 is $\{WL_i, YW\}$, the scope of *mostly* in P_2 is $\{YW, WL_i\}$.

Having represented the meaning of P_1 and P_2, it becomes a simple matter to represent the meaning of rd_1, and rd_2. Taking rd_1, for example, we have to add the following steps to the test procedure which defines P_1.

9. For each $Name_i$, find the degree to which $Name_i$ is young:

$$\delta_i \;=\; {}_{\mu}YOUNG[Age = B_i] \;,$$

where δ_i may be interpreted as the grade of membership of $Name_i$ in the fuzzy set, YM, of young men.

10. Compute the relative sigma-count of men who have property P_1 among young men:

$$\begin{aligned}\rho &= \Sigma Count(P_1/YM)\\ &= \frac{\Sigma Count(P_1 \cap YM)}{\Sigma Count(YM)}\\ &= \frac{\Sigma_i \tau_i \wedge \delta_i}{\Sigma_i \delta_i} \;.\end{aligned}$$

11. Test the constraint induced by $MOST$:

$$\tau \;=\; {}_{\mu}MOST[Proportion = \rho] \;. \qquad (3.6)$$

The test score expressed by (3.6) represents the overall test score for the disposition

$$d = \text{young men like young women}$$

if d is interpreted as rd_1. If d is interpreted as rd_2, which is a more likely interpretation, then the procedure is unchanged except that τ_i in (3.5) should be replaced by

$$\tau_i = {}_{\mu}MOST[Proportion = \sigma_i] ,$$

where

$$\begin{aligned} \sigma_i &= \Sigma Count(YW/WL_i) \\ &= \frac{\Sigma_j \alpha_j \wedge \beta_{ij}}{\Sigma_j \beta_{ij}} . \end{aligned}$$

4. Representation of Dispositional Commands

A *dispositional command*, dc, is an imperative exemplified by *avoid overeating, be honest, keep under refrigeration, do not exert yourself if you feel tired*, etc. Any command, whether dispositional or not, may be defined through its propositional content (Searle, 1979) or, equivalently, its compliance criterion (Zadeh, 1982). Naturally, if a command is dispositional, so is its defining compliance criterion. For example, the dispositional command

$$dc = \textit{avoid overeating} , \tag{4.1}$$

is associated with the dispositional compliance criterion

$$dcc = \textit{A avoids overeating} , \tag{4.2}$$

where A is an indexical representing the addressee of the command.

As an illustration, consider the conditional dispositional command

$$dc = \textit{do not go late bed } \text{if } \textit{you feel tired.} \tag{4.3}$$

which is assumed to be defined by the dispositional compliance criterion

$$dcc = \textit{A does not go late bed when he/she feels tired.} \tag{4.4}$$

As a first step in representing the meaning of the dispositional command (4.3), we have to make explicit the implicit fuzzy quantifiers in (4.4). We assume that the intended meaning of (4.4) is defined by prefixing (4.4) with the fuzzy quantifier *most of the time*:

$$cc = \textit{most of the time A does not go late bed when he/she feels tired.} \tag{4.5}$$

To represent the meaning of (4.5), we associate with the explicit compliance criterion (4.5) the *EDF*

$$EDF = RECORD\ [Day;\ Time;\ \mu Tired] + LATE\ [Time;\ \mu] + MOST\ [Proportion;\ \mu]\ . \tag{4.6}$$

In (4.6), *RECORD* may be interpreted as a diary—kept during the period of interest—in which *DAY* ranges over successive days; *Time* is the time of going to bed; and *μTired* is the degree to which *A* feels tired. In *LATE*, μ is the degree to which the value of *Time* fits the description *late*; and in *MOST*, μ is the degree to which the value of *Proportion* fits the fuzzy quantifier *most*.

The steps in the test procedure which represent the meaning of (4.5) may be described as follows.

1. Let Day_i denote the *ith* day in the Diary, $i = 1, \ldots, m$. For each Day_i determine the time of going to bed, and the degree to which *A* felt tired:

$$t_i = {}_{Time}RECORD\ [Day = Day_i] \tag{4.7}$$

$$\mu_i = {}_{\mu Tired}RECORD\ [Day = Day_i]\ . \tag{4.8}$$

2. Find the test score for the elastic constraint induced by the predicate *late*:

$$\tau_i = {}_{\mu}LATE\ [Time = t_i]\ . \tag{4.9}$$

3. Find the relative sigma-count of days when *A* did not go late to bed when *A* felt tired:

$$\sigma = \frac{\Sigma_i\ \mu_i \wedge (1 - \tau_i)}{\Sigma \mu_i}\ . \tag{4.10}$$

4. Compute the test score for the elastic constraint induced by the fuzzy quantifier *most*:

$$\tau = {}_{\mu}MOST\ [Proportion = \sigma]\ . \tag{4.11}$$

The value of compatibility expressed by (4.11) represents the overall test score yielded by the test procedure which represents the meaning of the conditional dispositional command (4.3).

It is of interest to note that maxims and, more generally, rules of conduct of a proverbial nature may be viewed as special cases of the concept of a dispositional command. In many cases, such rules of conduct result from the application of a dispositional version of *modus ponens* (see 7.6) to a disposition. As an illustration, from the disposition *regular exercise is good for your health* we can infer through the application of *modus ponens* the maxim *to be in good health, exercise regularly.*

In summary, representation of the meaning of a dispositional command involves three basic steps. First, the meaning of the command is defined by a dispositional compliance criterion. Second, the implicit fuzzy quantifiers in the dispositional compliance criterion are made explicit. And third, the meaning of the resulting compliance criterion is represented by a test procedure which computes the overall score of compliance in relation to an explanatory database which contains a record of executions of the given command.

5. The Concept of a Subdisposition

Among the examples of dispositions cited in the preceding sections there are some which may be described more precisely as *subdispositions*.

A subdisposition, as its name suggests, is a weaker variant of a disposition in the sense that it connotes an increase in relative cardinality or probability without implying that their values are close to unity. In this sense, *slimness is attractive* is a subdisposition if its intended reading is (c) (see Section 1) *the proportion of attractive people in the population of slim people is substantially larger than in the general population*, rather than (a) *most slim people are attractive* or (b) *most attractive people are slim*.

We shall employ the sentence *slimness is attractive* to illustrate how the meaning of a subdisposition may be represented. Specifically, let U be a universe of discourse (a collection of persons), and let S and A be, respectively, the fuzzy subsets of U representing slim and attractive people, respectively.

In terms of these sets, the proportion of attractive people in the general population is expressed by the relative sigma-count $\Sigma Count(A/U)$, while the proportion of attractive people among the slim is given by the relative sigma-count $\Sigma Count(A/S)$.

By viewing the relative sigma-count $\Sigma Count(A/U)$ as a reference point in the unit interval [0,1], the relative sigma-count $\Sigma Count(A/S)$ may be represented as a point in the interval $[\Sigma Count(A/U), 1]$ which is a convex combination of its endpoints. More specifically,

$$\Sigma Count(A/S) = \rho + (1-\rho)\, \Sigma Count(A/U)\,, \tag{5.1}$$

where ρ is a number in the unit interval which, as a weighting coefficient, serves as a measure of the closeness of the point $\Sigma Count(A/S)$ to unity. In this sense, ρ plays a role which is similar to that of the measure of increased belief in confirmation theory and, in particular, in its applications to the management of evidence in expert systems (Buchanan and Shortliffe, 1984).

In terms of ρ, the subdisposition *slimness is attractive* may be represented as a proposition which provides a fuzzy characterization of its value, i.e.,

$$slimness\ is\ attractive \rightarrow \sigma\ is\ MEDIUM\,, \tag{5.2}$$

where *MEDIUM* is a fuzzy number which for simplicity may be assumed to have a triangular membership function as shown in Figure 3. By contrast, if *slimness is attractive* were interpreted as a disposition (in the sense of (a)) then in terms of of ρ its meaning would be represented by

$$\textit{slimness is attractive} \rightarrow \sigma \textit{ is HIGH} \ , \tag{5.3}$$

where *HIGH* is a fuzzy number which, as an approximation, may be assumed to be triangular (Figure 3).

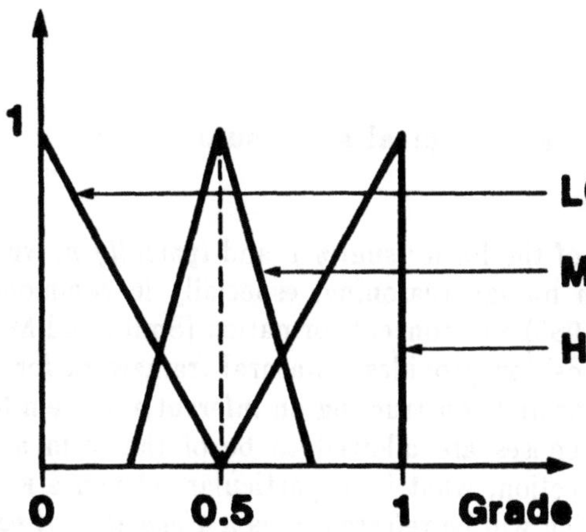

Figure 3. Representation of *HIGH*, *MEDIUM* and *LOW* as triangular possibility distributions.

As an illustration of the use of (5.1) in commonsense reasoning, consider the following premises and a related question:

$$\begin{array}{l} \textit{slimness is attractive} \\ \underline{\textit{Carol is slim}} \\ \textit{Carol is ?} \end{array} \tag{5.4}$$

The answer to the question may be deduced from (5.1) and (5.2) as follows. Let α be a fuzzy number which characterizes the relative sigma-count of attractive people in the general population, i.e.,

$$\Sigma Count(A/U) \text{ is } \alpha \ . \tag{5.5}$$

Furthermore, let β be the fuzzy number *MEDIUM*. Then from (5.1) it follows that the relative sigma-count $\Sigma Count(A/S)$ is characterized by a fuzzy number γ which is related to α and β by the expression

$$\gamma = \beta \oplus (1 \ominus \beta) \otimes \alpha \ , \tag{5.6}$$

where $\oplus$, $\ominus$, and $\otimes$ represent, respectively, addition, subtraction and multiplication in fuzzy arithmetic (Kaufmann and Gupta, 1985). If this number is interpreted as a fuzzy probability (Zadeh, 1984), then the answer to the question may be stated as follows:

(a) If all that is known is that Carol is slim, then the fuzzy probability that she is attractive is α. (5.7)

(b) If it is known in addition that there is a relationship between slimness and attractiveness which is expressed by the subdisposition *slimness is attractive*, then the fuzzy probability that Carol is attractive increases to γ, as expressed by (5.6).

6. Dispositional Valuations—Typical and Usual Values

Qualified propositions of the form *usually p* and *typically p*, where *p* is a proposition, play a basic role in human reasoning, especially in commonsense reasoning, default reasoning (Reiter, 1983) and concept formation (Smith and Medin, 1981).

The concept of a disposition provides a natural framework for representing the meaning of such propositions and constructing an inferential system for commonsense reasoning in which the premises are allowed to be of the form *usually p* (Zadeh, 1984c, 1985). In this connection, what is of particular relevance is a variant of the concept of a disposition which will be referred to as a *dispositional valuation*.

As its name implies, a dispositional valuation is a dispositional statement regarding the values of a variable. For example:

(a) *a cup of coffee costs about fifty cents*

(b) *it takes about an hour to drive from Berkeley to Stanford*

(c) *a typical TV set weighs about fifty pounds.*

In (a), the variable is the price of a cup of coffee and its dispositional value is *about fifty cents*; in (b), the variable is the duration of driving from Berkeley to Stanford and its dispositional value is *about an hour*; and in (c), the variable is the weight of a typical TV set and its dispositional value is *about fifty pounds*.

In more general terms, in the case of dispositional valuations of the form

$$usually\ (X\ is\ F)$$

$$typically\ (X\ is\ G)$$

F and G play, respectively, the roles of *usual* and *typical* values of the variable X. Thus,

$$usual\ value\ of\ X = U(X) = F \tag{6.1}$$

$$typical\ value\ of\ X = T(X) = G = \{t\} \tag{6.2}$$

with the understanding that, usually, F is a fuzzy subset of the universe of discourse, U, in which X takes its values, while G is a singleton $\{t\}$. What this implies is that

(a) with each fuzzy subset, F, of U we can associate the degree to which F qualifies to be regarded as a usual value of X; and (b) with each element, t, of U we can associate the degree to which t qualifies to be regarded as a typical value of X. In this sense, $T(X)$ and $U(X)$ are, respectively, first and second order predicates defined on U.

To define what is meant by a usual value it is necessary to define the meaning of the qualified proposition *usually* (X *is* F). Viewed as a fuzzy quantifier, *usually* may be interpreted as a *conditioned* quantifier whose definition involves a *conditioning variable* whose role is to characterize a set of *normal* or, equivalently, *non-exceptional* conditions which circumscribe the validity of a dispositional valuation. More concretely,

$$\textit{usually}\ (X\ \textit{is}\ F) \rightarrow \text{if } Z\ \textit{is}\ R\ \textit{then most}\ X'\textit{s are}\ F\ , \tag{6.3}$$

where Z is the conditioning variable and R is a prescribed set which is the complement of the set of exceptions. From this definition, it follows that if X takes the successive values $x_1, \ldots, x_n$ in U, then the meaning of *usually* (X *is* F) may be represented by the test procedure

$$\tau = {}_{\mu}MOST\ [Proportion = \Sigma CountF/R)]\ , \tag{6.4}$$

which computes the overall test score τ. If there is no conditioning variable or, equivalently, if $R = U$, then *usually* is an *unconditioned* quantifier defined by

$$\textit{usually}\ (X\ \textit{is}\ F) \rightarrow \textit{most}\ X'\textit{s are}\ F\ . \tag{6.5}$$

As an illustration, consider the disposition *snow is white*. Assume that its intended interpretation is *usually* (*snow is white*), in which *usually* is a fuzzy quantifier conditioned on the freshness of snow, i.e.,

usually (*snow is white*) → if *snow is fresh then in most instances it is white*

If in the explanatory database the successive samples of snow, say $s_1, \ldots, s_n$, are fresh to the degree $f_1, \ldots, f_n$ and white to the degree $w_1, \ldots, w_n$, respectively, then from (6.4) and (2.6), it follows that

$$\tau = {}_{\mu}MOST\ [\Sigma_i\ f_i \wedge w_i/\Sigma_i\ f_i]\ . \tag{6.6}$$

This expression defines the meaning of *snow is white* as a test procedure which computes its compatibility, τ, with a given explanatory database.

Turning to the concept of a typical value, we note that it is a common practice to treat *typical* and *usual* as almost synonomous notions. In what follows, we define a typical element (or value) in a way that sets it apart from the concept of a usual value. However, we also show that the concepts are related and exhibit the relation between them.

Let A be a fuzzy set in U (e.g., U = *cars* and A = *station wagons*). The definition of a typical element of A may be expressed in verbal terms as follows:

t is a *typical* element of A if and only if (a) t has a high grade of membership in A ,
and (b) most elements of A are similar to t . (6.7)

It should be remarked that this definition should be viewed as a *dispositional definition*, in the sense that it may fail, in some cases, to reflect our intuitive perception of the meaning of typicality. Furthermore, in (6.7), A plays a role similar to that of R in the definition of a usual value (6.3).

To put the verbal definition expressed by (6.7) into a more precise form, let S be a similarity relation defined on U which associates with each element u in U the degree to which u is similar to t.[2] Furthermore, let $S(t)$ be the *similarity class* of t, i.e., the fuzzy set of elements of U which are similar to t. What this means is that the grade of membership of u in $S(t)$ is equal to $\mu_S(t,u)$, the degree to which u is similar to t (Zadeh, 1971). (Figure 4.)

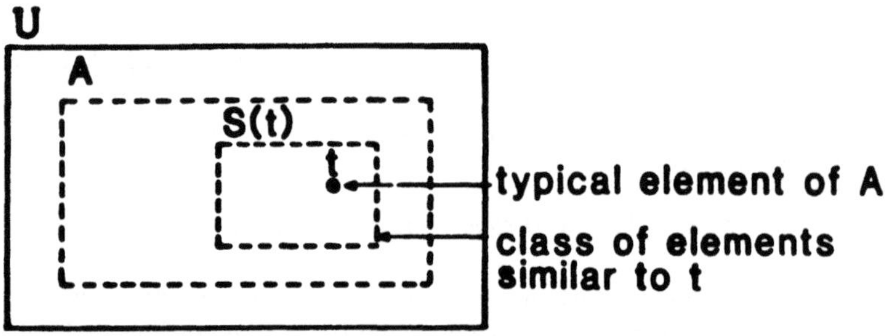

Figure 4. Definition of Typicality

Let *HIGH* denote the fuzzy subset of the unit interval which is the extension of the fuzzy predicate *high*. Then, the verbal definition (6.7) may be expressed more precisely in the form:

$$t \text{ is a } \textit{typical} \text{ element of } A \text{ if (a) } \mu_A(t) \text{ is } HIGH, \qquad (6.8)$$

$$\text{and (b) } \Sigma Count(S(t)/A) \text{ is } MOST\,.$$

The fuzzy predicate *high* may be characterized by its membership function μ_{HIGH} or, equivalently, as the fuzzy relation *HIGH* [*Grade*; μ], in which *Grade* is a number in the interval [0,1] and μ is the degree to which the value of *Grade* fits the intended meaning of *high*.

An important implication of this definition is that typicality is a matter of degree. Thus, it follows at once from (6.8) that the degree, τ, to which t is typical or, equivalently, the grade of membership of t in the fuzzy set of typical elements of A, is given by

$$\tau = {}_{\mu}HIGH[Grade = \mu_A(t)] \wedge {}_{\mu}MOST[Proportion = \Sigma Count(S(t)/A)]\,. \qquad (6.9)$$

In terms of the membership functions of *HIGH*, *MOST*,S and A, (6.9) may be written as

$$\tau = \mu_{HIGH}(\mu_A(t)) \wedge \mu_{MOST}\left[\frac{\Sigma_u\, \mu_S(t,u) \wedge \mu_A(u)}{\Sigma_u\, \mu_A(u)}\right], \qquad (6.10)$$

[2] For consistency with the definition of A, S must be such that if u and u' have a high degree of similarity, then their grades of membership in A should be close in magnitude.

where μ_{HIGH}, μ_{MOST}, μ_S and μ_A are the membership functions of *HIGH, MOST, S* and *A*, respectively, and the summation Σ_u extends over the elements of *U*.

It is of interest to observe that if $\mu_A(t) = 1$ and

$$\mu_S(t,u) \;=\; \mu_A(u)\,, \tag{6.11}$$

that is, the grade of membership of *u* in *A* is equal to the degree of similarity of *u* to *t*, then the degree of typicality of *t* is unity. This is reminiscent of definitions of prototypicality (Rosch, 1978) in which the grade of membership of an object in a category is assumed to be inversely related to its "distance" from the prototype.

Another point of interest is that in the definition of prototypicality given in Zadeh (1982), a prototype is interpreted as a so-called σ-summary. In relation to the definition of typicality expressed by (6.7), we may say that a prototype is a σ-summary of typical elements of *A*. In this sense, a prototype is *not*, in general, an element of *U* whereas a typical element of *A* is, by definition, an element of *U*. As a simple illustration of this difference, assume that *U* is a collection of movies, and *A* is the fuzzy set of Western movies. A prototype of *A* is a summary of the summaries (i.e., plots) of Western movies, and thus is not a movie. A typical Western movie, on the other hand, is a movie and thus is an element of *U*.

As was stated earlier, in our formulation of the concept of typicality the notion of a typical value is distinct from that of a usual value. However, there is a close connection between the two which is brought out by the following result.

Assume for simplicity that $\mu_A(t) = 1$. Then, the degree to which *t* is a typical element of *A* is given by

$$\tau \;=\; {}_{\mu}MOST\;[Proportion \;=\; \Sigma Count(S(t)/A)]\;\;. \tag{6.12}$$

Comparing this expression with (6.4), we note that if we equate *Z* to *X*, *R* to *A*, and *F* to *S*(*t*), the expression for the degree to which *F* is a usual value of *X* becomes identical with (6.12). Consequently, we can assert that

$$\begin{aligned}&\textit{the degree to which } t \textit{ is a typical element of } A = \\ &\textit{the degree to which } S(t) \textit{ is a usual value of } X.\end{aligned} \tag{6.13}$$

In other words, the similarity class of a typical element of *A* may be interpreted as a usual value of the variable which ranges over *U* and is conditioned on *A*.

7. Fuzzy Syllogisms

A concept which plays an essential role in reasoning with dispositions is that of a *fuzzy syllogism* (Zadeh, 1983c). As a general inference schema, a fuzzy syllogism may be expressed in the form

$$\begin{array}{l} Q_1\, A's \textit{ are } B's \\ Q_2\, C's \textit{ are } D's \\ \hline Q_3\, E's \textit{ are } F's \end{array}\;, \tag{7.1}$$

where Q_1 and Q_2 are given fuzzy quantifiers, Q_3 is fuzzy quantifier which is to be determined, and A, B, C, D, E, and F are interrelated fuzzy predicates.

In what follows, we shall present a brief discussion of two basic types of fuzzy syllogisms. A more detailed description of these and other fuzzy syllogisms may be found in Zadeh (1983c, 1984b).

The *intersection/product syllogism* may be viewed as an instance of (7.1) in which

$$C = A \textit{ and } B$$
$$E = A$$
$$F = B \textit{ and } D\ ,$$

and $Q_3 = Q_1 \otimes Q_2$, i.e., Q_3 is the product of Q_1 and Q_2 in fuzzy arithmetic. Thus, we have as the statement of the syllogism:

$$\begin{array}{l} Q_1\, A\text{'s are } B\text{'s} \\ \underline{Q_2\, (A \text{ and } B)\text{'s are } C\text{'s}} \\ (Q_1 \otimes Q_2)\, A\text{'s are } (B \text{ and } C)\text{'s}\ . \end{array} \tag{7.2}$$

In particular, if B is contained in A, i.e., $\mu_B \le \mu_A$, where μ_A and μ_B are the membership functions of A and B, respectively, then A and $B = B$, and (7.2) becomes

$$\begin{array}{l} Q_1\, A\text{'s are } B\text{'s} \\ \underline{Q_2\, B\text{'s are } C\text{'s}} \\ (Q_1 \otimes Q_2)\, A\text{'s are } (B \text{ and } C)\text{'s}\ . \end{array} \tag{7.3}$$

Since B *and* C implies C, it follows at once from (7.3) that

$$\begin{array}{l} Q_1\, A\text{'s are } B\text{'s} \\ \underline{Q_2\, B\text{'s are } C\text{'s}} \\ \ge (Q_1 \otimes Q_2)\, A\text{'s are } C\text{'s}\ , \end{array} \tag{7.4}$$

which is the *chaining syllogism* expressed by (1.4). Furthermore, if the quantifiers Q_1 and Q_2 are monotone increasing, i.e., $\ge Q_1 = Q_1$ and $\ge Q_2 = Q_2$, then (7.4) becomes the *product syllogism*

$$\begin{array}{l} Q_1\, A\text{'s are } B\text{'s} \\ \underline{Q_2\, B\text{'s are } C\text{'s}} \\ (Q_1 \otimes Q_2)\, A\text{'s are } C\text{'s}. \end{array} \tag{7.5}$$

An important special case of the product syllogism which obtains when $A = U$

and $Q_1 = Q_2 = $ *usually*, is a variant of the ***dispositional modus ponens***. On observing that the proposition *Q A's are B's* may be expressed equivalently as *Q* (*if X is A then Y is B*), the variant in question may be stated as the syllogism

$$\frac{\begin{array}{l} usually\ (X\ is\ B) \\ usually\ (if\ X\ is\ B\ then\ Y\ is\ C) \end{array}}{usually^2 (Y\ is\ C)} \tag{7.6}$$

when $usually^2$ is the product of the fuzzy number *usually* with itself.

In the case of the *consequent conjunction syllogism*, we have

$$\begin{aligned} C &= A \\ E &= A \\ F &= B \text{ and } D \ . \end{aligned}$$

In this case, the statement of the syllogism is:

$$\frac{\begin{array}{l} Q_1\ A's\ are\ B's \\ Q_2\ A's\ are\ C's \end{array}}{Q_3\ A's\ are\ (B\ and\ C)'s\ ,} \tag{7.7}$$

where Q_3 is a fuzzy number (or interval) defined by the inequalities

$$0 \circledvee (Q_1 \oplus Q_2 \ominus 1) \leq Q_3 \leq Q_1 \circledwedge Q_2 \ , \tag{7.8}$$

where $\oplus$, $\ominus$, $\circledvee$ and $\circledwedge$ are the operations of addition, subtraction, min, and max in fuzzy arithmetic.

As a simple illustration, consider the dispositions

$$\begin{aligned} d_1 &= students\ are\ young \\ d_2 &= students\ are\ single\ . \end{aligned}$$

Upon restoration, these dispositions become the quantified propositions

$$\begin{aligned} p_1 &= most\ students\ are\ young \\ p_2 &= most\ students\ are\ single\ . \end{aligned}$$

Then, applying the consequent conjunction syllogism to p_1 and p_2, we can infer that

$$Q\ students\ are\ single \text{ and } young$$

where

$$2\ most \ominus 1 \leq Q \leq most\ . \tag{7.9}$$

Thus, from the dispositions in question we can infer the disposition

$$d = students\ are\ single \text{ and } young$$

on the understanding that the implicit fuzzy quantifier in d is expressed by (7.9).

8. Negation of Dispositions

In dealing with dispositions, it is natural to raise the question: What happens when a disposition is acted upon with an operator, T, where T might be the operation of negation, active-to-passive transformation, etc.? More generally, the same question may be asked when T is an operator which is defined on pairs or n-tuples of dispositions.

As an illustration, we shall focus our attention on the operation of negation. More specifically, the question which we shall conside. briefly is the following: Given a disposition, d, what can be said about the negation of d, *not* d? For example, what can be said about *not (birds can fly)* or *not (young men like young women)*.

For simplicity, assume that, after restoration, d may be expressed in the form

$$rd = Q\ A's\ are\ B's\ . \tag{8.1}$$

Then,

$$not\ d = not\ (Q\ A's\ are\ B's)\ . \tag{8.2}$$

Now, using the semantic equivalence established in Zadeh (1978), we may write

$$not\ (Q\ A's\ are\ B's) \equiv (not\ Q)\ A's\ are\ B's\ , \tag{8.3}$$

where *not* Q is the negation of the fuzzy quantifier Q in the sense that the membership function of *not* Q is given by

$$\mu_{not\ Q}(u) = 1 - \mu_Q(u)\,,\quad 0 \le u \le 1\ . \tag{8.4}$$

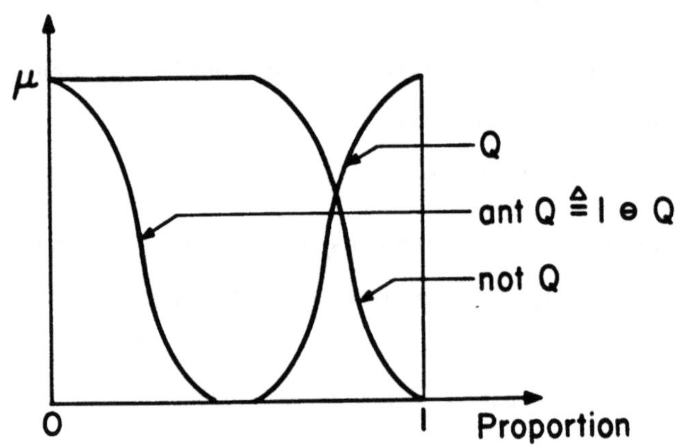

Figure 5. Representation of Q, *not* Q, and *ant* Q

Furthermore, the following inference rule can readily be established (Zadeh, 1983a):

$$\frac{Q\ A's\ are\ B's}{\geq (ant\ Q)\ A's\ are\ not\ B's}, \tag{8.5}$$

where *ant Q* denotes the *antonym* of Q, defined by (see Figure 5)

$$\mu_{ant\ Q}(u) = \mu_Q(1-u)\,,\quad 0 \leq u \leq 1\ . \tag{8.6}$$

On combining (8.3) and (8.5), we are led to the following result:

$$not(QA's\ are\ B's) = \geq (ant\ (not\ Q))\ A's\ are\ not\ B's \tag{8.7}$$

which reduces to

$$not(Q\ A's\ are\ B's) = (ant\ (not\ Q))\ A's\ are\ not\ B's \tag{8.8}$$

if Q is monotone increasing (e.g., $Q = most$).

As an illustration, if $d =$ *birds can fly* and $Q = most$, then (6.8) yields

$$not\ (birds\ can\ fly) = (ant\ (not\ most))\ birds\ cannot\ fly\ . \tag{8.9}$$

It should be observed that if Q is an approximation to *all*, then *ant(not Q)* is an approximation to *some*. For the right-hand member of (8.9) to be a disposition, *most* must be an approximation to *at least a half*. In this case *ant (not most)* will be an approximation to *most*, and consequently the right-hand member of (8.9) may be expressed—upon the suppression of *most*—as the disposition *birds cannot fly*.

REFERENCES AND RELATED PUBLICATIONS

Barwise, J., and Cooper, R., Generalized quantifiers and natural language, *Linguistics and Philosophy 4* (1981) 159-219.

Bellman, R.E., and Zadeh, L.A., Local and fuzzy logics, in: *Modern Uses of Multiple-Valued Logic*, Epstein, G., (ed.). Dordrecht: Reidel, 103-165, 1977.

Brachman, R.J., and Smith, B.C., *Special Issue on Knowledge Representation*, SIGART 70, 1980.

Buchanan, B.G., and Shortliffe, E.H., *Rule-Based Expert Systems*. Reading: Addison-Wesley, 1984.

Carnap, R., The methodological character of theoretical concepts, in: *Minnesota Studies in the Philosophy of Science*, Fiegl, H. and Scriven, M. (eds.), 38-76, 1956.

Cresswell, M.J., *Logic and Languages*. London: Methuen, 1973.

Cushing, S., *Quantifier Meanings—A Study in the Dimensions of Semantic Compentence*. Amsterdam: North-Holland, 1982.

Dubois, D., and Prade, H., *Fuzzy Sets and Systems: Theory and Applications*. New York: Academic Press, 1980.

Goguen, J.A., The logic of inexact concepts, *Synthese 19* (1969) 325-373.

Hacking, I., *Logic of Statistical Inference*. Cambridge: Cambridge University Press, 1965.

Hempel, C.G., Maximal specificity and lawlikeness, *Philosophy of Science 35* (1968) 116-133.

Hintikka, J., Statistics, induction and lawlikeness, *Synthese 20* (1969) 72-83.

Jeffrey, R.C., *The Logic of Decision*. New York: McGraw-Hill, 1965.

Kaufmann, A., and Gupta, M., *Introduction to Fuzzy Arithmetic*. New York: Van Nostrand, 1985.

Keenan, E.L., Quantifier structures in English, *Foundations of Language 7* (1971) 255-336.

Kyburg, H.E., *Probability and Inductive Logic*. New York: McMillan, 1970.

Mamdani, E.H., and Gaines, B.R., *Fuzzy Reasoning and its Applications*. London: Academic Press, 1981.

McCarthy, J., Circumscription: A non-monotonic inference rule, *Artificial Intelligence 13* (1980) 27-40.

McCawley, J.D., *Everything that Linguists have Always Wanted to Know about Logic*. Chicago: University of Chicago Press, 1981.

McDermott, D.V., and Doyle, J., Non-monotonic logic, I. *Artificial Intelligence 13* (1980) 41-72.

McDermott, D.V., Non-monotonic logic, II: Non-monotonic modal theories, *J. Assoc. Comp. Mach. 29* (1982) 33-57.

Mellor, D.H., *Matter of Chance*. Cambridge: Cambridge University Press, 1971.

Mellor, D.H., *Science, Belief and Behavior*. Cambridge: Cambridge University Press, 1980.

Miller, G.A., and Johnson-Laird, P.N., *Language and Perception*. Cambridge: Harvard University Press, 1976.

Moore, R.E., *Interval Analysis*. Englewood Cliffs: Prentice-Hall, 1966.

Osherson, D.N., and Smith, E.E., On the adequacy of prototype theory as a theory of concepts, *Cognition 9* (1981) 59-72.

Peterson, P., On the logic of *few, many* and *most*, *Notre Dame J. Formal Logic 20* (1979) 155-179.

Reiter, R., and Criscuolo, G., Some representational issues in default reasoning, *Computers and Mathematics 9* (1983) 15-28.

Rescher, N., *Plausible Reasoning*. Amsterdam: Van Gorcum, 1976.

Rosch, E., Principles of categorization, in: *Cognition and Categorization*, Rosch, E. and Lloyd, B.B., (eds.). Hillsdale, NJ: Erlbaum, 1978.

Searle, J., *Expression and Meaning*. Cambridge: Cambridge University Press, 1979.

Smith, E., and Medin, D.L., *Categories and Concepts*. Cambridge: Harvard University Press, 1981.

Suppes, P., *A Probabilistic Theory of Causality*. Amsterdam: North-Holland, 1970.

Yager, R.R., Quantified propositions in a linguistic logic, in: *Proceedings of the 2nd International Seminar on Fuzzy Set Theory*, Klement, E.P., (ed.). Johannes Kepler University, Linz, Austria, 1980.

Yager, R.R., Reasoning with fuzzy quantified statements—I, *Kybernetes 14* (1985) 233-240.

Zadeh, L.A., Similarity relations and fuzzy orderings, *Information Sciences 3* (1971) 177-200.

Zadeh, L.A., Fuzzy sets and their application to pattern classification and clustering analysis, in: *Classification and Clustering*, Ryzin, J., (ed.). New York: Academic Press, 251-299, 1977.

Zadeh, L.A., PRUF—A meaning representation language for natural languages, *Inter. J. Man-Machine Studies 10* (1978) 395-460.

Zadeh, L.A., A note on prototype theory and fuzzy sets, *Cognition 12* (1982) 291-297.

Zadeh, L.A., Test-score semantics for natural languages and meaning-representation via PRUF, *Proc. COLING 82*, Prague, 425-430, 1982. Full text in: *Empirical Semantics*, Rieger, B.B., (ed.). Bochum: Brockmeyer, 281-349, 1982.

Zadeh, L.A., A computational approach to fuzzy quantifiers in natural languages, *Computers and Mathematics 9* (1983a) 149-184.

Zadeh, L.A., A fuzzy-set-theoretic approach to the compositionality of meaning: Propositions, dispositions and canonical forms, *Journal of Semantics 3* (1983b) 253-272.

Zadeh, L.A., Fuzzy logic as a basis for the management of uncertainty in expert systems, *Fuzzy Sets and Systems 11* (1983c) 199-227.

Zadeh, L.A., A theory of commonsense knowledge, in: *Aspects of Vagueness*, Skala, H.J., Termini, S. and Trillas, E., (eds.). Dordrecht: Reidel, 1984.

Zadeh, L.A., Fuzzy Sets, Usuality and Commonsense Reasoning, *Berkeley Cognitive Science Report No. 32*, 1985. To appear in *Matters of Intelligence*, L. Vaina (ed.), Dordrecht: Reidel, 1986.

Zadeh, L.A., Syllogistic reasoning in fuzzy logic and its application to reasoning with dispositions, in *Proceedings of the 1984 International Symposium on Multiple-Valued Logic*, 148-153, Winnipeg, 1984.

AN APPROXIMATE REASONING FRAMEWORK FOR AGGREGATE PRODUCTION PLANNING

I.B. Turksen
University of Toronto
Toronto, Ontario M5S 1A4, CANADA

ABSTRACT

An approximate reasoning framework is proposed based on the developments in interval valued fuzzy set representation and inference. A detailed study of the aggregate production planning problem and its comparison with a classical operations management model known as HMMS model show that the approximate reasoning framework produces a favourably robust result.

1. AGGREGATE PRODUCTION PLANNING.

In computer integrated manufacturing, the aggregate production planning is an important top level planning activity. The aggregate production planning is mainly concerned with decisions on production and workforce levels in addition to managing inventories. More specifically, aggregate production planning attempts to balance costs: costs of regular payroll associated with rates of production; costs of changes in the size of the workforce (hiring/layoff costs); under or over utilization costs of the workforce (idle time or overtime costs); and inventory costs (carrying charges, backorder costs, lost sales costs); etc. For the factories of the future some of this classic terminology need to be adjusted to account for robots replacing humans and inventories adjusted to just-in-time schemes at the low detail level production planning. However, with these adjustments, the essential nature of the problem does not change but becomes much more complex requiring approximate reasoning methodology that will be introduced shortly.

In our quest for precision in operations research methodologies, a multitude of mathematical models of aggregate production planning were developed in the last thirty years of operational management literature. (Bowman, 1956; 1963; Hanssman and Hess, 1960; Holt et. al., 1955, 1956, 1960; Jones, 1967, etc.). However, the implementation of any of these models in industry is practically non-existent (Buffa and Taubert, 1972; Lee and Khumawala, 1974). This may be attributed to the fact that these operations

NATO ASI Series, Vol. F49
Computer Integrated Manufacturing
Edited by I. Burhan Turksen

research models enforce an artificial precision inappropriate to many real world situations, in particular to such complex systems involving humans and robots. It appears, most managers use their own heuristics which do not guarantee mathematical optimality. Available empirical evidence (Bowman, 1963; Vergine, 1966; Eilon, 1973) suggests that these managers' (experts') judgemental models do remarkably well. It seems that these experienced managers (experts) capture the knowledge of aggregate production planning in the form of rules which contain linguistic variables that are many-valued rather than two-valued. [In this context, many-valued representation of linguistic variables are expressed in terms of generalized truths and falsities that take on values in the interval [0,1] whereas two-valued representation of linguistic variables are expressed in terms of simple truths and falsities that take on values in {0,1} lattice.] A collection of such expert rules is given in Table 1a. For example rule number 4 should be read as:

IF the current period sales forecast, S_t , is HIGH,

AND the previous period inventory level, I_{t-1}, is AVERAGE

AND the previous period work force level, W_{t-1}, is HIGH

THEN the current period production rate, P_t, should be HIGH and the change in the work force level, ΔW_t should be AVERAGE.

Until recently, a theory for handling linguistic terms such as HIGH, AVERAGE, etc. and approximate reasoning that uses such linguistic terms did not exist. Fuzzy set theory (Zadeh, 1965, 1970, 1973, 1975, 1977, 1981, 1987) is concerned precisely with these kinds of problems. The development of this theory began about a quarter of a century ago to provide a sound basis for representing an object's membership in a set analogous to the way humans normally think. The basic idea is that an object may belong to a set with a grade of membership which may range from full to none membership. This differs from ordinary set notion which require objects to either belong or not. By allowing objects' attributes to be specified in grades of membership, knowledge is represented in a pattern which is more natural for human understanding.

Shortly after Zadeh's proposal for approximate reasoning, Mamdani and Assilian (1975) successfully applied fuzzy sets theory to derive a linguistic control strategy for a steam engine and boiler combination using the heuristic rules stated by a human operator. Other studies (Kickert and Lemke, 1976; Ostergaard, 1977; Turksen, 1988) have reported similar success in applications on processes that obey complex laws. Within their contexts, these studies demonstrate that knowledge can be represented and modelled at the meta level within the domain of linguistic variables and at the computational level within the domain of membership values.

2. APPROXIMATE REASONING.

Informally, approximate reasoning is the process or processes by which a possible imprecise conclusion is deduced from a collection of imprecise premises (Zadeh, 1977). There are two alternative but equivalent ways (Tong and Efstathiou, 1982) of approximate reasoning: truth value restriction (Zadeh, 1975; Baldwin and Guild, 1980) and compositional inference (Zadeh, 1975). It appears that most researchers prefer compositional inference in their work.

As an extension of the classical logic, there are at least two ways of reasoning in fuzzy logic: modus ponens, modus tollens, etc. Moreover, there are two additional plausible ways of reasoning that are available in fuzzy logic which are not possible in the classical logic: denial and confirmation (Bandler and Kohout, 1984). However, so far, only modus ponens appears to have been favoured by most researchers in fuzzy logic under the name of generalized modus ponens (GMP).

In the abstract GMP often takes the following structure in certain expert system prototypes: (Clearly, the following is just a skeleton. Other relevant aspects of an expert system developement is not considered in this paper.)

(i) Factual Knowledge (Observed System State)

X_1 is $\mathbf{A}_1^*$ AND X_2 is $\mathbf{A}_2^*$ AND ... AND X_n is $\mathbf{A}_n^*$.

A user of an expert system inputs the factual knowledge choosing the appropriate linguistic variables, i.e.,

$\mathbf{A}_1^*, \mathbf{A}_2^*, \ldots, \mathbf{A}_n^*,$

that describe the observed states of the system. Users may be given the option to state the meaning of these linguistic variables by specifying membership values and/or functions. Alternately they may be given the option to rely on the definitions provided by the domain expert.

(ii) Knowledge-Base (Expert Rules):

IF X_1 is $\mathbf{A}_1$ AND X_2 is $\mathbf{A}_2$ AND...AND X_n is $\mathbf{A}_n$, THEN Y is $\mathbf{B}$.

A domain expert provides such rules during the knowledge-acquisition phase, where $\mathbf{A}_1$, $\mathbf{A}_2$, ... $\mathbf{A}_n$ and $\mathbf{B}$ are the appropriate set of linguistic variables specified by a domain expert. In addition, the domain expert must also provide the meaning of these linguistic variables by specifying their membership values and/or functions over the domain of discourse. Rules are then encoded into the knowledge-base by a knowledge engineer during the design and development phase.

(iii) Expert System Advice

Y is (should be) **B***

The expert system response is displayed to a user after the inference subsystem operates on the "RULES" and the given "FACTS" in accordance with GMP.

In current literature, it may be observed that the meaning of linguistic variables are generally represented by "point-valued" membership assignments. However, it has been observed that membership values obtained from domain experts usually turn out to be "interval-valued" in most measurement experiments (Adlassnig, 1986; Turksen, 1986b;).

2.1 Interval-Valued Fuzzy Sets.

We have suggested previously that the linguistic connectives, AND, OR, IF...THEN, should be interpreted in a way that leads to an interval-valued representation of the linguistic compositions (Turksen, 1983).

Recently it was shown that there is a theoretical foundation for a class of interval-valued fuzzy sets (IVFS's) (Turksen, 1986b). This class of IVFS's is based on the disjunctive and the conjuctive normal forms, DNF and CNF, respectively, of a fuzzy composition as an extension of the canonical forms in Boolean logic. For example, DNF and CNF for the three basic linguistic compositions are:

(1) AND Composition:

$DNF(A \text{ AND } B) = A \cap B$

$CNF(A \text{ AND } B) = (A \cup B) \cap (A \cup B^c) \cap (A^c \cup B)$;

(2) OR Composition:

$DNF(A \text{ OR } B) = (A \cap B) \cup (A \cap B^c) \cup (A^c \cap B)$,

$CNF(A \text{ OR } B) = A \cup B$

(3) IF...THEN (→) Composition:

$DNF(A \to B) = (A \cap B) \cup (A^c \cap B) \cup (A^c \cap B^c)$,

$CNF(A \to B) = A^c \cup B$;

where $\cap$, $\cup$, and superscript c are used in the set notation to correspond to the intersection, union, and complementation operators, respectively. In the membership domain, the corresponding operators, are the various t-norms, t-conorms and pseudo or strong negation operators, respectively.

It should be observed that a DNF is formed with disjunction of some of the primary conjuctions:

$A \cap B, A \cap B^c, A^c \cap B, A^c \cap B^c$;

and a CNF is formed with conjuctions of some of the primary disjuctions:

$A \cup B, A \cup B^c, A^c \cup B, A^c \cup B^c$.

2.2 Linguistic Compositions.

The essential result of the previous work (Turksen; 1983, 1986) is that for certain classes of t-norms and t-conorms each DNF is contained in its corresponding CNF for every linguistic composition in general, and in particular, for the three linguistic compositions specified above.

As a consequence, it is suggested that every linguistic composition be represented by an IVFS, S(•), such that

$$DNF(\bullet) \leq S(\bullet) \leq CNF(\bullet).$$

For example, for the three compositions considered above, S(•) would be defined as follows:

$$S(A \text{ AND } B) = \{s | A \cap B \leq s \leq (A \cup B) \cap (A \cup B^c) \cap (A^c \cup B)\} \quad (a)$$
$$S(A \text{ OR } B) = \{s | (A \cap B) \cup (A \cap B^c) \cup (A^c \cap B) \leq s \leq A \cup B\} \quad (b)[1]$$
$$S(A \to B) = \{s | (A \cap B) \cup (A^c \cap B) \cup (A^c \cap B^c) \leq s \leq A^c \cup B\} \quad (c)$$

It should be observed that the "point-valued" definition of "A AND B" in current literature takes only the form $A \cap B$ which is the lower bound of **S(A AND B)**. On the other hand, the "point-valued" definition of "A OR B" in current literature takes only the form $A \cup B$ **which is the upper bound of S(A OR B)**. The case of $A \to B$ is defined in many ways in current literature. It is sufficient to say that **S(A → B)** contains most of the "point-valued" definitions.

Consider, for example, the following "point-valued" expressions of fuzzy implication (Bandler and Kohout, 1980):

R_5 Lukasiewicz: $a \to_5 b = \min(1, 1-a+b)$

$R_{5.5}$ Kleene-Dienes-Lukasiewicz: $a \to_{5.5} b = 1-a+ab$

R_6 Kleen-Dienes: $a \to_6 b = (1-a) \vee b$

R_7 Early Zadeh: $a \rightarrow_7 b = (a \wedge b) \vee (1-a)$

R_8 Willmott: $a \rightarrow_8 b = (a \rightarrow_7 b) \wedge (b \vee (1-b))$

It can be shown that:

$DNF(A \rightarrow B|\vee, \wedge) \leq (R_6, R_7, R_8) \leq CNF(A \rightarrow B|\vee, \wedge)$.

$DNF(A \rightarrow B|\hat{+}, .) \leq (R_{5.5}) \leq CNF(A \rightarrow B|\hat{+}, .)$

$DNF(A \rightarrow B|\mathbf{V}, \mathbf{A}) \leq (R_5, R_{5.5}, R_6) \leq CNF(A \rightarrow B|\mathbf{V}, \mathbf{A})$

$DNF(A \rightarrow B|S_W, T_W) \leq (R_5, R_{5.5}, R_6) \leq CNF(A \rightarrow B|S_W, T_W)$

where the collection of intersection operators Min($\wedge$), Product(.), **Bold** intersection($\mathbf{A}$), Drastic intersection(T_W) are the well known t-norms, and Max($\vee$), Sum($\hat{+}$), Bold union($\mathbf{V}$), Drastic union(S_W) are the well known t-conorms.

These four conjugate pairs of t-norms T and t-conorms S are defined as:

(1) $T(a, b) = a \wedge b = \min(a, b)$
$S(a, b) = a \vee b - \max(a, b)$

(2) $T(a, b) = a \,.\, b = ab$
$S(a, b) = a \hat{+} b = a + b - ab$

(3) $T(a, b) = a \mathbf{A} b = \max(0, a + b - 1)$
$S(a, b) = a \mathbf{V} b = \min(1, a + b)$

(4) $T_W(a, b) = \{a \wedge b$ if $a \vee b = 1$
$\{0$ otherwise
$S_W(a, b) = \{a \vee b$ if $a \wedge b = 0$
$\{1$ otherwise

2.3 Framework.

The proposed approximate reasoning framework is essentially designed to accomplish the following:

(i) an interval valued representation of rules and observed system facts;

(ii) a search for the rules closest to the observed system facts based on a similarity measure;

(iii) an inference based on the implication of the rules found in (ii) and the observed system facts; and

(iv) an advice based on the linguistic approximation of the result found in (iii).

Before a more detailed account of the framework is given, we need to review briefly the following:

(1) Representation: An interval-valued fuzzy set **A** is bound by the lower bound A_L and the upper bound A^U; i.e.,

$$\mathbf{A} = \{A | A_L \le A \le A^U\}$$

The bounds A_L and A^U are determined by DNF and CNF, respectively, as specified by [1].

(2) Centre. The centre of an interval-valued fuzzy set **A** is a point-valued fuzzy set A, i.e.,

$$C(\mathbf{A}) = \{(A)_i\} = 1/2\{(A_L)_i + (A^U)_i\} \qquad [2]$$

(3) Similarity. There are a number of similarity measures. A comparative analysis may be found in Zwick, Carlstein and Budescu (1987). Any one of these similarity measures could be used as part of this approximate reasoning algorithm, for example, the Minkowski r-metric, i.e., the similarity between two interval valued fuzzy sets $\mathbf{I}_1$ and $\mathbf{I}_2$ is measured by the distance between their centres $C(\mathbf{I}_1)$ and $C(\mathbf{I}_2)$ defined as:

$$d(\mathbf{I}_1, \mathbf{I}_2) = [\Sigma_i(|C(\mathbf{I}_1)_i - C(\mathbf{I}_2)_i|^r]^{1/r}$$

where r is any arbitrary positive constant with the special cases of interest being r = 1,2, ∞, **where r = 2** defines the well known Euclidian distance measure.

(4) Knowledge-base. The knowledge-base contains all the rules specified by a domain expert as discussed in Section 2, i.e.,

IF X_1 is $\mathbf{A}_1$ AND X_2 is $\mathbf{A}_2$ AND...AND X_n IS $\mathbf{A}_n$, THEN Y is **B**.

(See Table 1a and 1b for two examples of a knowledge-base for aggregate production problems.)

(5) Facts. Observed system facts are specified by a user as also discussed in Section 2, i.e.,

X_1 is $\mathbf{A}_1$* AND X_2 is $\mathbf{A}_2$ * AND...AND X_n is $\mathbf{A}_n$*

There are various choices of design alternatives to be made by a knowledge engineer of an expert system during the design and development phase. For example, "should all of the rules in the knowledge-base be left in linguistic form or should their interval-valued representation be developed at the beginning and kept in the knowledge base in that form?"

Since such questions of concern affect the choice of design, we will now be presenting only a basic structure of this approximate reasoning framework without detailing such design alternatives.

2.3.1 Structure.

The structure is expressed in the fuzzy set notation. However, the computations are to be carried out in the membership space with the interval-valued representation of these fuzzy set which is dependent on a choice amongst the many conjugate pairs of t-norms, T, and t-conorms, S.

(1) Build the fuzzy knowledge-base with the interval-valued representation of each rule using knowledge representation schemes such as production rules, frames, etc.,

(i) Compute the interval-valued representation of the "IF" part of each rule, i.e.,

$$\mathbf{A}_j = \mathbf{A}_{1j} \text{ AND } \mathbf{A}_{2j} \text{ AND...AND } \mathbf{A}_{nj},\ j = 1, \ldots, m$$

$$= \{A | A_L \leq A \leq A_U\}$$

where m is the number of rules.

This is achieved by the recursive application of the representational form [1](a).

(ii) Compute the interval-valued representation of each rule, i.e.,

$$\mathbf{R} = \text{IF } \mathbf{A} \text{ THEN } \mathbf{B}$$

$$= \{R | R_L \leq R \leq R_U\}$$

This is achieved by the application of the representational form [1](c).

(2) Determine the interval-valued representation of the observed system facts; i.e.,

$$\mathbf{A}^* = \mathbf{A}_1{}^* \text{ AND } \mathbf{A}_2{}^* \text{ AND...AND } \mathbf{A}_n{}^*$$

This is also achieved by the recursive application of the representational form [1](a).

(3) Determine the expert system advice:

(i) Choose the rule $\mathbf{R}_j$ such that the linguistic variable $\mathbf{A}_j$ of the jth rule is the closest to the linguistic variable $\mathbf{A^*}$ of the observed system state based on a suitable similarity measure; e.g., the Minkowski metric. Identify the rule so selected as $\mathbf{R^*}$.

(ii) If $\mathbf{R^*}$ is not unique then go to step (iv); otherwise continue with step (iii).

(iii) Determine the expert system advice:

(a) Compute the interval-valued GMP.

$$\mathbf{D} = \mathbf{A^*} \circ \mathbf{R^*} = \{D \mid D_L \subseteq D \subseteq D_U\}$$

where $D_L = A_L^* \circ R_L$, $D_U = A_U^* \circ R_U^*$, and "o" is the compositional rule of inference; and then,

(b) Do a linguistic approximation using a similarity measure, i.e., determine the linguistic variable that is closest to the computed $\mathbf{D}$ in (a) above.

Linguistic variable so determine would be inserted into the appropriate slot of the expert system advice given to a user.

(iv) If, for k non-unique R*'s, $\mathbf{A}_j$, $j = i_1, \ldots, i_k$, have the same minimal similarity measure to $\mathbf{A^*}$; then,

(a) Compute the interval-valued

$\mathbf{S^*} = \mathbf{R^*}_{i2}$ OR $\mathbf{R^*}_{i2}$ OR ... OR $\mathbf{R^*}_{ik}$; using formula [1](c) and then,

(b) Compute the interval-valued

$\mathbf{D} = \mathbf{A^*} \circ \mathbf{S^*}$; and then go to step (iii)(b) above.

Naturally, there is an essential choice to be made at the outset of any approximate reasoning. As indicated above, this is the choice of operators to be used during the computations in the membership space. A large collection of conjugate-pair of operators are at the disposal of the knowledge engineers. Initially, there are of course the well known conjugate pairs of t-norms and t-conorms: max-min, sum-product, bold union and intersection, and drastic union and intersection which were specified at the end of section 2.2.

Once the knowledge engineers make a decision on various design alternatives then the details of the framework would become more specific. Finally a crucial choice is the selection of an appropriate language such as LISP or PROLOG or one of other computer languages.

3. AGGREGATE PRODUCTION PLANNING PROTOTYPE

It is well known, with the research of the last thirty years on operations management, that the current period sales forecast, the previous period work force and the inventory level are essential independent variables for the determination of production and work force levels for the current period of concern in the planning. The following definitions are usually stated in current literature of operations management:

S_t = the sales forecast for period t;

W_{t-1} = the work force level in period t-1;

I_{t-1} = the inventory level at the end of period t-1;

ΔW_t = the change in the work force level at the beginning of period t;

P_t = the production rate for period t.

In quantitative models of operations management, the relationships between these variables are expressed in the form of linear and/or non-linear equations. In approximate reasoning, the relationships between these variables are expressed in propositions containing linguistic (fuzzy) variables in forms of linguistic (fuzzy) statements. Such statements form the core of a meta language of approximate reasoning providing a representational structure for the human expert knowledge.

In the context of modelling management reasoning for the aggregate production planning, linguistic conditional statements provide a structure for a direct modelling of expert rules as shown in Table 1. In addition, propositions about observed system states and conclusions from syllogisms are also represented in such linguistic expressions. The terms: high, H, low, L, average, A, etc., in such linguistic expressions are known as linguistic (fuzzy) variables. It is obvious that such linguistic expressions are a part of the natural language. The attractiveness and potential power of approximate reasoning rests on the observations that (i) the thought process of managers (experts) incorporate such linguistic variables in expressing their knowledge of system interactions; and (ii) linguistic rule based models of planning has the potential of producing a better cognitive simulation of managerial thought processes than the ones which use precise mathematical models of operations research which, it appears, so far had no appeal for managers.

3.1 Knowledge Base for a Prototype.

The knowledge base for an aggregate production planning prototype should contain:

(i) a rule base which consists of management decision rules, and;

(ii) a data base which consists of all other information needed to support the inference system in order to infer an expert system advise.

3.1.1 Rule Base.

Management decision rules for aggregate production planning should be expressed in the meta language representation as:

IF X_1 IS $\mathbf{A_1}$ AND X_2 is $\mathbf{A_2}$ AND X_3 is $\mathbf{A_3}$

THEN Y is (should be)**B**

where X_1 is the base variable of sales forecast and X_2 is the inventory level I_{t-1} and X_3 is the work force level W_{t-1}. Furthermore $\mathbf{A}_i$'s are the linguistic variables describing these independent base variables with the allowable linguistic variables given in Table 2.

In an previous study, Rinks (1982) has investigated a knowledge base with 40 such rules (see Table 1b). We have hypothesized however that 27 such rules should be sufficient for aggregate production planning. Our hypothesis is based on the knowledge

(i) that aggregate production is a robust system for both planning and execution. This has been suggested by earlier studies (Holt, Modigliani and Muth, and Simon 1960; Eilon, 1973); and

(ii) that approximate reasoning based on fuzzy logic is a robust inference mechanism in the sense that a system observation $A^1 \neq A$ can be combined with a rule $A \rightarrow B$ to provide a conclusion B^1.

Thus our hypothesis suggest that it would be sufficient to use just the basic ancor variables known as "high", "low", and "medium" for all of the three base variables, sales forecast, inventory level, and work force level to determine both the rate of production and the level of work force for the planned period of production.

Based on this hypothesis, we have decided that the linguistic management rules depicted in Table 1a should form a reasonable rule base in contrast to Table 1b provided by Rinks (1982). (A comparison of these two rule bases will be made toward the end of this paper). It is to be noted that in Table 1, the independent system variables of aggregate production planning are:

(i) the sales forecasts for period t: There will be a need to determine two sales forecasts one for production decision rules, S_t^P; and one for work force decision rules, S_t^W. These will be further explained in the "Data Base" section;

(ii) the inventory level at the end of the period t-1, I_{t-1};

(iii) the work force level in period t-1, W_{t-1};

and the dependent decision variables are:

(iv) the production rate for period t, P_t, and

(v) The change in work force level at the beginning of period t, ΔW_t.

3.1.2 Data Base.

Data base consists of three basic components

(i) weights for the sales forecast

(ii) membership-values for the linguistic variables allowed to be used in the system

(iii) descriptions of observed system characteristics.

In general (i) and (ii) would be permanent components of the data base which may require adjustments from time to time to accommodate environmental changes. Whereas (iii) would be temporary within the span of a particular system diagnosis or system planning for the period of concern.

3.1.2.1 Forecasting Weights

In order to determine the effects of the future sales forecasts on the current period planning decisions, the decision - maker needs to choose two weighted forecasting functions, one for the work force level and the other for the production rate (Rinks, 1982). Let F_j denote the sales forecast for the jth time period obtained by a forecasting method, say regression analysis, etc., then for the use in determining the work force level by a rule, we need to compute

$$S_t^W = \Sigma\, a_i\, F_{t+i-1};$$

and for the use in determining the production rate by a rule, we need to compute

$$S_t^P = \Sigma\, b_i\, F_{t+i-1}.$$

For these weighted forecasting functions, the decision maker needs to choose a_i's and b_i's appropriately. It is suggested by Rinks (1982) that these weights be given by the following formulae:

$$a_i = [\frac{1}{4+i}] \quad \sum [\frac{1}{4+i}] \quad , i = 1, ..., n;$$

$$b_i = [\frac{1}{2}]^i \quad \sum [\frac{1}{2}]^i \quad , i = 1, ..., n.$$

In order to be able to compare our results with Rinks' "point-valued" fuzzy set approach, we have used the same weights. Naturally, other weighting schemes are just as plausible and infact they should be specified by the manager (expert) in charge of planning or the one who provides the rules for the rule base. The ultimate choice will depend on the cost structure of the firm. It is to be noted, however, in choosing a particular weighting scheme, the decision-maker is implicitly making trade-offs amongst the various costs.

3.1.2.2. Membership Values.

The linguistic variables e.g.: high, H; low, L; and average, A; etc., as shown in Table 1, for the aggregate production planning variables, i.e., the independent base variables, S_t, I_{t-1}, W_{t-1} and the dependent base variables, P_t and ΔW_t are defined by exponential functions shown in Table 2 as suggested by Ostergaard (1977) as found experimentally by Norwich and Turksen (1984) and as found experimentally and validated by regression analysis by Zysno (1987) and as used by Rinks (1982). Table 2 contains all the membership functions of the allowable linguistic terms used in the prototype study. Several of the membership functions are depicted in Figure 1. An important feature of the exponential functions used in practice is that all base variables are scaled to be placed in the interval [-1,1]. When the same linguistic variable is used to describe several base variables, this proves to be extremely advantageous for the adoption of a linguistic variable to the particular context of the base variable. For example, we show the upper and lower bounds of the production variable for the case of the the "Paint Factory" to be discussed in the sequel. It needs to be pointed out that, although the membership functions are defined analytically for continuous values of the base variables, the linguistic variables are actually represented by finite support sets for computational efficiency. Furthermore, they need not be defined analytically. They could be taken as given by experts or as found by a measurement experiment (Norwich and Turksen 1984; Zysno, 1987).

3.1.2.3 Observed System.

Descriptions of the observed system characteristics are given with the statements of the form X_1 is $\mathbf{A}^*_1$ and X_2 is $\mathbf{A}^*_2$ and X_3 is $\mathbf{A}^*_3$ where X_1, the sales forecast, is either, S_t^P or S_t^W, for the production rate or the work force level decision, respectively, X_2 is the inventory level I_{t-1} and X_3 is the work force level W_{t-1}. Furthermore the $\mathbf{A}^*_i$'s are the linguistic variables describing the observed base variables with the allowable linguistic variables given in Table 2. If the user would like to use a linguistic variable which is not in Table 2 then the system would ask for its membership description from the user.

3.2 Application to Paint Factory

In order to test the performance of this approximate reasoning approach to aggregate production planning, we used the paint factory data depicted graphically in Figure 2 and compared it with the classic Holt, Modigliani, Muth, and Simon, HMMS, (1955, 1956, 1960) analysis. The reasons for choosing the paint factory data for comparative purposes are two fold: (1) the case uses "real world" data and is sufficiently well documented to perform such a comparative analysis; and (2) the HMMS paint factory linear decision rule solutions have become the standard by which other aggregate planning models are compared (Gordon, 1966; Jones, 1967; Taubert, 1967; Lee and Khumawala, 1974).

Important assumptions of the comparative analysis are: (i) the quadratic cost function derived for the paint factory is the "true" cost function, (ii) the decisions computed from the derived linear rules and (iii) perfect sales forecasts i.e., actual orders (see Figure 2) are available. Other studies, in particular Jones (1967) and Taubert (1967), Rinks (1982) have also made these assumption and compared their results with the results of the linear decision rules.

3.2.1 The Paint Factory

According to HMMS formulation, the objective of the production planner is to minimize the total cost of regular payroll, overtime, hirings and layoffs, inventories, shortages, and machine set up during a given planning interval of N periods. After making the necessary assumptions and statistically estimating the cost coefficients, the mathematical formulation of the problem for the HMMS paint factory was stated as (Holt, Modigliani, Muth and Simon, 1960):

$$\text{Minimize} \quad C_N = \sum_{t=1} C_t \qquad [3]$$

where

$$\begin{aligned} C_t = {} & 340W_t && \text{Regular payroll costs} \\ & + 64.3(W_t - W_{t-1})^2 && \text{Hiring and layoff costs} \\ & + 0.20(P_t - 5.67W_t)^2 + 51.2P_t - 28W_t && \text{Overtime costs} \\ & + 0.0825(I_t - 320)^2 && \text{Inventory costs} \end{aligned}$$

and subject to material balance equations:

$$I_t = I_{t-1} + P_t - S_t; \quad t=1,\ldots, N.$$

The objective function as can be seen, is a quadratic function in the decision variables. With the assumption of quadratic costs, the HMMS solution methodology forms a set of linear equations by taking partial derivatives with respect to each period work force and each period inventory and setting them equal to zero. Then, solving this set of linear equations, two linear decision rules are obtained that indicate for any planning period the production rate and size of work force that yield the lowest cost.

The HMMS solutions for the paint factory are the following two linear decision rules:

$$P_t = \begin{array}{l} 0.463F_t \\ 0.234F_{t+1} \\ 0.111F_{t+2} \\ 0.046F_{t+3} \\ 0.013F_{t+4} \\ -0.012F_{t+5} \end{array} + \begin{array}{l} -0.008F_{t+6} \\ -0.010F_{t+7} \\ -0.009F_{t+8} \\ -0.008F_{t+9} \\ -0.007F_{t+10} \\ -0.005F_{t+11} \end{array} + 0.993W_{t-1} + 153 - 0.464I_{t-1} \qquad [4]$$

$$W_t = 0.743W_{t-1} + 2.09 - 0.011I_{t-1} + \begin{array}{l} 0.0101F_t \\ 8.0088F_{t+1} \\ 0.0071F_{t+2} \\ 0.0054F_{t+3} \\ 0.0042F_{t+4} \\ 0.0031F_{t+5} \end{array} + \begin{array}{l} 0.0023F_{t+6} \\ 0.0016F_{t+7} \\ 0.0012F_{t+8} \\ 0.0009F_{t+9} \\ 0.0006F_{t+10} \\ 0.0005F_{t+11} \end{array} \qquad [5]$$

The actual order pattern for the six-year period that HMMS analyzed, 1949-54, was extremely variable -- including both the 1949 recession and the Korean War. In this respect, the paint factory data is a rather rigorous test environment. The order pattern is graphically depicted in Fig. 2.

The results of applying the linear decision rules [4] and [5], using 12-month perfect sales forecasts, is a 60-month total cost of $2,053,196. Since it is assumed that the quadratic cost structure is the "true" cost structure, the 60-month total cost of $2,053,196 is the base against which the approximate reasoning approach is compared.

3.2.2 Parametrization

In order to implement the approximate reasoning approach, it is necessary that the model variables first be parametrized. In actuality, this step describes the universes over which the values (fuzzy subsets) of the linguistic variables are defined. Two parameters are required for each variable in the model -- the lower bound and the upper bound that the variable is expected to take on. In effect, the lower bound corresponds to the scaled value of -1 used in the definition of the basic fuzzy subsets in Table 2 and, the upper bound corresponds to the scaled value of +1 used in the definition.

Table 3 lists the values used to parametrize the approximate reasoning model for the paint factory data. It should be noted that the information required for determination of the parameter values is usually available to managers making aggregate planning decisions. In the absence of historical data, the manager would use his judgment to make the determinations.

4. COMPARISON OF APPROACHES

In Table 4a for the years 1949-1953, we show the comparison of HMMS-LDR model of aggregate production planning against the Rinks' "Point-Valued Fuzzy Set" (PVFS) model and our "Interval-Valued Fuzzy Set" (IVFS) model. This comparison is made on the base of the 40 rules provided by Rinks (1982). It should be observed that IVFS approach gives a total cost of $\$2,109 \times 10^3$ as compared to PVFS approach of $\$2,156 \times 10^3$. Thus IVFS approach is closer to LDR result of $\$2,053 \times 10^3$ and infact it is within 2.7% of LDR result. In Table 4b, for the years 1949-1950, we show the comparison of HMMS-LDR model against our IVFS approach. This comparison is made on the bases of the 27 anchor rules that we have hypothesized in the previous section. It should be observed that IVFS model did much better than LDR model 721×10^3 versus 731×10^3 which is within -1.36% of LDR model. In this case, it was not possible to compare it against Rinks (1982) PVFS case since there were no results available from Rinks PVFS model for the case of 27 rules and for the years 1949-1950.

In these comparisons, total costs achieved by the approximate reasoning models were based on

(i) twelve period weighted forecasts for S_t^P and S_t^W, and

(ii) the base variable values corresponding to the membership value of 1 for each of the decision variables. (An average was taken over the base variable values whenever there were more than one value of the base variable corresponding to the membership value of 1) and then using the quadratic cost function of the "Paint Factory".

Based on these comparative results shown in Table 4, it is reasonable to say that approximate reasoning based on interval valued fuzzy set approach provides reasonably "good" and "robust" results in addition to providing a communication link between managerial experts and expert systems based on linguistic expressions of the natural languages.

REFERENCES

1. Adlassnig, K.-P. "Fuzzy Set Theory in Medical Diagnosis", IEEE-Transactions on SMC, 16, 2, 260-265, (1986).

2. Baldwin, J.F. & Guild, N.C.F. Feasible algorithms for approximate reasoning using fuzzy logic. Fuzzy Sets and Systems, 3, 225-251, (1980).

3. Bandler, W. & Kohout, L.J. The four modes of inference in fuzzy expert system, in R. Trappl (Ed.), Cybernetics and Systems Research 2, Elsevier Science Publishers (North Holland), 581-586, (1984).

4. Bellman, R.E. and L.A. Zadeh. Decision Making in a Fuzzy Environment, Management Science, 17, B141-B164, (1970).

5. Bowman, E.H. Production scheduling by the transportation method of linear programming. Operations Research, 4, 100-103, (1956).

6. Bowman, E.H. Consistency and optimality in managerial decision making. Management Science, 9, 310-321, (1963).

7. Buffa, E.S. and W.H. Taubet. Production-Inventory Systems: Planning and Control, Richard I. Irwin, Inc., Homewood, Illinois, (1972).

8. Eilon, S. Five Approaches to Aggregate Production Planning. AIIE - Transactions, 7, 2, 118-131, (1975).

9. Gordon, J.R.M. A multi-model analysis of an aggregate scheduling decision. Ph.D. dissertation, Sloan School of Management. MIT, (1966).

10. Hanssman, F. and S.H. Hess. A linear programming approach to production and employment scheduling. Management Technology, 1, 46-51, (1960).

11. Holt, C.C., F. Modigliani and H. Simon. A linear decision rule for production and employment scheduling. Management Science, 2, 1-30, (1955).

12. Holt, C.E., F. Modigliani and J.F. Muth and H. Simon. Derivation of a linear decision rule for production and employment scheduling. Management Science, 2, 159-177, (1956).

13. Holt, C.C., F. Modigliani, J.F. Muth and H. Simon. Planning Production, Inventories and Work Force, Prentice Hall, New York, (1960).

14. Jones, C.H. Parametric production planning. Management Science, 13, 843-866, (1967).

15. Kickert, W.J.M. and H.R. Van Nauta Lemke. Application of a fuzzy controller in a warm water plant. Automatica, 12, 301-308, (1976).

16. Lee, W.B. and B. Khumawala. Simulation testing of aggregate production planning models in an implementation methodology. Management Science, 20, 903-911, (1974).

17. Mamdani, E.H. and S. Assilian. An experiment in linguistic synthesis with a fuzzy logic controller. International Journal of Man-Machine Studies, 7, 1-13, (1975).

18. Norwich, A.M.; I.B. Turksen. A Model for the Measurement of Membership Systems and the consequences of its empirical implementation, Fuzzy Sets and Systems , 12, 1-25, (1984).

19. Ostergaard, J.-J. Fuzzy logic control of a heat exchanger process. In M. Gupta, G. Saridis, and B. Gaines (Eds.), Fuzzy Automata and Decision Processes, North-Holland Publishing Col, New York., (1977).

20. Rinks, D.B. A heuristic approach to aggregate production scheduling using linguistic variables: methodology and applications. In R.R. Yager (Ed.), Fuzzy Sets and Possibility Theory, Recent Developments, Pergamon Press, New York, 562-581, (1982).

21. Taubert, W.H. A search decision rule for aggregate scheduling problem. Management Science, 14, 343-359, (1967).

22. Tong, R. & Efstanthiou, J. A critical assessment of truth functional modification and its use in approximate reasoning. Fuzzy Sets and Systems, 7, 103-108, (1982).

23. Turksen, I.B. Inference Regions for Fuzzy Propositions. Fuzzy Set Theory and Applications, P.P. Wang (Ed.), 137-148, (1983).

24. Turksen, I.B. and D.D.W. Yao. Representation of connectives in Fuzzy Reasoning: The View Through Normal Forms. IEEE Trans. Systems, Man and Cybernetics, 14, 1, 146-150, (1984).

25. Turksen, I.B. Fuzzy representation and inference with normal forms. In J. Kacprzyk and R.R. Yager (Eds.), Management Decision Support Systems Using Fuzzy Sets and Possibility Theory, 213-228, (1985a).

26. Turksen, I.B. Fuzzy sets and systems and its applications in production research. In H.-J. Bullinger, and H.J. Warnecke (Eds.), Toward the Factory of the Future, (1985b).

27. Turksen, I.B. Interval-valued fuzzy implication. W. Bandler and A. Kandel (Eds.), Recent Developments in the Theory and Applications of Fuzzy Sets, NAFIPS' 86, 576-592, (NAFIPS '86 Outstanding Paper Award), (1986 a).

28. Turksen, I.B. Interval valued fuzzy sets based on normal forms. Fuzzy Sets & Systems, 20, 2, 191-210, (1986 b).

29. Turksen, I.B. An Approximate Reasoning Framework for Production Planning, Fuzzy Sets and Systems, 26, 1, 23-37 (1988).

30. Vergine, R.C. Production scheduling under seasonal demand. Journal of Industrial Engineering, 17, 260-266, (1966).

31. Zadeh, L.A. Fuzzy Sets, Information control, 8, 338-353, (1965).

32. Zadeh, L.A. (with R.E. Bellman). Decision making in a fuzzy environment. Management Science, 17, B-141-B-164, (1970).

33. Zadeh, L.A. Outline of a new approach to the analysis of complex systems and decision processes. IEEE Trans. Systems, Man and Cybernetics, SMC-3, 28-44, (1973).

34. Zadeh, L.A. The concept of a linguistic variable and its application to approximate reasoning -I, II, III. Information Sciences, 8, 199-249; 301-357; 9, 43-80, (1975).

35. Zadeh, L.A. A Theory of Approximate Reasoning, memorandum No. UCB/ERL M77/58, (1977).

36. Zadeh, L.A. Possibility Theory and Soft Data Analysis, Mathematical Frontier of the Social and Policy Sciences, L. Cobb and R.M. Thrall (Eds), 69-129, Boilder, CO: Westview Press, (1981).

37. Zadeh, L.B. A Computational Theory of Dispositions, International Journal of Intelligent Systems, 2, 39-63, (1987).

38. Zwick, R., Carlstein, E. & Budescu, D.V. Measures of Similarity Between Fuzzy Concepts: A Comparative Analysis IJAR (to appear), (1987).

39. Zysno, P. Representing human concepts by fuzzy sets: Systematical conditions and empirical results. Reprints of Second IFSA Congress, 618-621, Tokyo, July 20-25, (1987).

TABLE 1a. Rules for Production Level and Change in Work Force.

Rule No.	Independent Variables			Decision Variables	
	S_t	I_{t-1}	W_{t-1}	P_t	ΔW_t
1	H	H	H	SH	RL
2	H	H	A	A	RH
3	H	H	L	A	RH
4	H	A	H	H	A
5	H	A	A	RH	RH
6	H	A	L	SH	VH
7	H	L	H	H	A
8	H	L	A	RH	RH
9	H	L	L	SH	H
10	A	H	H	SH	RL
11	A	H	A	A	A
12	A	H	L	A	RH
13	A	A	H	SH	RL
14	A	A	A	A	A
15	A	A	L	A	RH
16	A	L	H	SH	RL
17	A	L	A	A	A
18	A	L	L	A	RH
19	L	H	H	SL	L
20	L	H	A	RL	RL
21	L	H	L	RL	A
22	L	A	H	SL	VL
23	L	A	A	RL	RL
24	L	A	L	L	A
25	L	L	H	SH	RL
26	L	L	A	A	A
27	L	L	L	A	RH

TABLE 1b. Rules for Production Level and Change in Work Force.

Rule No.	Independent Variables			Decision Variables	
	S_t	I_{t-1}	W_{t-1}	P_t	ΔW_t
1	H	AMA	H	H	A
2	H	AMA	A	RH	RH
3	H	AMA	L	SH	VH
4	SH	L	H	H	A
5	SH	L	A	RH	RH
6	SH	L	L	SH	VH
7	SH	SH	H	SH	RL
8	SH	SH	A	A	A
9	SH	SH	L	A	RH
10	A	A	H	SH	RL
11	A	A	A	A	A
12	A	A	L	A	RH
13	SL	SL	H	SH	RL
14	SL	SL	A	A	A
15	SL	SL	L	SL	RH
16	RL	L	H	SH	RL
17	RL	L	A	A	A
18	RL	L	L	A	RH
19	L	ALA	H	SL	VL
20	L	ALA	AL	RL	RL
21	L	ALA	L	L	A
22	SL	H	H	SL	VL
23	SL	H	A	RL	RL
24	SL	H	L	RL	A
25	H	AMA	SH	H	SH
26	H	AMA	SL	SH	H
27	SH	L	SH	H	SH
28	SH	L	SL	SH	H
29	SH	SH	SH	A	A
30	SH	SH	SL	A	SH
31	A	A	SH	A	SL
32	A	A	SL	A	SH
33	SL	SL	SH	A	SL
34	SL	SL	SL	A	A
35	RL	L	SH	A	SL
36	RL	L	SL	A	A
37	L	ALA	SH	RL	L
38	L	ALA	SL	L	SL
39	SL	H	SH	RL	L
40	SL	H	SL	RL	A

TABLE 2. Membership Functions of the Allowable Linguistic Terms.

Linguistic Terms	Acronym	Membership Function Expression[a]
VERY HIGH	VH	HIGH * HIGH
HIGH	H	$1-e^{\left[-\left(\frac{0.5}{\lvert 1-x\rvert}\right)^{2.5}\right]}$
RATHER HIGH	RH	$1-e^{\left[-\left(\frac{0.25}{\lvert 0.7-x\rvert}\right)^{2.5}\right]}$
SORT OF HIGH	SH	$1-e^{\left[-\left(\frac{0.25}{\lvert 0.4-x\rvert}\right)^{2.5}\right]}$
AVERAGE	A	$e^{[-5\lvert x\rvert]}$
SORT OF LOW	SL	$1-e^{\left[-\left(\frac{0.25}{\lvert -0.4-x\rvert}\right)^{2.5}\right]}$
RATHER LOW	RL	$1-e^{\left[-\left(\frac{0.25}{\lvert -0.7-x\rvert}\right)^{2.5}\right]}$
LOW	L	$1-e^{\left[-\left(\frac{0.5}{\lvert -1-x\rvert}\right)^{2.5}\right]}$
VERY LOW	VL	LOW * LOW
AT LEAST AVERAGE	ALA	$e^{[-5\lvert x\rvert]}$ $\quad -1 \leq x \leq 0$ 1 $\quad 0 < x \leq 1$
AT MOST AVERAGE	AMA	1 $\quad -1 \leq x \leq 0$ $e^{[-5\lvert x\rvert]}$ $\quad 0 < x \leq 1$

TABLE 3. Bounds on the Base Variables of "Paint-Factory".

Variable	Lower Bound	Upper Bound
W_{t-1}	60.	115.
ΔW_t	-10.	10.
P_t	250.	750.
I_{t-1}	150.	490.
FS_t	250.	750.

TABLE 4a. Total Operating Costs 1949-1953 - Paint Factory Case (40 Rules, Max-min Operators)

Costs ($000)	LDR		RINK's Point Valued Fuzzy Set Model		Turksen's Interval Valued Fuzzy Set Model	
Regular Payroll	1879		1857		1812	
Hiring & Layoff	20		42		42	
Overtime	129		198		201	
Personnel		2028		2097		2055
Inventory		25		59		54
Total Cost		2053		2156		2109

TABLE 4b. Total Operating Costs 1949-1950 - Paint Factory Case. (27 Rules, Max-Min Operators)

Costs ($000)	HMMS-LDR Model	Turksen's IVFS Model
Total Costs	734	721

FIG. 1. Membership Functions for Several Linguistic Terms. In this example the base variable represents the production variable with the lower bound of 250 and upper bound of 750 units of production for the case of "Paint Factory" data. For Bounds on other base variables see Table 3.

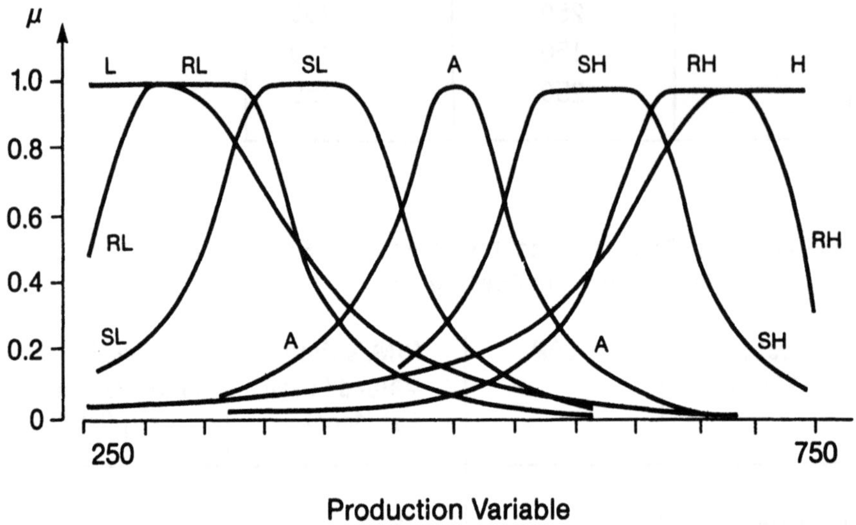

FIG. 2. Perfect Sales Forecast i.e., actual orders.

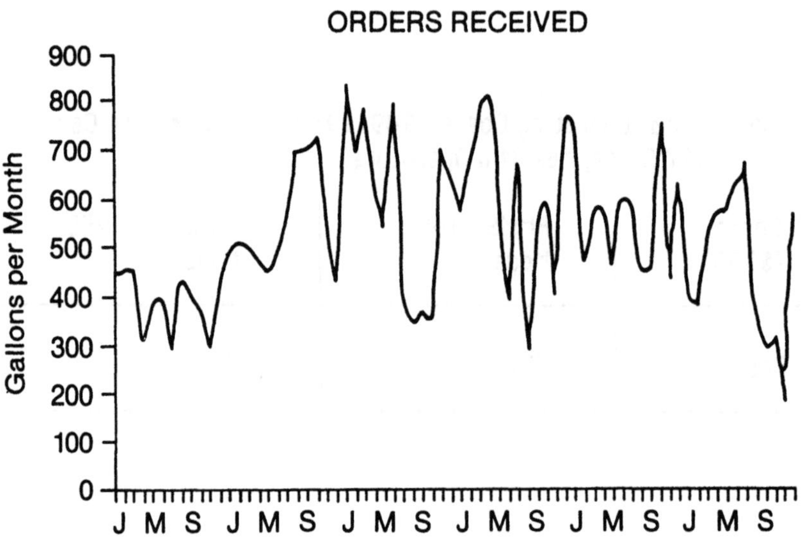

FUZZY ROBOT VISION AND FUZZY CONTROLLED ROBOT

K.Hirota, S.Hachisu
Dept. of Instrument & Control Eng.,College of Eng.,Hosei Univ.
Kajino-cho 3-7-2, Koganei-city, Tokyo 184, Japan

Y.Arai
CSK Research Institute, CSK Computer Building,
Higashi Ikebukuro 3-22-17, Toshima-ku, Tokyo 170, Japan

ABSTRACT

A robot-arm system which is able to recognize a moving patterns and to manipulate a moving object on a belt-conveyor at a various speed is built. This system consists of two parts.

The first part is related to recognizing patterns. In this part, a method of constructing a discriminant-tree is proposed, where three newly defined measures called effectiveness, importance and applicability are introduced. The robot-arm-system is able to recognize the shape and the size of moving patterns on a belt conveyor based on the discriminant-tree.

The second part is concerned with replacing a moving object (i.e. grasping a moving object and putting it on an indicated moving mark) based on fuzzy-inference rules with the aid of image processing technique. The main idea is based on the concept of probabilistic sets in extended fuzzy expression. Ambiguous instructions in terms of membership and vagueness are generated by the robot itself using imagery data from a CCD-camera. Each of these instructions consists of three fuzzy items. In replacing part, two of three fuzzy items are input (ambiguous) information, and one is output information. One of the input information is the fuzzy speed of the moving object/mark, and another is the fuzzy distance between the robot-hand and the object/mark. Output fuzzy information shows the distance between the present position of the object/mark and the next position of it. This output information is **calculated based on Fuzzy inference method.**

NATO ASI Series, Vol. F49
Computer Integrated Manufacturing
Edited by I. Burhan Turksen

The whole system is controlled by only one 16-bit-personal-computer, and works in real time. It also allows a human-like, flexible movement. The advantages of the proposed method are the reduction of processing time and the availability of low level devices, which have not been realized by other methods.

1.Introduction

A repetitive-movement-method of a specified orbit has been a fundamental approach in a conventional robot control. Recently however, it is hoped to realize a flexible, intelligent robot which is able to cope with various situations. Among a lot of approaches to this problem, methods dealing with human ambiguity have been studied positively in order to make the man-machine-communication smoothly (Sugeno and Nishida, 1985, Terano and Masui, 1985).

Fuzzy theoretical approach by Zadeh (Zadeh,1965,1973) is well known as a method dealing with human ambiguity. Mamdani and Assilian have opened a new field to control problems known as fuzzy control by applying fuzzy theoretical decision making technique (Mamdani and Assilian,1974). Recently an advanced version of fuzzy control is also studied by taking a learning factor into consideration (Yamazaki and Sugeno,1984).

In fuzzy control approaches, the system runs based mainly on linguistic control rules in which empirical control strategies of experts (operators) are described by using linguistic variables. Fuzzy control is now in practical use in industrial level, e.g. plant control and automatic train operation (Larsen,1980, Yasunobu,Miyamoto and Ihara,1983). We have also applied this method to a robot-arm control problem and developed a fuzzy controlled robot arm (equipped with a CCD camera) which grasps an object at rest (Hirota, Arai and Pedrycz,1985, Hirota, Arai and Ikezu, 1985, Hirota,1985, Hirota and Arai,1985).

Search has been one of fundamental methods from the early stages in the field of artificial intelligence (Barr and Feigenbaum, 1981, Rich,1983). Applications of it could be seen, for example, in natural language understanding, information retrieval, games, expert systems, scene analysis, and robotics. In these studies, the following two facts are important; 1.A search space is clarified from a state space, and a search tree which leads to solutions effectively is generated. 2.Solutions are obtained suitably based on the search tree.

In dynamic pattern recognition, one of the most important things is a discriminant-tree. The discriminant-tree distinguishes patterns by using several features. Each of end nodes(leaves) corresponds to one pattern category. The method using discriminant-tree has been a fundamental approach in real time pattern recognition. In general, however, how to construct the discriminant-tree depends on experiences of researchers. Although various features of patterns have been studied (Kono,1980,1981), construction methods of discriminant-trees have not been studied enough until now.

In this paper a new approach concerning how to construct an efficient discriminant-tree is proposed first. Necessary information to choose minimum features-set which are used to construct an efficient discriminant tree are frequency of appearance of each pattern category, features themselves, and computing time of extracting each feature. Shape recognition using the discriminant-tree and size recognition based on fuzzy logic are applied to distinguish moving patterns on a belt conveyor. In the proposed system, an object or a mark moves on a belt conveyor, and the robot grasps an indicated object and put it on an indicated mark.

In the next place an outline of robot arm control is mentioned. In order to express ambiguous information the notion of vagueness, which has been proposed in an extended fuzzy expression of probabilistic sets (Hirota,1979), is used as well as that of membership in conventional fuzzy theory. Ambiguous information used in this system consists of the following three items; the velocity of moving object/mark, the distance between moving object/mark and robot hand (both of which are given by an imagery data from a CCD camera), and the next position of the object/mark (which is estimated by fuzzy inference method based on the former two information).

It should be noted that only one 16-bit personal computer controls the whole system (i.e. real time image processing of CCD camera data, fuzzy inference, and robot control of 2 robot arms are done by this 16-bit personal computer alone). Though a lot of processing time and memory are required in general to realize a robot control system with a visual sensor (e.g.CCD camera), only lower level devices were successfully utilized in this report to construct a real time robot system equipped with visual function. It is realized by applying fuzzy inference method with vagueness to rather insufficient visual information from a CCD camera, which have not been realized by other methods.

2.Mark/Object recognition using image processing technique

2.1 recognition of size

Size of each mark/object may be expressed by an area of the observed mark/object in a physical sense. But the size in usual sense is rather a conceptual one.

As a size index we consider not only an area of the observed mark/object but also an area of the circumscribed rectangle of the mark/object. An area density (i.e. a ratio of numbers of pixels of observed mark/object to that of circumscribed rectangle) is used as the size index in the latter sense. Precisely speaking three kinds of sizes, which are big, medium and small, are used. Big and small are expressed by fuzzy labels shown in Fig.1.

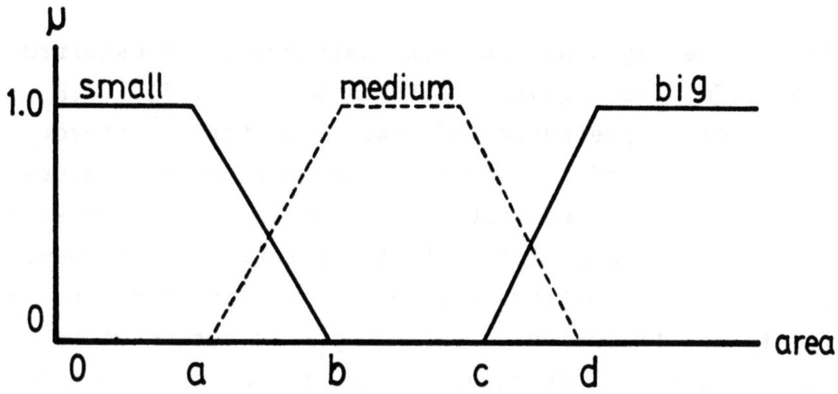

Fig. 1. Fuzzy labels of size.

The horizontal axis shows area, and the vertical axis shows membership value. If the area of the mark/object is less than 'a' or more than 'd', then the size is considered to be small or big, respectively. If the area of it is more than or equal to 'b' and less than or equal to 'c', then the size is medium. If the area is more than 'a' and less than 'b', then the membership value μ will be checked first. If the value is greater than the area density α , then the size is considered to be medium, and if not, then small. This is because the mark with small α seems to be big in comparison to the one with big α , even if they have the same area. In much the same way as this, the size is regarded as big in the case μ is greater than α , and as medium in the converse situation, if the area takes a value between 'c' and 'd'.

That is to say, the size is determined with the aid of the area density in the case that the membership value is less than 1 (i.e. in an ambiguous case).

2.2 algorithm of features-selection in shape recognition

We propose an algorithm of how to select effective and minimum features-set under the constraint that some special pattern categories are given.

Let some special pattern categories considered be C1,...Ci,...Cm, and given features be $F_1,\ldots F_j,\ldots,F_n$. Let frequency of appearance of each pattern category be $p_1,\ldots p_i,\ldots p_m$, and computing time of each feature be $t_1,\ldots,t_j,\ldots,t_n$. These four information are necessary to select effective and minimum features-set. The frequency of appearance of each pattern category is not necessary to be a probability distribution, i,e, $\sum p_i$ is not equal to 1 in general. Because the frequency of appearance is used only as a measure of evaluating efficiency.

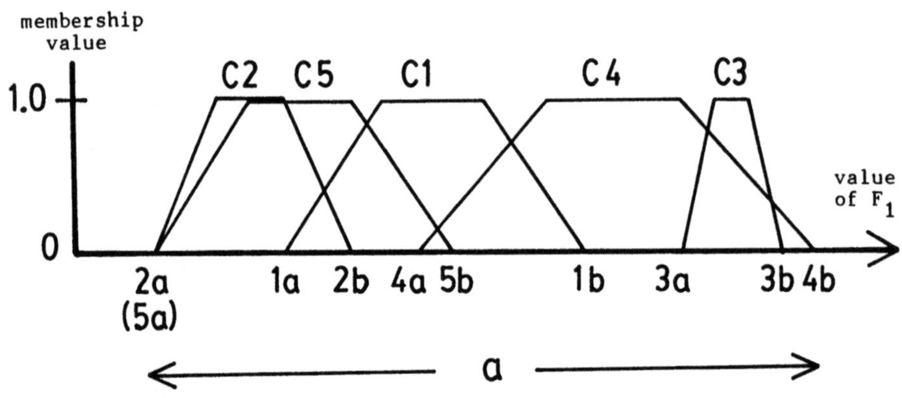

Fig. 2. Distribution map of F_1

In order to select necessary and sufficient features-set, we must know the distribution of each feature for each pattern category. For example, let Fig.2 be such distribution of one feature F_1 for given five pattern categories C1,...,C5. Each pattern category has a membership value on feature-axis because of observation errors or variety of patterns in each pattern category.

According to the distribution map of each feature, we are able to know that which pattern categories are distinguished from others. For example, in Fig.2, we can say that C1 is able to be distinguished from C3, but is not from C2,C4 and C5, since the support set of membership function of C1 is overlapped with that of C2,C4 and C5.

Table 1. Discriminant Table of Feature F_1

	C5	C4	C3	C2	C1
C1	0.2	0.29	1	0.6	
C2	0	1	1		
C3	1	0.2			
C4	0.86				
C5					

Table 2. Effectiveness of Feature F_1

	C5	C4	C3	C2	C1
C1	.005	.0145	0.25	.045	
C2	0	0.25	.575		
C3	0.45	.005			
C4	.129				
C5	$E_1 = \Sigma e' = 1.7235$				

A discriminant table, which is shown in Table 1, is able to be made where the value ($\in [0,1]$) indicates the separation degree of two pattern categories, i.e. 1's complement of intersect-value of two membership functions. It should be noted that the degree of validity of discrimination of two pattern categories depends on the distance between the two membership functions. Let the total area be 'ax1=a' in Fig.2. And let the area of the valley between two membership functions be b, the shape of which is either a reversed rectangle or a trapezoid. Then a measure e' is defined by the following equation (1),

$$e' = b \ / \ a. \tag{1}$$

A result of calculation of measure "e'" is arranged in a tabular form as, for example, shown in Table 2. The sum of each "e'" is regarded as an evaluation function of the effectiveness of each feature. So it is called an "effectiveness" of corresponding feature F_j.

In the next place, let us consider a measure "importance" of each feature. If two features have the same effectiveness, then a feature with shorter computing time is better. It should also be noted that frequency of appearance of each pattern plays an important role, because the feature which is able to discriminate the pattern category with high frequency of appearance from others is more important.

Table 3. Importance of Feature F_1

P_a \ P_b		5	4	3	2	1
		C5	C4	C3	C2	C1
1	C1	0.03	.0725	1	.135	
2	C2	0	1.5	2.875		
3	C3	3.6	.035			
4	C4	1.161				
5	C5					

$\Sigma i' = 10.4085$
$t_1 = 1$
$I_1 = 10.4085$

In the above example of five pattern categories C1,---,C5, let the frequency of appearance $p_1, \ldots, p_5$ of each pattern category be 1,2,3,4,5. Then for each of ten entries e's in Table 2, we can calculate the importance i' of the feature F_j based on

$$i' = e' \times p , \tag{2}$$

where p is the sum of frequency of appearance of corresponding two pattern categories. It is shown in a tabular form as shown in Table 3. We consider that both of the total sum of all i's and the computing time t_j of a feature F_j are closely related to the importance of the feature F_j. And a measure "importance" of the feature F_j is defined by

$$i_j = i' / t_j. \tag{3}$$

If every pair of two pattern categories is able to be discriminated, then we consider that the features' set used is sufficient. In order to get such a sufficient features' set, we superimpose discriminant tables as shown in Table 1 in the order of "importance", and update the entry value to the biggest one. When every entry of the table is filled with "1" (i.e. variant discrimination), then the features-set obtained is a minimum sufficient one. (It should be noted that the superimposition process may be stopped if the minimum value of all entries is greater than 1-E for some small E>0.)

Thus we are able to choose efficient, minimum features-set from given features. Every pair of pattern categories is discriminated by applying this features' set, and other features are redundant ones.

2.3 construction algorithm of discriminant tree in shape recognition

In order to construct an efficient discriminant tree, it is necessary to use, first and foremost, the feature with shorter computing time and with powerful discrimination ability.

Now, in the above example of five pattern categories, we assume that efficient, minimum features-set is $\{F_1,---,F_j,---,F_k\}$, and corresponding computing time is $\{t_1,---,t_j,---,t_k\}$ (,where $k \leqq n$).

Again consider the discriminant table shown in Fig.1. We are able to guess the possibility of corresponding pattern categories from the computed value of the feature F_j (j= 1 in this case) for the observed pattern, i.e.

1	2a-1a ---->	C2 or C5
2	1a-2b ---->	C1 or C2 or C5
3	2b-4a ---->	C1 or C5
4	4a-5b ---->	C1 or C4 or C5
5	5b-1b ---->	C1 or C4
6	1b-3a ---->	C4
7	3a-3b ---->	C3 or C4
8	3b-4b ---->	C4

Next, we determine the feature which is applied to the root of a discriminant tree. In order to do this procedure, a new measure called "applicability" is introduced. The applicability is defined based on the degree of discrimination by taking discrimination degree, frequency of appearance and computing time into consideration. It indicates how the feature is effectively applied for the purpose of discrimination.

Actual calculation example in the above case is shown in the following.

At first, we rewrite each entry in the discriminant table (Table 1) by the product of the value of each entry and the frequency of appearance of both pattern categories (cf. Table 4). Note again that the frequency of appearance of C1-C5 is 1,2,3,4,5 in this example. The total sum of all components in Table 4 is divided by the computing time t_j (= t_1 = 1 in this case) of the feature F_j. And the value is defined as the "applicability" a_j of the feature F_j,

Table 4. Applicability of Feature F_1

P_a \ P_b		5 C5	4 C4	3 C3	2 C2	1 C1
1	C1	1	1.16	3	1.2	
2	C2	0	8	6		
3	C3	15	2.4			
4	C4	17.2				
5	C5					

$\Sigma a' = 54.96$

$t_1 = 1$

$A_1 = 54.96$

$$a_j = 54.96 \ / \ 1 = 54.96 \ . \qquad (4)$$

In the same way as this, applicability for the other features is calculated. And the feature which takes the greatest value of applicability is applied to the root of the discriminant tree.

The tree is expanded to the next stage, and several successors (new nodes) are generated (cf. Fig.3). The measure of applicability is also calculated in the same way.

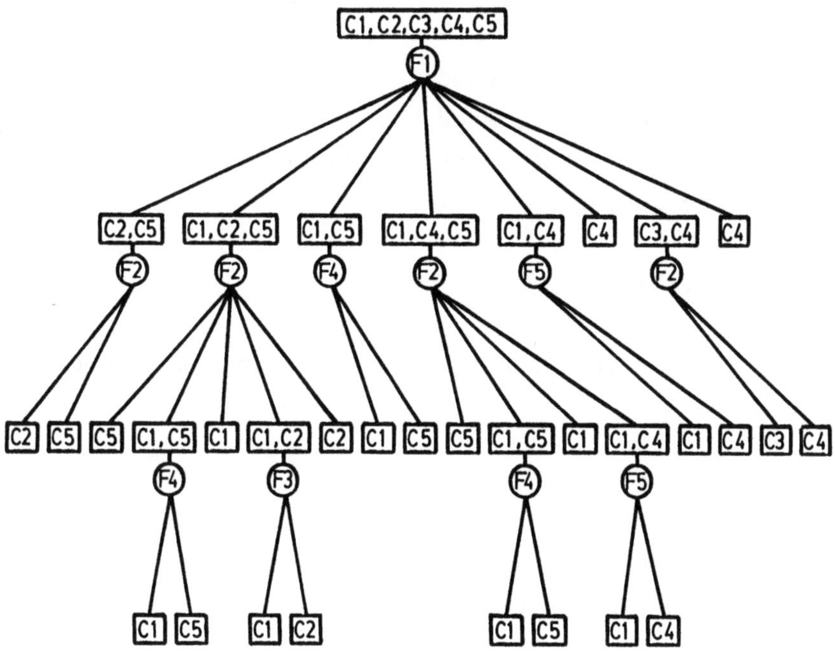

Fig. 3. An example of discriminant tree

A feature with the greatest applicability is applied to the node in each stage and the node is expanded to the next stage. This procedure will be continued, till every successor (or new node) consists of a single pattern category. And finally an efficient discriminant tree is obtained.

3. Dealing with ambiguous information

In this paper, both of the membership notion and the vagueness notion, which have been proposed in extended fuzzy expression of probabilistic sets (Hirota,1979), are used in order to deal with ambiguous information. Membership notion is applied to fuzzy inference method based on compositional rule of inference introduced by Zadeh (Zadeh,1973), and vagueness concept is used to ensure a safety movement of the robot system. This section briefly refers to deal with ambiguous information in inference procedures.

3.1 utilization of membership information

The inference method used in this report is based on the compositional rule of inference proposed by Zadeh (Zadeh,1973). Let F and G be fuzzy subsets on different support sets. Consider the following fuzzy proposition P;

$$P : \text{IF } F \text{ THEN } G \quad \triangleq \quad F \rightarrow G \,. \tag{5}$$

Then the fuzzy relation R is given by using a propositional function

$$R = f (F , G) \,. \tag{6}$$

If this relation R is known and the antecedent F^o is given then the conclusion G^o is expressed as

$$G^o = F^o \circ R, \tag{7}$$

where o stands for a compositional operation, and the Max-Min operation is usually used. A cartesian product of two membership functions is used as a propositional function f

$$R = f (F , G) = F \otimes G \,. \tag{8}$$

Table 5. Fuzzy labels of V,L,P

(a) Fuzzy labels of Speed (V)

	low	----	----	----	----	----	----	----	----	----	----	----	----	----	high
	0	1	2	3	4	5	6	7	8	9	10	11	12	13	14
V1	1	1	1	1	.5	.1	0	0	0	0	0	0	0	0	0
V2	0	.2	.4	.8	1	.8	.4	.2	0	0	0	0	0	0	0
V3	0	0	0	0	0	.2	.4	.8	1	.8	.4	.2	0	0	0
V4	0	0	0	0	0	0	0	0	0	.2	.4	.8	1	1	1

(b) Fuzzy labels of Distance between object/mark and robot-hand (L)

	near	----	----	----	----	----	----	----	----	----	----	----	----	----	----	far	
	0	1	2	3	4	5	6	7	8	9	10	11	12	13	14	15	16
L1	1	.8	.2	0	0	0	0	0	0	0	0	0	0	0	0	0	0
L2	.1	.6	1	.6	.1	0	0	0	0	0	0	0	0	0	0	0	0
L3	0	0	.1	.6	1	.6	.1	0	0	0	0	0	0	0	0	0	0
L4	0	0	0	0	.1	.6	1	1	1	.6	.1	0	0	0	0	0	0
L5	0	0	0	0	0	0	0	0	.1	.6	1	1	1	.6	.1	0	0
L6	0	0	0	0	0	0	0	0	0	0	0	0	.1	.6	1	1	1

(c) Fuzzy labels of (estimated) moving-Distance (P)

	a little away	----	----	----	----	----	----	----	----	----	----	----	----	----	far a way		
	0	1	2	3	4	5	6	7	8	9	10	11	12	13	14	15	16
P1	1	0	0	0	0	0	0	0	0	0	0	0	0	0	0	0	0
P2	.1	.6	1	.6	.1	0	0	0	0	0	0	0	0	0	0	0	0
P3	0	.2	1	.2	0	0	0	0	0	0	0	0	0	0	0	0	0
P4	0	0	.1	.6	1	.6	.1	0	0	0	0	0	0	0	0	0	0
P5	0	0	0	0	.1	.6	1	.1	.6	.1	0	0	0	0	0	0	0
P6	0	0	0	0	0	0	.1	.6	1	.6	.1	0	0	0	0	0	0
P7	0	0	0	0	0	0	0	0	.1	.6	1	1	.6	.1	0	0	0
P8	0	0	0	0	0	0	0	0	0	0	0	.1	.6	1	1	1	1

In this case each antecedent of fuzzy proposition (fuzzy inference rule) consists of two factors; the velocity of the moving object/mark and the distance between the robot hand and the object/mark, which are expressed as fuzzy sets V and L, respectively. For each support set, normalized fuzzy labels are prepared as shown in Table 5 (a)(b). Conclusion (output information) of each fuzzy inference rule indicates the distance between the present position and the next (estimated) position of the moving object/mark and is expressed by a fuzzy set denoted by P. Normalized fuzzy labels are also prepared for the support set of P as shown in Table 5 (c).

Estimated P_k which is derived from the state (V_i, L_j) is defined in a tabular form shown in Table 6, i.e. control rules applied is expressed as the following composite proposition,

$$\begin{array}{l} \text{IF } V \text{ is } V_1 \text{ AND } L \text{ is } L_1 \text{ THEN } P \text{ is } P_1 \text{ ELSE} \\ \text{IF } V \text{ is } V_1 \text{ AND } L \text{ is } L_2 \text{ THEN } P \text{ is } P_1 \text{ ELSE} \\ \qquad \vdots \qquad\qquad \vdots \qquad\qquad \vdots \\ \text{IF } V \text{ is } V_4 \text{ AND } L \text{ is } L_5 \text{ THEN } P \text{ is } P_7 \text{ ELSE} \\ \text{IF } V \text{ is } V_4 \text{ AND } L \text{ is } L_6 \text{ THEN } P \text{ is } P_8 \text{ .} \end{array} \tag{9}$$

This composite proposition makes a fuzzy relation R which is represented as a three dimensional array of size 15X15X17. This array is calculated in advance based on (8) and stored in memory of the computer.

Using this array and eq. (7), we can infer an output information P^o from observed information (V^o, L^o).

Table 6. Rule map of fuzzy inference

	near	----	----	----	----	far
	L1	L2	L3	L4	L5	L6
low V1	P1	P1	P1	P1	P1	P1
: V2	P1	P2	P2	P3	P4	P5
: V3	P1	P2	P4	P5	P6	P7
high V4	P1	P3	P5	P6	P7	P8

a little a way			------------				far a way
P1	P2	P3	P4	P5	P6	P7	P8

3.2 utilization of vagueness information

In extended fuzzy expression of probabilistic sets (Hirota, 1979), usual membership function is called a 1-st monitor and the vagueness function is also called a 2-nd monitor, both of which are expressed as [0,1] valued functions on a support set. The vagueness ,i,e. the 2-nd monitor, is interpreted as an index which indicates the degree of ambiguity (vagueness) of the value of the 1-st monitor (membership function), and has several similar points in approaches of other extended notion such as TYPE 2 (Mizumoto and Tanaka,1976). Extended inference method added vagueness information has been proposed (Hirota, 1979), but vagueness information is not utilized in the inference block of the proposed system because of the reduction of processing time. But the value of vagueness is utilized to realize a safety-movement when the robot hand moves according to the result of inference. Since the inferred conclusion P is interpreted as "the approximate position where the robot should grasp the object or put in on the mark", there is no need to make an orbit correction in microscopic time intervals. That is if the conclusion is rather a reliable one then the robot should move according to the inference result, but if it is not then the robot should infer again during the movement. This way of movement is usually observed in human hand behavior.

Definitely speaking, it will be natural to think that the reliability of inference result decreases as the velocity of the object/mark becomes higher or the distance between the object/mark and the robot hand becomes longer, so a vagueness map shown in Table 7 is prepared. In the case of (V_3,L_4), for example, vagueness is 0.3 from Table 7, so the robot stops and infers again at the 70 (=(1-0.3)x100) % point of the inferred orbit.

Table 7. Vagueness map

	L1	L2	L3	L4	L5	L6
V1	0	0	0	0	0	0
V2	0	0	0	.2	.3	.4
V3	0	.2	.2	.3	.4	.5
V4	0	.3	.3	.4	.5	.5

4. Robot Control Method

The control algorithm of the presented robot system consists of the following 6 blocks (the recognition block in Fig.4 is included in the recognition part and the control part consists of the following 6 blocks);

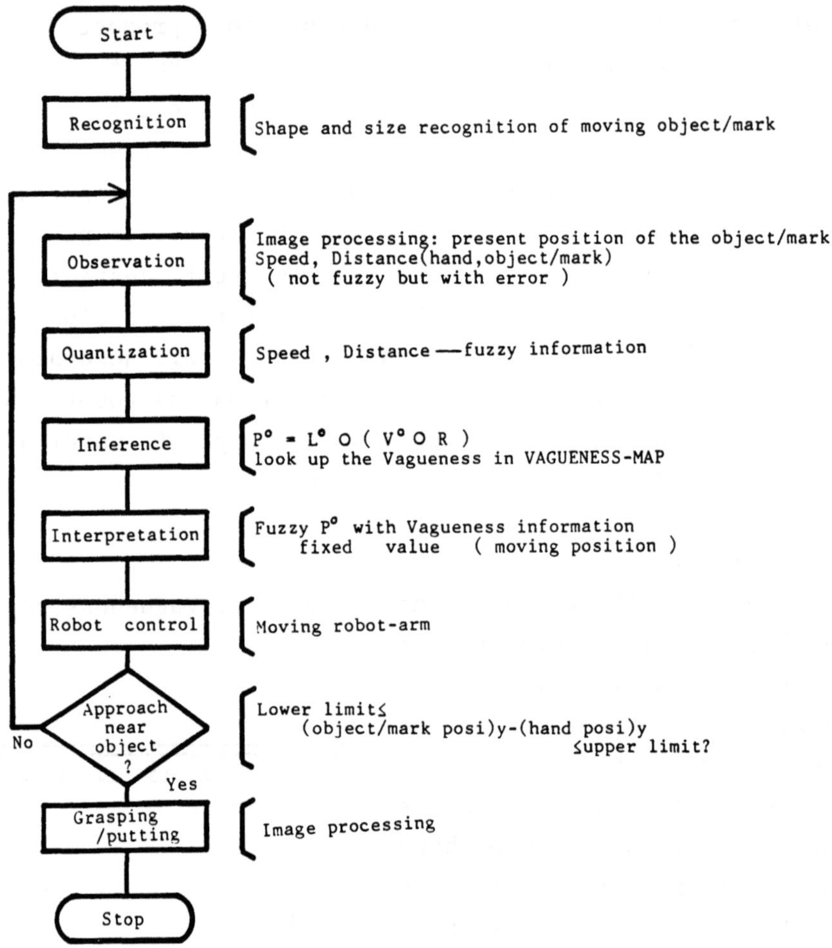

Fig. 4. A flow chart of robot-arm system

(1) observation block

(2) quantization block

(3) inference block
(4) interpretation block
(5) (robot) control block
(6) grasping/putting block .

In the first block, an image is observed by a CCD camera and necessary (but rather ambiguous) information is extracted. Then the robot hand is guided near the object/mark through blocks (2) to (5). If the hand approaches closely enough to the object/mark then it grasps the object or puts the grasped object on the mark (block (6)). Fig.4 shows a flow chart of the whole algorithm and Fig.5 shows the outline of the whole system. The algorithm of each block will be described briefly in the following.

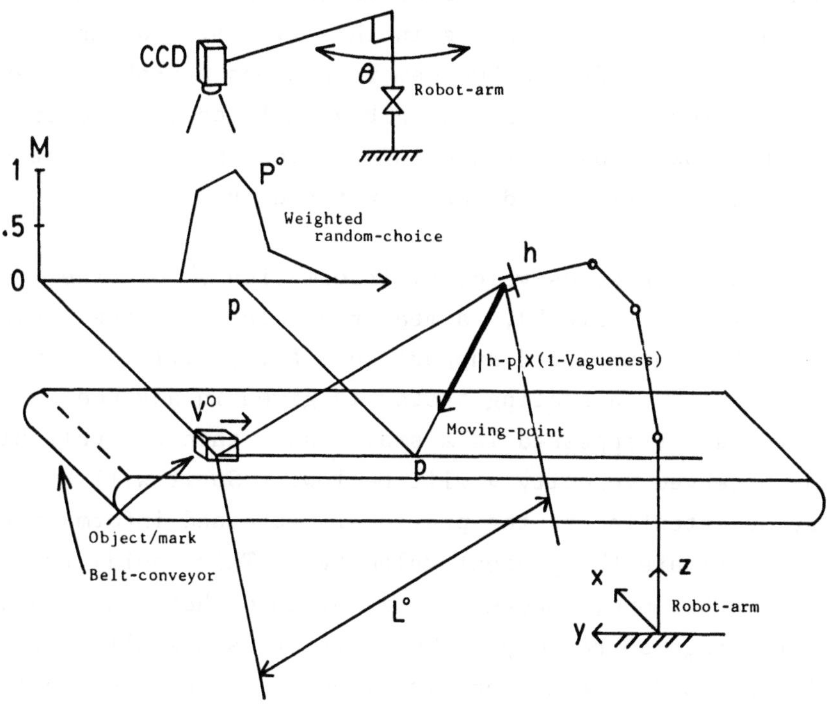

Fig. 5. Outline of the robot-arm system

4.1 observation block

A moving object/mark is observed by a CCD camera and the necessary information is extracted. Here, the necessary information for inference procedures is the approximate velocity v of the object/mark and the approximate distance l between the robot hand and the object/mark. Two observed information (which are non-fuzzy information but with error) will be passed to the next block.

4.2 quantization block

Observed non-fuzzy information is interpreted as a fuzzy information taking error into consideration in this block.

The velocity of a moving object/mark may change continuously, and the distance between a robot hand and an object/mark is always changing in this system. So ambiguity of input information from a CCD camera becomes a problem even if the observation error is a negligibly small one. However it is enough to know almost certain information in the present state because the real time feedback is not required.

As for ambiguous velocity information V^{o} of the moving object/mark (cf. Fig.5), a measured value is mapped into an element of a normalized support set {0,1,...,14}. In this mapping process, a scaling factor is given in advance. If a time interval t appeared is a small one, then the reliability of the observed velocity v will be low. So it will be better to pay an attention to the previously obtained information in order to estimate the present velocity. This holds true even if t is a big one, because it is assumed that the velocity does not change so rapidly. Hence if t is a small one, then 0.8 is assigned to a membership value of a mapped element and 90% values of one step before are allocated to other elements.

If t is large enough, then 1.0 is assigned to a membership value of a mapped element, and 60% values of 1 step before are allocated to other elements.

As measured value of L^o (cf. Fig.5) is also mapped into an element of a normalized support set { 0,1,...,16 }. A scaling factor is also given in advance.

Now let us consider about the distance between a robot hand and a moving object/mark. Suppose that measured values are the same. If the hand is located in front of the moving object/mark, then it will be natural to consider that the moving object/mark exists in a closer position compared with the case that the hand is located backward the moving object/mark. Hence, if the hand exists in front of the moving object/mark, then 0.6 is assigned to the membership value of a mapped element, and 1.0/0.4 are allocated to that of one smaller/two smaller element, respectively. Conversely in the case that the hand is located backward the object/mark, then 0.2/1.0/0.5 are allocated to membership values of mapped element/one bigger element/two bigger element, respectively. And if the hand exists near the object/mark, then 1.0/0.5/0.5 are assigned to membership values of mapped element/one smaller element/one bigger element, respectively.

Non fuzzy numbers are usually used in most fuzzy control systems in order to simplify algorithms, but in this paper ambiguity is fully adopted in the information used in inference procedures.

4.3 inference block

In this block, control information P^o is derived by an inference procedure (cf.(7)) based on two ambiguous information (V^o,L^o) which are given in the quantization block, and a vagueness value is also determined using Table 7.

As already mentioned in section 3, a fuzzy relation R is calculated in advance based on 24 control rules, and control information P^o (cf. Fig.5) is derived by an inference procedure based on (7). The control information P^o is, of course, a fuzzy information and it indicates the approximate hand position in the next step.

Uncertainty of inferred P^o is expressed by a concept of vagueness. The value of vagueness in (V^o , L^o) is obtained directly from a vagueness map which is prepared in advance. (Precisely speaking, V^o and L^o are transformed into fuzzy labels V_1 V_4 and L_1 L_6 firstly, then the vagueness map shown in Table 7 is referred.)

4.4 interpretation block

Inferred fuzzy information P^o with vagueness is transformed into an actual moving position of the robot hand in this interpretation block. It consists of the following three parts (cf. Fig.5); (1) a fixed value p is selected from a support set of P^o, (2) corresponding three dimensional coordinates are calculated, (3) actual moving position of the robot hand is decided based on vagueness information.

A weighted random choice method is used in part (1). A region is picked up from the support set of P^o such that the membership value of P^o in the region is greater than a fixed threshold value (=0.3 in this case). A weighted random number is generated in the region and it is adopted as p, where the membership function of P^o in the region is used as the weight of generated random numbers. (Another method, i.e. center of gravity method, is also programmed in this system.) It should be noted that the movement of the object/mark is one dimensional, i.e. the object/mark moves on a belt conveyor. So the calculation in part (2) is simplified. Vagueness information is used in part (2) to realize a safety movement. If the vagueness value is equal to 0.3, for example, then the robot hand moves till 70% point of the directed orbit. And if it is 0, then the hand moves to just the directed point.

4.5 (robot) control block

Movement parameters of each articulation (motor) of the robot hand are calculated in order to realize the directed movement in the interpretation block. And the robot hand moves according to the parameters.

4.6 grasping/putting block

If the robot hand moves closely enough to the moving object or mark, then it grasps the object or it puts the grasped object on the mark with the aid of imagery information from a CCD camera.

5. Experimental result

A robot arm system with a visual function has been built. The system is capable of recognizing moving objects or marks and is able to grasp an indicated moving object on a belt conveyor at a various speed based on a fuzzy inference method with fuzzy information of speed and distance which are given by a CCD camera. After grasping an object, the robot is also able to put it on an indicated moving mark. Total system is illustrated in Fig.5 and Photo 1.

Photo 1. Overview of the robot system with CCD camera

It consists of two vertical articulated robots with 5 joints (MITSUBISHI RM-501, so called Move Master 2), a CCD camera (SONY, XC-37), an image frame memory of 256 X 256 pixels with 6 bits (i.e. 64) gray levels (EDEC ED-1161, so called image PC, with 128 KB i.e. 2 frame-memories), an image monitor CRT, and a 16 bit personal computer (NEC PC-9800, with 8086 & 8087 cpu's, 5 MHz clock, 384 KB RAM, KB, CRT,8 inches FD). Objects or marks are conveyed by a belt conveyor (SANWA, MN-200-20 (various speed type)). The size of objects is limited to 5X5X5 cm because of the grip size limitation. One of two robots is used to grasp a moving object or to put it on a moving mark on a belt conveyor. Another is utilized as a CCD camera pedestal and it follows the moving object/mark on a belt conveyor.

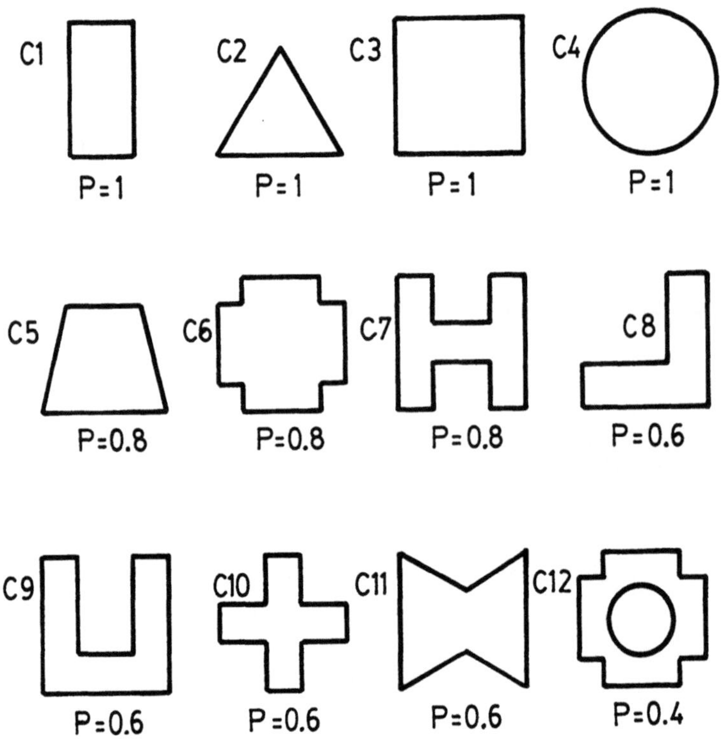

Fig. 6. Twelve patterns used in the shape recognition experiment

The program is coded in N88-NIHONGO BASIC(86) (MS-DOS version) (about 850 steps) and Assembler language (about 2KB including data set). Main flow is coded in BASIC language, whereas image processing part and fuzzy inference part are coded in assembler language in order to improve the processing speed. The upper limit of the speed of the belt conveyor is about 5 cm/sec in this system. If the speed is less than it then the robot is able to grasp or put the object surely.

Patterns used in the experiment are illustrated in Fig.6, where twelve pattern categories C1-C12 are considered and p indicates the frequency of appearance of each pattern category. Categories C1 to C4 indicate typical examples of the shape of both objects and marks, whereas C5 to C12 indicate that of only marks. Directionality of each object or mark is restricted to 1 direction, i.e. if generated shapes by rotation of the observed pattern (object or mark) are different, then they are regarded as belonging to other pattern categories. Intuitively given features are shown in Table 8, where computing time of each features is also given (the value [sec] indicates the real processing time by the system used in the experiment);

- F_1: the ratio of vertical length to horizontal length for the circumscribed rectangle of the observed pattern
- F_2: the variance of projection-image on x-axis for the observed pattern
- F_3: the variance of projection-image on y-axis for the observed pattern
- F_4: the ratio of mean value to maximum value for the projection image on x-axis
- F_5: the ratio of mean value to maximum value for the projection image on y-axis
- F_6: the ratio of the number of pixels of observed pattern to that of circumscribed rectangle
- F_7: the ratio of square of circumference of observed pattern to area of it

F_8: the percentage expression of the position of the center of gravity of observed pattern in x-axis

F_9: the percentage expression of the position of the center of gravity of observed pattern in y-axis

Table 8. Given features and their computing time in recognition process

	F_1 Aspect ratio	F_2 Variance of marginal distribution on x-axis	F_3 Variance of marginal distribution on y-axis
Computing time[sec]	0.25	0.808	0.807

	F_4 x-mean: max length	F_5 y-mean: max length	F_6 Area density
Computing time[sec]	0.778	0.60	0.59

	F_7 Circum-area ratio	F_8 CG offset in x-axis direction	F_9 CG offset in y-axis direction
Computing time[sec]	1.04	0.59	0.59

Distribution maps (cf.Fig.2) of these features are given in advance, where three patterns of different size are observed ten times for each pattern category and the lower bound or the upper bound of the interval is decided by the minimum or the maximum of these data, respectively. Discrimination tables of these features are superimposed in the order of the measure "importance". In the seventh step of this superimposition procedure all entries are filled with "o", so the necessary minimum features-set is $\{ F_1, F_6, F_9, F_7, F_5, F_8, F_4 \}$,($F_1$ is the highest and F_4 is the lowest from a viewpoint of "importance"), and F_2 and F_3 are redundant ones. Applicability of each feature in this set is calculated, and the greatest one F_1 is adopted as a root of the discriminant tree. The root F_1 has twenty one successors. Applicability of features are calculated for each of successors other than end nodes, and this procedure is repeated till arriving end nodes. Finally an efficient discriminant tree is obtained, where the maximum depth is five and the total number of branches is 455.

Objects (C1-C4) and marks (C1-C12) on a belt conveyor are correctly recognized by using this discriminant tree, where the average recognition time is 1.2 sec in this system. A typical example of robot movement is shown in a tabular form in Table 9, and each corresponding photo is shown in Photo 2.

Table 9. An example of experimental result

Step	1 →	2 →	3 →	4 →
Hand posi. Object posi. Next posi. Vagueness	(160, 0, 498) (404, 343, 100) (404, 209, 100) 0.3	(281, 146, 186) (405, 260, 100) (405, 125, 100) 0	(355, 125, 145) (406, 192, 100) (406, 169, 100) 0	(356, 169, 145) (408, 123, 100) (408, 35, 100) 0.2
→ 5 →	6 →	7 →	8	
(348, 62, 145) (407, 58, 100) (407, -81, 100) 0	(357, -81, 145) (407, 5, 100) (407, -16, 100) 0	(357, -16, 145) (407, -29, 100) (407, -85, 100) 0	(357, -85, 145) (407, -82, 100)	(407, -85, 120) (407, -85, 100) grasping

Step	9 →	10 →	11 →	12 →
Hand posi. Mark posi. Next posi. Vagueness	(160, 0, 498) (405, 323, 100) (405, 211, 100) 0.3	(281, 148, 186) (403, 258, 100) (403, 206, 100) 0	(353, 206, 145) (405, 209, 100) (405, 173, 100) 0	(355, 173, 145) (406, 160, 100) (406, 119, 100) 0
→ 13 →	14 →	15 →	16	
(356, 119, 145) (407, 110, 100) (407, 56, 100) 0	(357, 56, 145) (416, 58, 100) (416, 20, 100) 0	(366, 20, 145) (416, 10, 100) (416, -46, 100) 0	(366, -46, 145) (416, -43, 100)	(416, -46, 120) (416, -46, 100) putting

posi.=(X,Y,Z)[mm]

hand posi. : present position of hand
object/mark posi. : present position of object/mark based on imagery data from CCD-camera
next posi. : estimated next position of object/mark based on P^o (calculated by fuzzy inference using I^o and V^o)

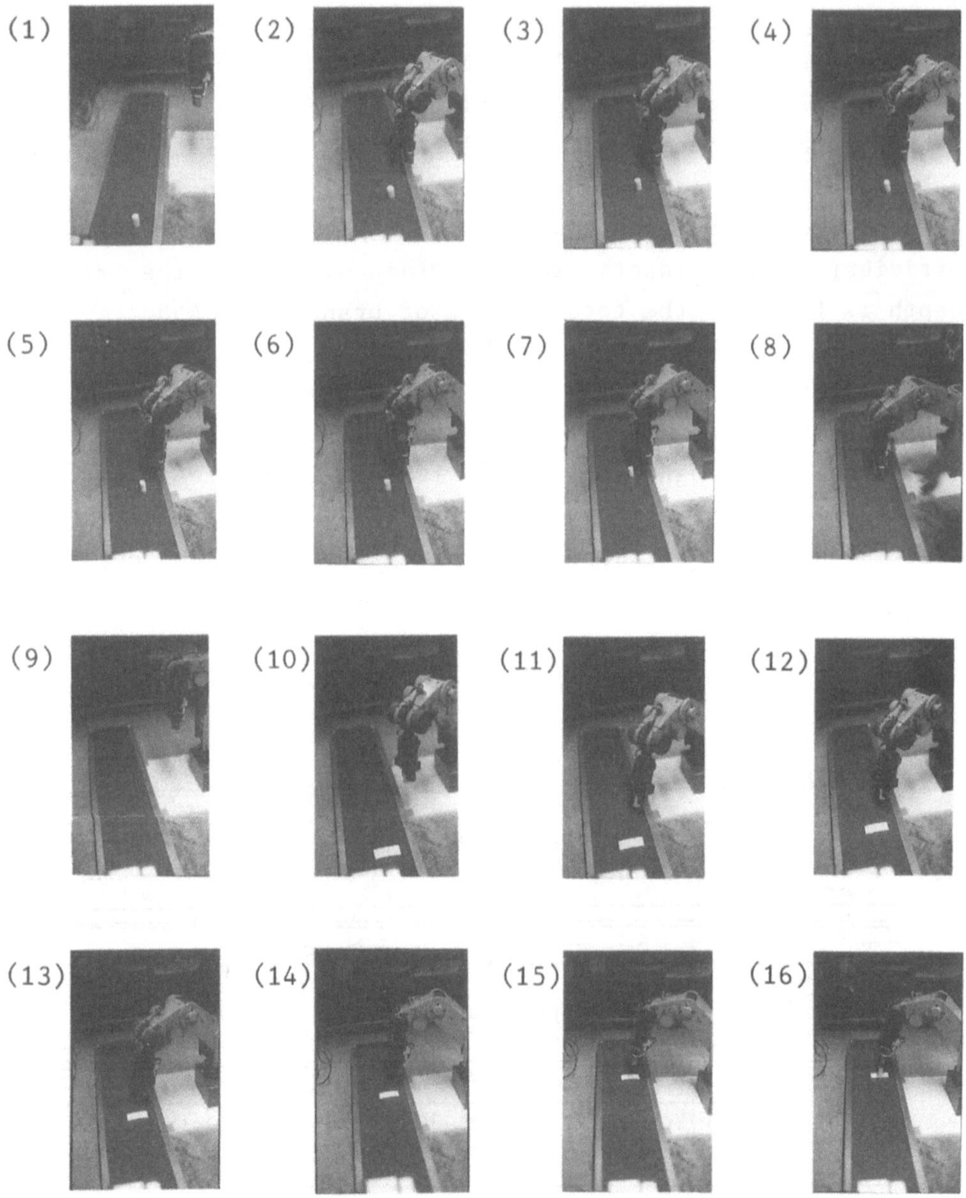

Photo 2. An example of robot movement

An example of experimental results is shown in Photo 2. In this example (1)-(8) and (9)-(16) correspond to grasping an object and putting an object on a mark, respectively, where the number of repetitions of inference procedure is eight in both cases. Corresponding numerical results is shown in Table 9. In this system, the range of closely enough approach is able to be set appropriately according to the velocity. The vagueness-value in the 1st step (1) is 0.3 and the robot stops at 70%

point on the inferred orbit. Then the system moves to the next step (2) and infers again. In the same way as this, after the eighth steps the distance becomes 3 mm (= -82 - (-85) mm) and the robot grasps the object because at this step 8 in Tab.9, the lower limit and upper limit of closely enough approach is 2 and 12 mm, respectively. The total time required is about 10 seconds. Putting process is continued in almost the same way. In general, grasping or putting operation is finished in our system after 3 to 15 steps repetitions, if the velocity of the belt conveyor is less than 5 cm/sec.

6. Conclusions

A robot arm system equipped with a CCD camera has been built. The robot is able to recognize an indicated object or mark on a belt conveyor at a various speed. It is also able to grasp an object or put it on an indicated mark. In recognition part, a new algorithm constructing an efficient discriminant tree is proposed. Fuzzy control method makes it possible to realize the whole system using only lower level devices. Especially it should be noted that only one 16-bit personal computer (5MHz clock) controls the whole system, i.e. real time image pattern recognition, fuzzy inference with vagueness, and 2 industrial-robots-control are done by the 16-bit computer alone. Practical implementation in real industrial lines is also in progress now.

References

1. Barr,A. and Feigenbaum,E.A. : The handbook of Artificial Intelligence, Kaufman, Los Altos,Calif,1981
2. Czogala,E. : Probabilistic Sets in Decision Making and Control (Verlag TÜV Rheinland,Köln),1984
3. Hirota,K. : Extended Fuzzy Expression of Probabilistic Sets, in :M.M.Gupta, R.K.Ragade and R.R.Yager Eds, Advances in Fuzzy Set Theory and Applications (North-Holland), 201-214,1979
4. Hirota,K. Arai,Y and Pedrycz,W. : Robot Control Based on Membership and Vagueness, in :M.M.Guputa, A.Kandel, W.Bandler and J.B, Kiszka Eds, Approximate Reasoning in Expert Systems (North-Holland), 621-635,1985
5. Hirota,K. Arai,Y and Ikezu,T. : Robot Control Based on Imagery Information with Ambiguity (in Japanese Language), Bulletin of the College of Engineering Hosei University, 21,55-76,1985
6. Hirota,K. : Fuzzy Controlled Intelligent Robot (In Japanese Language), McGraw Hill, (Tokyo),1985
7. Hirota,K. and Arai,Y. : Intelligent Arm Robot Based on Probabilistic Fuzzy Control, First IFSA Congress (in Spain),1985
8. Kono,H. : Pattern Discrimination of Convex Plane Shapes Using the Pattern-Classification Value (in Japanese Language), Transaction of the Society of Instrument and Control Engineers (Japan), 16-2,222-227,1980
9. Kono,H. : An Approach to Profile Pattern Recognition of complicated Plane Shapes (in Japanese Language), Trans. of SICE,17-3,381-388,1981
10. Larsen. : Industrial Application of Fuzzy Logic Control, Int.J. of Man-Machine Studies, 12, 3-10,1980
11. Mizumoto,M. and Tanaka,K. : Some Properties of Fuzzy Sets of Type 2, Information and Control, 31-4, 312-340,1976
12. Mamdani,E.H. and Assilian,S. : A Case Study of the Application of Fuzzy Set Theory to Automatic Control, Proc. IFAC Symp. on Stochastic Control,1974
13. Rich,E. : Artificial Intelligence, McGraw-Hill,1983
14. Sugeno,M. and Nishida,M. : Fuzzy Control of Model Car, Int.J. of Fuzzy Sets and Systems, 16-2, 103-113,1985
15. Terano,T. and Masui,S. : Study on Fuzzy Robot (in Japanese Language), 3-rd Knowledge Engineering Symposium (Tokyo), 137-142,1985
16. Yamazaki,T. and Sugeno,M. : Self-Organizing Fuzzy Controller (in Japanese Language),Trans. of the Society of Instrument and Control Engineers (Japan), 20-8,720-726,1984
17. Yasunobu,S. Miyamoto,S. and Ihara,H. : A Fuzzy Control for Train Automatic Stop Control (in Japanese Language), Trans. of the Society of Instrument and Control Engineers (Japan), 19-11,873-880,1983
18. Zadeh,L.A. : Fuzzy Sets, Inf. and Control,8,338-353,1965
19. Zadeh,L.A. : Outline of a New Approach to the Analysis of Complex System and Decision Process,IEEE Trans. of SMC-3, 28-44,1973

OPAL

A KNOWLEDGE-BASED SYSTEM

FOR INDUSTRIAL JOB-SHOP SCHEDULING

E.BENSANA* - G.BEL* - D.DUBOIS**

* CENTRE D'ETUDES ET DE RECHERCHES DE TOULOUSE
Département d'Etudes et de Recherche en Automatique
BP 4025
31055 TOULOUSE Cedex (France)

** LABORATOIRE LSI
Université Paul Sabatier
31062 TOULOUSE Cedex (France)

ABSTRACT

A job-shop scheduling software currently under development is described, based on Artificial Intelligence programming techniques. The idea is to be able to make three kinds of knowledge cooperate in the derivation of a feasible schedule : theoretical knowledge (issued form scheduling theory) which achieves the management of time ; empirical knowledge about priority rules and their influence on production objectives; and practical knowledge (provided by shop-floor managers) about technological constraints to be satisfied in a given application. The latter is usually not considered in pure Operations Research algorithms. The system is actually implemented in COMMON LISP and runs on a Texas Explorer LISP Machine and a SUN workstation. Computational results are reported.

This work is part of a multiteam project on "Production Management" supported by the G.I.P. PROMIP, Toulouse (France).

NATO ASI Series, Vol. F49
Computer Integrated Manufacturing
Edited by I. Burhan Turksen

INTRODUCTION

Scheduling is a field of investigations where advances in theory and solving techniques did not produce expected results at the industrial level due to specific features of case-studies which prevent the application of general approaches. The scheduling problem in a workshop can be stated as follows :

- given a set of production facilities and technological constraints,
- given production requirements expressed in terms of quantity, product quality, and time constraints,
- find a feasible sequencing of processing operations on the various facilities, which satisfies the production requirements.

Three approaches to scheduling can be contemplated :

- optimisation [6]
- constraint-based analysis [13]
- heuristic dispatching techniques with simulation based efficiency-checking [3,5] .

The two first approaches apply to pure academic problems with specific structures. In real environments, one is often led to resort to heuristic techniques whose efficiency is often structure or objective-dependent, and is thus hard to predict .

To make a step towards flexibility a new approach which takes advantage of Artificial Intelligence techniques is under development. The expert-system methodology looks attractive because it may help general-purpose knowledge cooperate with dedicated information provided by shop-floor managers. Moreover, the scheduling strategy may explicitly depend upon the production objectives and account for changes in the workshop configuration. Also, the scheduling software could be more easily carried from a workshop to another, by suitable modification of its knowledge base. Using AI in the manufacturing environment seems to become a major trend (e.g. Kempf [19], Rayson [24]), although people from O.R. do not entirely share the over-optimistic views of AI advocates (Grant [17]).

AI scheduling systems have already been developed based on theorem-proving and predicate calculus [4,22], heuristic, constraint-guided search and frame-based knowledge representation [16], [27], learning techniques[23]. Another constraint-guided search system SOJA [21], involves an operation selection step which reduces the size of the problem. Let us mention lastly PLANEX [8], stemming from works on the GARI system [7] in process planning. See Steffen [26] for an exhaustive review of recent AI-based scheduling systems. Very often these systems prove to be very much time consuming. One of the reasons for such lengthy computations is the use of a single knowledge representation framework at the conceptual level (constraints expressed as schemata in ISIS [16], Horn clauses in logic programming oriented systems [22], [4]). While a single knowledge representation framework at the implementation level is good to prevent interfacing problems, it is not so interesting at the conceptual level where objects, rules, constraints, procedures, etc... should remain distinct, as means of structuring and efficiently using the available knowledge.

The OPAL system, presented here, uses several conceptual entities and tools in order to account for different kinds of knowledge involved in the formulation of the problem, description of the data, and solution methods. These representation principles are presented in the next section which contains the description of the manufacturing data-base. Section 2 presents the architecture of the OPAL system. The different sub-sections are devoted each to a specific module : the constraint propagation and analysis module, the decision-support module and the supervisor. The role of the supervisor in reducing the computation time of a feasible schedule is emphasized, especially the link between the amount of elementary decisions proposed by the decision-support module, and the remaining amount of freedom in scheduling the various operations. Further developments of OPAL are outlined at the end. An illustrative example is explained to show how the system works, and computational results on a realistic sized case-study are reported.

1 - KNOWLEDGE REPRESENTATION

The idea of using Artificial Intelligence techniques in the scheduling area is not really surprizing, since AI face some types of problems already encountered in Operations Research, e.g. combinatioral problems and their solution by ordered search methods. However, it is believed by the authors that the classical expert-system methodology based on production rules is not directly applicable to scheduling problems because expertise in the field takes various forms .

1.1 Required knowledge for a scheduling problem

A natural distinction is between static knowledge (the data) and dynamic knowledge (the solving methods). The static knowledge gives all information about the workshop (available machines and other resources ...) and the production objectives (part types with their processing sequences, quantities, due dates...). The dynamic knowledge describes the available expertise about how to derive feasible or interesting schedules. It may involve three components:

- theoretical expertise which is very much formalized and in the hands of O.R. specialists. It deals with the management of time and resources.

- empirical expertise obtained in using heuristic dispatching rules in simulations.

- practical dedicated expertise provided by shop-floor managers who supervise the production methods, and are aware of technological or human-originated constraints to be respected by the scheduler. In the case of new and/or very complex manufacturing systems, this type of knowledge can be very poor, and only expresses constraints on the manufacturing process.

Viewed as an optimization problem, scheduling only requires the use of theoretical expertise, which is not realistic because it is hard to formulate the wishes of a shop-floor manager in terms of a single objective function. Another O.R. technique is constraint analysis [13]; it tries to charaterize a set of feasible schedules which do not violate due-dates. This approach is much more flexible as it does not prescribe a unique way of discriminating among feasible schedules. The idea is then to integrate in the same system theoretical knowledge in constraint analysis and ill-structured knowledge about dispatching rules or technological and human-originated constraints, with a view to select, among feasible schedules, those which match shop-floor manager wishes according to several possible criteria (quality of finished products, reasonable in- process inventories, set-up time minimization, machine utilization...).

1.2 Choice of a knowledge representation setting

The existence of several types of knowledge with different levels of structuration precludes the use of a unique knowledge representation setting. More particularly the pure Expert-system methodology (Rule base + Factual Base + Inference Engine) looks poorly fitted to the problem. The OPAL system, presented in further sections, involves several knowledge representation techniques :

* the description of the workshop, of processing sequences of parts, and the production requirements is done by means of an object-centered approach, i.e., structured objects related by inheritance links especially.

This type of representation makes OPAL accessible for various different uses from scheduling to simulation [18] or real time monitoring of the workshop [25].

* the description of the schedule in progress is done by means of a precedence graph closely related to the object base, and visualized by means of a Gantt diagram.

* constraint propagation and analysis techniques would easily be expressed in PROLOG because they are basically logical conditions on the overlapping of time windows. This PROLOG module is developed with its full power by a companion team [14]. Presently only a simplified constraint analysis module is working in OPAL. It is implemented by means of four production rules and a small inference engine, written in COMMON-LISP, for the sake of computational efficiency.

* empirical and practical knowledge can be represented by means of "if... then..." rules. They represent pieces of advice which help in the construction of a schedule. However the vagueness which usually pervades this type of knowledge has required the use of fuzzy set theory to model the "if" part of the rules, which express the situations where rules apply.

What is obtained is a hybrid system which mixes ill-structured dedicated knowledge with well-structured generic knowledge issued from scheduling theory. A suitable programming environment where these types of knowledge can cooperate is needed so as to produce realistic, ready to use,production plans.

1.3 Knowledge encoding and the data base

A data base for computer-aided manufacturing requires a knowledge representation technique which is flexible (accessibility from various tasks), modular (for easy modification), efficient and user-friendly (for man-machine communication). The requirements naturally lead to structured declarative languages[20]. We have used COMMON-LISP with DEFSTRUCT functions. It is then possible to specify structured entities having attributes, create instances of these entities by assigning values to attributes. An entity can be a particular case of another, and inherits its attributes. Classes of identical objects can be defined. The choice of entities must be general enough to allow access to the data base at various stages of production management, i.e. from processing sequence generation to scheduling and real time control, or simulation.

The workshop is usually dedicated to the processing of prescribed classes of workpieces. The workpieces are clustered into families called "piece". Entity "piece is characterized by a set of possible processing sequences, represented as a directed tree of groups of operations. The ordering of operations in each group is to be decided at the scheduling level. The tree structure encompasses assembly processes. Entity "job" captures some of the attributes of "piece", necessary to specify the production requirements. The physical components of the workshops are captured by entities "machine", "auxiliary facility" deriving from a more general entity "resource" : "machine" is specialized into dedicated entities ("lathes", "drills", etc...). "Auxiliary facility" expresses resources required for part transportation, machine set-up etc... (transition tasks). Processing sequences are made of operations. Each entity "operation" is related to the entity "phase" whose attributes are the set of machines capable of performing the operation, together with the corresponding processing time, set-up time and other technological information. Based on the same approach other features of the workshop are represented such as "teams of operators" their capabilities and "time tables" which induce machine availability constraints, "production horizon", "quantities of products", "earliest starting times of jobs", "due-dates", etc... Data are stored as instances of entities, part of which are described above. These entities are accessible to the user (they are "named entities"), as well as other kinds of entities such as "nodes" (which allow the search-graph to be developped), "rules" of the D.S. module. Other internal entities ("unnamed") are not accessible to the user, but are useful for efficiency. Figure 1 presents part of the inheritance relations ("is-a") for some entities.

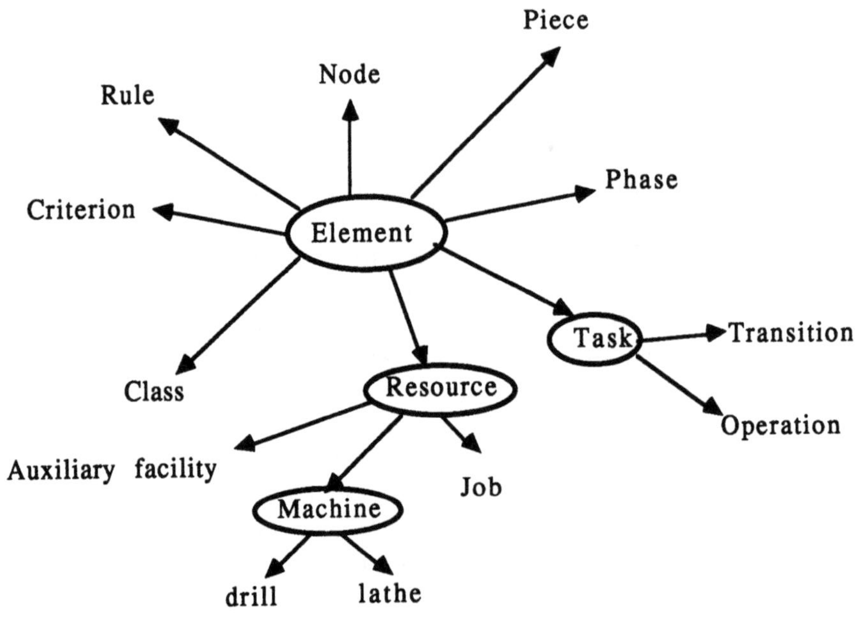

FIGURE 1 : hierarchy

The knowledge representation approach used here is compatible with the one described in the OASYS system [18] for workshop description and simulation under development by the CERT Toulouse with industrial partners.

It is important to figure out that while OPAL uses different sources of knowledge (from theory to practice) and different conceptual entities to formalize these various types of knowledge (semantic nets for the description of the problem, precedence graphs and Gantt Diagrams for schedules, rules for pieces of advice...), there is a unique representation language at the implementation level : namely COMMON-LISP involving DEFSRUCT macrofunctions. Every piece of knowledge is encoded into the form of an object in the sense of object-centered languages. Especially , dynamic knowledge including rules,criteria,and the search graph describing the state of the solving process, as well as the precedence graph describing the schedule in progress are encoded as objects. Hence there are no interfacing problems at the programming level in OPAL.

2 - ARCHITECTURE OF OPAL

The OPAL system, already outlined in [1] [2], is a general software tool for short-term scheduling of a workshop dedicated to small of medium sized production of discrete parts. However it does not deal with the assignment problem of operations to machines .Production requirements must be expressed in terms of earliest starting times and due dates. The generality of the software is due to its knowledge representation capabilities, for instance :

- OPAL accepts the notion of a "batch" and accounts for the possibility of "batch overlapping", i.e. several successive machines processing the same batch.

- processing sequences can be underspecified, i.e. only a partial ordering of operations is available for each job. In this case the job is considered as a logical resource which has to be scheduled. Assembly operations are permitted.

- transportation and set-up times are considered explicitly.

- the existence of time-tables, which implicitly specify an availability calendar for machines is also accounted for.

OPAL is built around four main modules monitored by a supervisor which guides the search process (see figure 2).

- A data base, already described in section 1, contains the description of the worshop (work-stations, part types...), of the scheduling problem (production, requirements, due-dates, machine availability...) and the current state of the scheduling plan.

- A constraint-based analysis (CBA) module calculates the consequence of time constraints on the sequencing of operations. It can generate new precedence relations between operations and propagates such new constraints.

- A decision support (DS) module provides advice on the sequencing of operations, based on practical or heuristic knowledge.

- A supervisor controls the dialogue between CBA and DS and constructs the schedule step by step according to the decisions made by these modules.

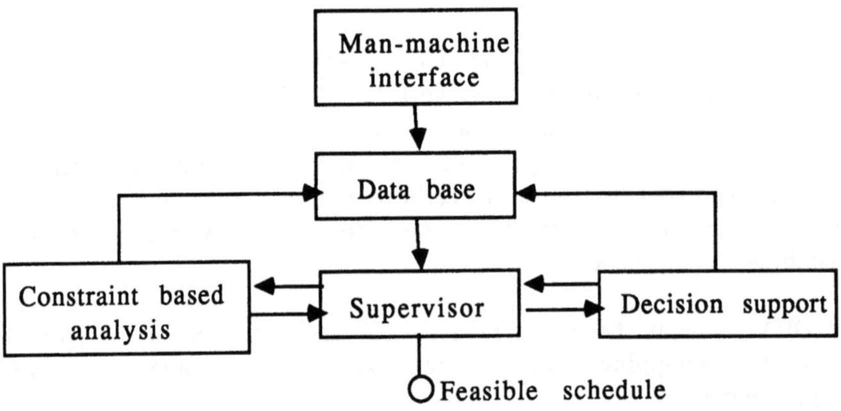

FIGURE 2 : Structure of the jobshop planner

2.1 - The constraint-based analysis (CBA) module

This module deals with a simplified model of the scheduling problem involving only the machines and the operations assigned to these machines, the partial ordering on operations deriving from processing sequences, or from previous decisions made by the DS module, and the time constraints. More specifically, a set K of jobs must be performed by a set M of machines. Each job k is characterized by a set Q_k of operations, and each operation O_{ik} in Q_k must be performed by a machine m_{ik}. This operation requires a processing time d_{ik}. Moreover Q_k is a partially ordered set in the sense that $O_{ik} < O_{jk}$ means that O_{ik} must be performed before O_{jk}. Each job k is assigned to a time horizon bounded by an earliest starting time e_k expressing the availability of raw material, and a latest ending time f_k expressing a due-date. By propagating due-dates and earliest starting times along the directed graph induced by precedence relationships among operations, a window can be assigned to each operation. This window is the time span in which processing must take place. Comparing the respective location of the windows of two not yet ordered operations on the same resource, it is possible to sometimes prove that one operation should precede the other, or that the windows do not allow for the processing of the operations. Such comparisons are systematically carried out by the module. The precedence constraints which are discovered this way are called "essential characteristics"[13],[15]

Let O_i and O_j be two still unsequenced operations to be performed on a machine M (for simplicity the subscript referrring to job number is now omitted). We say that (O_i, O_j) is a pair of conflicting operations, for short a *conflict* on M.

Let est_i and lft_i be the bounds of the window allowed for O_i and d_i the duration of O_i (and the same for O_j).

We assume that $est_i < est_j$.
The test is then : are the intervals overlapping ?
- if they don't O_i must precede O_j
  - if they do different situations may appear [13,12]:

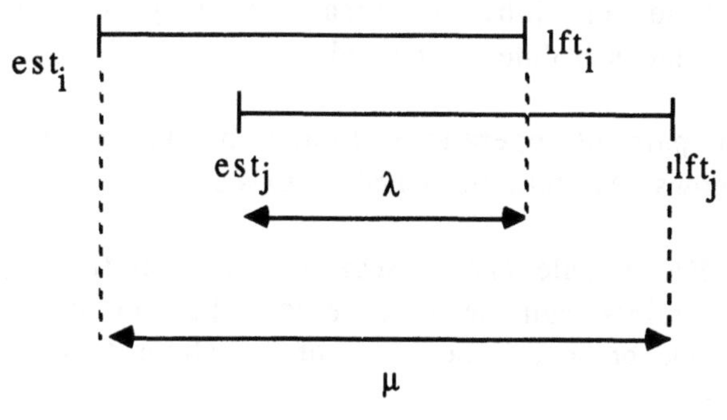

We compute

$$\lambda = lft_i - est_j$$
$$\mu = lft_j - est_i$$
$$\sigma = d_i + d_j$$

* if $\sigma > \lambda$
  and : - if $\sigma \leq \mu$ then O_i must precede O_j
  - if $\sigma > \mu$ then there is no feasible schedule for O_i and O_j.

* if $\sigma \leq \lambda$
  and: - if $\sigma > \mu$ then O_j must precede O_i
  - else no precedence relation can be stated. The conflict remains.

When a new precedence constraint is discovered, say O_i precedes O_j, windows are updated to account for it. Namely,

est_j becomes $\max(est_j, est_i + d_i)$
and lft_i becomes $\min(lft_i, lft_j - d_j)$

These updated values of earliest starting (resp: latest finishing) time est_j (resp:lft_i) are then propagated forward in time on operations $O_l > O_j$ (resp: backward in time on operations $O_l < O_i$).Modified windows are compared again. It is an iterative process which commutes between precedence constraint generation and window updating. As the procedure proceeds, windows cannot but shrink, as more precedence relations are discovered. The procedure stops in one of the following states :

* <u>success</u> : all pairs of operations can be sequenced. For each resource (machine or job) a linear ordering of operations is available. A feasible schedule is derived.

* <u>failure</u> : a pair of operations cannot be sequenced because of conflicting windows. No feasible schedule exists.

* <u>wait</u> : the CBA module cannot sequence all operations and no more precedence constraints can be discovered. The DS module must be called to select one or several new pair of operations to be ordered.

A more elaborated CBA Module is under study, including more refined window comparison tests is described in [14]. However the present module already integrates two kinds of information :

- machine availability time-tables which splits the time scale into non-adjacent windows,

- time spent in transition operations (transportations time, set-up times). Set-up times are included once two operations are known to be consecutive.

Moreover, the basic CBA procedure is also extended so as to allow for batch overlapping : if batches can be dissociated into sub-batches you can try to process a batch simultaneously on several machines to save time; batch overlapping means that it is not necessary to wait for the operation O_i on machine m_k to be entirely done for a batch before starting operation O_{i+1} on the same batch, on machine m_{k+1}. Suppose a batch is composed of n sub-batches, two cases may occur depending on the respective values of d_i and d_{i+1} (processing times of the batch for operation O_i and O_{i+1} of the processing sequence). The schedule of O_{i+1} is located so that machine m_{k+1} can process the batch without any interuption. When $d_i < d_{i+1}$, a sufficient condition is that the last sub-operation on machine m_{k+1} starts right after the end of the last sub-operation on machine m_k(plus transportation time). When $d_i > d_{i+1}$, the same reasoning is applied to the first sub-operation of the batch. Hence the following updating formulas, where $d'_i = d_i/n$, and transportation times are neglected:

* if $d_i > d_{i+1}$, then $est_{i+1} \geq est_i + d_i - (n-1)\, d'_{i+1}$
and $lft_i \leq lft_{i+1} - d'_{i+1}$

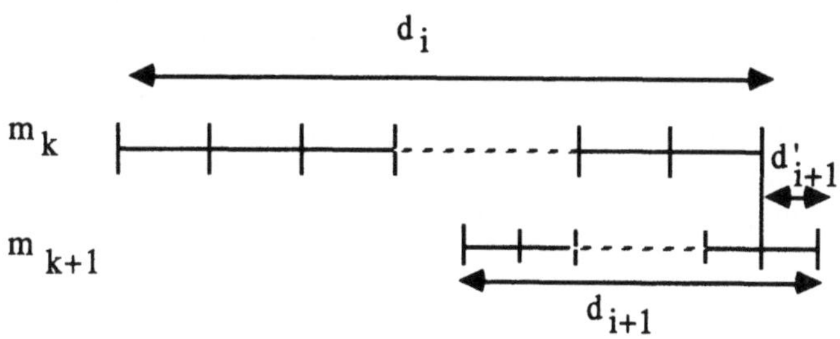

* if $d_{i+1} > d_i$, then $lft_i \leq lft_{i+1} - d_{i+1} + (n-1)d'_i$
and $est_{i+1} \geq est_i + d'_i$

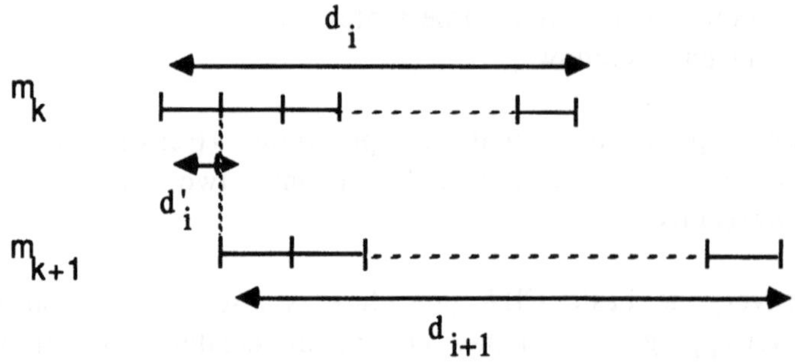

These inequalities are the general constraint propagation formulas in OPAL. Letting n = 1 forbids the overlapping effect for this batch (this is the default option in OPAL). Note that the overlapping effect only occurs along a processing sequence, for consecutive operations on different machines. Especially, the sequencing tests for discovering new precedence constraints, as described earlier, on a single machine are valid regardless of the value of n.

2.2 - The decision-support module

When the CBA module reaches a "wait" state, the decision pertaining to operation ranking is no longer dictated by feasibility considerations with respect to due-dates. Such decisions can be made according to other kinds of criteria of a technological nature (e.g. it is better not to cut a workpiece made of metal M before a workpiece made of metal M'), or related to productivity (facilitate material flow, avoid filling up machine input buffers, avoid long set-up times...). Advice provided by the shop-floor manager is welcome, in order to derive realistic schedules which tend to satisfy technological requirements; productivity aspects can be dealt with thanks to advices from priority-rule specialists who know which rule favors which objective in a given situation.

2.2.1 - Preliminary selection of a set of potential decisions

When the CBA module stops and the supervisor finds that the problem is not entirely solved, it activates the decision-support module by feeding its factual base with the set S of unordered pairs of operations, to be ordered in a further step. In realistic-sized problems this set may contain hundreds of pairs. Note that if there are n_m operations on machine m, the cardinality of S never exceeds :

$$\sum_m (n_m - 1) \frac{n_m}{2}$$

Each pair (i,j) in S is called a "conflict" because time windows pertaining to O_i and O_j are located so that it is not known which operation O_i or O_j is performed first.

Conflicts are equally considered by the CBA module ; however the decision-support system tries to classify the conflicts in order to keep only interesting ones, i.e. those which correspond to the shop-floor manager's point of view and are relevant to his scheduling criteria. We form the fuzzy set of interesting conflicts C by giving to each conflict a value to express how it is interesting to solve this conflict.

Selection criteria are for instance based on the following points of view.

- the machine on which conflicts occur, or the jobs involving O_i and O_j
  - the temporal location of conflicts
  - the influence of solving a conflict on the quality of the schedule
  - the influence of solving a conflict on the resolution speed.

Examples of selection criteria presently available in OPAL can be formulated as follows :

* Select the machine m where the greatest number of conflicts occur, and select the conflicts on this machine .

* Select the conflicts (i,j) where either O_i or O_j is an operation of a particular job of interest (e.g. the one which must be achieved without violating the due date at any rate).

* Select the machine where operations have small slack times, i.e. the machine where the schedule in progress is the less flexible, and choose C to be the set of conflicts on this machine.

* Select the conflicts which are located at early points on time. This criterion leads to solve the scheduling problem following the ordering of operations in the processing sequences.

* Select the conflicts which are linked to many other conflicts.The conflicts linked to (i,j) are in the set $\{(i,k) / k \neq j\}$. This point of view tends to favor the solving of conflicts which may make the constraint analysis most productive.

Let C_k be the set of conflicts selected according to criterion k choosen by the user. If the latter has choosen several criteria, then the set of conflicts on which the decision rules are applied is $C = \cap_k C_k$. The default option (if $\cap_k C_k = \emptyset$ or the user has not choosen any criterion) is to take C=S, that is all unsolved conflicts are interesting a priori . In the OPAL system, the selected subsets C_k are fuzzy, and their intersection is performed by means of the minimum operation, as usual in fuzzy set theory[11].

2.2.2 - Rule modeling and the voting procedure

Further selection in C is achieved by using a collection of pieces of advice expressed as "if...then" rules. Rules differ by their origin and by their range of application (general or application-dedicated). Moreover their efficiency is more or less well known and depends upon the prescribed goal, or the state of completion of the schedule. They can express antagonistic points of views. Lastly they are usually pervaded by imprecision and fuzziness, because their relevance in a given situation cannot be determined in an all-or-nothing manner[10].

To take these features into account, each rule r is assigned a grade of relevance $\pi_r(k)$ with respect to goal k. $\pi_r(k)$ can be viewed as the grade of membership of rule r to the fuzzy set of relevant rules for goal k. The aim of these coefficients is basically to create an ordering on the set of rules. They quantify usefulness, rather than uncertainty, i.e. are grades of utility on the scale [0,1]. The "if" part of a rule describes the situation according to some point of view. It is evaluated on the basis of an index which charaterizes the value v_i of some attribute of operation O_i with respect to the value v_j of the same attribute for operation O_j (static index) ; or it can measure the relative consequence of "O_i precedes O_j" with respect to "O_j precedes O_i" (dynamic index). Matching the <u>if</u> part of rule r against the fact (i,j,m) pertaining to operations O_i and O_j boils down to locate the ratio $v_i/(v_i + v_j)$ with respect to prescribed thresholds. In order to avoid thresholding effects inducing "bang-bang" types of responses., three fuzzy sets H = High ratio, M = Medium ratio, S = Small ratio are defined which form a fuzzy partition of the unit interval [11] that is (see figure 3) :

$$\forall x \in [0,1] \; \mu_H(x) + \mu_M(x) + \mu_S(x) = 1$$

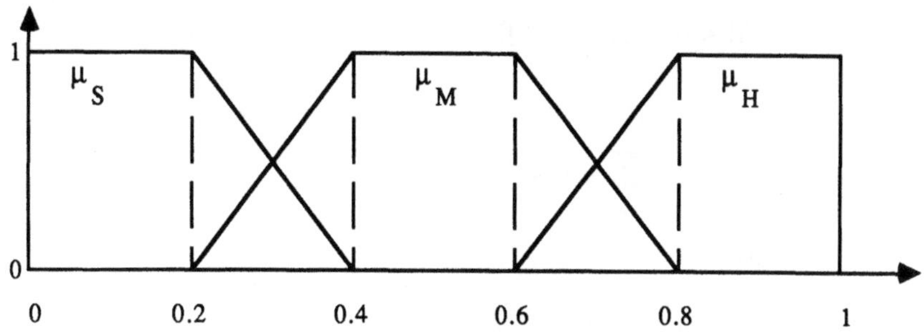

FIGURE 3 : Fuzzy sets Small, Medium and High

The relevance of rule r with respect to fact (i,j,m) is evaluated as the grade of membership of the ratio $x_{ij}=v_i/(v_i+v_j)$ to each of the fuzzy categories H,M,S. Hence relation R appearing in the rule is a fuzzy relation describing the extent to which v_i is greater (H), smaller (S) or equal (M) to v_j. In the case of a static attribute, v_i can be for instance the processing time of O_i (then rule r is for instance a "shortest processing time" priority rule). In the dynamic case the attribute can be a slack time or set-up time etc....

The "then" part of all rules is identical. It provides an advice about whether O_i should precede O_j (i<j), O_j should precede O_i (j<i) or if the rule does not know (i~j). This advice is expressed by three numbers $\mu_r(i<j)$, $\mu_r(j>i)$ and $\mu_r(i\sim j)$ which aggregate the grades of relevance of rule r with respect to goal k and fact (i,j,m) in a conjunctive manner :

$$\mu_r(i < j) = \min (\mu_S(x_{ij}), \pi_r(k))$$

$$\mu_r(i \sim j) = \min (\mu_M(x_{ij}), \pi_r(k))$$

$$\mu_r(i < j) = \min (\mu_H(x_{ij}), \pi_r(k))$$

The rules relevant for goal k are all triggered and applied to all facts of the form (i,j,m) for a preselected resource m. Their results are aggregated through a voting-like procedure. Rule r has $\mu_r(i<j) + \mu_r(i\sim j) + \mu_r(j<i)$ ballots which are shared between the three possible options. The global weights :

$$P(i < j) = \sum_r \mu_r(i < j),$$

$$P(i < j) = \sum_r \mu_r(i < j)$$

and

$$P(i \sim j) = \sum_r \mu_r(j \sim i)$$

are calculated and represent the advice provided by the DS module regarding the pair of operations (i,j). Namely, when $\pi_r(k) \in \{0,1\}$ and S,M,H are standard sets , P(k<j) is the number of triggered rules favoring i<j. In the general case, it is the cardinality of the fuzzy set [11] of rules which prefer i<j. The proportions of relevant triggered rules preferring i<j, j<i, i~j are respectively obtained as relative fuzzy cardinalities[28]:

$$\rho(i<j) = P(i<j)/(\Sigma_r \ \pi_r(k))$$

$$\rho(j<i) = P(j<i)/(\Sigma_r \ \pi_r(k))$$

$$\rho(i_\sim j) = P(i\sim j)/(\Sigma_r \ \pi_r(k))$$

The preference index for decision i<j is defined as min($\rho(i<j)$, 1- $\rho(i\sim j)$) ; in terms of fuzzy logic , it expresses to what extent most of the triggered rules prescribe i<j *and* most are not indifferent about the conflict (i,j).

On the whole the DS module works as follows :

1) preliminary selection : a subset C of non-ordered pairs (i,j) is selected as the set of potential candidates for a precedence decision, i. e. each conflict c is given a value Ic (between 0 and 1). Ic express how c belongs to the fuzzy set of interesting conflicts.

2) voting procedure : apply all rules to all pairs of operations in C, the ordering of which is not decided yet.

3) precedence constraint selection : choose the relation i<j such that the conflict solved is interesting and the preference index is maximal, i.e. the relative weight of indifference $\rho(i_\sim j)$ does not dominate. When $\rho(i\sim j)$ is close to 1, it means that one is unable to decide between i < j and j < i because the rules are indifferent. Contrastedly, when $\rho(i\sim j)$ is close to 0, but $\rho(i<j)$ is close to $\rho(j<i)$, the set of rules is strongly conflicting.

$r(i<j)$ is computed by $\min(Ic\ ,\ \rho(i<j))$ with $r(i<j) = \min\ (P(i<j)\ ,\ 1 - P(i\ j))$

2.2.3 - Description of the rule base

In the present version of OPAL the decision support module has about 15 rules of thumb, which are rather generic and relevant for any scheduling problem. It can be extended by adding application-dedicated knowledge in each case-study. Present rules belong to 3 categories :

* Priority rules issued from the study of queuing systems
* Rules linked to the utilisation of auxiliary resources (other than machines)
* Rules pertaining to dynamic slack times of operations.

Exemples of rules are as follows :

SPT :	$O_i < O_j$ if the duration d_i significantly smaller than d_j
SST :	$O_i < O_j$ if the slack time $lft_i - est_i - d_i$ is significantly smaller than $lft_j - est_j - d_j$
EST :	$O_i < O_j$ if the earliest starting time est_i is significantly smaller than est_j
LFT :	$O_i < O_j$ if the latest finishing time lft_i is significantly smaller than lft_j
EDD :	$O_i < O_j$ if the due date of the job involving O_i is significantly smaller than the due date of the job involving O_j.
SRPT :	Let R_i be the sum of the processing times of operations occuring after i in the job involving i then $O_i < O_j$ if R_i is significantly smaller than R_j
RST :	$O_i < O_j$ if the sum of <u>updated</u> slack times of O_i and O_j when O_i precede O_j is significantly greater than the similar quantity when O_j precedes O_i
RTCO :	$O_i < O_j$ if the set-up time from O_i to O_j is significantly smaller than the set- up time from O_j to O_i.
FASC :	$O_i < O_j$ if lft_i-est_i is significantly smaller than lft_j-est_i and d_i+d_j is close to lft_i-est_j (this rule is similar to RST but more demanding).

Rules can be classified according to whether the result they produce changes or not as the schedule is being constructed (dynamic versus static), or whether they depend only upon operations, or on the ordering of operations. Namely :

SST, EST, LFT are dynamic rules involving operations alone
SPT, EDD, LRPT are static rules involving operations alone
RST, FASC are dynamic rules depending on the ordering of { O_i , O_j }
RTCO is a static rule which depends on the ordering of { O_i , O_j }

Rules are not always applicable. The if-part of a rule precises the subset of conflicts to which the rule applies. For instance the RTCO rule is meaningless if there are other operations located between O_i and O_j. The above rules are given as examples, they can be easily modified, deleted and other rules can be added if needed.

2.3 - The supervisor

The supervisor module plays the role of an inference engine and guides the search process by alternatively calling the CBA and the DS modules. Everything behaves as if 2 sets of rules were available and triggered in turn ; the CBA module has the higher priority however. The schedule is gradually built by adding precedence constraints between operations. The search graph is developed as follows : each time the CBA module stops, a new node is generated and the current state of the schedule is stored, including pending decisions. The supervisor then triggers the DS module which selects a precedence constraint (branching) ; this selection re-activates the CBA module which checks for forced decisions. Back-tracking occurs if the explored path leads to a failure state.
The search process is globally of the depth-first type, because a solution in progress is not given up unless proved unfeasible. However, it is locally best-first since the DS module always solves the best conflict. It is a "greedy" algorithm with backtrack. Because the number of pending precedence relationships is finite, the procedure finds a feasible solution, if any, after a finite number of steps. This way of controlling the search process enables both types of knowledge to cooperate : the CBA module helps pruning sterile nodes, and the DS module takes into account application-oriented objectives and guides the search towards realistic schedules. The overall strategy of solving problems presented to the system is outlined in the flow chart on figure 4.

When no feasible schedule exists, the data must be modified in order to recover feasibility. This can be done by relaxing suitable constraints as done in the ISIS system [16] :

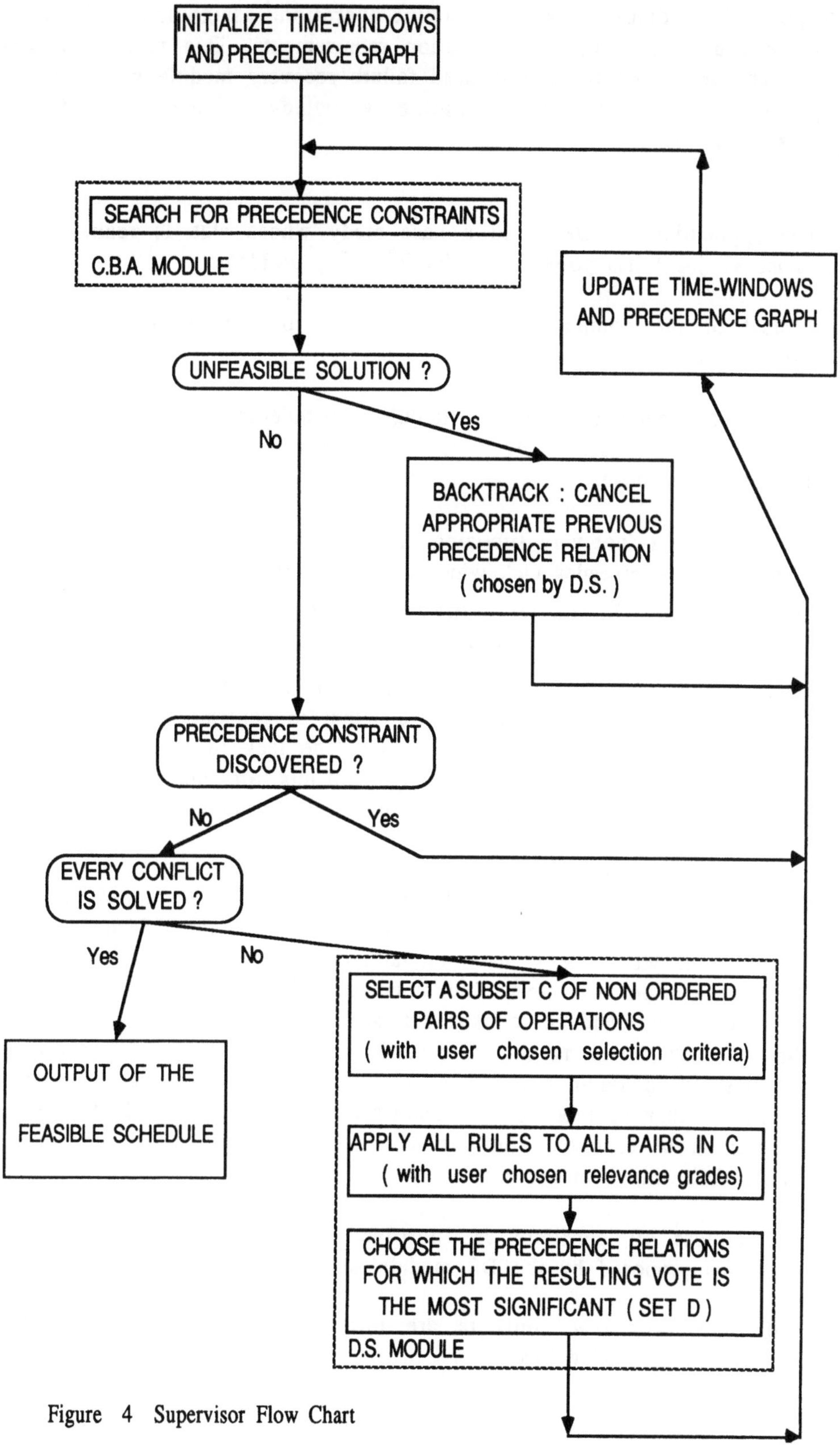

Figure 4 Supervisor Flow Chart

postpone due dates, change machine-operation assignments, improve machine availability, modify production requirements. This relaxation step should be carried out by a rule-based failure recovery module which is not implemented yet. The corresponding knowledge depends upon the application at hand.

The application of OPAL to a case study whose size is realistic (23 machines 47 jobs) has pointed out the following problems :

- most of the computing time is devoted to constraints propagation (80 %),

- decisions made by the DS module (precedence orderings) are too elementary to help speed up the solution process, when the due-dates are loose.

The last point is not very suprizing. Indeed, a decision of the DS module is of the form "O_i precedes O_j", which does not forbid other operations to be located between O_i and O_j ultimately. When there are hundreds of unsolved conflicts, deciding "O_i precedes O_j" may have only a very limited influence on the construction of the final schedule. Especially, the time window updating it entails will be unsufficient to generate new precedence constraints by means of constraint analysis. The lack of significance of a single precedence decision when many conflicts are unsolved leads to useless constraint analysis steps which are time-consuming.

Hence we have felt the need to enhance the decision power of the DS module when the problem is underconstrained, so as to speed up the construction of a feasible schedule. In its present state, the DS module has two working modes : one is the minimal mode where only one precedence constraint is chosen at a time. The other mode is called the normal mode where several precedence constraints are simultaneously chosen. The number of such decisions should depend upon the amount of constraints acting on the schedule in progress, at a given point of the search. We have linked the number of precedence constraints selected by the DS module to the proportion of unsolved conflicts by the CBA module. The idea is to model the following heuristic rule :

* if almost all conflicts are unsolved, the set D of precedence constraints proposed by the DS module should be maximum

* if only very few conflicts are left unsolved, then the minimal mode should be applied ($|D| = 1$)..

To characterize the search speed rate, we use a parameter, denoted $\alpha \in [0,1]$ representing the maximal proportion of conflicts which the DS module is allowed to solve in a single step. If there are NBC_i unsolved conflicts at node i of the search graph, no more than $\alpha.NBC_i$ precedence relations can be selected by the DS module. An empirically reasonable value of α is 0.2. If α is too large, the decision may become unfeasible, and creates backtracking. If α is too low, the CBA module is not efficiently used, and the duration of the search becomes prohibitive.

Note that NBC_i reflects the current number of unsolved conflicts during the search process. Indeed an underconstrained sheduling problem may become very much constrained if the decisions of the DS module entail the discovery of many precedence constraints by the CAB module. The size of the set of decisions made by the DS module is empirically calculated by a linear interpolation and truncated from below, to always allow for the minimal mode as a default option, i.e., it is :

$$|D| = \max(1, E(\alpha NBC_i))$$

where E extracts the integer part of a real number.

The DS module produces at most |D| precedence constraints. Indeed, the set of chosen precedence constraint must be consistent. A consistency check is carried out by looking for possible cycles in the precedence graph of the schedule augmented by the precedence constraints in D. For each detected cycle, the least preferred precedence relation is discarded from D, until no cycle is left.

When backtracking occurs in the normal mode, it is difficult to understand what is the cause of the failure. As a consequence the first step after backtracking is always taken in the minimal mode.

This heuristic technique which relates the behavior of the DS module to the results of the CBA module enables computing times to be significantly reduced, which is crucial for the use of OPAL in industrial environments.

As an example, table1 compares results obtained on a 23 machines, 47 batches example involving 93 elementary operations to be scheduled. The rule base is kept invariant, as well as the conflict selection criteria, only α is varying.

α (per cent)	N° of nodes in the search graph	Computer time (sec) (TEXAS EXPLORER)
0 = minimal mode)	80	1543
10	34	587
15	30	522
20	24	442
25	20	451
30	18	523

Table 1 : execution time

3. USING OPAL

This section addresses the aspects of OPAL which directly concern the users, namely what types of problems can be solved by OPAL in a workshop, how does the user interact with the system. A toy illustrative example is given lastly.

3.1 The role of OPAL

The OPAL system aims at constructing a predictive schedule on a given time horizon for which production requirements are known. It is supposed to be periodically run, the period being smaller than the time horizon. The production schedule is then passed over to the real-time level which tries to follow it (see figure 5).

Note that slack times are calculated by OPAL, and this information enables the real-time level to cope with small perturbations causing delays. When significant perturbations occur, which may lead to a violation of the due-dates, the OPAL system is run again , on the basis of the current workshop state, and possibly new orders.

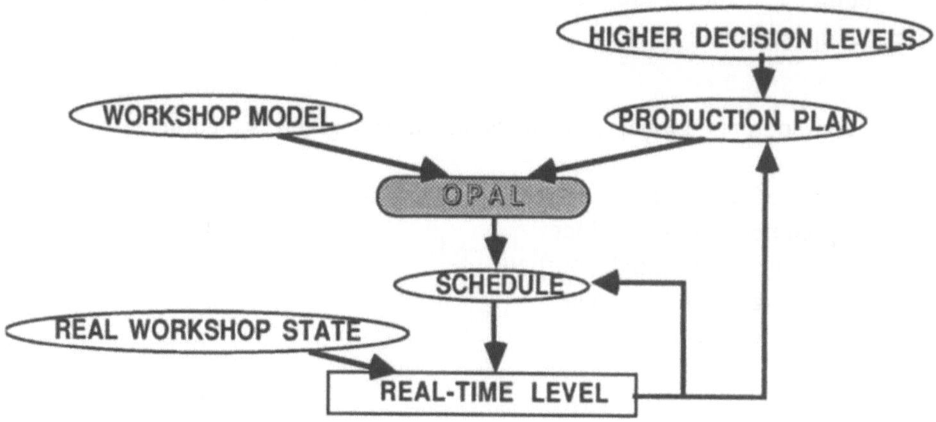

Figure 5 : OPAL in the global scheduling problem

3.2 The user interface module

3.2.1 - Input data

OPAL requires 3 input files to describe the workshop and the production requirements :

- one file to describe the resources of the workshop (essentially the objects belonging to the class 'machine')

- another for the processing sequences (objects belonging to the classes 'piece' and 'phase'). In the present state of OPAL, this includes the operation/machine assignment.

- the third file contains data about the production requirements, i.e. for each piece the number of batches to be processed and for each batch its size in number of elementary pieces, its due date , its starting time.

Given this information OPAL performs some basic computations :

* first create and complete the data base it will work on (create the objects 'job' and 'operation', links between objects)

* verify that prescribed production requirements are feasible
  - the workload of a resource does not exceed its availability
  - the time horizons for jobs are not too short

* build a window (earliest starting time, latest finishing time) for each operation.

OPAL then outputs global statements about the given situation :

- number of machines
- total number of operations
- total number of initial conflicts
- job with greatest due date
- most loaded resource...

3.2.2 The input of expert knowledge

Before the beginning of the search, the user is asked to choose among the rules, the criteria to be used in the DS module, the working mode used for the supervisor . There are 5 conflict preselection criteria and 15 rules of advice, in the present system.The criteria can be aggregated by means of logical "and" and "or" connectives, internally interpreted as fuzzy set-theoretic operators. The assignment of relevance weights to the rules is done in linguistic terms, chosen in a 5-element, ordered term set ("irrelevant", "weak", "medium", "strong", and "totally relevant"), mapped into the unit interval. All these options are selected using mouse sensitive menus, which are user-friendly. This possibility makes the system quite versatile, in the sense that the user may significantly act on the solving strategy and the properties of the obtained schedules . It is also possible to add new rules , or new conflict preselection criteria; however there is no easy interface mechanism to do so in the present state of OPAL.

During the search OPAL displays information about the state and speed of resolution of the problem, under the form of a Gantt diagram and a plotting of the number of solved conflicts against the number of CBA calls respectively, but does not allow the user to interfere.

3.2.3 - Outputs of OPAL

When a feasible schedule is found, OPAL

* displays information about the resolution :

- total CPU time
- number of generated nodes
- number of performed backtracks
- depth of the solution in the graph

* reminds the expertise used :

- rules
- criterions
- and values of α .

* gives global information about the solution, its flexibility, the time required for set-up. One can visualize the schedule by displaying :

- a Gantt chart for each resource, including slack times and time windows for each operation.
- the graph of precedence between tasks

All this information can be saved in a file and then be printed. The user can modify if needed the solution found by OPAL, by pairwise exchange of operations on the precedence graph. The system automatically recomputes the new windows for all operations.

3.2.4 - Illustrative example

Consider a simple three machine - two job problem without time table constraint. Processing sequences are identical (it is a flow-shop problem) and the processing times are given in table 2 . Jobs are called x and y and x_i (resp y_i) is operation i on job x (resp y).

Machines \ Jobs	1	2
M_1	4	4
M_2	3	5
M_3	3	4

Table 2 : Processing times

The problem is to do both jobs on a time horizon of 17 units of time.

Figure 6 represents a part of the corresponding data-base where one can see that parts, machines and rules are represented by structured objects . "Time-table1" refers to an object which contains the time intervals when the machine is working .

EXAMPLE OF MACHINE

```
(setq M1
      (make-machine
            :name "M1"
            :comment "drill number 1"
            :time-table  Timetable1))
```

EXAMPLE OF PIECE

```
(setq P1
      (make-piece
            :name "X"
            :operation-sequence
            (list
            (make-phase
                  :name "x1"
                  :alternatives
                  (list  (make-option
                              :machine M1
                              :unit-operation-time  4
                              :set-up-time  0)))
            (make-phase
                  :name "x2"
                  :alternatives
                  (list  (make-option
                              :machine M2
                              :unit-operation-time  3
                              :set-up-time  0)))
            (make-phase
                  :name "x3"
                  :alternatives
                  (list  (make-option
                              :machine M3
                              :unit-operation-time  3
                              :set-up-time  0)))))
```

EXAMPLE OF CRITERION

```
(setq C1  (make-criterion
            :name "C1"
            :comment "chronological order"
            :definition
            '(lambda (c) (- 1 (/ (date c) end))))
```

EXAMPLE OF RULE

```
(setq RST (make-rule
                    :name "RST"
                    :comment " less slack time reduction"
                    :if-part
                    '(lambda (c) (not (job (resource c))))
                    :then-part
                    '(lambda (c)
                            (exp-3-pref
                               (sigma-slack (2-op c) (1-op c))
                               (sigma-slack (1-op c) (2-op c))))))
```

Figure 6 : Example of objects in the data-base

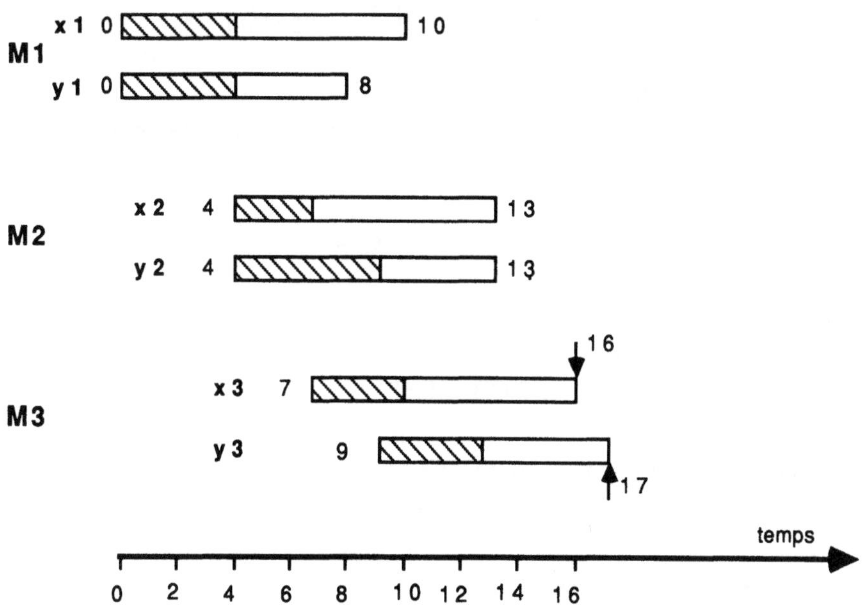

FIGURE 7

The time-window allocated to each operation appears on figure 7. It is calculated on the basis of earliest starting dates $e_x = e_y = 0$ and due dates $f_x = 16$ and $f_y = 17$. It can be checked that there are three conflicts $(\{x_i,y_i)(i = 1,3)\}$ one per machine, and that none of them can be solved by comparing the time windows. For instance :

$est_{x1} = est_{y1}=0$ $\quad lft_{x1} = 8 \quad lft_{y1}=10$ and

$d_{x1} + d_{y1} < \min(lft_{x1} - est_{y1}, lft_{y1} - est_{x1}) = 8$ ....

The decision support module must then decide which conflict to solve.

Conflict selection rule : Choose the conflicts in chronological order

Ic1 = 1 - 0/17 = 1
Ic2 = 1 - 4/17 = 0.35
Ic3 = 1 - 7/17 = 0.18

Rule base : The main objective is: keep as much slack time as possible. We thus consider RST (π = 0.9) which takes care of this objective and SST (π =0.5) which tries to rank first the operations with less current slack time.

* Ballots of rules about $C_1 = (x_1, y_1, M_1)$

rule SST : importance 0.5

$\mu_{SST}(x_1 < y_1) = 0$, $\mu_{SST}(x_1 \sim y_1) = 0.32$, $\mu_{SST}(y_1 < x_1) = 0.68$

rule RST: importance 0.9

$\mu_{RST}(x_1 < y_1) = 0$, $\mu_{RST}(x_1 \sim y_1) = 0$, $\mu_{RST}(y_1 < x_1) = 1$

* Ballots aggregation

$P(x_1 < y_1) = 0$ $P(x_1 \sim y_1) = 0.32$ $P(y_1 < x_1) = 1$

Hence the ratings

$r(x_1 < y_1) = \min(1, 0, 0.68) = 0$

$r(y_1 < x_1) = \min(1, 1, 0.68) = 0.68$

The same computation applied to $C_2 = (x_2, y_2, M_2)$ leads to

$r(x_2 < y_2) = 0$ $r(y_2 < x_2) = 0.35$

To C3 = (x_3, y_3, M_3) it leads to

$r(x_3 < y_3) = 0$ $r(y_3 < x_3) = 0.18$

The highest rating is thus for $y_1 < x_1$. This decision modifies time windows as indicated on figure 8.

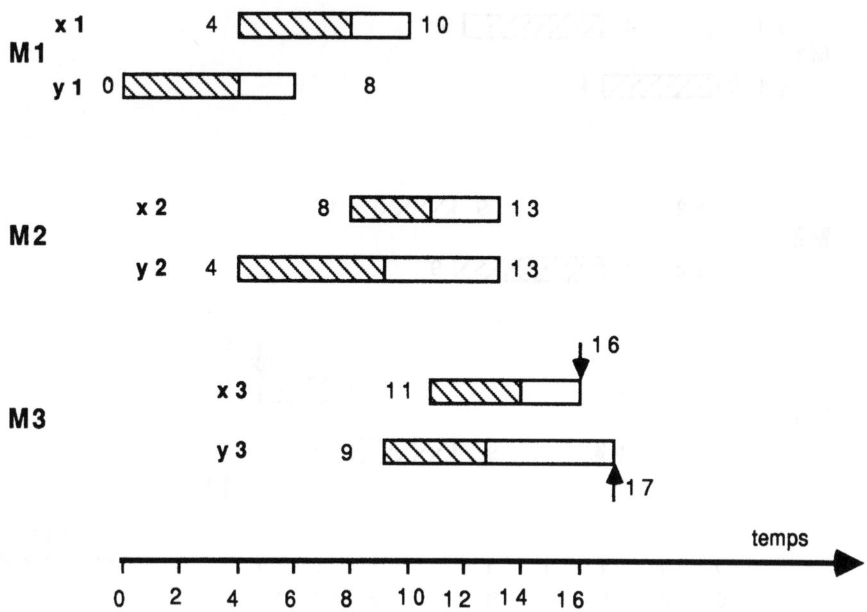

FIGURE 8

But noticing that

$$lft_{x2} - est_{y2} = 9 > d_{x1} + d_{x2} = 8 > lft_{y2} - est_{x2}$$

The constraint analysis module can find that $y_2 < x_2$ and then after the modifications of the time windows find that $y_3 < x_3$. Hence the final schedule (figure 9) where slack times appear and time windows have been updated again.

As this problem is underconstrained we can obtain another schedule if we use other rules .For example with the conflict selection criterion : "most-critical-product" and the voting rule : "RST" we obtain the following schedule on figure 10 which differs from the preceding one :

- the operation sequences are different
- the slack-times are greater

This shows the versatility of OPAL, i.e. its ability to build various solutions following different criteria .

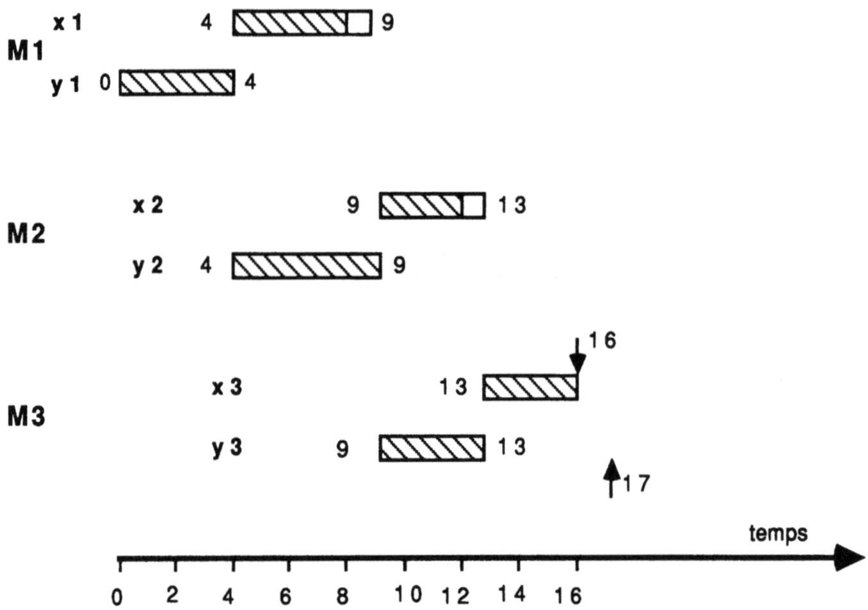

FIGURE 9 : first solution

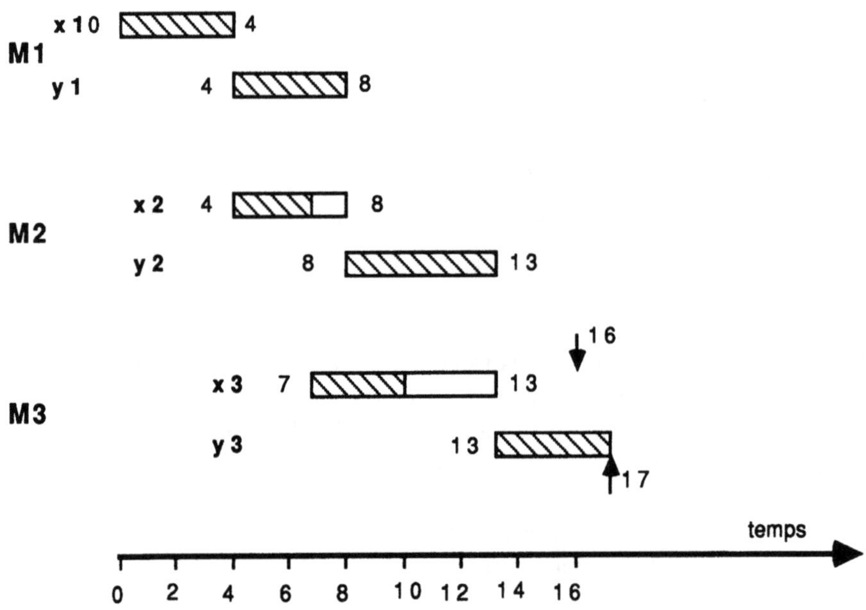

Figure 10 : second solution

4 - FURTHER STAGES IN THE DEVELOPMENT OF OPAL

The system, in its present version, can be improved in several respects. Several research lines are surveyed in the following.

4.1 - Improvements of the DS module

In its present version the DS module only stores pieces of advice which may not be imperatively followed, contrary to the GARI system[8]. One may think of a diversification of the types of rules by allowing the formulation of constraints which must be simultaneously respected. These ones would be dealt with separately from the pieces of advice whose present mode of aggregation expresses the possibility of making trade-offs. The format of the rules may be generalized to allow for a more flexible way of expressing pieces of advice. See Dubois[9] for a formal proposal along these lines.Lastly one may think of producing consistent macroscopic decisions (i.e. where consistency checking is useless)) i.e. i immediately precedes j, or choosing the absolute location of an operation in the sequence associated with a machine.

4.2 - Improvements of the CBA module

When the OPAL system finally stops in a failure state, it means that no feasible schedule exists which does not violate the prescribed due dates. Hence these temporal constraints have priority over the pieces of advice in the DS module. This is not always acceptable in real situations. Especially two kinds of failures are worth distinguishing : a limited violation of due dates versus a strong violation. The total failure diagnosis produced by the system in case of limited violation is over-rated with respect to the way due-dates are actually specified ! In practice such a specification is quite rough, and there is implicit flexibility in the due-dates, which OPAL does not account for. Such pieces of information could be used at the level an error recovery system which would be called upon in case of complete failure. However it is more satisfactory to use the due-date flexibility within the procedure which builds a schedule, in order to avoid total failure diagnoses whenever a small admitted violation of delays is enough to solve the case. The error analysis would be triggered only in case of strong violation of due dates. A mathematical setting for fuzzy constraint-based analysis is proposed in Dubois [9], including an algorithm.

4.3 - Guide-lines for an error recovery system

A strong violation of due dates implies a new statement of the scheduling problem by modifying the initial set of data. This situation may occur either when a graph search has been uselessly scanned or even when preliminary feasibility checks on machine utilization fail. In both case, the problem is to relax constraints. What are the degrees of freedom ?

a) - modify the production requirements by discarding one of the job considered as not too urgent (e.g. the one with the farthest imperative due dates)

b) - modify the production process by resorting to alternative processing sequences which influence the machine workload balance. It amounts to a new assigment of operations to machines.

c) - modify the resource availability : it may involve different types of actions : change the time table of machines, their speed if relevant (e.g. by changing the assigment of human operators), add new machines.

For instance, b) is interesting if the workloads of the machines are not balanced at all. However, in case of dedicated machines, the operation assignment is not always easy to modify. One may think of a set of advices which either act on a set of data in order to recover a first sight feasibility (to be eventually verified by producing a schedule), or acts on a unfeasible schedule in order to make it feasible (by looking at what happens when a job is deleted, or when time tables are modified - the reassignment of operations to other machines requires some sequencing steps while other modifications can be done while keeping the same sequencing).

5 - CONCLUSION

OPAL is a short term scheduling system which makes various types of knowledge cooperate towards the construction of realistic production plans at the level of elementary operations. The system guides the search in order to avoid unfeasible schedules but also to provide production plans which a shop-floor manager is likely to accept. The use of AI programming techniques helps getting a highly flexible software which can adapt to real problems. One may think that this type of knowledge base system, which is not a regular expert system, but integrates well-structured as well as ill -structured knowledge, will become very common in all fields where some theoretically founded methods already exist. The use of the AI approach must not lead to reject traditional methods which already gave good results, although being unsufficient to completely solve a real problem. AI enables the amount of useful knowledge for automated or semi-automated solving to be augmented, where pieces of ill-structured but important information were neglected so far. In that sense Artificial Intelligence does not make traditional O.R methodologies obsolete, but it may bring them closer to real problems. This is the basic idea behind the OPAL system.

The authors are grateful to the two anonymous referees for their constructive suggestions that helped improve the presentation of the paper. More details about the OPAL system appear in the first author's Ph.D. dissertation [30] (in French).

REFERENCES

[1] BEL G., BENSANA E., DUBOIS D.
Un système d'ordonnancement prévisionnel d'atelier utilisant des connaissances théoriques et pratiques
Actes 6ème Journées Internationales - Les Systèmes Experts et leurs applications, Avignon, Mai 1986, pp 757-770.

[2] BENSANA E., CORREGE M., BEL G., DUBOIS D.
An expert system approach to industrial job shop scheduling
Proc. IEEE International Conference on Robotics and Automation, San Francisco, April 1986, pp.1645-1650

[3] BLACKSTONE J.H., PHILIPPS D.T., HOGG G.L.
A state-of the-art survey of dispatching rules for manufacturing job-shop operation
International Journal of Production Research, 1982 , vol 20, pp 27-45

[4] BULLER W.I., NOF S.Y, WHINSTON A.B.
Artificial Intelligence in manufacturing planning and control
AIEE Transactions, 12, 351-363, 1980

[5] CANALS D.
Ordonnancement d'atelier par simulation :
étude des règles de priorité et aide au lancement
Thèse de Docteur Ingénieur, ENSAE, Juillet 1986

[6] BELLMAN R. ,ESOGBUE A.O., NABESHIMA I.
Mathematical aspects of scheduling and applications
Pergamon Press, 1982

[7] DESCOTTE Y., LATOMBE J.C.
GARI : a problem solver that plans how to machine mechanical parts
7th International Joint Conference on Artificial Intelligence
Vancouver,1981.

[8] DESCOTTE Y., DELESALLE M.
Une architecture de système expert pour la planification d'activité
6ème Journées Internationales - Les Systèmes Experts et leurs applications, Avignon 1986,903-906.

[9] DUBOIS, D.
Fuzzy knowledge in an artificial intelligence system for job-shop scheduling.
to appear in "Fuzzy Methodologies for Industrial and Systems Engineering"
(G.W. Evans, W.Karwowski,M.R. Wilhelm, Eds) North Holland, 1988

[10] DUBOIS D., BEL G., FARRENY H., PRADE H.
Toward the use of fuzzy rule-based systems in the monitoring of manufacturing systems
In : "Software for Discrete Manufacturing"
(J.P.Crestin, ED.) North Holland Pub.Co., 1986, 525-535

[11] DUBOIS D., PRADE H.
Fuzzy sets and systems : theory and applications
Academic Press, New York, 1980

[12] DUBOIS D., PRADE H.
The advantages of fuzzy approach in OR/MS demonstrated on two examples of resource allocation problems
In : Progress in Cybernetics and Systems Research,
vol.8, Hemisphere Pub.,New York, 1982, pp.491-497

[13] ERSCHLER J.
Analyse sous contrainte et aide à la décision pour certains problèmes d'ordonnancement
Thèse d'Etat, LAAS, Toulouse 1976

[14] ERSCHLER J., ESQUIROL P.
Decision-aid in job shop scheduling : a knowledge based approach
IEEE International Conference on Robotics and Automation
San Francisco, April 1986, pp.1651-1656

[15] ERSCHLER J. , ROUBELLAT F., VERNHES J.P.
Finding some essential characteristics of the feasible solutions for a scheduling problem
Operations Research, 24, n°4, 1976, pp.774-782

[16] FOX M.S., ALLEN B., STROHM G.
Job-shop scheduling : an investigation in constraint-directed reasoning
Proc. 2d American Artificial Intelligence Conference,
Pittsburg, 1982, 155-158.

[17] GRANT T.J.
Lessons for OR from AI : a scheduling case study
Journal of the Operational Research Soc., 37, 41-57, 1986

[18] HOLLINGER D., BEL G.
An object-oriented approach for CIM systems specification and simulation
CAPE86 2d International Conference on Computer Applications in Production and Engineering - Copenhangen, May 1986

[19] KEMPF K.G.
Manufacturing and artificial intelligence
Robotics 1, 1985, 13-25.

[20] LEE Y.C., FU K.S.
A relational approach to the integrated data-base management system for computer-aided Manufacturing
In "Automated Manufacturing" (L.B.Gardener, Ed.), American Society for Test. and Mat., Philadelphia, 1985, 150-152.

[21] LE PAPE C., SAUVE B.
SOJA : un système d'ordonnancement journalier d'atelier
5ème Journées Internationales "Les systèmes experts et leurs applications" Avignon, 1985, 849-867

[22] PARRELLO B.D., BARRAT W.C., WOS L.
Job-shop scheduling using Automated Reasoning :
a case study of the Car-Sequencing Problem
Journal of Automated Reasoning 1986 Vol 2 , n1 , pp1-43

[23] PROTH J.M. et al
Utilisation de l'intelligence artificielle dans un problème d'ordonnancement
Actes Congrès AFCET Automatique, Besançon 1983, pp 53-61.

[24] RAYSON P.T.
A review of expert systems principles and their role in manufacturing systems
Robotica 3, 279-287, 1985

[25] SAHAROUI A., ATABAKHCHE H., COURVOISIER M., VALETTE R.
Joining Petri nets and knowledge based systems for monitoring purpose
Proc. IEEE Int. Conf. on Robotics and Automation, 1987

[26] STEFFEN M.S.
A survey of AI based scheduling systems
Fall Industrial Engineering Conference, Boston, Dec. 1986

[27] SMITH S.F., FOX M.S., PENG SI OW
Constructing and maintaining detailled production plans : investigations into the development of knowledge based factory scheduling systems
AI Magazine, Fall 1986, pp 45-61

[28] ZADEH L.A
A computational approach to fuzzy quantifiers in natural languages
Computers & Math. with Appl., 9, 149-184, 1983.

[30] BENSANA E.
Artificial Intelligence techniques for job shop scheduling
Thèse de doctorat de l'ENSAE, november 1987

PART 4: MODELS TOWARD INTEGRATION

DESIGNING THE LAYOUT AND THE CONTROL SYSTEM OF FMS

Prof. Dipl.-Ing. A. Reinhardt
University of Kassel (GhK) - FG Produktionssysteme
Monchebergstr. 7 D-3500 Kassel

ABSTRACT

An epistemological concept for the development of technical designing tools was developed in the early seventies and implemented in a simulator. It is based on an object-subject model that originate from sensory and intellectual perception and action cycle. Models depict a picture of the plant. They describe distinctive features and states of an installation. The simulator is a tool for experimental studies in the design of future factories. Simulation experiments could be used as a test bed of integration.

NATO ASI Series, Vol. F49
Computer Integrated Manufacturing
Edited by I. Burhan Türksen

1 Production on Schedule

Being capable of a flexible and prompt reaction to market demands at reduced costs is an essential aim of all thoughts concerning production. Product development and manufacturing on schedule, just-in-time during the whole process, is a basic requirement for achieving this aim /1/.

The development of Flexible Manufacturing Systems (FMS) in the sixties was one approach to a solution of the task. A flexible manufacturing system was then a relatively clear system of interlocked individual machines. Today, a whole factory must be considered as an FMS, where production facilities are connected via data and material flow.

It is a common fact today in the electronical engeneering and the automotive industry that the development of products and plants must go together in order to place a product on the market quickly and ahead of the competitors.

Methods and tools of the CAD/CAM-line assist the development of a product. In this field, the companies have reached different levels. Some have already been working with CAD for a few years and try to include CAM. Integration of these methods is a task of immediate interest. Progress is considerably delayed by a lack of experts.

At present however, the development of factories is still carried on conventionally with few computer aided tools. In this field, simulation as a computer aided method is getting more and more important.

2 Aspects of conventional Factory Design

Factory design is presently carried on by individual teams that work on parts of the development and not on the process as a whole. Simply speaking, each team develops its concept of the factory without bearing in mind the entire model.

The designers of a plant draw the layout on a drawing board thus creating a static model. The dynamics and accordingly the performance of a plant can hardly be calculated with conventional methods. Different performance specifications are written in informal description languages for simulation models, for realtime software, and for the development of the electronics and the mechanics of the plant.

Simulation orders are mainly placed with universities, because here the few existing experts are found. Models are based on performance specifications and conferences. Each task leads to a new simulator. Experiment results are transmitted to the plant designer and the designer for the control software as standards for the performance. The validity of these models, and accordingly, the results are often not reliable.This is why some software developers do not accept these standards. Today, animation is often considered a supplement for the management and not the most important instrument of validation. The method and technics of simulation need strong impulses. There are only few universities that go deeply into research and teaching in this area.

Software for realtime control is developed and implemented according to the standards of performance specifications and many discussions about the functions of plant components. The developer of the realtime software is not aware that his software is a detailed model of the plant and that the simulation model should at last contain the same algorithms and data structures, if both teams talk about the same installation.

The "model" of a plant constructor is the real installation, the mechanics and controlling system on the level of signals with programmable logic controller (PLC). The constructor develops and implements the mechanics of the installation or of parts of it on the basis of the performance specification, the technical description, the layout drawing and many discussions with his client. The entire installation cannot be tested, until the realtime software is available.

A test of the installation, the mechanics and the control software takes place on initiation. At this stage, 90 % of the costs have already arisen. In endless work during days, nights and week-ends, bugs are being eliminated. The installation is going to work, because it must work. The simulation model is no longer valid, because the installation has been changed at will. Nobody knows, whether it has ever been valid. The high costs for the whole process of planning and development can only be reduced by new computer aided methods with homogenous and transparent data bases in all steps of the development.

3 Integrated Methods and Models

3.1 Software Islands

There is hardly a company to be found that does not employ computer aided methods in its management, its technical development, its process control, etc. What is missing is the integration of all these computer aided methods. We need the integration of all tools that support job processing in the organizational field and the technical development, as well as in manufacture and assembly.

The leaflets and presentations of hardware and software suppliers boost such concepts of computer integrated manufacturing (CIM). Although computers, respectively display units, are located closely together, the methods presented there are not at all connected or only to a small degree. All computer suppliers offer today their own networks which work reliably now. The development of communication protocols, such as TOP (Technical Office Protocol) and MAP (Manufacturing Automation Protocol), give hope for computer independant networks, that is, stable networks with distributed data bases as a technical fundament for the integration of different methods /2/.

Each program contains a model or a view of the working area for which it has been designed. Looking at the implementation languages for single methods, we note that in the field of administration COBOL, in the field of technical development FORTRAN and for manufacturing control FORTRAN or Assembler are employed. Each software designer has his own point of view and his own language /3/. A fundamentally new approach is necessary.

3.2 An Integration Concept

Based on the discussion of fundamental philosophies in Berlin during the sixties, an epistemological concept for the development of technical designing tools was developed in the early seventies and implemented in the simulator SIMFLEX/2 .

The concept is based on the relation object-subject-model (fig.1).

In the object-subject-relationship, cognitions originate in the cycle of sensory perception, intellectual perception, and action. In this cycle, thought models are established or experience on the structure and function of a system is gained, which may be embodied in language and drawings as static models. These models are the basis of communication during the process of developing an installation. The application of formal systems leads to detailed and unequivocal models, which can be transferred into dynamic models by means of computers.

In the <u>subject-model-relationship</u> cognitions may be gained considering the model as a substitute for the system. Conventional computer models with numerical input-output-relationships have a limited relevance for the complex systems considered here, because they disregard sensory perception as an essential component of cognition. A model can only be applied efficiently for gaining knowledge, if graphical representation allows a minimum of sensory perception. Therefore, animation cannot be considered as a supplement, but as an essential element of the method. Furthermore, the process of gaining knowledge is characterized by interventions in the process (action) and observation of the reaction (sensory perception). Models must allow graphic-interactive experiments, in order to gain a major part of knowledge in advance that otherwise could only be obtained at considerable costs on initiation.

In the <u>model-object-relationship</u>, the computer model serves as a controlling system. Models are pictures of the object area, that is, a picture of the plant. They describe distinctive features and states of the installation. They contain the algorithms for process control decisions. Models have graphic-interactive qualities and, accordingly, essential qualities of a control station for the installation control. Models substitute the conventional control software of factories. Consequently, they may be connected via corresponding control electronics, and then we see, whether or not our model is valid for the installation considered.

An initial object-orientated implementation of the concept in SIMULA was presented in 1976. As SIMULA is not implemented on many computers, a prototype in PASCAL was designed with support from the industry, presented on the industrial fair at Hannover in 1982, and introduced for industrial planning tasks.

The implemented concept consists of units for the control and processing of modelled or real material flows, and of software tools for handling the units.

4 Composing Models of Units

A unit library in SIMFLEX/2 contains a set of predefined units for the control and processing of material flow by means of transportation and working operations. The set of units can be increased according to the task. There are basic types of units whose behaviour is programmed by data structures and algorithms in the logical modelling system (LMS). By specifying differentiating attributes of geometry and unit parameters, the user can create variant types and file them in the library.

In the field of material flow, there are concrete and abstract types of units. Concrete types deal with individual objects and show single operations of motion in mechanics. Their geometry is shown in two-dimensional graphics (fig.2). Transportation and processing orders can be assigned to them by abstract units. In this case, they report the result of an operation back to the orderer. Abstract types of units take on global control functions for individual concrete units or for unit pools. They deal with objects of the material flow as quantities, not as individuals.

4.1 Concrete Types of Units (fig.2)

Load and unload stations (unit no. 1-2) imitate the functions of sources and sinks for dynamic objects in the system.

Rolling conveyors (unit no. 3) assume the functions rolling and storing, respectively buffering of objects. With these functions, objects are filed and removed in sequential access.

Turn tables (unit no. 4) take objects from one to three incoming streams and put them in one to three outgoing streams.

Vehicles (unit no. 5-6) represent the functions of rail-borne vehicles which take over objects at stops, move on rails to goal stations, and hand over their objects.

The type "shelf" (unit no. 7) takes on storage functions for the construction of simple shelves or entire warehouses. Storage and removal is carried out by vehicles with lifting and lowering functions.

Machines or work stations (unit no. 8-10) imitate the functions of take-over and hand-over of objects in the material flow, as well as processing within a given time.

Pallet turn tables, respectively rotating buffers, (unit no. 16) imitate the function of object filing with random access. Empty places are put into the position for take-over and occupied places in the position of handing over by rotation of the buffer.

Automatic guided vehicles (unit no. 17) imitate the function of inductively controlled vehicles which search a route in a net of nodes and routes from their actual position to a given point carrying objects.

Nodes and routes (unit no. 11-31) are way elements for the development of transportation networks for inductively controlled vehicles.

4.2 Abstract Types of Units

Job managers create jobs as dynamic objects for system loads, make them available at the load stations, which schedule them into the system, and wait for ready-messages.

Route managers control parts of the transportation network. They find out routes for announced objects, clear them for transport and await ready-messages.

Vehicle managers hold pools of free-moving vehicles. They distribute drive orders to the individual vehicles.

Working managers give out single working orders to working stations in pools.

Warehouse managers control the storage and removal of objects in pools of shelves, respectively warehouses.

4.3 Tools

The user can specify variants of the basic types by geometrical, technical, and control parameters in dialogue and file them in a unit library. For this purpose, he can apply the Graphic Unit Designing Tool (GBS, fig.3). Design dialogues are controlled by the menu (fig.4).

The layout of an installation can be designed with a Graphic Topology Designing Tool (GTS). Individual units are called in from the library and are placed in a factory key plan, provided with specific unit parameters. This layout description is filed in a layout library (fig.3). A set of products is specified as a system load. Different layout variants (fig.5,6) may be created for a set of products in one step. Disturbance processes may be specified for individual units. Here too, the dialogues are controlled by a menu (fig.7).

The Logical Modelling Tool (LMS) imitates the dynamic process by means of a layout description from the library (fig.8). Movements and changes of state in the process are displayed by the Graphic Dialogue Tool (GDS) on a CAD-screen (fig.9). The results of the operational data acquisition are displayed by the statistic dialogue tool (SDS) on a second screen (fig.10).

Using the dialogue tools GDS and SDS, the user may observe and control the flow of the modelled process (fig.11). Either both dialogue modules together or each as a single module constitute the control station of the modelled system.

If the mechanical facilities are connected to the LMS by control electronics, the user is able to control the real process by the control station.

5 The Structure of a System Model

A modelled system consists of individual units, which have different tasks and communicate with each other in order to transport and process jobs on schedule. Logical units in the systems can have mere management and control jobs (abstract units), or they can imitate physical processes of transportation and working logically with their control logic reduced to a local range of view.

If mechanical model facilities are connected, physical processes are carried out by mechanical units.

The mechanical units are connected via electronic units (PLC) with the logical modelling system.

5.1 Features of Model Units

From the electromechanical level and the concrete level of modelling up to the abstract level of modelling, the following features may be distinguished (fig.12):

- mechanical behaviour
- electronical behaviour
- logical behaviour
- geometrical features
- historical behaviour
- identifying attributes
- communication protocol

5.2 Communication of Units

Each individual unit has relations with other units within the limits of its task. Units can be neighbours in the material flow, and they communicate with input and output units. A unit can be assigned to a control unit. A unit receives transport or working orders from and reports to the control unit.

A control unit is in contact with other control units and with the units of the tranport and working level assigned to it. A control unit, e.g. in charge of a pool of working stations, calls in orders from the order management, passes them on to individual working stations and sends transportation requests to the transport control. The latter fixes the route giving transportation orders to individual units of the material flow.

Each unit is known by its identifying attributes. It holds a list of requirements, where other units file their announcements. Its state is visible for all other units of the environment. A unit sends annonuncements to another unit and activates it, if it is idle, and takes on a defined state. An activated unit examines the list of announcements and following given strategy parameters decides its further activities on the basis of the environment situation. It schedules starting events and final events and takes on a defined state.

Operations of communication consist of announcing and activating. They are implemented according to the computer structure as procedure calls, task operations or network operations.

In the concept, the connection of the modelling system and a software system for planning and controlling production (PPS) is intended. The control unit "job manager" can receive orders from the PPS and the unit "operational data acquisition" reports ready-messages.

The software of the modelling system runs as one task or as a multitasksystem on a computer or as a multitasksystem on different nodes of a computer network. A unit can be integrated as a whole in one task and consequently in one computer, but it can also be distributed with its features among several tasks (fig.13).

6 Realtime Control of the Electromechanical Model

The concept of transition from a modelling system to realtime control is being implemented actually within the scope of laboratory experiments. First results have already been presented on industrial fairs. For these experiments, factories were designed as physical models. The individual units of an installation were controlled by miniaturized PLC. The PLC are interlinked among each other and the modelling system by a local area network (LAN) with RS-232-protocol.

6.1 Model Events and Process Signals

During the process of the model the dynamics of individual units are implemented by activities that are started by initial events and terminated by final events. The events are scheduled by event operations and managed by the internal calendar of events. The virtual time proceeds from event to event. The time between events can be projected on the real time. Thus the proceeding of the modelled process can run over a delay parameter either quickly, slowly or in realtime. The time of the model remains virtual.

Turning to realtime control the internal time mechanism is connected to the realtime clock of the computer system. Initial events are sent via realtime interface as control signals to the external process. Final events are managed in a list as signals expected from the external process. If the external process supplies the signal, it is returned with the real time to the internal time mechanism as final event. If the signal does not arrive within the time expected, a failure is reported to the control station.

6.2 Electronical Units

A miniaturized PLC is provided for each mechanical unit of the physical model installation subject to control. On initialization, PLC are adapted to internal logical types of units and represent them in the real world. Each PLC takes the signals bound for it from the LAN and transfers them into electrical signals which start drives in mechanical units. A mechanical unit indicates via signal transmitter the end of action to the PLC. The controller sets the drives into defined states, returns the final signal to the LMS and waits for new instructions. For a test of the external control systems, each PLC can imitate single activities by means of time-counters. In this case we talk about electronical simulation.

6.3 Electromechanical Units

The mechanical modelling system simulates all technical functions of the material flow that are necessary for the transport of objects. It consists of a construction set of prefabricated units like load stations, rolling conveyors, turn tables, machines, vehicles (fig.14), etc. Each unit includes electric drives for the conveyor system and for signal transmitters. The units are designed according to the requirements specified in the layout drawing (fig.15) and connected to the logical modelling system that now operates as realtime control via electronical control. Then the units are activated. The control station of the simulator is now control station for the real installation (fig.9,10).

7 State of Development and Aims

After 15 years of developing, the simulator SIMFLEX/2, the essential part of the concept, is employed as a mature tool in industry. It runs on CAD-computers with high performance like ND 500 in the computer network COSMOS, and as well on micro-computers with correspondingly less performance.

It remains important to see that the simulator is only one step on the way to a tool for factory design. Before contracting for the simulator a seminar for users is held. During the seminar the user gains detailed knowledge about the simulator and the method. He should realize first of all that using the simulator means a fundamental change in the work of developing installations.

The designer of installations, the developer of software, the constructor of an installation, and the owner are going to carry out experiments with the simulator as a basis for all of their decisions from the first step of planning until the installation is started. In this manner, simulation can become the basis for the integration of all methods of factory planning and controlling.

The security of planning is going to rise considerably, the time of planning, the cost of development, and especially the cost of software are going to reduce to the same extent as the cost of computer hardware is reducing today.

However, the main issue of the task is developing an entire model of the factory in the minds of the people involved, as well as developing know-how for the implementation of the entire model and all tools, and finally, convincing all persons concerned in a company to work with the tools with respect to the entire model.

8 Literature

/1/ Mortimer, John: Just-in-Time, IFS (Publications) Ltd, Springer Verlag, Berlin, 1986

/2/ Hollingum, Jack: Implementing an Information Strategy in Manufacture, IFS (Publications) Ltd, Springer Verlag, Berlin, 1987

/3/ Reinhardt, Adolf: Die Fabrik im Simulator, Vieweg-Verlag, Wiesbaden, 1987

9 Appendix: Figures

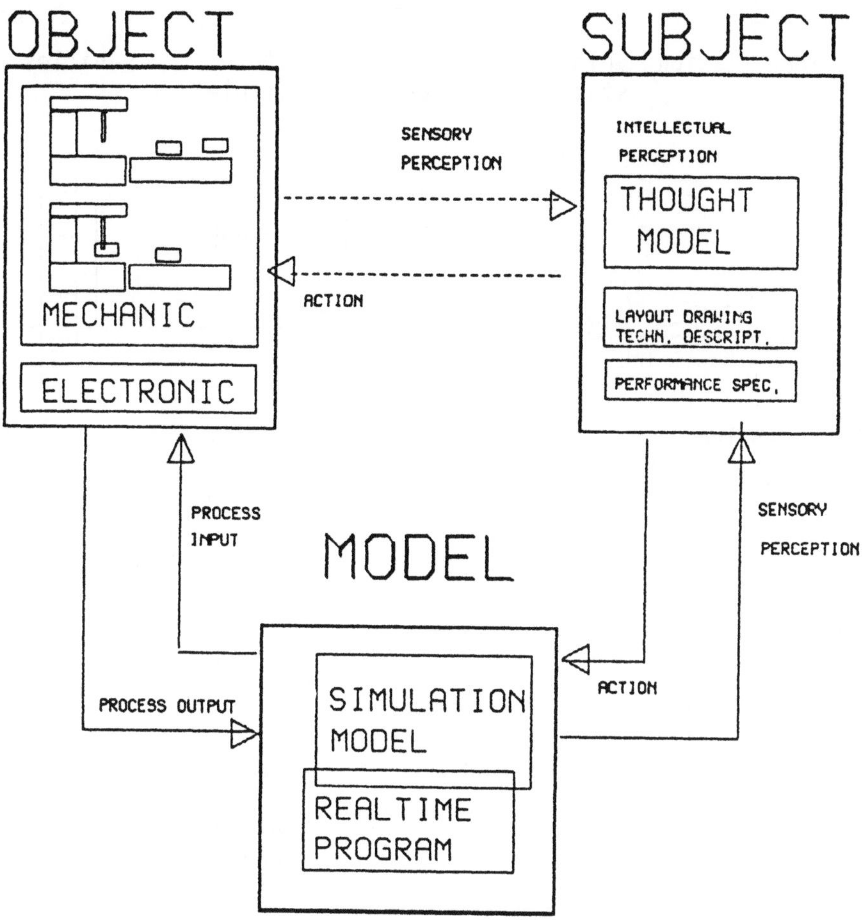

Fig. 1. The Relationship Object-Subject-Model

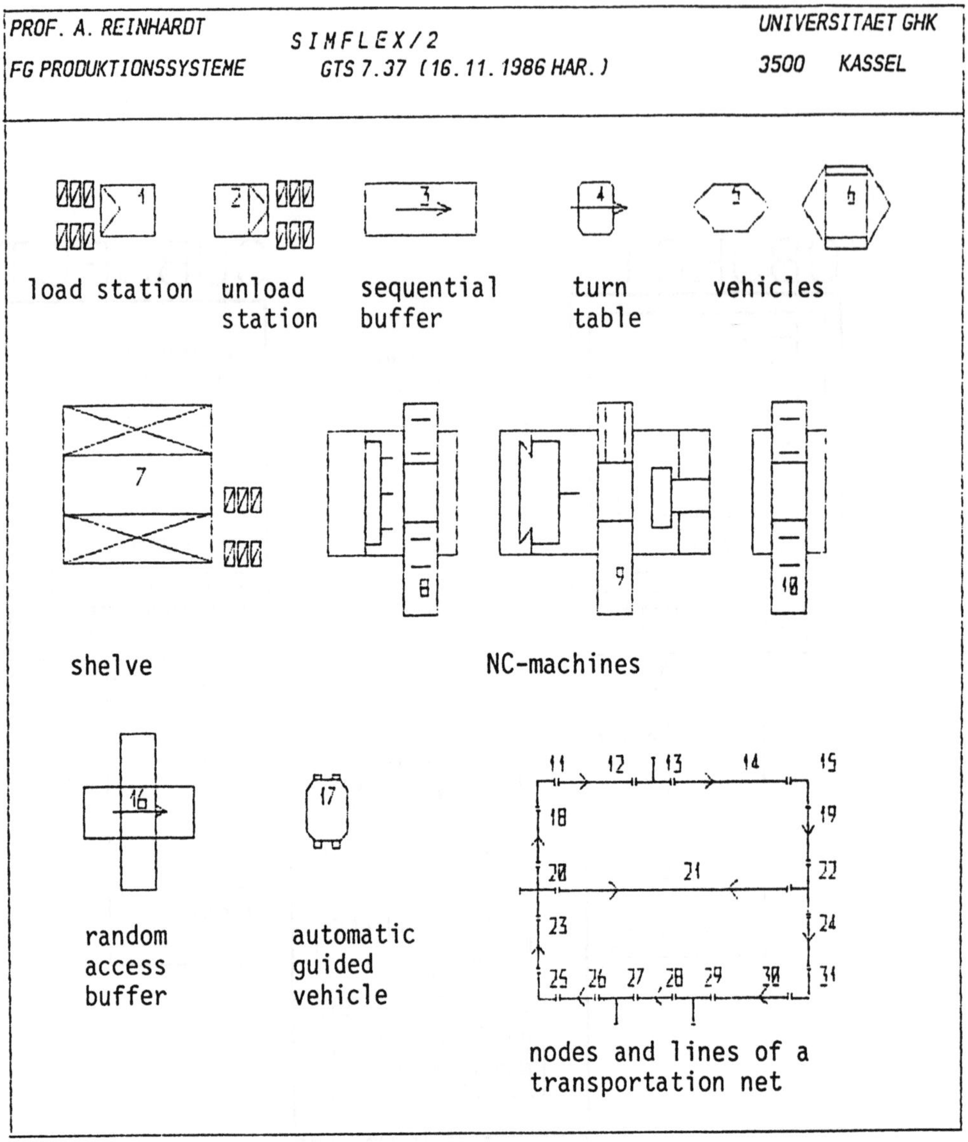

Fig. 2. Graphic Unit Library

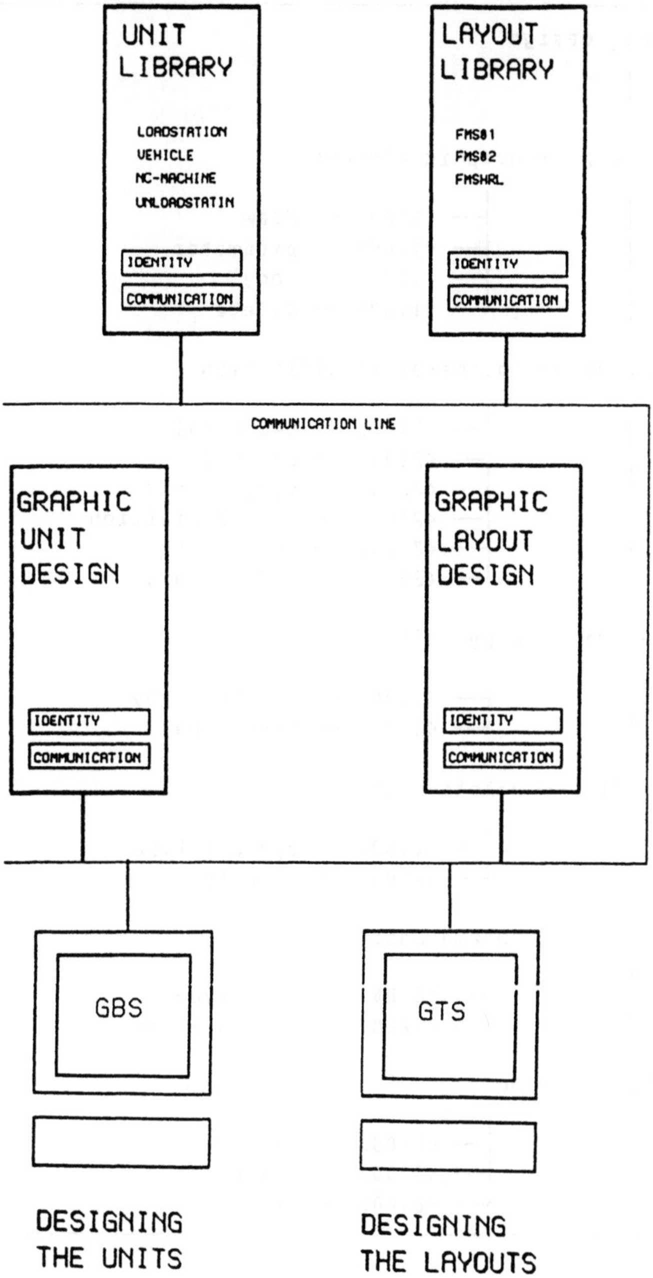

Fig. 3. Unit and Layout Design

```
<0000> unit-design
        |
       <1000 > — graphic-elements
        |         ├— <1100> — draw
        |         ├— <1200> — parameter
        |         ├— <1300> — copy
        |         └— <1400> — delete
       <2000> — parameter-specification
        |         ├— <2100> — technical
        |         ├— <2200> — control
        |         ├— <2300> — system-load
        |         ├— <2400> — data-aquisition
        |         ├— <2500> — unit-states
        |         └— <2600> — modify-text
       <3000> — copy-unit
        |         ├— <3100> — variant-type
        |         └— <3200> — base-type
       <4000> — delete-unit
        |         ├— <4100> — variant-type
        |         └— <4200> — base-type
       <5000> — rename-unit
        |         ├— <5100> — base-type
        |         └— <5200> — variant-type
       <6000> — mode
                  ├— <6100> — exit
                  ├— <6200> — break
                  └— <6300> — fix
```

Fig. 4. Dialogue for Unit Design <0000>

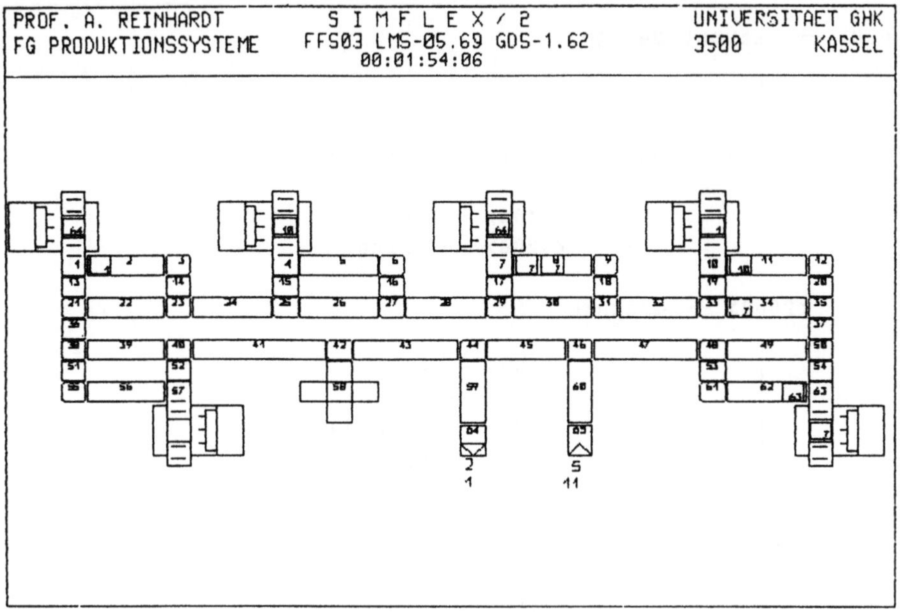

Fig. 5. Variant Layout FMS03

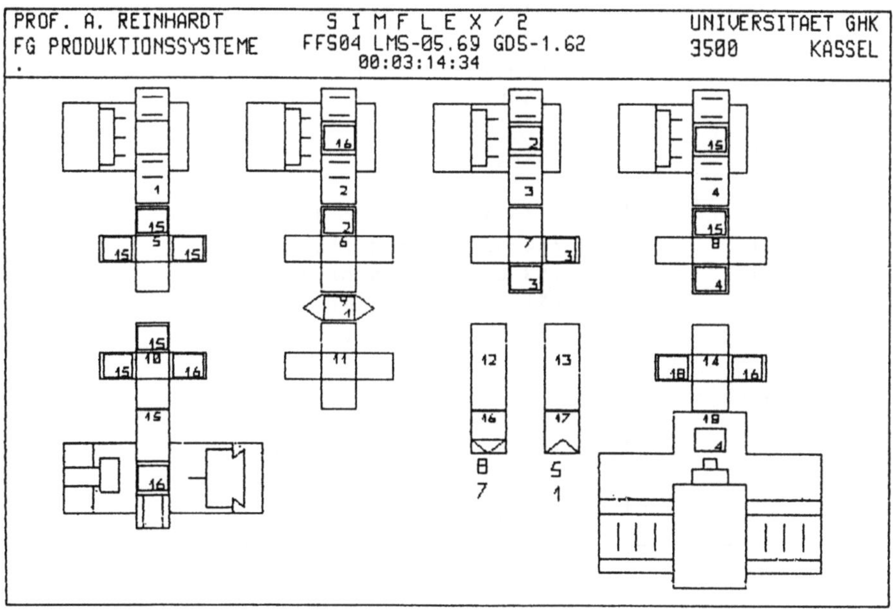

Fig. 6. Variant Layout FMS04

```
<0000> ── layout-design
          │
          <1000> ── experiment-specification
          │          ├── <1100> ── read
          │          └── <1200> ── write
          <2000> ── graphic
          │          ├── <2100> ── unit
          │          ├── <2200> ── layout
          │          ├── /2300/ ── renumber
          │          ├── /2500/ ── grid on/off
          │          └── <2600> ── window
          <3000> ── logic
          │          ├── <3100> ── connect
          │          ├── <3200> ── route
          │          ├── <3500> ── technical-parameter
          │          └── <3600> ── control-parameter
          <4000> ── order-description
          │          ├── <4100> ── info.-order
          │          ├── <4200> ── info.-sequence
          │          ├── <4300> ── modify
          │          └── /4500/ ── check
          <5000> ── exp.planning
          │          ├── <5100> ── system-load
          │          ├── <5200> ── disturbance
          │          ├── <5300> ── external-events
          │          ├── <5400> ── runtime
          │          └── <5500> ── data-aquisition
          <6000> ── mode
                     ├── /6100/ ── test on/off
                     ├── /6200/ ── reinitialisation
                     └── /6300/ ── exit
```

Fig. 7. Dialogue for Layout Design <0000>

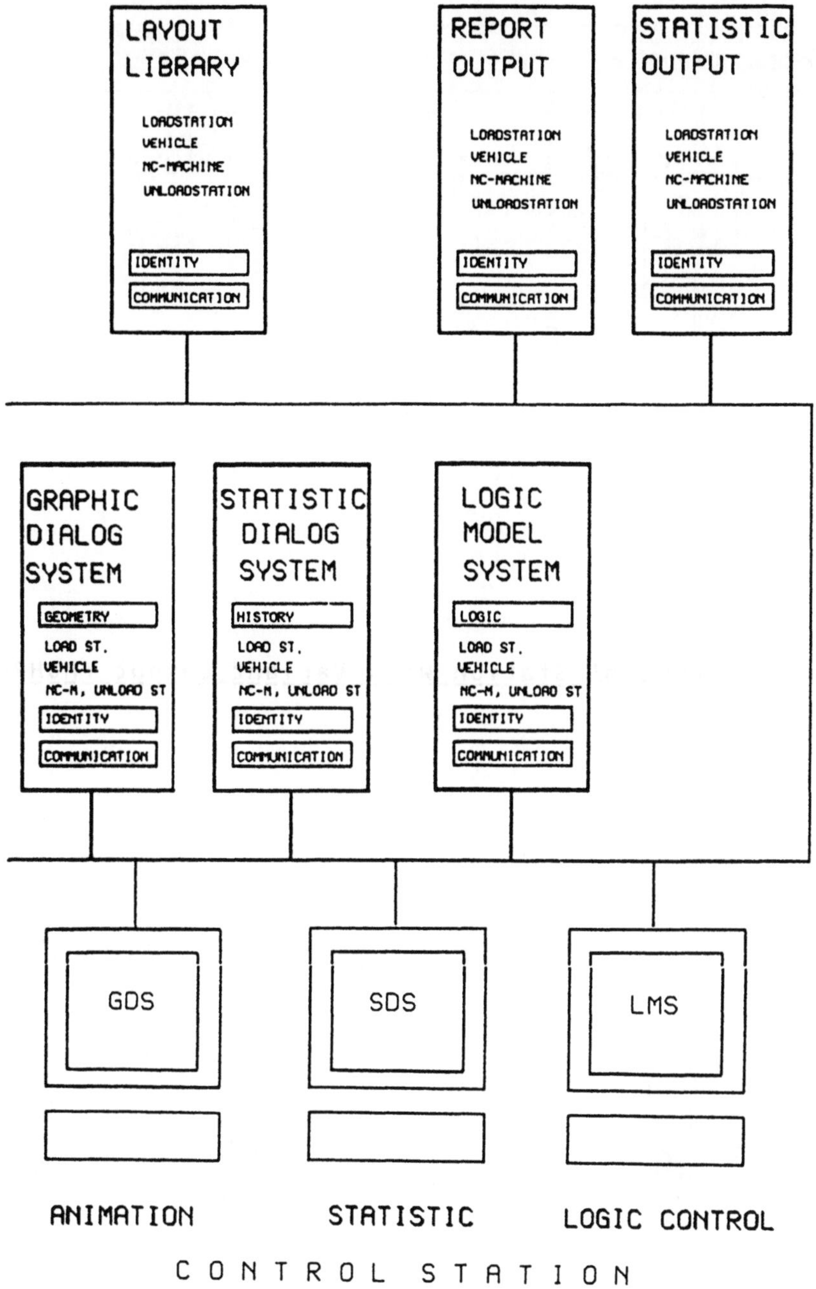

Fig. 8. Control Station for Dynamic Experiments

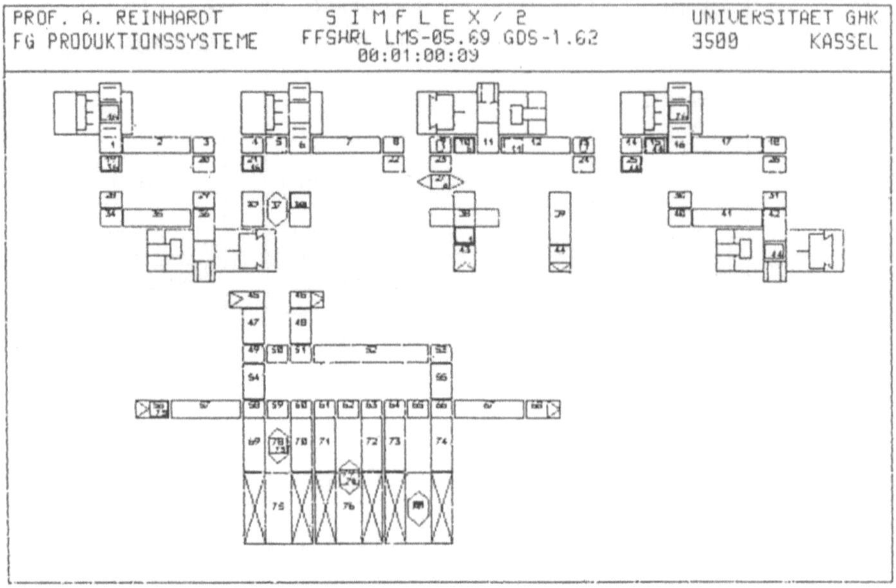

Fig. 9. Control Station with Variant Layout FMSHRL

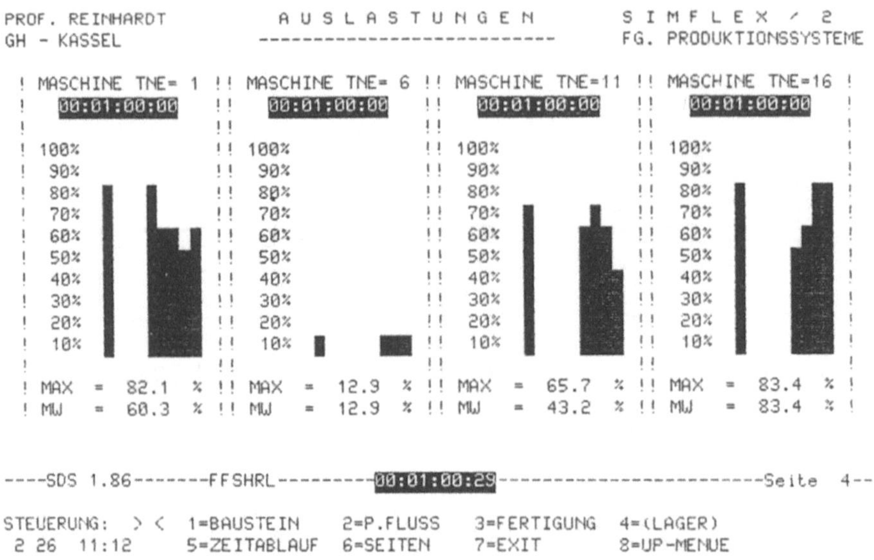

Fig. 10. Control Station with Statistics for FMSHRL

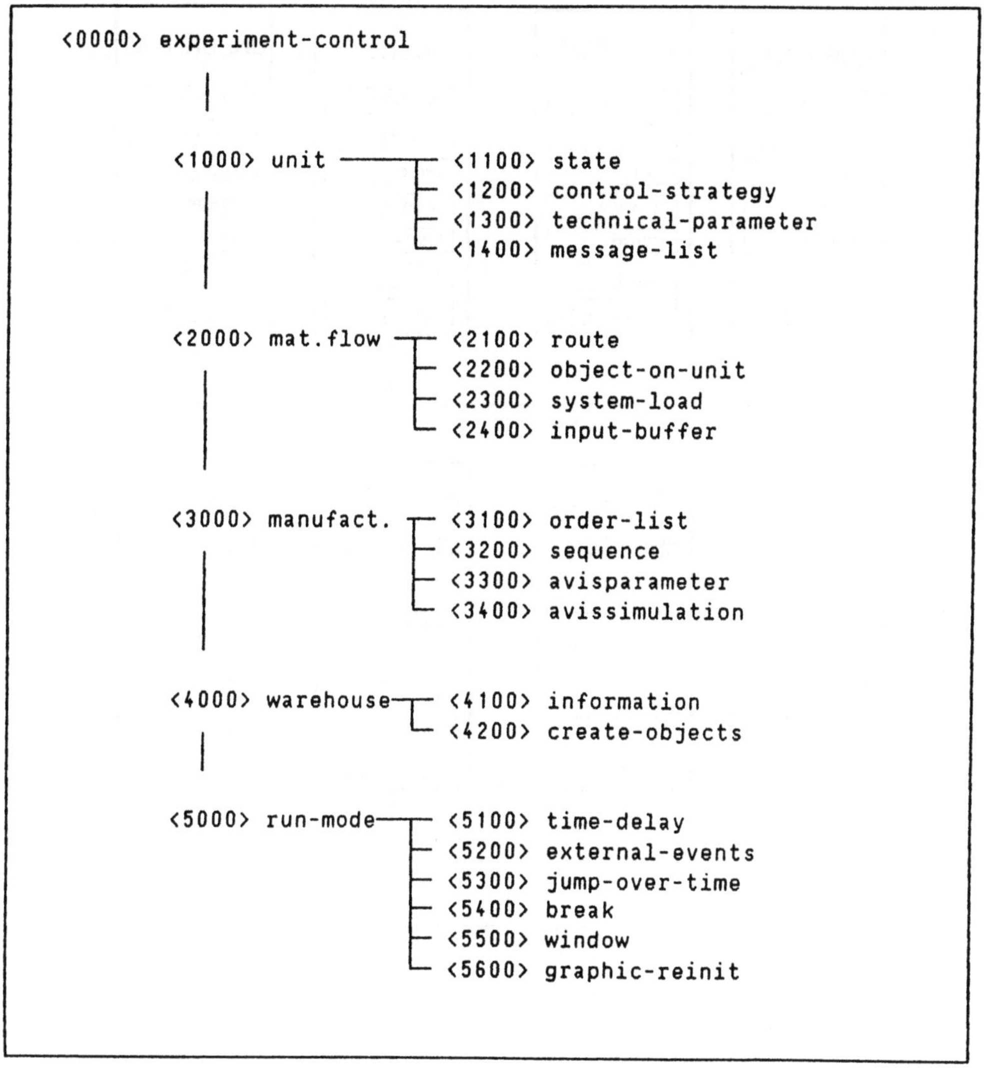

Fig. 11. Dialogue for Experiment Control <0000>

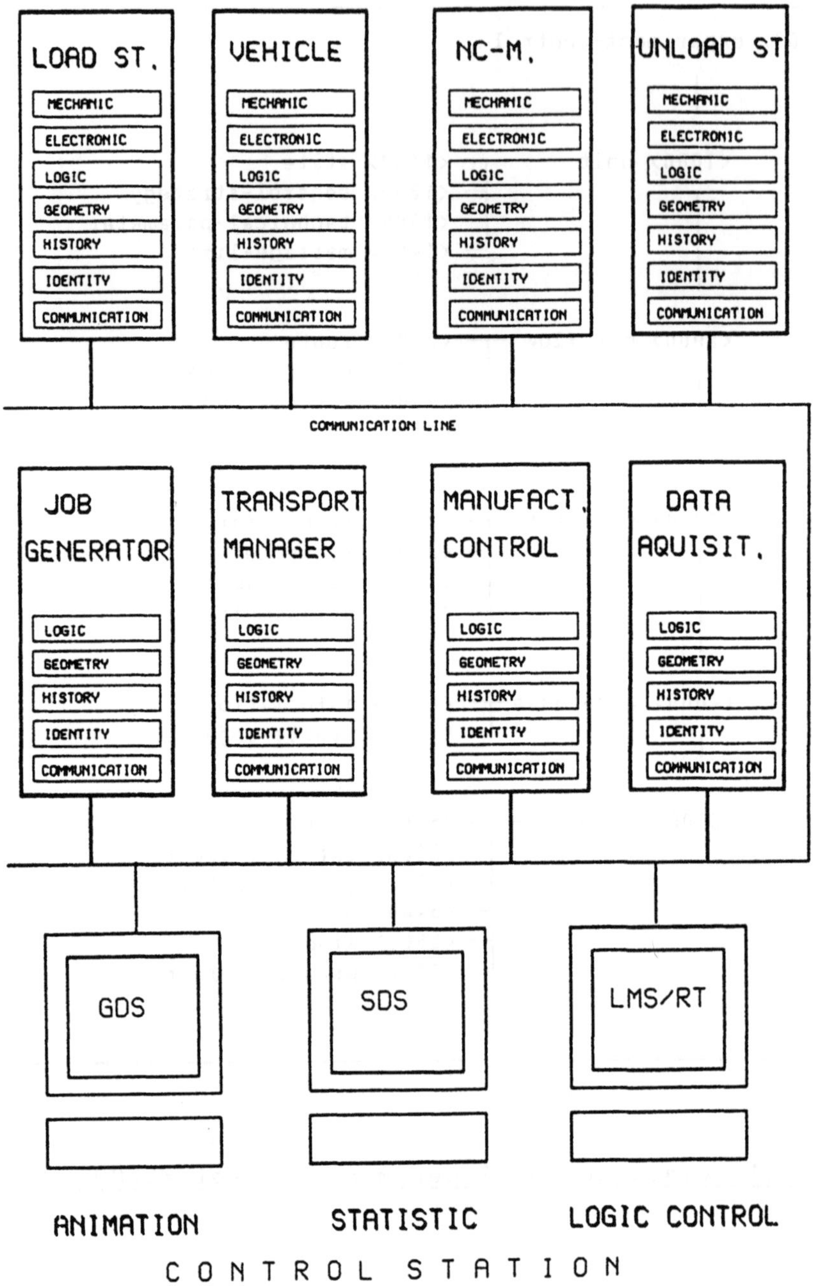

Fig. 12. Object Oriented Modelled Units

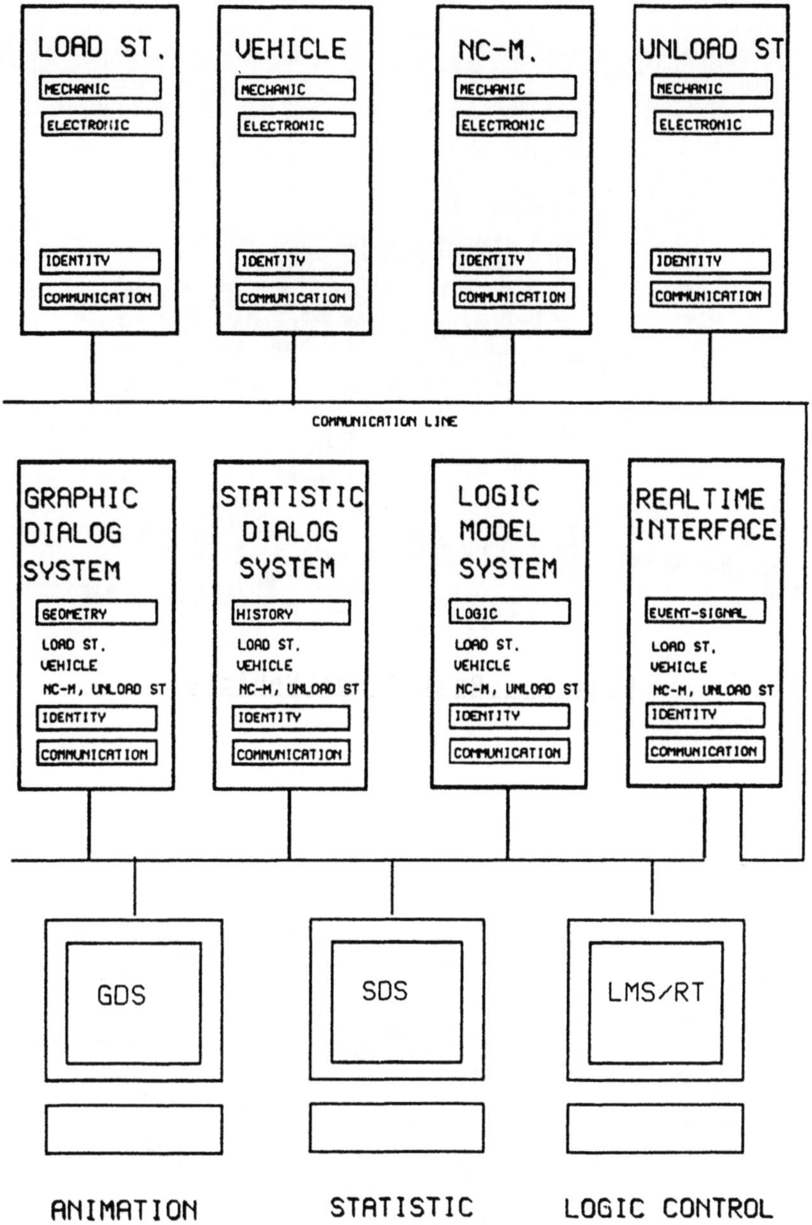

Fig. 13. Realtime Control System

Fig. 14. Rail-Borne Vehicle

Fig. 15. Layout of a Physical Model

FEEDBACK CONTROL THEORY APPROACH FOR SCHEDULING FLEXIBLE MANUFACTURING SYSTEMS

K. RAVI KUMAR
University of Southern California, USA

ABSTRACT

The area of feedback control has a rich history of applications in the various fields of engineering, notably electrical and electronic systems. On the other hand, adaptation of formal feedback control techniques to business planning and control systems have been less than successful. Typically, these unsuccessful applications have been in the environment of man-machine production systems.

In this paper, we are going to address various aspects of business planning and control as applicable to highly automated systems. Flexible Manufacturing Systems are an example of such computer integrated manufacturing systems. Our hypothesis is that feedback control methodologies are easier to implement in this environment and can contribute to great increases in productivity in terms of increased throughput rates as well as decreased tardiness.

We present our implementation of Intelligent Feedback Methodology (IFM-1), the structure of its design and the experimental design for testing the efficacy of such an approach to control an FMS. Preliminary results indicate the robustness of this methodology.

1.0 INTRODUCTION

The area of feedback control has a rich history of applications in the various fields of engineering. The classic example is a thermostat used as a means to control room temperature within a particular temperature range. More complex controllers have been successfully developed to manage multi-axis machining centers and robots in manufacturing environments. On the other hand, adaptation of formal feedback control techniques to business planning and control systems have been less than successful. Typically, these unsuccessful applications have been in the environment of

This work represents part of ongoing research conducted by the author and two students, James R. Gross and Dennis E. Kroll.

NATO ASI Series, Vol. F49
Computer Integrated Manufacturing
Edited by I. Burhan Turksen

man-machine production systems. For example, the evolution of materials requirement planning (MRP) systems to closed-loop MRP systems (and MRP II), while laudable for its conceptual breakthrough, has not provided the promised benefits of better control.

One of the pioneering efforts in the application of feedback control theory to the business planning problem is by Holt and Simon (1954). They provided an analysis of the aggregate production planning and control problem using quadratic cost functions and indicated that they give rise to unstable classes of production feedback rules. While the instability characteristics of such feedback can be remedied [Vassian (1955)], the practical application of such a methodology was infeasible due to its sensitivity to demand forecasts. It is important to note that the spirit of this research was to explore the direct application of mathematical control theory to a highly simplified production system.

In this paper, we are going to present a methodology for applying feedback control to address various aspects of business planning and control as applicable to highly automated and integrated systems. Flexible Manufacturing Systems are an example of such computer integrated manufacturing systems. They typically consist of computer controlled work stations linked with an automated flow of information and material. Our hypothesis is that feedback control methodologies are easier to implement in this environment and can contribute to great increases in productivity in terms of increased throughput rates as well as decreased tardiness. The primary motivation for the hypothesis is the increased scope of controllability of the system elements as well as the speed of information flows within the system.

We present, in Section 2, aspects of control theory constructs that are used to motivate our methodology as well as how they relate to the FMS control problem. In Section 3, we elucidate our modular construction of the feedback methodology, including requirements of each module. Section 4 contains an implementation of the methodology, termed IFM-1, using simulation as the vehicle for system analysis and show statistically that the methodology is robust in its endeavor to increase productivity of the FMS, productivity being measured by throughput and tardiness. We present our conclusions and areas for future investigation in Section 5.

2.0 FMS SCHEDULING CONTROL

A Flexible Manufacturing System may be defined as an integrated system of computer controlled machine tools and other work stations with an automated flow of information, workpieces and possibly tools in such a way that automated production of a group of complex workpieces in small and medium lot sizes is possible. It is a miniature computer integrated manufacturing facility.

As an example, a firm producing parts for the aerospace industry has installed a flexible machining system consisting of 8 CNC machining centers capable of profile milling, drilling, boring, reaming and tapping operations, 2 CNC coordinate measuring machines capable of verification of machined parts, an automated wire-guided cart system for material transport using robocarriers and computer control for traffic zones as well as palletised parts pickup/delivery at carousels and load/unload at workstation on/off shuttles.

FMS scheduling control involves several decisions to be made concerning the introduction of parts, routing of parts between workstations to complete the necessary operations, timing of preventive maintenance, reschedules upon workstation breakdown as well as coordination of material handling devices in the system. Lenz (1982) describes the primary decision areas to be : a) maintenance of a desirable mix of the various parts being produced by the system, b) sequencing the operations to be performed on the parts and selecting workstations necessary to make full use of the flexibility of the workstations and c) selecting the transporters (automated guided vehicles, robocarriers, etc.) necessary and routing the traffic to make best use of the flexibility of the material handling system (see also Stecke and Solberg, 1981).

A typical list of functions that an FMS control system has to attend to is as follows:

1) workpiece and fixture control and records maintenance
2) material transport system control
3) workstation control and records maintenance
4) NC part program control and maintenance
5) tool control and tool life control and maintenance
6) management performance reporting

and 7) visual display of key items and exception status.

Buzacott and Yao (1986) provide a summary of analytic models for FMS analysis. They point to the set of models developed at Purdue University,

Draper Labs, MIT, Harvard University and University of Toronto, their foundations being set in the one or more of the following approaches: network of queues, perturbation analysis, mathematical programming and simulation. Among the inadequacies identified with these models is their inability to take transient behavior into account. More seriously, most of these analysis tools have not been configured to support real-time, on-line scheduling control with decisions being based on the actual state of the system.

From a control theory perspective, the objective of a control system is to guide the manufacturing system over time maintaining some desired conditions within the manufacturing system. Tocher (1970) delineates five necessary characteristics of feedback control problems:

1) set of times at which action is possible
2) choice among a set of actions
3) an evaluative model to predict future history of system for each choice of action
4) a criteria or objectives to base choice of action

and 5) feedback on actual system performance on objectives or criteria.

The issue of time is of importance since there are two distinct categories: times at which actions are taken and times over which actions are planned. The latter is the planning horizon and the former the action implementation horizon. Typical planning assumes that the two are the same but for dynamic planning, the planning horizon is usually much larger than the action implementation horizon. This type of planning is also termed "rolling horizon" in the literature on materials requirement planning. Obviously, the choice of action, in dynamic planning, is based on a comparison of objectives/criteria evaluated over the planning horizon across the various alternatives.

The dimension of the action space as well as the size of the choice set is of great importance in the design of the control system. Clearly, if the entire action space is the choice set at each control decision time, complete enumeration poses tremendous problems in practice since the number of choices are so large. Therefore, for practical purposes, either there must exist a fast algorithm for arriving at the action choice or the choice set must be a dimensionally smaller subset of the action space. This dictates a need for some method by which a reduction of the action space to the choice set or to the actual choice is accomplished. Clearly, the feedback from actual system performance can give some indication as to the efficacy of certain choices made in the past and thus, aid in the reduction of the choice set.

The work done at MIT in the Laboratory for Information and Decision Systems is particularly of interest since it incorporates the essence of control theoretic constructs including feedback. Particularly, Kimemia (1982) proposes a hierarchical control structure for an FMS. The hierarchy consists of three levels, namely flow control, routing and sequence control. The flow controller regulates the production rate of each part type via a linear program [Kimemia and Gershwin (1983)] using information feedback on demand changes and workstation reliability. This control boils down to a functional form dependent on machine state and actual output levels. This automatic control typifies what Tocher (1970) calls off-line control implemented automatically. The routing and sequence controllers are based on algorithms which use system status as feedback to decide operation routing and sequencing decisions. It should be noted that the FMS configurations considered by them do not entail any automated material handling control decisions at all. Also, the feedback is basically of the system status which triggers either fixed rules or affects decision parameters of a fixed model. Another example of such an approach is the work of Akella et al. (1985) in which a hierarchical control structure similar to Kimemia (1982) is operationalized in the context of a card insertion line at IBM, Lexington. The control algorithm replans every time a work station malfunctions and not on a continuous basis as in Kimemia and Gershwin (1983). All of these works do not, in any fashion, incorporate any forward-looking algorithms or use any measure of tardiness as feedback.

The feedback methodology that we propose (see Figure 1) is closely related to the dynamic planning concept proposed in Tocher (1970). The FMS system that is to be controlled within our framework is defined in terms of stations (work and load/unload), number of on/off shuttle locations at each station (for storage prior to and after processing), pallet types and part types associated with them and material handling equipment characteristics such as carts, track segments and travel times. The FMS employs an on-line computer control system to apply a set of predetermined rules (direct control parameters) for part introduction, dispatching and material handling control.

The direct control parameters are derived from the action control plan/policy adjustment. The action implementation horizon is limited by the time required for the necessary feedback computations to be performed.

The action times are defined by either fixed times and/or by the occurrence of certain events in the system (such as machine failure, which are expected to be rare). At each such one, a sequence of specific decisions for the (shorter term) decision horizon as well as a system trajectory for the (longer term) planning horizon is generated. Obviously, such a plan can be visually displayed for approval by an FMS system manager, if one exists. The approval is based on system performance criteria such as throughput rate and system tardiness. The generation of the control plan is achieved

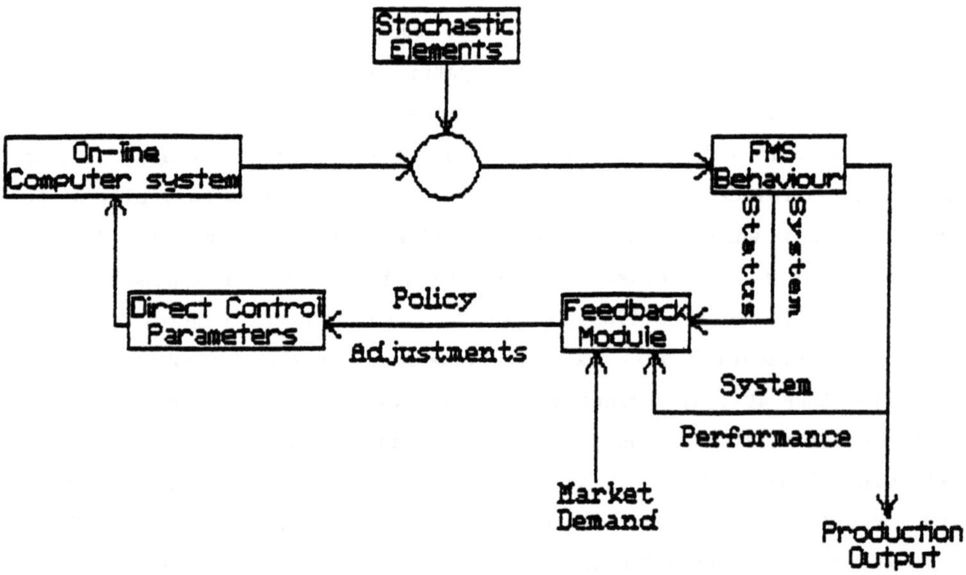

Figure 1. Proposed Feedback Control Model

by the use of a feedback module operating off-line. The module consists of several subsystems which take, as input, market demand feedback, system status feedback and system performance feedback regarding throughput and tardiness. These feedbacks are used to generate new policy adjustments regarding part balance, operation sequencing, selection of work stations, transporter selection and traffic control.

In the next section, we will describe how the control plan is generated and the adaptive nature of the proposed feedback methodology.

3.0 FEEDBACK ANALYSIS MODULAR CONSTRUCTION

The objective of the feedback module in Figure 1 is to be able to periodically analyze the system status over time to detect whether the assigned manufacturing goals are being achieved. It is also charged with the responsibility to suggest adjustments to the current control policy parameters imbedded in the software of the on-line control computer of the FMS. The feedback analysis module, thus, contributes to the adaptive nature of the control system. The conceptual framework for the feedback analysis is presented in Figure 2. The subsystems involved are the forecast module, the control policy nomination module, the FMS performance modelling system, and the control policy adjustment selection module.

The forecast module takes on the task of projecting useable values of all relevant non-control parameters. For example, machine reliability factors could be predicted using feedback data from past manufacturing system trajectory. Part order requirements could be generated by combining regular needs from assembly and forecasts of spare part needs. Tool wear could be predicted using tool life cycle trajectories generated by mapping historical data. Both work station and tool reliability can be used for preventive maintenance. The techniques that can be used for forecasting can range from statistical methodology to artificial intelligence. Statistical techniques encompass time-series analysis methods, ranging from moving averages to Box-Jenkins autoregressive models, and causal methods such as multiple regression. From the standpoint of adapting to new information, statistical techniques have a great advantage in that the theoretical foundations have already been laid. Expert systems and pattern recognition techniques are potential candidates for use in the artificial intelligence variety. The theory of learning in expert systems and pattern recognition is, at best, in its infancy and the outlook for practical implementation is not very good at this point in time.

The policy nomination module, at each decision time, is in charge of offering a choice set of potential policies to be evaluated by the performance modelling module for adjusting the current operating policy. In the case of use of mathematical programming techniques (such as linear, integer and non-linear programming models) for performance optimization, the nomination task is simply to offer the entire action space. The solution methodologies (branch and bound, steepest gradient, simplex/interior etc.) are not affected by the number of feasible points but rather the size of the constraint set and the dimension of a typical policy. An inherent problem with non-linear models is that the algorithms may end up in a local optimum. In any of the above cases, it is clear that the policy adjustment procedure does not take advantage of past history; clearly, the performance evaluation model is not dynamically updated even though its parameters may.

If the performance evaluation model is based on simulation, then the task of the policy nomination module is particularly critical since evaluating the entire action space may be practically impossible (given real-time environment). It would be particularly helpful to build a forecast model of performance given past history. This model would

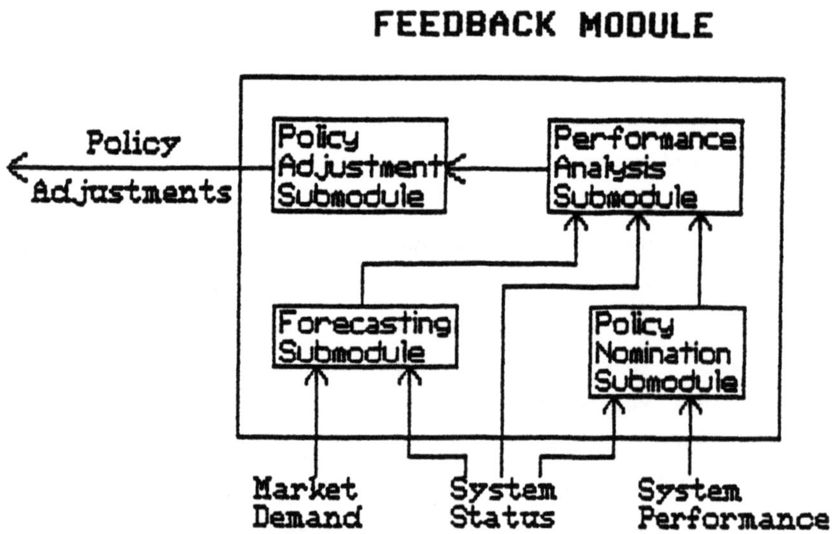

Figure 2. Proposed Adaptive Feedback Analysis

pre-process the action space quickly to predict potential policies (to be evaluated in detail) that would offer the best system performance. As in the forecast module, techniques to do this pre-processing could range from statistical methodology to artificial intelligence. Multiple regression models with performance being the dependent variable and the policy parameters, system status and other fixed system parameters as independent variables may be used to rank the action space. Similarly, expert systems with rules based on the system status and fixed system parameters could point to choice sets of potential policies to be evaluated. The flexibility of these pre-processors in being able to use feedback of current performance history adds to the adaptive nature of the system. It is also quite possible that such aggrégate models of performance may be used directly without going through the detailed performance module under certain environmental conditions.

The FMS performance modelling system can be based, as previously mentioned, on approaches such as networks, mathematical programming, perturbation analysis, simulation and symbolic representation. Network of queues has the advantage of modelling uncertainty in the scheduling environment but suffers from the drawback that it does not deal with transient real-time system information. The technique of using dynamic networks to model FMS scheduling shows a lot of promise due to its ability to model part sequencing and the existence of fast network algorithms to solve such models (Kumar and Kroll [1987]). Petri nets are another technique to model detailed event-switching in FMS operations. While it has been shown to work in real-time (Shaw and Whinston, 1985), its ability to cope with uncertainty is suspect. Mathematical programming, while being precise in formulation and is potentially optimal in analyzing performance, cannot take into account uncertainty. Also, the detailed modelling of an FMS invariably leads to integer programming formulations. While solution techniques are getting more sophisticated and fast, nevertheless typical integer formulations are difficult to solve. Easier to implement currently are simulation modelling and perturbation analysis techniques. While the general variety of simulation models can evaluate performance measures given a particular policy, perturbation techniques help in directing search for optimal policies without simulating all the alternatives. Both of these techniques have the advantage of rich modelling capacity and the duplication of actual FMS control logic, involving machine instructions and data architecture, within the model framework. Symbolic representation of the system has been shown possible by the ISIS project (Fox et al., 1983) involving artificial intelligence techniques. Whether incorporation of uncertainty, dynamic planning and fast execution times is possible with such techniques remains to be seen. These techniques will certainly have the advantage of rich modelling features and explanatory powers.

The control policy adjustment selection module is in charge of selecting the adjustment to current operating policy. It evaluates the performance measures, as reported by the FMS modelling module, of the nominated policies in terms of their desirable characteristics and chooses that policy meeting current objectives. This decision can be made very simple through choice of a single objective measure but, in practice, it can be difficult due to the presence of multiple objectives. It is quite possible that decision analysis can be used effectively by incorporating not only average measures of performance but also measures of risk (or variance). One alternative in designing this module is to allow a window for interacting with the FMS manager/scheduler to obtain management preferences. Once the selection has been made, the new direct control parameters have to be computed and communicated to the on-line FMS control system software.

These four modules comprise our model of a feedback system that is to be interfaced with the regular on-line computer control of an FMS. The feedback module has the advantage of using current information about the system status and performance as well as the external environment to suggest changes in control policy reflecting current objectives. One of the major benefits will be the ability to continuously track and adapt the predictive capability of the model of the system as well as learning about the efficacy of control policies in given environments.

In the next section, we will describe a particular implementation of this feedback analysis concept and report on its effectiveness as compared with fixed control policies.

4.0 EFFECTIVENESS OF FEEDBACK ANALYSIS CONCEPT

To test the robustness of the feedback concept, a very rudimentary prototype was developed. The prototype design is influenced by considerations of providing a sample of all the modules in the feedback system while keeping the development time as short as possible. The outcome is Intelligent Feedback Methodology Version 1 (IFM-1) implemented on a VAX 11/750 super-minicomputer operating under VMS 4.0 operating system.

A central portion of IFM-1 includes a modified version of the MAST FMS simulator, a proprietary software product of CMS Research Inc. (Lenz 1982). The selection of MAST package was based on its wide application to existing FMSs and ease of interface. While the development time of IFM-1 was certainly helped by the use of MAST, it was confining with respect to part introduction.

The inputs to IFM-1 include data files describing the FMS itself, order release data, control policy data and experiment parameter data. The system description data file includes information on 1) the station types including description of number of on/off shuttle positions, location in terms of material handling path and time taken to transfer to and from a shuttle and 2) the part types including description of alternative sequencing of operations, alternative station selection and operation times, priority of part type and pallet requirements. The order release file contains information on the description of part lots including lot size of part type, due date of lot and release sequence. IFM-1 introduces each part to the system in accordance to this data allowing no intervention during operations. For experimental purposes, configurations of FMS include six to eight workstations, two material handling carts, single pallet type, three part types with alternative routes for each part type varying from four to twenty seven.

The control policy file lists the entire policy action space. A specific control policy, in this implementation, consists of a choice of decision rule for selecting the next part for processing and a choice of system congestion control by means of individual limits on parts of each type as well as on total number of parts in the system at any time. In our experimentation, the part selection rules have been restricted to five, namely, first in-first out, shortest processing time/total processing time, due date priority, critical ratio and fixed part type priority. The system congestion limits vary from two to nine for individual part types and from eight to ninety-nine for total part types. Thus a vector of dimension eight constitutes a specific policy that can be used to control the FMS model. While this is quite simplistic, it should be noted that this is a prototype and a more realistic policy space would just increase the dimension of the policy vector. Finally, the experiment data file contains information on the length of time for start-up of the FMS and data base collection (to be used by the policy nomination module), decision horizon and the number of policy alternatives to be evaluated by the system performance module. In our experimentation, the start-up time for the FMS has been two hours, an allowance of six hours for initial data base collection for all policies, a decision horizon time of six hours and policy alternatives ranging from eight to all available.

The forecast module, in this implementation, takes on the very simple form of prescribed error to be added to the "actual" value of the variable.

The only environmental variable being forecast is the demand order and the forecast is similar to the order release file (described above) in data structure. Three distinct "actual" demand order patterns namely step function, ramp function and oscillating function were used for two of the part types in the experiment, keeping the total number of parts to be produced a constant.

The control policy nomination module has been configured using multiple regression as the pre-processing methodology. FMS performance objective is taken to be average part tardiness and is regressed against the independent variable vector comprising of part selection and part limits (representing the control policy), number of parts of each type in the system currently, average lead times associated with parts in the system and average number of parts of each type per order (representing feedback information on system status and performance). In the experiment, the independent variables are a vector of dimension seventeen.

The FMS performance evaluation module is, as mentioned before, a simulation model of the actual system and has as the core the MAST simulator package. Specific events modelled by MAST include material handling movement from station to station, on and off shuttles, completion of part operations and part introduction/removal from the system.

The control policy adjustment module, in this implementation, tries to trade off the objectives of minimizing average part tardiness as well as maximizing the total production rate. In this implementation, the simple selection rule used is choosing that rule with lowest average part tardiness and yet within 10% of the highest production rate evaluated over all nominated policies.

In trying to establish the effectiveness of the prototype feedback control module, one needs to establish a reference system. A very stringent test is proposed wherein the best possible fixed part selection policy, assuming perfect forecast of demand pattern, is chosen. Measures of the effectiveness of the feedback methodology is defined by the ratio of the average part tardiness in the feedback model to that in the reference model run and the ratio of the average production rate in the feedback model to that in the reference model run.

Figures 3 and 4 represent the results of the experimental runs expressed as bar charts. They illustrate the overall improvement that the feedback control provides inspite of the stringent reference system chosen. Figure 3 shows that 37 of the 54 runs had the tardiness ratio less than

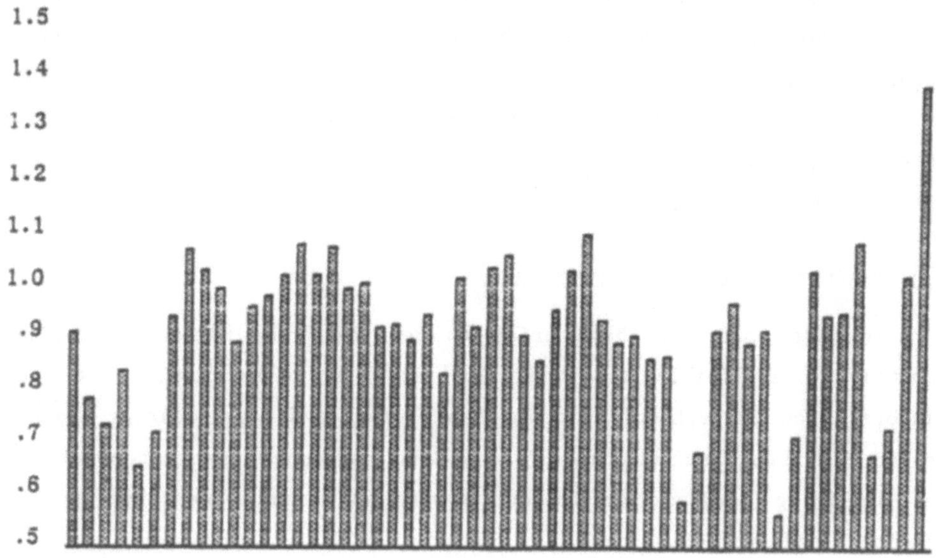

Figure 3. Tardiness Ratio

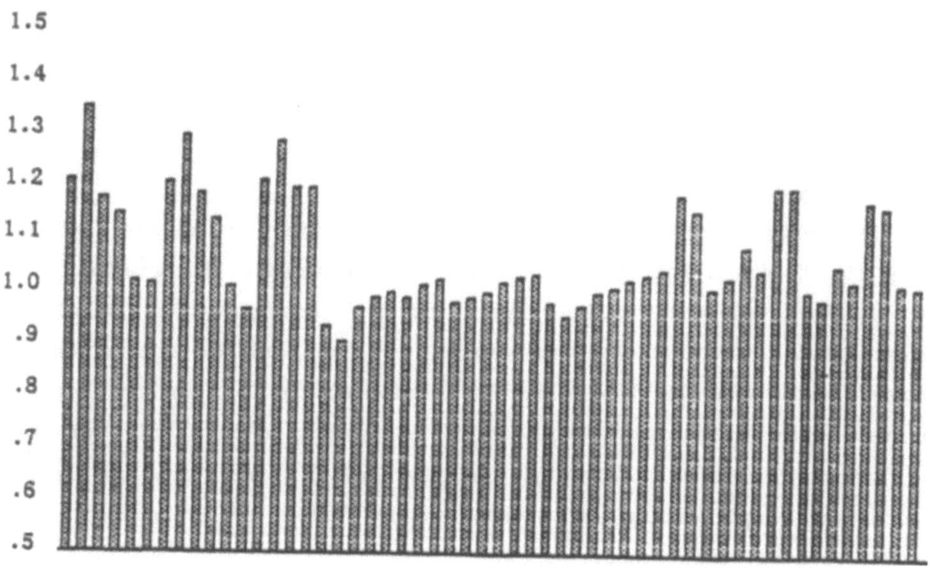

Figure 4. Production Rate Ratio

unity showing superior performance of the feedback system. Performing the Wilcoxon signed ranked test shows that the hypothesis that the reference system is better than the feedback system is rejected at the significance level of 0.005. Similarly, Figure 4 illustrates that 38 of the 54 runs had a production rate ratio greater than unity supporting the feedback system. The Wilcoxon test rejects this hypothesis also at the 0.005 level of significance.

This experiment establishes the effectiveness of the feedback system concept implemented through a simple prototype system and compared to a stringent reference.

5.0 CONCLUSIONS AND FUTURE RESEARCH

In this paper, we have tried to resurrect the notion of feedback control as applicable to production systems. Specifically, we have elaborated on the scheduling control of integrated manufacturing systems, called Flexible Manufacturing Systems, and developed the conceptual base for utilizing feedback to improve system performance measured by part tardiness and production rate.

To evaluate the effectiveness of the concept, a prototype version of the feedback control system was built around a simulation model as the performance analysis module. The prototype was used to generate performance results for a variety of FMS configurations and fixed control parameters. It was shown that the performance of the prototype under feedback control was statistically superior to that of a stringent reference system.

This effort is just the beginning of a variety of research thrusts along the feedback control approach. The performance characteristics of the feedback approach with respect to FMS configurations should be investigated. Each of the submodules in the feedback module can be enhanced. For example, the policy nomination module is a potential target for developing intelligent decision tools to aid in the choice process. The forecast module requires implementation of prediction models and algorithms. Implementing this system in an industrial environment would provide valuable experience.

REFERENCES

1. Akella, R., Y. Choong and S. B. Gershwin: Real-time production scheduling of an automated cardline. Annals of Operations Research, Vol. 3, pp. 403-425, 1985.
2. Buzacott, J. A. and D. D. Yao: Flexible manufacturing systems: A review of analytical models. Management Science, Vol. 32, No. 7, 1986.

3. Fox, M. S., B. P. Allen, S. F. Smith and G. A. Strom: ISIS: A constraint-directed reasoning approach to job-shop scheduling. Research Report, Robotics Institute Intelligent Systems Laboratory, Carnegie-Mellon University, Pittsburgh, Pennsylvania, 1983.
4. Holt, C. C. and H. A. Simon: Optimal decision rules for production and inventory control. Proceedings of the Conference on Operations Research in Production and Inventory Control, Case Institute of Technology, Cleveland, Ohio, January 1954.
5. Kimemia, J. G.: Hierarchical control of production in flexible manufacturing systems. Ph.D. Dissertation, Report No. LIDS-TH-1215, Laboratory for Information and Decision Systems, Massachusetts Institute of Technology, Cambridge, Massachusetts, 1982.
6. Kimemia, J. G. and S. B. Gershwin: An algorithm for the computer control of a flexible manufacturing system. IIE Transactions, Vol. 15, pp. 353-362, 1983.
7. Kumar, K. R. and D. E. Kroll: Dynamic network modeling of an FMS. Modern Production Management Systems, Elsevier Science Publishers B. V. (North-Holland), pp. 19-30, 1987.
8. Lenz, J.: Design and control of flexible manufacturing systems with MAST. CMS Research, Oshkosh, Wisconsin, 1982.
9. Shaw, M. J. and A. B. Whinston: A distributed knowledge-based approach to flexible automation. Working Paper, College of Commerce and Business Administration, University of Illinois, Champaign, Illinois, 1985.
10. Stecke, K. E. and J. J. Solberg: Loading and control policies for a flexible manufacturing system. International Journal of Production Research, Vol. 19, No. 5, pp. 481-490, 1981.
11. Tocher, K. D.: Control. Operational Research Quarterly, Vol. 21 No. 2, pp. 159-180, 1970.
12. Vassian, H. J.: Application of discrete variable servo theory to inventory control. Operations Research, Vol. 3, No. 3, pp. 272-289, 1955.

1. [illegible]

2. [illegible] Control [illegible] Technology, Cleveland, Ohio, January 1972.

3. [illegible] Ph.D. Dissertation [illegible] Report No. [illegible] Laboratory for Information and Decision Systems, Massachusetts Institute of Technology, Cambridge, Massachusetts, [illegible]

OPTIMUM STORAGE LAYOUT IN FLEXIBLE MANUFACTURING

A.S. KIRAN and B.C. TANSEL
Department of Industrial & Systems Engineering
University of Southern California

ABSTRACT

Two layout problems in flexible manufacturing have been considered. First, the layout of Work In Process (WIP) storage area has been modelled as a generalized assignment problem. Secondly, the optimum location is investigated for the pick up point of the WIP storage unit on a material handling network. Algorithms have been provided for efficient solution of these two problems.

NATO ASI Series, Vol. F49
Computer Integrated Manufacturing
Edited by I. Burhan Turksen

1. INTRODUCTION

Design of the physical layout for a manufacturing system is one of the most important issues that must be resolved in early stages of the system design. In Computer Integrated Manufacturing (CIM), expensive hardware that is used for processing and handling parts necessitates careful consideration of alternative layouts with respect to cost and operational performance. Cost consequences of the decisions related to number and layout of machines, buffer (WIP storage) areas, and load/unload stations can be observed not only during the implementation but also operations of the system. Good solutions to layout related design problems also provide a necessary foundation for effective utilization of the system.

Although there is a substantial literature available for the facilities layout problem (see [2,3,4,9]) there are very few works published on the layout problems in flexible manufacturing. Cell level machine and robot layout problems are defined by Sarin and Wilhelm [10]. They propose mixed-integer programming models for task assignment, selecting number and type of robots, and determining system configuration. Jayaraman [6] developed a CAD based system for evaluation of different cell designs. A Q-GERT based approach for cell design is proposed in [12].

The system level problems related to Material Handling System (MHS) are described in [8]. Afentakis [1] developed a graph theoretic model and proposed heuristic methods for machine and MHS layout, Heragu and Kusiak [5] proposed clustering based heuristics for machine layout problems.

In this brief review we shall summarize models and solution methods for two layout related problems that arise in flexible manufacturing context. The paper is organized as follows: in the next section, we will consider layout for the WIP storage areas in flexible manufacturing cells. The problem is modeled as a discrete storage layout problem and solution methods are illustrated on numerical examples.

A more general storage layout problem is considered in section 3. Location for pick point of storage unit in a flexible manufacturing system (FMS) is analyzed as a discrete location problem. Algorithms for solving this location problem and its variations are also provided. Summary and conclusions are given in section 4. More detailed discussions on the problems provided in sections 2 and 3 can be found in [11] and [7], respectively.

2. STORAGE LAYOUT IN FLEXIBLE MANUFACTURING CELLS [11]

In this section we consider a flexible manufacturing cell where there are a small number of machines, usually two or less, a load/unload station, a material handling mechanism such as a robot, and some area within the cell for short term storage of WIP parts. The operation of the cell is explained as follows: After being fixed on a pallet fixture, parts are located in the storage area before their first operation. Material handling equipment such as a robot picks up a part from the storage area and places it on a machine table. After completion of the operation the part may be taken to another machine for its next operation, or to load/unload station. If the machine which will perform the next operation is not available the part may be taken to the storage area where it has to wait until the required station becomes available. The storage area must be located in the cell because of the constraints on the robot's work envelope. In this context we analyze the problem of determining optimum areas to store WIP parts.

Let us consider a flexible manufacturing cell where there are m stations (e.g. machines, load/unload stations) to produce p parts. Assume that stations in the cell are at known locations. Each part will be assigned a storage area of size A_i, $i=1,\ldots,p$ within the cell. We shall assume that the area that can be allocated to parts can be divided into unit regions each of size equal to an area where one pallet can be stored.

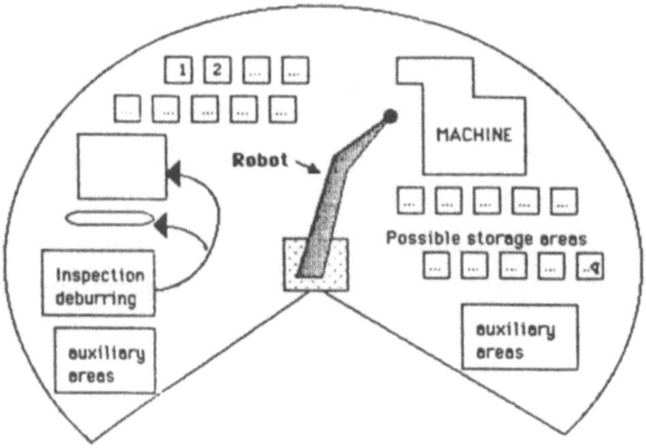

(a) Cell with a spherical robot

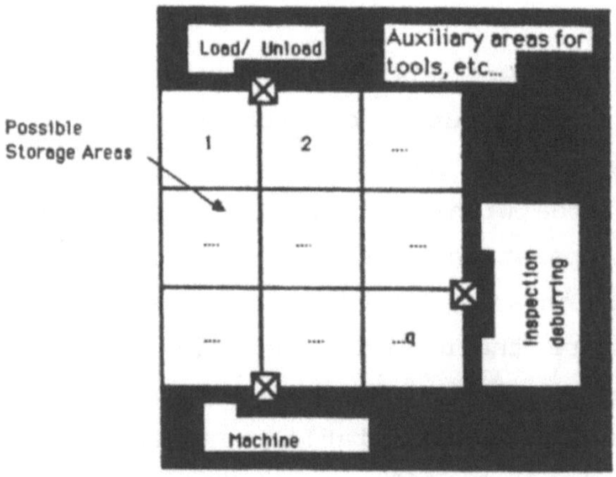

(b) Cell with a Gantry type robot

Figure 1

Figure 1 shows some possible configurations.
Let Q be the region available for storage, divided into q unit regions. In the case of Figure 1.b each area can be considered as a grid square which may simplify the representation of Q.

Let w_{ij}, i=1,...,p, j=1,...,m be the number of pallet loads of part i to be moved between storage areas of part i and station j. W_{ij}'s can be calculated using part routings and machine utilizations (see [11] for details). We state the problem as follows: given the item movement matrix, $W=[w_{ij}]_{pxm}$, and the areas $A_1,\ldots,A_p$, find A_i unit regions for the storage of each item i, i=1,...,p so that the material handling cost is minimized.

Layout for n items

Let $S_i \subset Q$ be the set of unit regions to be allocated to item i. Note that $|S_i| = A_i$. Even though w_{ij} pallet loads will be moved between region S_i and station j, we do not know which unit region generates each move. A reasonable assumption is that each unit region in S_i is equally likely to generate such a move. With equal likelihood assumption w_{ij}/A_i is the number of movements of item i between any unit region in S_i and the station j.

Let d(j,k) be the distance between unit region k, k=1,...,q and the station j. Given an appropriate definition of the distances, the material handling cost for the movement of part i between its storage region S_i and station j is given by

$$\sum_{k \in S_i} W_{ij} \quad d(j,k)/A_i \tag{1}$$

and the total cost is then defined by

$$f(S_i,\ldots,S_p) = \sum_{i=1}^{p} f_i \ (S_i) \tag{2}$$

where

$$f_i(S_i) = \sum_{j=1}^{m} \sum_{k \in S_i} W_{ij} \quad d(j,k)/A_i \tag{3}$$

The optimization problem of interest is to find regions $S^*_i,\ldots,S^*_p$ such that

$$\begin{aligned} P: f_1(S^*_1,S^*_2,\ldots,S^*_p) &= \min f(S_1,\ldots,S_p) \\ \text{s.t.} \quad & S_i \subset Q \qquad i=1,\ldots,p \\ & |S_i| = A_i \qquad i=1,\ldots,p \end{aligned}$$

This formulation is quite similar to the one given in Francis and White [3, Chap. 6] in the context of assigning dedicated storage spaces to p items in a warehouse. The problem can be easily converted to a generalized assignment problem.

For i=1,...,p and k=1,...,q, let x_{ik} be a zero/one variable such that $x_{ik} = 1$ if unit region k is assigned to item i and zero otherwise. If will be convenient to assume

$$\sum_{i=1}^{p} A_i = |Q|$$

so that every unit region is assigned to some item. If this is not the case, we may define a p + 1st dummy part which will take the unused space: i.e.

$$A_{p+1} = q - \sum_{i=1}^{p} A_i \quad \text{and} \quad W_{p+1,j} = 0 \qquad j=1,\ldots,m$$

Now

$$f_i(S_i) = \sum_{j=1}^{m} \sum_{k=1}^{q} W_{ij} \; x_{ik} \; d(j,k)/A_i \tag{4}$$

and the total cost is

$$f(S_i,\ldots,S_p) = \sum_{i=1}^{p} \sum_{j=1}^{m} \sum_{k=1}^{q} W_{ij} \; x_{ik} \; d(j,k)/A_i \tag{5}$$

Note that with an interchange of summation, we get

$$f(S_i,\ldots,S_p) = \sum_{i=1}^{n} \sum_{k=1}^{q} x_{ik} \sum_{j=1}^{m} W_{ij} \; d(j,k)/A_i \tag{6}$$

If we define

$$c_{ik} = \sum_{j=1}^{m} W_{ij} \; d(j,k)/A_i \tag{7}$$

then The Generalized Assignment Problem (AP) below is equivalent to P

AP:

$$\text{minimize} \sum_{i=1}^{p} \sum_{k=1}^{q} c_{ik} \; x_{ik} \tag{8}$$

s.t.

$$\sum_{i=1}^{p} x_{ik} = 1 \qquad k=1,\ldots,q \tag{9}$$

$$\sum_{k=1}^{q} x_{ik} = A_i \qquad i=1,\ldots,p \tag{10}$$

AP can be solved using any transportation algorithm.

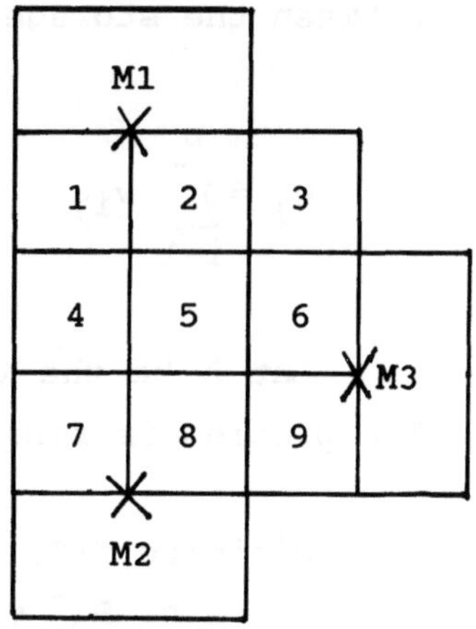

Figure 2: Example 1

Example 1: Let us consider a cell with a Gantry robot similar to Figure 1.b. Let m = 3 (i.e. M1, M2, M3), $p = 2$, $A_1 = 2$, $A_2 = 3$ and $Q = \{1,2,...,9\}$.

Let
$$W = \begin{bmatrix} 300 & 320 & 160 \\ 320 & 0 & 320 \end{bmatrix}$$

and

$$D = [d(j,k)]_{mxq} = \begin{bmatrix} 1 & 1 & 2 & 2 & 2 & 3 & 3 & 3 & 4 \\ 3 & 3 & 4 & 2 & 2 & 3 & 1 & 1 & 2 \\ 4 & 3 & 2 & 3 & 2 & 1 & 3 & 2 & 1 \end{bmatrix}$$

First we calculate

$$C = [C_{ik}]_{pxq} = WD = \begin{bmatrix} 1900 & 1740 & 2200 & 1720 & 1560 & 2020 & 1700 & 1540 & 2000 \\ 1600 & 1280 & 1280 & 1600 & 1280 & 1280 & 1920 & 1600 & 1600 \end{bmatrix}$$

Applying the transportation algorithm, we have $x_{15} = x_{18} = 1$, $x_{22} = x_{23} = x_{26} = 1$ and $f(S_1^*, S_2^*) = 6,940$. The solution is locating part 1 at unit areas 5 and 8 and locating part 2 at unit areas 2, 3 and 6.

Aggregated Approach

In the previous model we assumed each part is assigned to a region S_i dedicated solely to its storage. We wish to relax this assumption now by allowing random storage of parts at any available storage area. Such a model is applicable to cases where the MHS is capable of recognizing each part type (e.g. using bar codes).

Since we are not interested in the individual item movements, we define the total number of pallet loads to be moved between the storage region and station j

$$W_j = \sum_{i=1}^{p} w_{ij} \qquad j=1,\ldots,m \tag{11}$$

Let A be the total storage area requirement for all parts. The problem is finding the storage region $S \subset Q$ such that

$$\begin{aligned} &\text{minimize } f(S) \\ &\quad \text{s.t. } S \subset Q \\ &\quad\quad |S| = A \end{aligned}$$

where

$$f(S) = \sum_{j=1}^{m} W_j \sum_{k \in S} d(j,k)/A \tag{12}$$

Note that, each unit area is assumed to be equally likely to generate these movements. Thus f(S) is the total distance items are moved between the storage and all stations.

Rewriting f(S) with an interchange of summation, we get

$$f(S) = \sum_{k \in S} (1/A) \sum_{j=1}^{m} W_j \ d(j,k) \tag{13}$$

and defining

$$f_k = \sum_{j=1}^{m} W_j \ d(j,k) \tag{14}$$

The problem can be rewritten as

$$\begin{aligned} &\text{minimize } (1/A) \sum_{k \in S} f_k \\ &\text{s.t.} \quad S \subset Q \\ &\quad\quad |S| = A \end{aligned}$$

The above problem can be solved by arranging unit areas in

ascending order of the f_k values. S^* is given by the first A unit areas.

<u>Example 2</u>: Let us consider the cell given in Example 1.

Let $W_1 = 620$, $W_2 = 320$, $W_3 = 480$, and $A = 5$. D is given in Example 1.

$f_1 = 3500$ $\quad f_2 = 3020$ $\quad f_3 = 3480$

$f_4 = 3320$ $\quad f_5 = 2840$ $\quad f_6 = 3300$

$f_7 = 3620$ $\quad f_8 = 3140$ $\quad f_9 = 3600$

$f_5 < f_2 < f_8 < f_6 < f_4 < f_3 < f_1 < f_9 < f_7$

The solution, $S^* = \{2,4,5,6,8\}$ is shown in Figure 3.

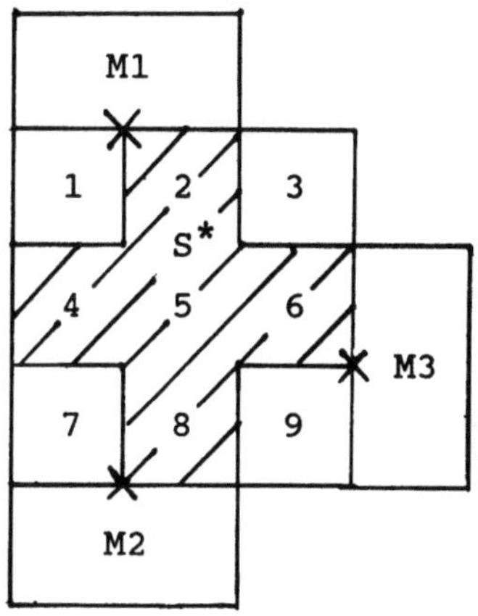

Figure 3: Solution of Example 2

3. OPTIMUM CENTRAL STORAGE LOCATION IN FLEXIBLE MANUFACTURING SYSTEMS [7]

In this section, we consider a central storage unit in a FMS. We assume that pallets, loaded with parts, are transported via conveyors, overhead monorail carts, tow-line carts, or automated guided vehicles that can move on a given Material Handling System (MHS). Transportation is often restricted by unidirectional movement of conveyors, carts or vehicles. The central storage unit consists of a carousel or a vertical storage unit with a single pick point, i.e., the interaction point for material exchange to and from the central storage unit. A pallet, transported by MHS, is brought to pick point where an automated mechanism such as a shuttle or rotary indexing mechanism will move the pallet to the storage. Hence the distances must be measured on the MHN as shortest path distances between the pick point and stations. We assume that all stations in the FMS are located along the tracks of the MHS.

The problem we consider here is the minimum cost location of the central storage unit: Given a MHS layout with m stations, determine a pick point to minimize a total cost function which is defined by total transport cost of the parts to and from the central storage. For the analysis purposes, the material handling system layout is represented by the "Material Handling Network" (MHN). An example of a MHS layout and corresponding MHN is shown in Figure 4. The MHN, G=(N,A) is defined by a node set that corresponds to stations and intersections of the MHS, and a directed arc set A that corresponds to material handling paths.

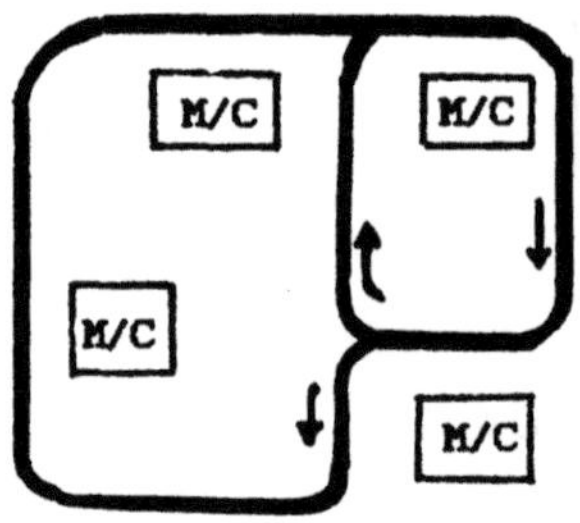

Figure 4.a: Material Handling System Layout

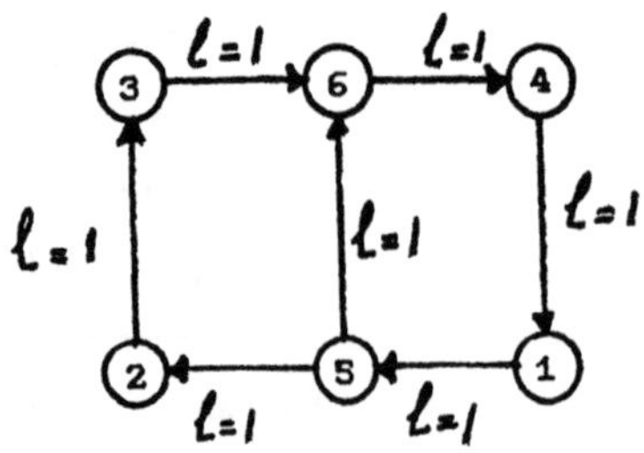

Figure 4.b: Corresponding Material Handling Network

Let d(x,j) be the shortest path distance from point x to node j (j=1,...,n) of G. We define a point x as a (possible) node on graph G assuming that node set may be expanded by adding a pick point, i.e., $N \cup \{x\}$. Arc set of the new graph may also be extended. If x is located on arc [a,b] the new arc set is defined as $\{A-[a,b]\} \cup [a,x] \cup [x,b]$.

For each node j, let u_j be the number of pallets to be transported <u>from station</u> j to storage, and let v_j be the number of pallets to be transported <u>from storage</u> to station j per unit time. Note that, $u_j \neq v_j$ in general.

For a pick point x the total cost of material flow is given by

$$f(x) = f_1(x) + f_2(x) \tag{15}$$

where $f_1(x)$ is the total cost of material flow from stations to storage, and $f_2(x)$ is the total cost of material flow from storage to stations, i.e.:

$$f_1(x) = \sum_{j=1}^{n} u_j \, d(j,x) \tag{16}$$

$$f_2(x) = \sum_{j=1}^{n} v_j \, d(x,j) \tag{17}$$

The problem now can be defined as to find x* such that $f(x^*) \leq f(x)$ for any point on G.

Node Optimality

The following theorem will be useful in solving the above problem:

Theorem 1 (Node Optimality): Let [a,b] be an arc and let $x \in [a,b]$, then $f(x) \geq \min \{f(a), f(b)\}$ (18)

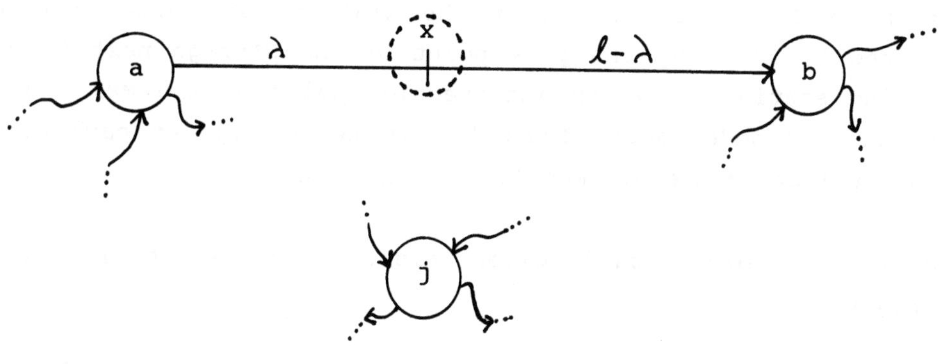

Figure 5: Theorem 1

Proof: Let ℓ be the length of arc [a,b]. Let λ be the distance between a and x (see Figure 5). For x = a or x = b, the proof is trivial. For $x \neq a$, b we have

$$f(x) = \sum_{j=1}^{n} u_j \,[d(j,a) + \lambda] + \sum_{j=1}^{n} v_j \,[\ell - \lambda + d(b,j)] \qquad (19)$$

rewriting (19) for f(a) and subtracting from (19) we have

$$f(x) - f(a) = \sum_{j=1}^{n} v_j \,[d(b,j) - d(a,j)] + \ell V + \lambda(U - V) \qquad (20)$$

where $U = \sum_{j=1}^{n} u_j$ and $V = \sum_{j=1}^{n} v_j$.

In (20) $d(a,j) \leq \ell + d(b,j)$ since arc [a,b] may not be on the shortest path from a to j. Hence

$$f(x) - f(a) \geq \sum_{j=1}^{n} v_j(-\ell) + \ell V + \lambda(U - V) = \lambda(U - V) \qquad (21)$$

A similar derivation gives

$$f(x) - f(b) \geq (V-U)\;(\ell - \lambda) \qquad (22)$$

(21) and (22) imply that either $f(x) \geq f(a)$ or $f(x) \geq f(b)$.

Theorem 1 reduces the candidate solution set to node set of the graph G. This means that the minimum total cost can be obtained by locating the pick point of the storage next to an existing station or at an intersection point of the MHN. The minimum cost node can be found by evaluating f(j) at each node j, and selecting the minimum f(j), j=1,...,n.

<u>Example 3</u>: Problem data is given below. Graph G=(N,A) is given in Figure 6.

j	1	2	3	4	5
u_j	5	10	15	10	0
v_j	15	5	10	15	0

$$D = \begin{bmatrix} 0 & 2 & 2 & 3 & 1 \\ 1 & 0 & 3 & 4 & 2 \\ 4 & 3 & 0 & 1 & 2 \\ 3 & 2 & 2 & 0 & 1 \\ 2 & 1 & 1 & 2 & 0 \end{bmatrix}$$

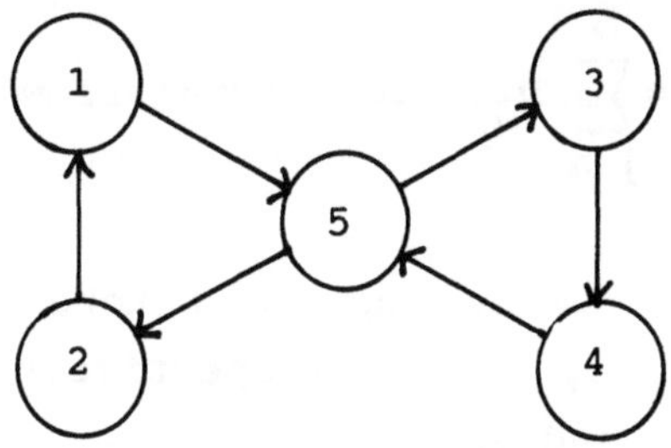

Figure 6: Graph G=(N,A) for Example 3

Solution:

j	$f_1(j)$	$f_2(j)$	$f(j)$
1	100	75	175
2	75	105	180
3	60	90	150
4	70	75	145
5	65	75	140

Optimum location is node 5 with f(5) = 140.

Unicyclic Case

In the case of unicyclic networks a more efficient solution method can be developed. Let us number the nodes of the G=(N,A) from 1 to n, starting at an arbitrary node, and following arcs (see Figure 6).

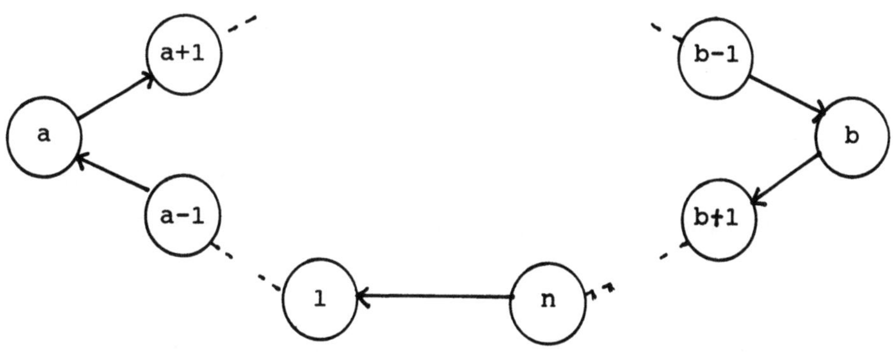

Figure 7: Numbering a Unicyclic Graph

Let a and be be two nodes such that 1 < a < b < n. We define

$$U_{ab} = \sum_{j=a}^{b} u_j \text{ and } V_{ab} = \sum_{j=a}^{b} v_j,$$ i.e., total flow from

stations a, a+1,..., b-1, b to storage, and total flow from storage to stations a, a+1,..., b-1, b respectively. Now, we can define $U_{ba} = U - U_{a+1,b-1}$ and $V_{ba} = V - V_{a+1,b-1}$

Property 1:

$$f(a) - f(b) = d(b,a)\ [U_{a+1,b} - V_{a,b-1}] + d(a,b)\ [V_{b,a-1} - U_{b+1,a}] \tag{23}$$

Proof: Rewriting (15) for a and b we have

$$f(a) = \sum_{j=1}^{n} u_j\ d(j,a) + \sum_{j=1}^{n} v_j\ d(a,j)] \tag{24}$$

$$f(b) = \sum_{j=1}^{n} u_j\ d(j,b) + \sum_{j=1}^{n} v_j\ d(b,j)] \tag{25}$$

Subtracting (25) from (24) gives (23) after cancellation of similar terms (see [7] for details).

Let c be the cycle length of MHN, i.e. c = d(a,b) + d(b,a). We can easily show that

Corollary: If a and b are adjacent then

$$f(a) - f(b) = c(u_a - v_b) + \ell_{ab}(V - U) \tag{26}$$

where l_{ab} is the arc length from a to b.

Equation (26) can be used in developing the following O(n) algorithm to find optimal pick point in the case of unicyclic MNHs.

Algorithm ONLUN (Optimal Node Location on Unicyclic Network)

INITIAL: Number nodes from 1 to n. Set i=1, f*(1)=0, MIN=0,

OPNODE=0

MAIN: $i \longleftarrow i+1$, compute

$$f^*(i) = f^*(i-1) + c(-u_i + v_{i-1}) - \ell_{i-1,i}\,(V - U)$$

If $f^*(i) < MIN$ then $MIN = f^*(i)$, and OPNODE = i.

If i = n stop, otherwise repeat MAIN.

<u>Example 4:</u> A unicyclic MHN is given in Figure 8. Nodes of the MHN is located with unit distance between adjacent nodes. Let

j	1	2	3	4	5	6
u_j	10	5	12	18	15	20
v_j	8	10	11	17	15	19

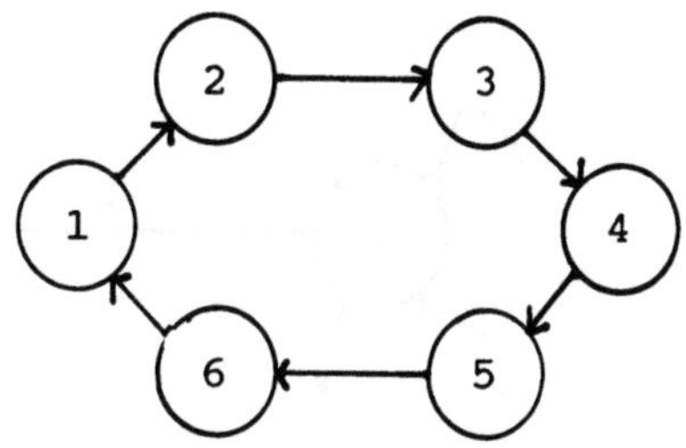

Figure 8: MHN for Example 4

The solution can be found by applying the algorithm.

i	f*(1)	MIN
1	0	0
2	18	0
3	6	0
4	-36	-36
5	-24	-36
6	-54	-54

Optimum solution is node 6.

<u>Location on Arcs</u>

The previous results indicate that to minimize the total material flow costs the central storage must be located such that to share the same pick point on the MHN with some other station. This may not be possible due to computer control difficulties or size constraints. In most of the real world situations, the pick point must be located sufficiently away from any existing station.

The following result simplifies the search for the optimum location.

<u>Property 2</u>: Let [a,b] be an arc with $\ell = \ell_{ab}$. Let $x \in [a,b]$ be a point with $d(a,x) = \lambda$. We have

$$f(x) = f_1(a) + f_2(b) + \lambda U + (\ell - \lambda)V \qquad (27)$$

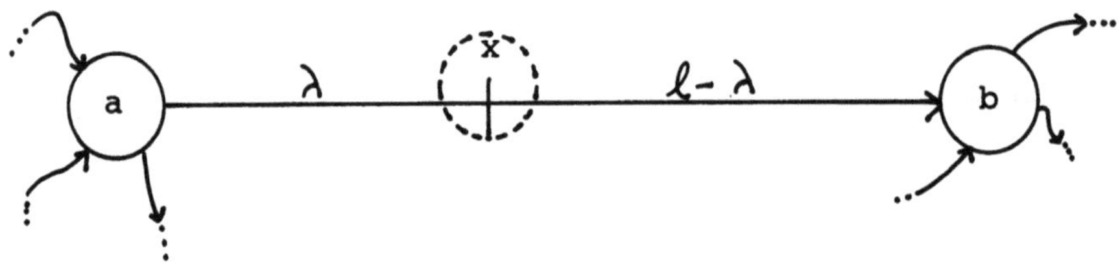

Figure 9: Property 2

<u>Proof</u>: Rewriting (15) for x we have

$$f(x) = f_1(x) + f_2(x) = \sum_{j=1}^{n} u_j\, d(j,x) + \sum_{j=1}^{n} v_j\, d(x,j)] \qquad (28)$$

But $d(j,x) = d(j,a) + \lambda$ and (29)

$$d(x,j) = d(b,j) + (\ell - \lambda) \qquad (30)$$

By substituting (29) and (30) into (28) we obtain (27). (27) further simplifies if U = V. This is not uncommon in practical cases due to the fact that any part coming to storage will eventually leave the storage. We refer to this case as the "balanced" case.

<u>Corollary 2:</u> In a balanced case, (27) becomes

$$f(x) = f_1(a) + f_2(b) + \ell V \qquad (31)$$

Corollary 2 indicates that in the balanced case the total cost of the pick point is independent of λ. A practical implication of this is to locate the pick point of the central storage at anywhere along the minimum cost edge. The minimum cost edge can be found by evaluating (31) for all arcs in the network.

In the general case of $U \neq V$, the minimum cost location can be found in $O(n^2 + |A|)$ time if the distance matrix is given. First let us rewrite (27) as

$$f(a,b) = f_1(a) + f_2(b) + \ell V + \lambda(U - V) \tag{32}$$

and observe that f(a,b) is monotonicly increasing (decreasing) in if $U > V$ ($U < V$). This indicates that on any arc [a,b] pick points must be chosen as close as possible to a if $U > V$ and as close as possible to b if $U < V$.

Hence a candidate point on each arc $[a,b] \in A$ of the graph G=(N,A) can be easily determined. The optimum pick point can be found by evaluating (32) at candidate points and choosing the minimum cost point. Example 5 illustrates the calculations.

<u>Example 5:</u> Recall the data given in Example 3. Let us consider the size constraints. The pick point must be located at least .2 unit distance away from nodes 1,2,3,4 and .1 unit distance away from node 5. Since $U = 40 < V = 45$ we shall consider pick points closer to b on directed arcs [a,b]. Calculations are shown below.

Arc [a,b]	Node b	$f_1(a)$	$f_2(b)$	ℓV	λ	f(x)
[1,5]	5	100	75	45	.9	215.5
[2,1]	1	75	75	45	.8	191
[3,4]	4	60	75	45	.8	176
[4,5]	5	70	75	45	.9	185.5
[5,2]	2	65	105	45	.8	211
[5,3]	3	65	90	45	.8	196

The optimum solution is on arc [3,4], .8 unit distance away from node 3.

4. SUMMARY AND CONCLUSIONS

We have examined the storage layout problems that arise in flexible manufacturing context. The objective function is defined as the minimization of material handling cost associated with the total part movement between stations and a central WIP storage area.

First, we have shown that known models of layout can be adapted to the context of flexible manufacturing. We have adapted the generalized assignment model to find optimum storage layout in flexible manufacturing cells. Examples are supplied to illustrate the approach.

Next, we have considered the problem of locating a pick point of the central storage unit in an FMS. We have shown that the pick point must be located at a node of the MHN to minimize the total cost function which is defined as the sum of the products of material flow and travel distances. In a real manufacturing environment, locating the pick point at existing nodes may not be possible due to constraints on computer controls and size limitations. We analyzed such cases in which the pick point must be located on an arc of the MHN. The analytical results obtained for this case are somewhat surprising. The actual location of the pick point on the arc does not affect the objective function under a reasonable "balancedness" assumption. This translates into practice as the selection of the minimum cost arc. An $O(n^2 + |A|)$ method has been developed to find the minimum cost arc where n and $|A|$ are the number of nodes and arcs, respectively. If MHN consists of a unicycle, the optimum node location can be found in $O(n)$ time.

REFERENCES

1. Afentakis, P., "A Model for Layout Design in FMS," pp. 127-139 in Flexible Manufacturing Systems: Methods and Studies, A. Kusiak (Ed.), North-Holland, 1986.
2. Domschke, W. and A. Drexl, Location and Layout Planning: An International Bibliography, Lecture Notes in Economics and Mathematical Systems, Springer-Verlag, 1985.
3. Francis, R.L. and J.A. White, Facility Layout and Location, Prentice-Hall, 1974.
4. Francis, R.L., L.F. McGinnis and J.A. White, "Locational Analysis," EJOR 12, pp. 220-252, 1983.
5. Heragu, S.S. and A. Kusiak, "Machine Layout Problem in Flexible Manufacturing Systems," Working Paper 08/86, Department of Mechanical and Industrial Engineering, University of Manitoba, 1986.
6. Jayaraman, R., "GALOP/2D--A Graphical System for Work-Cell Layout Evaluation," Research Report RC 11148, IBM Thomas J. Watson Research Center, 1985.
7. Kiran, A.S. and B.C. Tansel, "Optimum Pickup Point Location in Flexible Manufacturing Systems," Working Paper 87-06, Dept. of Industrial & Systems Engr., University of Southern California, 1987.
8. Kusiak, A., "Material Handling in Flexible Manufacturing Systems," Material Flow, 2, pp. 79-95, 1985.
9. Kusiak, A. and S.S. Heragu, "The Facility Layout Problem," EJOR, v. 29, No. 3, pp. 229-251, 1987.
10. Sarin, S.C. and W.E. Wilhelm, "Prototype Models for Two-Dimensional Layout Design of Robot Systems," IIE Transactions, 16, No. 3, pp. 106-125, 1984.
11. Tansel, B.C. and A.S. Kiran. Optimum Storage Location in Flexible Manufacturing Cells," Working Paper 87-05, Dept. of Industrial & Systems Engr., University of Southern California, 1987.
12. Wilhelm, W.E., M. Kalkunte, C. Cash, "A Modelling Approach to Aid in Designing Robotized Manufacturing Cells," Working Paper 1986-002, Department of Industrial & Systems Engineering, The Ohio State University, 1986.

Minimal Bottleneck Cell Approach for Generating Part-Machine Families in Cellular Manufacturing

by

Anthony Vannelli[†,*]

K. Ravi Kumar[††]

Abstract

Graph theoretic models are presented for finding part-machine families subject to production and management constraints. The new models allow one to reorganize the existing parts and machines into disaggregated cells. Two efficient heuristic techniques are developed which identify the most inexpensive bottleneck machines to duplicate while achieving capacity balance of the disaggregated cells. This approach has the flexibility of letting the designer control the number of cells and cell size thus generating a variety of cellular manufacturing system designs to choose from.

---

† Assistant Professor, Department of Electrical Engineering, University of Waterloo, Waterloo, Ontario N2L 3G1 Canada

†† Associate Professor, Department of Decision Systems, College of Business Administration, University of Southern California, Los Angeles, CA 90089, U.S.A.

* Part of the research of the first author (A. Vannelli) is funded by the Natural Sciences and Engineering Research Council of Canada - Grant No. U0381

NATO ASI Series, Vol. F49
Computer Integrated Manufacturing
Edited by I. Burhan Turksen

1. Introduction

The lagging growth in productivity by North American manufacturers has captured much attention during the last decade. A number of analysts attribute this problem, at least in part, to a certain reluctance on the part of management to invest in the capital equipment and innovative systems necessary to automate production at competitive and heightened levels. What is often not noted, however, is that all such investments necessarily would not carry equivalent benefits. Of the three principle manufacturing methods; namely, flow production, mass production and batch production, North American manufacturers would gain the most from stimulating automation in batch manufacturing as this form of production accounts for our sixty percent of all manufacturing activities.

In both flow and mass production, where single products in large volumes are manufactured, activities follow a predetermined sequence and use specialized machinery. Moreover, these manufacturing modes are highly automated. Batch (or job shop) production, which applies to the manufacture of several different products with relatively low volume and standardization, follows no invariable sequence, requires more general purpose equipment, involves greater unit costs and is far less automated. To compound these problems further, it has been found that historically, batch manufacturing has encouraged the submission of special orders and this has led to increased product variation and a growing emphasis on shop flexibility. The outcome of all this has included increased machine and tool investments per unit produced, complex scheduling and machine loading, higher setup time and costs and increased quality control costs.

In an effort to alleviate the problems inhirent in batch oriented manufacturing it was realized that many of the potential benefits of mass production may be derived for the otherwise batch orjob shop systems of manufacturing through a rationalization of production using the concept of *Group Technology (GT)* or *Cellular Manufacturing*. The concept of GT was originated by Mitrofanov (1959) and was developed into an integrated discipline by Burbidge (1975).

Group Technology may be defined as the manufacturing philosophy which recognizes that individual parts have similar characteristics by which they may be grouped into *families* or *cells*. It is a means of organizing and using information concerning component similarities to enhance the efficiency and effectiveness of small lot (or batch) manufacturing. In the last decade, the grouping of components based upon their design or production attributes has proven to be a powerful instrument in making the economies of scale normally associated with mass production available to the job shop batch processor. GT enables the benefits and technology of large scale production systems to be extended to the manufacture of small and medium sized batches through the creation of dedicated production cells.

The fundamental problem in GT is the identification of part-machine families. Burbidge (1975) initially generated part-machine families based on existing route sheet information. This problem has attracted considerable interest since Burbidge's (1975) original approach (Basale and Tripathi, 1981; Kusiak, 1984; Chakravarthi and Shtub, 1984; King and Nakornchai, 1982; Vannelli and Kumar, 1986; Kumar and Vannelli, 1986).

This paper develops graph theoretic models and heuristic solution techniques for redesigning the traditional production system into a disaggregated production system using Group Technology techniques. We focus on the disaggregation process by evaluating critical strategic decisions involving duplication of bottleneck machines as well as balancing of capacity between various cells. We propose a flexible heuristic technique which helps the designer to deal with these issues and give examples of how it can be used in practice. The heuristic technique is an extension of the *minimal bottleneck cell* approach that was introduced by the authors in Vannelli and Kumar (1986).

This paper is divided into five sections. In Section 2, we look at three graph theoretic model formulations of part-machine family problems that one considers in cellular manufacturing. We describe the fundamental bottleneck machine and part problem. We focus on the problem of determining the least expensive bottleneck machines to duplicate so that one can balance the capacity between generated cells. Two greedy solution techniques are developed for solving these two problems in Section 3. These heuristic techniques offer management and plant designers the flexibility of controlling several operational factors while disconnecting the plant into two or more subplants. The financial and production difficulties can be better controlled in such a cellular environment. The proposed solution techniques are tested on a representative problem in Section 4. Conclusions are presented in Section 5.

2. Part-Machine Family Model Formulations

In this section, we outline several important criteria that management and plant designers consider in using Group Technology concepts. The identification of the part-machine families remains the fundamental problem in attempting to disaggregate the manufacturing setup into two or more disconnected manufacturing cells. There are several possibilities in finding part-machine families which disaggregate the plant. In the following subsections, we outline three graph theoretic models which can be used to find disaggregated part-machine families. These models are based on the *minimal bottleneck cell* approach which was introduced in Vannelli and Kumar (1986). Subsequent models using this approach appear in Kumar and Vannelli (1986, 1987).

2.1 Identification of Minimal Bottleneck Machines and Parts

The identification of bottleneck machines is a well known problem in the design of efficient cellular manufacturing systems (Burbidge, 1975; King, 1982). Bottleneck machines are machines that process a large number of parts. However, parts themselves can be bottlenecks if they require a large number of machines to process them. Ideally, one would like to identify the *minimal* number of bottleneck machines and parts which disaggregate the plant into a fixed number of subplants satisfying size constraints; i.e., the subplant i contains no more than m_i machines and p_i parts. The disaggregation is achieved by subcontracting parts to outside vendors and purchasing duplicate bottleneck machines which are placed in other subplant cells. We illustrate these concepts on the following 3 machine-3 part example given in Fig. 1.

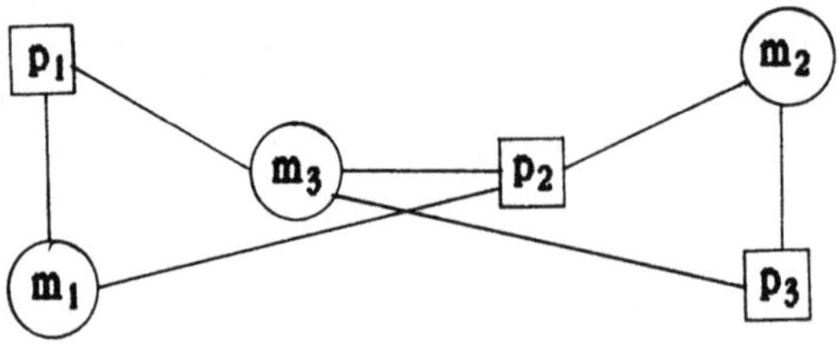

$m_i \triangleq$ machine #i
$p_i \triangleq$ part #i

Figure 1.

Note that part p_2 is processed on all three machines and machine m_3 is processing all three parts. By purchasing another machine m_3 and subcontracting part p_2, the plant can be disconnected into the two subplants given in Fig. 2.

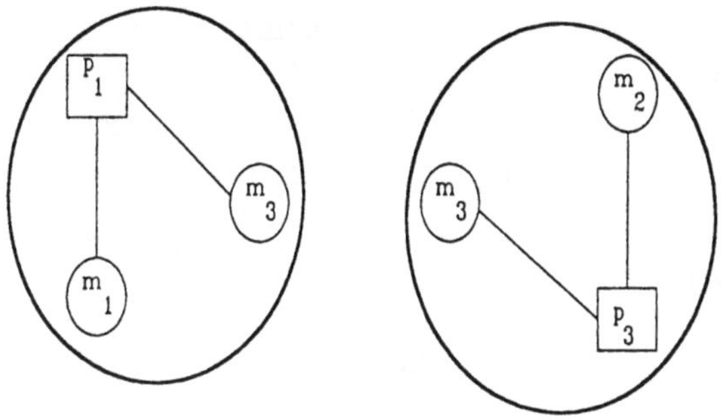

subplant 1 Figure 2 subplant 2

The problem of finding the minimal number of bottleneck machines and parts can be shown to be equivalent to the *tearing problem* in networks (Steward, 1965). This problem can be shown to be NP complete (Sangiovanni-Vincentelli, Chan and Chua, 1977). However, an efficient heuristic technique (Vannelli and Kumar, 1986-) can be used to find an approximate solution. This technique is outlined in the next section.

2.2 Identification of Bottleneck Machines Only

The graphical model of the part-machine family problem given in Section 2.1 considers the fundamental problem of finding the smallest number of bottleneck parts and machines. However, financial considerations may dictate that it is not feasible to use subcontraction, as a cellular design manufacturing tool. Unreliable or unavailable vendors for certain or all part types may dictate that it is more feasible in the long run to buy extra machinery only to achieve plant disaggregation. We illustrate this concept by considering the 3 machine - 3 part example give in Fig. 3.

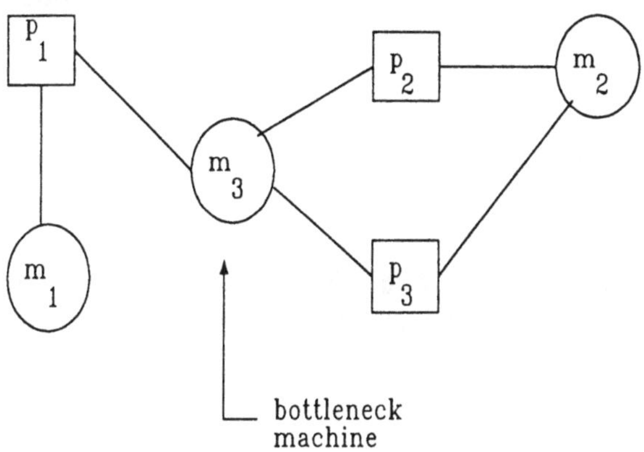

Figure 3

Note that machine m_3 is a bottleneck machine. The duplication of this machine would disconnect the operations into two disconnected subplants as shown in Fig. 4.

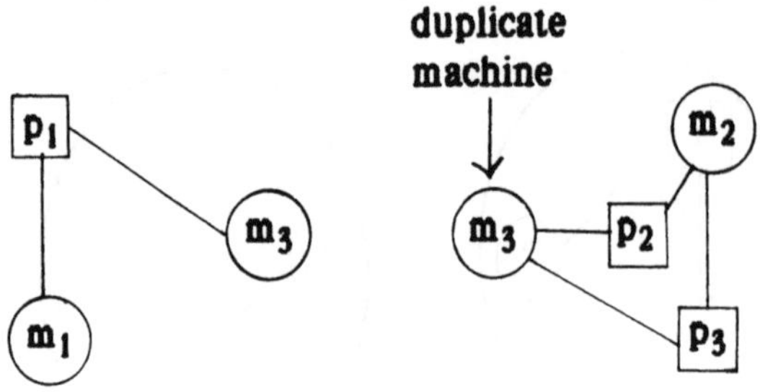

Figure 4.

2.3 Identification of Inexpensive Bottleneck Machines to Achieve Capacity Balancing

The graphical models described in the last two subsections can be expanded to account for other realistic production constraints. For example, consider a production system that produces 5 parts on 4-machines. The machine-part requirement is depicted in Fig. 5. Also shown in Fig. 5 are the various machine duplicating costs

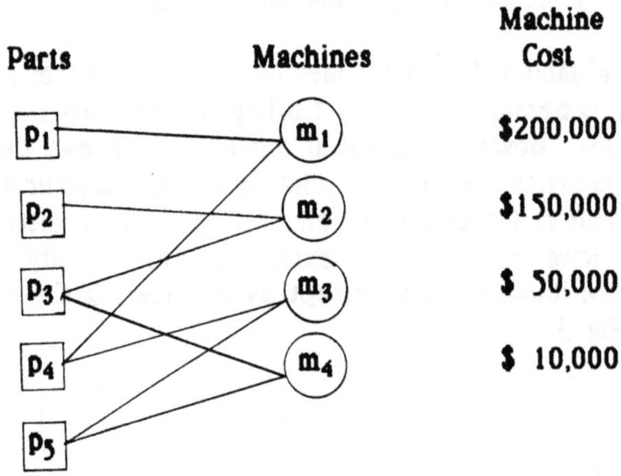

Figure 5. 5 part - 4 machine Example

An arc between a part node and a machine node indicates the required part processing time of a machine. The machine capacity requriement of each part being produced is given in Fig. 6.

Processing Weights

PARTS

MACHINES	p_1	p_2	p_3	p_4	p_5
m1	5	0	0	3	0
m2	0	1	2	0	0
m3	0	0	0	2	2
m4	0	0	3	0	2

Figure 6. Capacity Requirements

For example, part p_3 requires 2 units of machining time on machine m_2 and 3 units of machine time on machine m_4. The difference in requirements indicates that although the actual volume of part p_3 being produced annually is fixed, the processing time for machine m_2 is 2/3 that on machine m_4.

In the next section, we show that there are two possible disconnected cells that one can generate for this example containing <u>one</u> bottleneck machine. These are shown in Figs. 7 and 8.

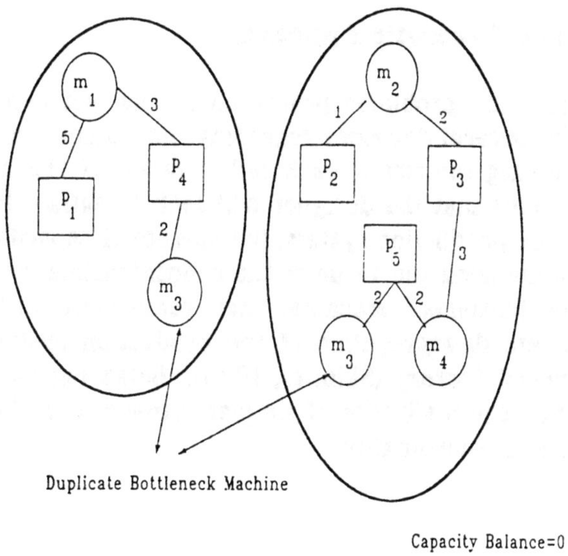

Figure 7. Machine m_3 as Bottleneck

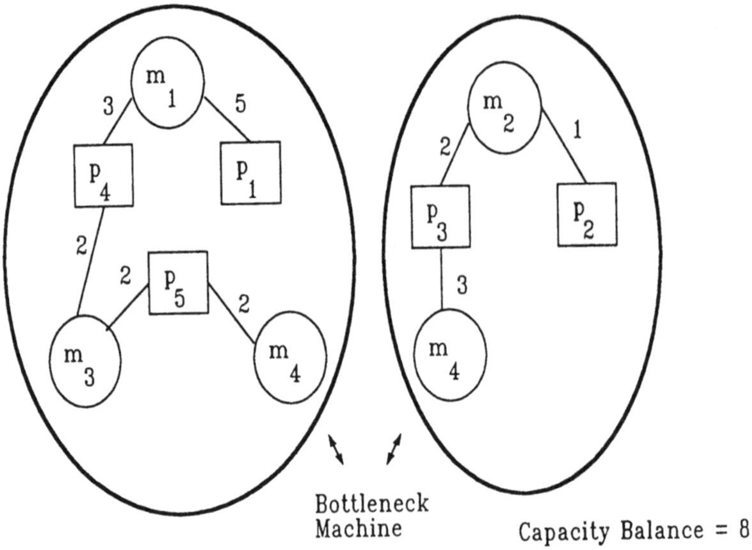

Figure 8. Machine m_4 as Bottleneck

Although the most inexpensive bottleneck machine to be duplicated is machine m_4 ($10,000) it yields a capacity balance of 8. By purchasing a slightly more expensive machine - machine m_3 ($50,000), the two disconnected cells are balanced in the total workload that the cells process. This example illustrates the tradeoffs in operational and cost criteria that management and engineers must consider.

3. Greedy Graph Theoretic Approaches

The disaggregation problems presented in the last section dictate that the designer can make several decisions regarding the number of cells that would be preferable and limiting the size of each cell in terms of the number of machines. Some of the objectives that the designer will seek to satisfy include the complete disaggregation of the production system, the balance of capacity among the various cells, as well as minimizing the value of the subcontracting cost of those parts and the value of those bottleneck machines that create material flow problems. The objective of complete disaggregation of the production system is to achieve the concept of a *focused factory* (Skinner, 1974). Balancing the work among all the cells is motivated by the realization of simpler managerial and control structures as well as easier material coordination.

We apply a variation of previous greedy graph theoretic heuristics (Vannelli and Kumar, 1986, Kumar and Vannelli, 1987) to solve the part-machine family probelm outlined in Section 2.3. We describe two *heuristic techniques* to solve this problem. There are two criteria that we attempt to optimize - capacity balance and duplicating cost of bottleneck machines. In the first heuristic, capacity balance is the primary objective and duplication is the secondary factor. We reverse these objectives in the second heuristic.

Heuristic 1

Step 1: Choose one initial seed (part or machine) and assign it to each of the predetermined number of cells. If the seed is a machine, add all the parts attached to the machine into the same cell as the machine.

Step 2: Find the boundary set of machines, one for each cell that is attached to the parts.

Step 3: (i) Choose at most one machine node from each of the boundary machine sets and add it to the corresponding cell with its part nodes such that the sum of the pairwise capacity differences among all cells increases minimally and that the upper bound on the number of machines is satisfied.

(ii) If there is a tie between several machine nodes, break the tie by choosing the machine node which minimally increases the cost of duplicating bottleneck machines in the bottleneck machine set.

Step 4: Go to Step 2 and continue until each node is assigned to a cell or bottleneck machine set.

As an illustration of Heuristic 1 consider the 5 part - 4 machine example with processing weights given in Figs. 5 and 6.

Step 1: We wish to find two part-machine families containing no more than three machines in each cell. Choose machines m_1 and m_2 as seeds for each cell. Add parts p_1 and p_4 to machine m_1 and parts p_2 and p_3 to machine m_2. This is shown in Fig. 9.

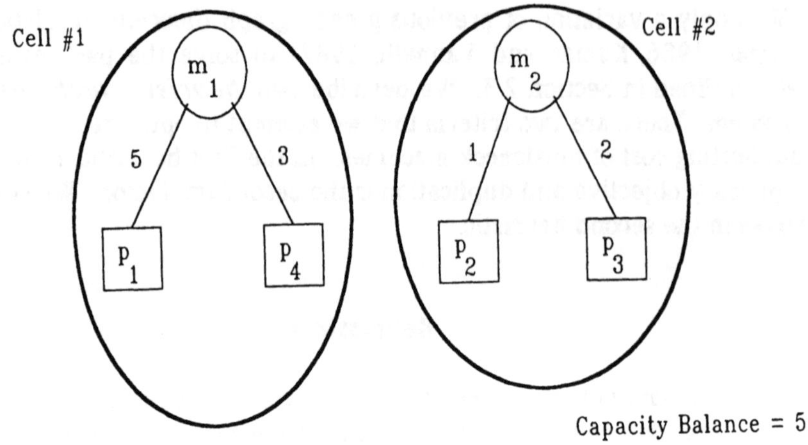

Figure 9

The capacity balance is |8 - 3| = 5 at this stage.

Step 2: **The boundary set of machines for Cell #1 is (m_3) and for Cell #2 is (m_4). This is shown in Fig. 10.**

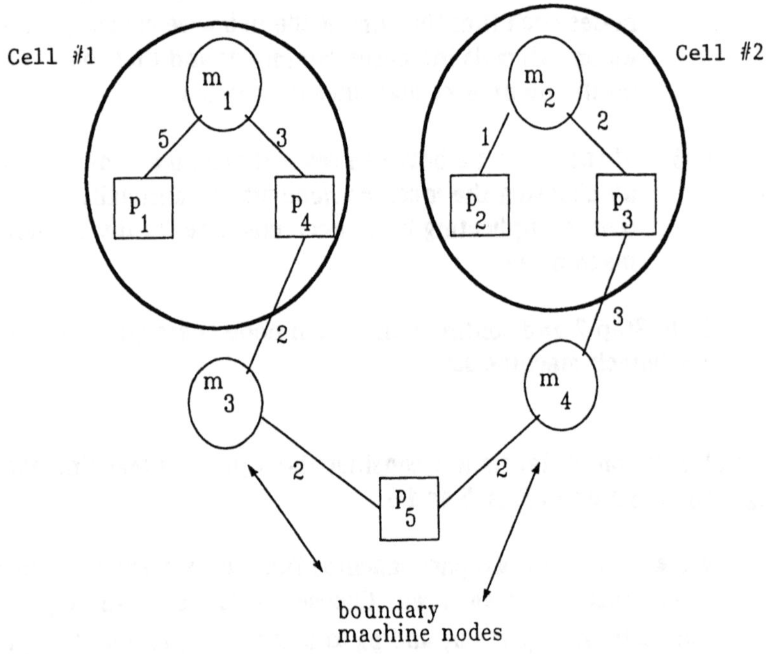

Figure 10

Step 3: Note that we cannot choose to add mchine m_3 to Cell #1 and machine m_4 to Cell #2 since a bottleneck part would be generated. There are only two feasible choices:

(i) machine m_3 and part p_5 is added to Cell #1 and machine m_4 is the bottleneck machine node to be duplicated. The resulting configuration is given in Fig. 8. The capacity balance is 8 with a cost of $10,000 for machine duplication.

(ii) machine m_4 and part p_5 is added to Cell #2 and machine m_3 is the bottleneck machine node to be duplicated. The resulting configuration is given in Fig. 7. The capacity balance is 0 with a cost of $50,000 for machine duplication.

In the second heuristic minimizing the cost of machine duplication is the primary factor and capacity balance is a secondary factor.

Heuristic 2

This is the same as Heuristic 1 except that the objective functions that are minimized in Step 3 are interchanged.

Applying Heuristic 2 on the example given in Figs. 5 and 6, we obtain the configuration given in Fig. 8 as the two disconnected cells which are obtained by minimizing bottleneck machine duplication cost.

4. Numerical Results

Heuristics 1 and 2 have been coded in APL and are presently running on an IBM 4341 computer at York University Computing Centre. We tested the preceeding heuristics on a 30 machine-41 part example shown in Fig. 11 (Kumar and Vannelli, 1987) and with the machine-part process weights given in Fig. 12.

Parts

	1	2	3	4	5	6	7	8	9	10	11	12	13	14	15	16	17	18	19	20	21	22	23	24	25	26	27	28	29	30	31	32	33	34	35	36	37	38	39	40	41
1										1																					1	1	1						1		1
2											1	1						1														1	1						1		1
3												1											1								1	1							1	1	
4							1								1									1	1			1							1						
5					1												1																	1		1					
6						1																				1															
7																1											1							1		1					
8	1							1	1					1															1												
9			1										1								1																				
10		1										1								1			1																1		
11										1																							1						1		1
12											1	1							1	1			1	1															1	1	
13																			1						1													1			
14				1	1													1								1											1				
15																	1																				1				
16						1																						1													
17							1									1											1									1					
18														1		1											1							1							
19	1		1																		1	1								1											
20	1		1																		1	1																			
21										1		1																			1	1	1							1	
22										1		1											1								1	1							1	1	
23		1										1											1								1								1		
24																																						1			
25				1																																					
26																1											1							1		1					
27								1							1																										
28								1																					1						1						
29	1		1						1				1								1									1											
30	1		1																		1	1																			

Fig. 11.

Parts

	1	2	3	4	5	6	7	8	9	10	11	12	13	14	15	16	17	18	19	20	21	22	23	24	25	26	27	28	29	30	31	32	33	34	35	36	37	38	39	40	41
1										3																					6	4	3						2		7
2											7	9						8														2	8						6		9
3												2											2								6	2							6	5	
4							7								1									5	1			9							5						
5					8												4																	3		7					
6						7																				8															
7																3											1							1		2					
8	4							4	2					3															3												
9			2										2								5																				
10		3										3								8			2																1		
11										1																							5						9		3
12											4	4							1	8			7	4															8	1	
13																			4						8													9			
14				9	1													6								6											7				
15																	5																				2				
16						7																						8													
17							7									4											9									8					
18														9		4											3							9							
19	7		7																		2	2								1											
20	2		4																		9	7																			
21										4		5																			3	4	3							5	
22										5		2											2								7	7							4	7	
23		9										1											8								3								3		
24																																						8			
25				9																																					
26																3											3							9		1					
27								2							1																										
28								7																					6						7						
29	8		6						7				1								3									9											
30	7		1																		8	4																			

Fig. 12.

The purchase cost of the machines are:

m1 = $ 2000	m2 = $13000	m3 = $34000	m4 = $ 5000
m5 = $ 7000	m6 = $ 9000	m6 = $22000	m8 = $32000
m9 = $40000	m10= $17000	m11= $15000	m12= $ 3000
m13= $10000	m14= $11000	m15= $23000	m16= $35000
m17= $29000	m18= $28000	m19= $21000	m20= $20000
m21= $33000	m22= $ 6000	m23= $ 5000	m24= $ 2000
m25= $22000	m26= $39000	m27= $12000	m28= $19000
m29= $16000	m30= $36000		

Using Heuristic 1; that is, with a first priority of minimizing capacity balance difference between cells, parts 1 and 30 as seeds, and an upper bound of 20 machines allowed per cell, the bottleneck machine set is found to be {2,4}.

Cell #1 contains:

machines - 1, 3, 10, 11, 12, 13, 21, 22, 23, 24.

and parts - 2, 10, 11, 12, 19, 20, 23, 24, 25, 31, 32, 33, 38, 39, 40, 41.

Cell #2 contains:

machines - 5, 6, 7, 8, 9, 14, 15, 16, 17, 18, 19, 20, 25, 26, 27, 28, 29, 30.

and parts - 1, 3, 4, 5, 6, 7, 8, 9, 13, 14, 15, 16, 17, 18, 21, 22, 26, 27, 28, 29, 30, 34, 35, 36, 37.

The capacity balance difference is 85 while the bottleneck machine purchase cost is $18,000.

Using Heuristic 2 with a priority of minimizing bottleneck machine cost first and the other criteria remaining the same, we find that the bottleneck machine set is found to be {4,5}.

Cell #1 contains:

machines - 1, 2, 3, 6, 10, 11, 12, 13, 14, 15, 16, 21, 22, 23, 24, 25.

and parts - 2, 4, 5, 6, 10, 11, 12, 17, 18, 19, 20, 23, 24, 25, 26, 28, 31, 32, 33, 37, 38, 39, 40, 41.

Cell #2 contains:

machines - 7, 8, 9, 17, 18, 19, 20, 26, 27, 28, 29, 30.

and parts - 1, 3, 7, 8, 9, 13, 14, 15, 16, 21, 22, 27, 29, 30, 34, 35, 36.

The machine purchase cost is $12,000 and the capacity balance difference is 136.

The encoded heuristic codes are very efficient. Although the heuristics are coded in APL, the two heuristics solve the problem in 8.5 and 9.6 sec. respectively. More important, the example illustrates the flexibility that a designer has in purchasing duplicate bottleneck machines to disaggregate plant operations.

5. Conclusions

In this paper we extended the minimal bottleneck cell approach (Vannelli and Kumar, 1986) to model and solve the part-machine family problem where one attemps to disaggregate existing production into two or more disconnected cells. The disconnected cells were generated by finding the least expensive bottleneck machines to duplicate. The number of cells as well as the size of the group cells are under the control of the plant designer. We also attempted to achieve such a disaggregation by balancing the capacity differences between cells.

Two heuristic approaches were developed for identifying these disaggregated part-machine families subject to these criteria. The heuristic methods were encoded in APL and tested on a representative problem. Any APL code can be accelerated by an order of magnitude by coding it into a higher level language. This is currently being undertaken for Heuristics 1 and 2. The efficiency and flexibility of the proposed heuristic methods may serve as an important decision aid in production system design.

References

Burbidge, J.L. (1975), The Introduction of Group Technology, Halsted Press, John Wiley and Sons, New York.

Basale, P.C. and Tripathi, D.K. (1981), "Group Technology in CRS through PFA," J. Institute of Engineers (India) - M.E., pp. 61, 204-206.

Chakravarthi, A.K. and Shtab, A. (1984), "Selecting Parts and Loading Flexible Manufacturing Systems,"Proceedings of First ORSA/TIMS Special Interest Conference on FMS, Ann Arbor, Michigan.

King, J.R. and Nakornchai, V. (1982), "Machine-Componnent Group Formation in Group Technology: Review and Extension," International Journal of Production Research, vol. 20, no. 2, pp. 117-133.

Kumar, K.R. and Vannelli, A. (1986), "Design of Flexible Production Systems: Capacity Balancing and Subcontracting Strategies", in Proceedings of Second ORSA/TIMS Conference on Flexible Manufacturing Systems: Operations Research Models and Applications", edited by K.E. Stecke and R. Suri, Elsevier Publishers, B.V., Amsterdam, pp. 203-208.

Kumar, K.R. and Vannelli, A. (1987), "Strategic Subcontracting for Efficient Disaggregated Manufacturing," to appear International Journal of Production Research. Working Paper #85-9. Department of Industrial Engineering. University of Toronto.

Kusiak, A. (1984), "The Part Families Problem in Flexible Manufacturing Systems", Working Paper #06-84, Technical University Nova Scotia, Department of Industrial Engineering.

Mitrofanov, S.P. (1959), The Scientific Principles of Group Technology, Leningrad.

Sangiovanni-Vincentelli, A. Chen, L.K. and Chua, L.O. (1977), "An Efficient Heuristic Cluster Algorithm for Tearing Large-Scale Networks," IEEE Transactions on Circuits and Systems", vol. CAS-24, no. 12, pp. 709-711.

Skinner, W. (1974), "The Focussed Factory", Harvard Business Review, May-June, pp. 113-121.

Steward, D.V. (1965), "Partitioning and Tearing Systems of Equations, "Journal of Numerical Analysis", vol. 2, pp. 345-365.

Vannelli, A. and Kumar, K.R. (1986), "A Method for Finding Minimal Bottleneck Cells for Grouping Part-Machine Families, "International Journal of Production Research, vol. 24, no. 2, pp. 387-400.

PRESENT STATE OF MANUFACTURING AUTOMATION

Y. Altintas
University of British Columbia
Vancouver, B.C., Canada

ABSTRACT

This paper discusses the present state of the Computer Aided Design and Computer Aided Manufacturing (CAD/CAM) technology. Particular attention is given to efficient, automated use of professional CAD/CAM systems, and to unsolved problems of unmanned machining technology. Automated design and manufacture of sculptured aircraft components is presented as an example of efficient use of CAD/CAM systems. In-process detection of tool failure for untended milling operations is presented in the paper to address the tool condition monitoring problems in unmanned machining systems.

1. INTRODUCTION

Commercial CAD/CAM systems have been commonly used since early 1980s. However, it is still not possible to claim that the systems are used with its full capacity. Various stages of product development cycles, such as Computer Aided Design (CAD), Computer Aided Engineering Analysis (CAE), Computer Aided Manufacturing (CAM), Computer Aided Process Planning (CAPP), Material Requirement Planning (MRP) systems have not been sufficiently integrated, although they strongly share information with each other. Many researchers and companies are working towards achieving fully integrated efficient systems. However, it is still possible to integrate and automate some of the product development stages using the existing CAD/CAM systems. A representative example, a design automation of a family of sculptured aircraft parts is given in the following section of the paper.

The recent trend in unmanned manufacturing is to achieve integrated and self-adjusting machining systems which are capable of machining varying parts without the supervision of operators. The absence of human supervision requires on-line monitoring of machining operation, ensuring a safe and efficient metal removing rate and taking corrective actions in the event of failures and disturbances. One of the most

NATO ASI Series, Vol. F49
Computer Integrated Manufacturing
Edited by I. Burhan Turksen

important monitoring requirements is a system capable of detecting tool breakages and wear on line. The author provides two methods of tool breakage detection for unmanned machining systems. The methods, one for the single batch size flexible production, the other for the medium size batch production, are presented in section 3 of the paper.

2. AUTOMATED DESIGN OF COMPLEX SCULPTURED COMPONENTS

It is common practice to use Computer Aided Design and Computer Aided Manufacturing Systems as conventional drafting and Numerical Control (NC) programming tools. The conventional method is that the operator designs the part and generates the tool paths interactively without utilizing the common geometric and manufacturing features in the family of parts. CAD/CAM systems have programming language interfaces to the graphics data base of the system. This enables the users to embed any interactive command in a user application software, therefore minimizing the operators' repetitive interactive use of the CAD/CAM system. The method enables the automation of various stages of product design and manufacturing cycle.

Automation of the design and NC tool path generation of complex family of sculptured aerodynamic components is shown in this paper as an example of the method proposed. Gas turbine blades, aircraft propellers and electro-chemical machining dies for a fighter aircraft jet's turbine blades are used as members of the part family. Part family members have similar but different number of airfoil sections along the blade's axis. The roots of the blades are standard for each major member. For example; the gas turbine blades manufactured by the particular company have standard roots which fit into the gas turbine carrier, so as the propeller roots which fit into the standard housing in a particular aircraft. A design manufacturing application program, which runs on GE-CALMA CAD/CAM system, was developed by the author to automatically design the mentioned similar parts. All the airfoil data supplied by the aerodynamicist is stored in a data base. The data base consists of part number, number of airfoils along the axes and their stagger angles around the airfoil centreoids, and the airfoils' two (x and y) coordinates. For each part number, the associated standard blade root exists which is also identified in the part's data base.

The program runs as follows. It reads and decodes all the part data from the data base. The first stage is to fit cubic splines around each airfoil and to find their centroids (Figure 1). All the airfoils are stacked and rotated around the centroid at

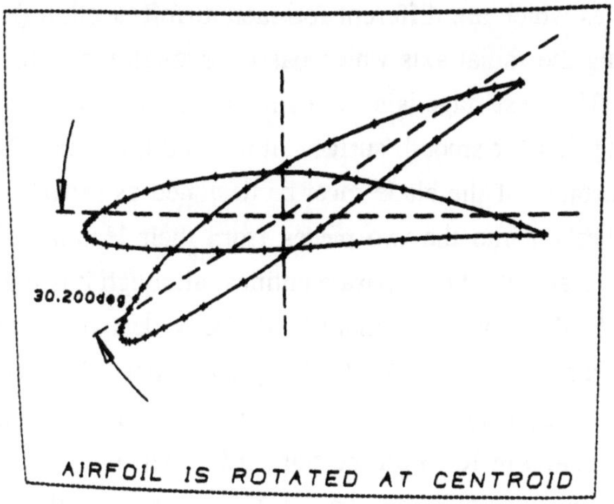

Figure 1. Fitted airfoil section with a calculated centroid.

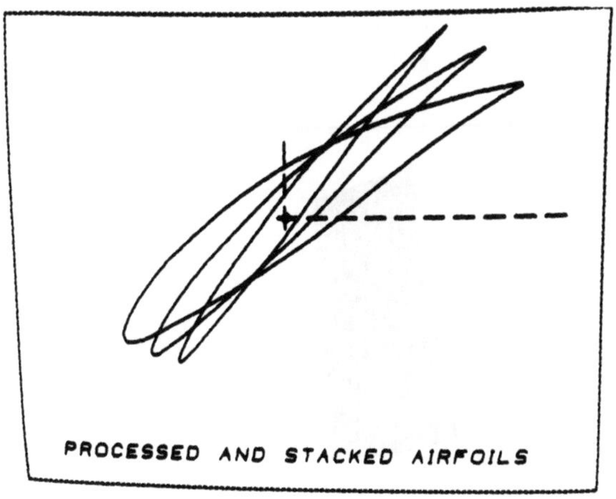

Figure 2. Stacked airfoils at their centroids.

amount of stagger angles which are different for each airfoil section (Figure 2). The airfoils are spaced along the radial axis which passes through the airfoil centroid and turbine's rotor center. The next stage is to fit a smooth surface over the spaced airfoil sections. Remembering that for smooth surface fitting and three axes NC machining, the upper and lower surfaces of the blade must be designed as separate entities. The airfoils are therefore broken into the two pieces from their leading edge tips. The upper and lower surfaces are fitted as separate entities, although it is not visible by the designer (Figure 3). At this point, the main blade body design is completed. The standard root was searched from the CAD/CAM system's part library (Figure 4) and assembled to the blade body (Figure 5). The program automatically generates four drawing views of the part and includes all the standard dimensions. The drawing view is stored under the drawing code number. The roughing and semi-finishing tool paths are generated and stored under a manufacturing view number. The program continues to design and produce tool paths for all the turbine blades and propellers existing in the part design data base. It can run overnight, or in a batch mode from an ASCII terminal thus leaving the more expensive graphics workstation for operators' use. The drawing and NC manufacturing views are further modified interactively by the draftsman and the process planner until they are satisfied. However, the program elimi-

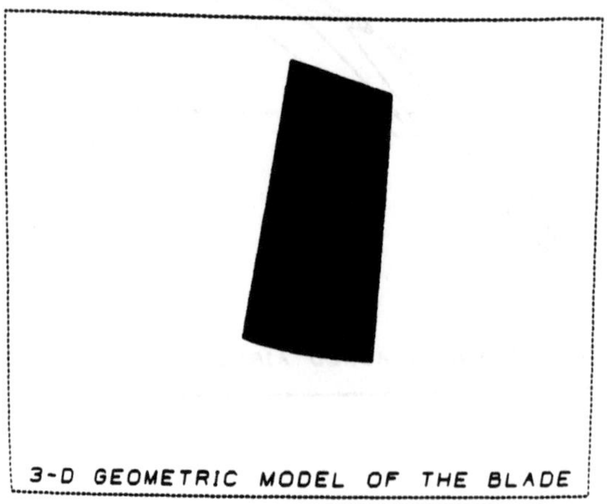

Figure 3. Main blade body with a fitted smooth surface.

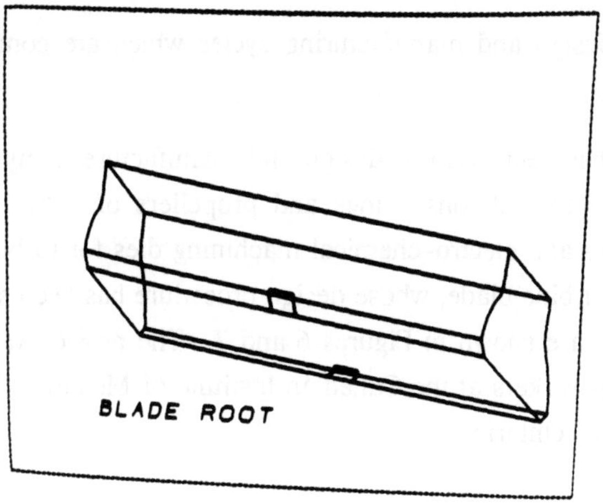

Figure 4. Standard blade root fetched from part library.

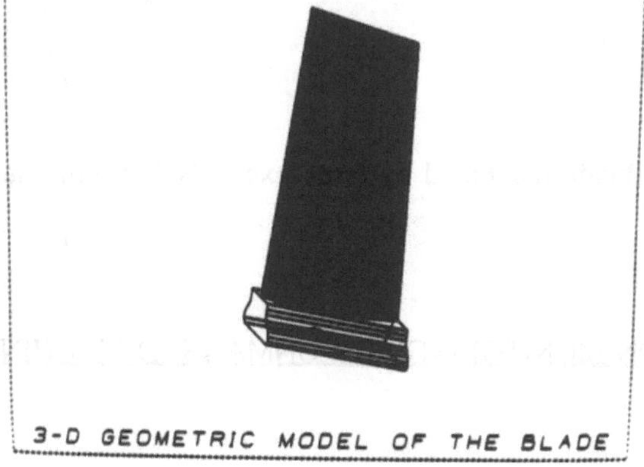

Figure 5. Three dimensional geometric model of the turbine.

nates the repetitive design and manufacturing cycles which are common for all the part family members.

The program has been used to design and manufacture many aircraft components (quarter scale flaps, alerons, wings and propellers of a Canadian passenger aircraft, turbine blades and electro-chemical machining dies for turbine blades, etc.). Two sample parts, a turbine blade, whose design procedure has been shown in Figures 1 to 5, and propellers are shown in Figures 6 and 7. The project was carried out by the author and his co-workers at the Canadian Institute of Metalworking at McMaster University, Hamilton, Ontario.

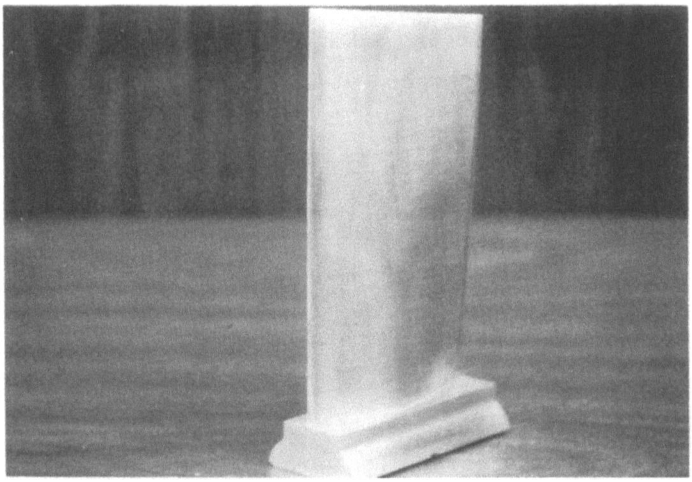

Figure 6. Turbine blade machined by three axes CNC milling machine.

TOOL CONDITION MONITORING FOR UNMANNED MACHINING CENTERS

The recent attempts in tool breakage detection involve the monitoring of acoustic emission, spindle motor current, vibration and cutting force signals [2]. It has been proven that it is quite difficult to relate the signals to the state of the cutting process excepting the cutting force measurements. The mechanics of cutting have been satisfactorily modeled to relate the cutting parameters (e.g. feedrate, speed, axial depth of

Figure 7. CNC machined aircraft propellers designed by the automated CAD/CAM software.

cut and immersion) with measured milling forces. It was shown that tool condition can be monitored using the equations of mechanics of milling and the measured cutting forces [3].

The first method, which was developed for single size batch flexible production, an adaptive tool breakage detection threshold algorithm was developed [4]. The threshold is adaptively updated as a function of the average cutting forces and in-process identified radial width of the cut [5]. The measured normalized average cutting forces are continuously compared with the adaptively identified tool breakage threshold at each tooth period. When the measured normalized force exceeds the present threshold value, a feed-hold and tool-change signal is sent to the CNC unit of the machine tool. Figure 8 shows the average cutting force history with a tool breakage event at tooth period 187. The face milling cutter has six teeth, and one of the teeth was broken during the machining. It is evident from Figure 9 that the measured normalized force difference amplitudes exceed the tool-breakage detection threshold values which stay at the stationary level. The algorithm is independent of the workpiece geometry, and does not need any calibration cut in order to identify the threshold values. The parameters, which are used to identify the threshold, depend on maximum allowable radial throws and the number of teeth on the cutter. The threshold identification parameters are calculated using an off-line computer algorithm

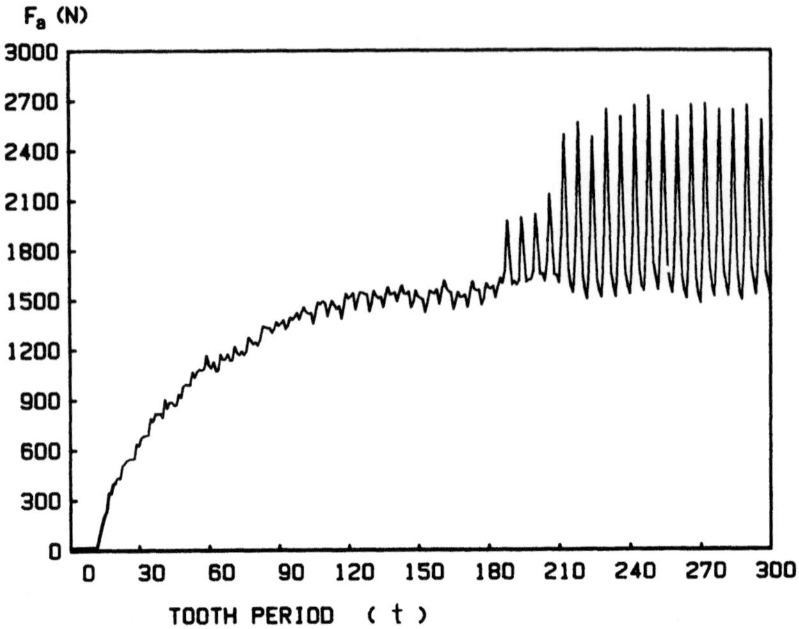

Figure 8. Quasi-mean resultant forces with a tool breakage event.

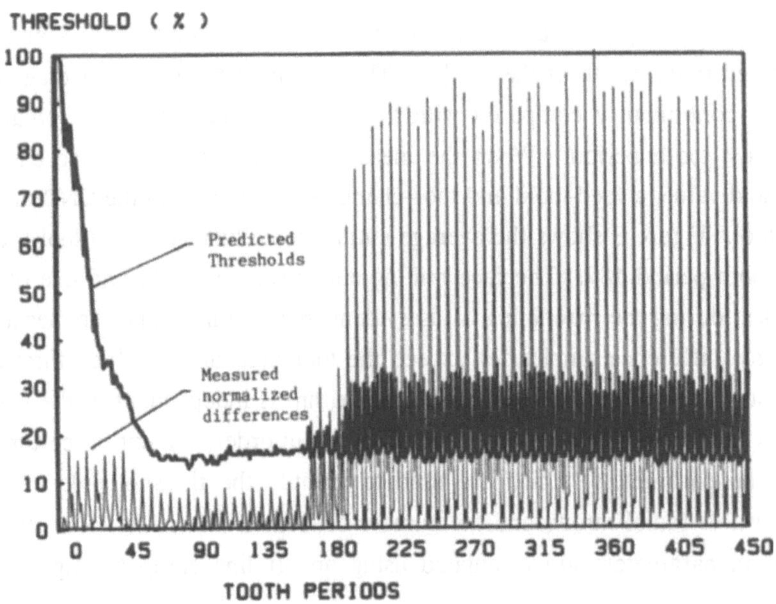

Figure 9. Detection of natural tool failure by threshold adaptation algorithm.

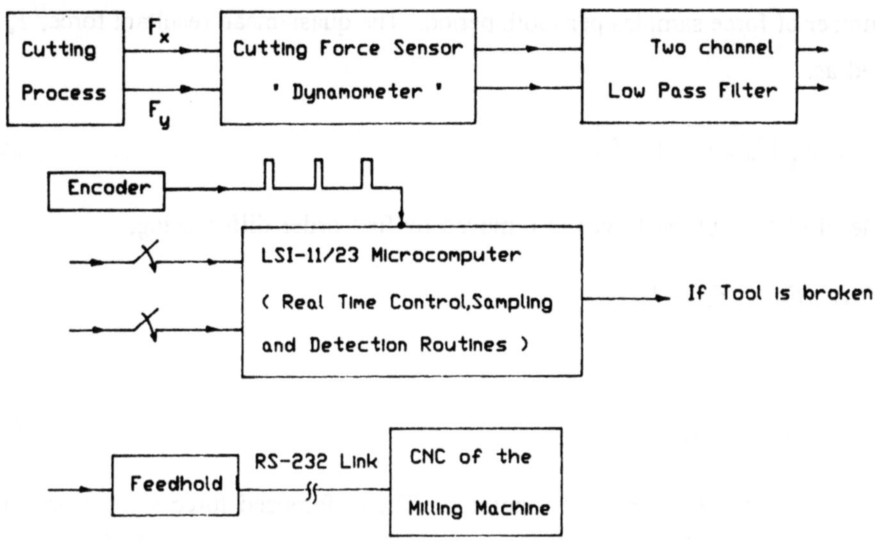

Figure 10. Block diagram of the experimental tool breakage detection system

developed from the mechanics of milling. The mentioned method has been tested in real time machining of many parts with severe geometric transients, such as holes, slots, and step changes in the axial depth of cuts. At all cases, the algorithm detected the tool-failure and sent a feed-hold signal to the CNC within one revolution of the spindle. The block diagram of the experimental set-up is shown in Figure 10.

The second method is based on the Time Series Monitoring of the average cutting forces per tooth period [6]. The cutting forces in milling is periodic at the tooth frequency. The periodicity in the cutting forces is first removed by averaging at each tooth period.

$$F_{ax}(t) = \frac{\sum_{i=1}^{k} F_x(i)}{k} \tag{1}$$

$$F_{ay}(t) = \frac{\sum_{i=1}^{k} F_y(i)}{k} \tag{2}$$

Where $F_x(i)$ and $F_y(i)$ are the instantaneous cutting forces, $F_{ax}(t)$ and $F_{ay}(t)$ are the average forces per tooth period (t) in the feeding and normal direction, and k

is the number of force samples per tooth period. The quasi–mean resultant force, F_a, is defined as,

$$F_a(t) = \sqrt{[F_{ax}(t)]^2 + [F_{ay}(t)]^2} \tag{3}$$

The dc level of the cutting forces is removed by first order differencing,

$$f_a(t) = F_a(t) - F_a(t-1)$$

or

$$f_a(t) = [1 - B]\ F_a(t) \tag{4}$$

where B is the backward time shift operator. The differenced force signals provide comparison between the performances of the successive teeth. A first order autoregressive filter is found to be sufficient to detect the abnormal changes in the cutting forces which are due to the sudden tool breakage event,

$$f_a(t) = \frac{1}{1 - \phi_1 B}\ \epsilon(t)$$

or

$$\epsilon(t) = f_a(t) - \phi_1 f_a(t-1) \tag{5}$$

When the residuals $\epsilon(t)$ of the filter exceeds a threshold value, which is predetermined with a calibration cut for particular production batch, the algorithm sends a feed–hold signal to the CNC, and tool–breakage alarm signal to the operator. The cutting forces with a tool breakage event is shown in Figure 8. The residual forces of the first order filter is shown in Figure 11 which clearly indicates the tool breakage at tooth period 187. Since the algorithm is based on a first order adaptive autoregressive filter, it is robust and can adapt itself to the changing geometric transients while detecting abrupt force changes caused by the tool breakage event.

4. CONCLUSION

The author discussed the present state of the CAD/CAM technology and unmanned manufacturing by providing the research projects as examples. The examples cover the state of the automation at the engineering design, manufacturing planning

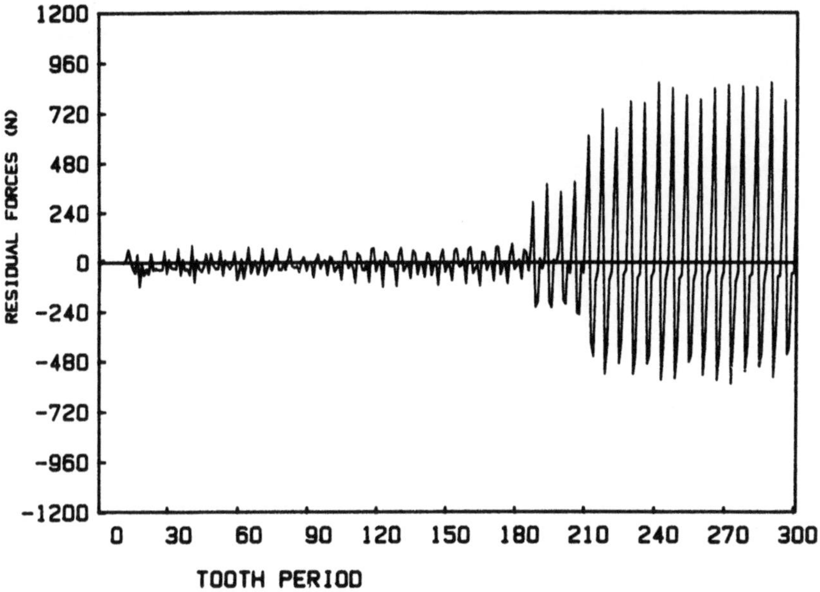

Figure 11. Residuals of AR(1) filter for tool breakage detection. Input data quasi mean resultant forces with tool breakage event, see Fig. 8. $[\epsilon(t) = (1 - \phi_1 B)\, f_a(t)]$

and unmanned production stages. The author is currently involved in the research investigation of constraint based adaptive control of unmanned machining centers for optimum metal removal, industrial cutting force sensor development, control of machining vibrations and design of intelligent, multi-processor based CNC machine tool controllers at the University of British Columbia.

BIBLIOGRAPHY

1. Altintas, Y., "Automated Design and Manufacture Turbine Blades", Canadian Industrial Computer Systems Conference, McMaster University, 1981.
2. Tlusty, J., Andrews, G., "Critical Survey of Sensors", Annals of the CIRP, Vol. 32/2/1983.
3. Altintas, Y., Yellowley, I., and Tlusty, J., "Detection of Tool Breakages in Milling", Proc. of ASME Winter Annual Meeting, Nov. 1985.
4. Altintas, Y., Yellowley, I., "In-Process Detection of Tool Failure in Milling Using Cutting Force Models", Proc. of ASME Winter Annual Meeting, Dec. 1987.
5. Altintas, Y., Yellowley, I., "The Identification of Radial Width and Axial Depth of Cut in Peripheral Milling", Int. J. of Machine Tool Manuf., Vol. 3/1987.
6. Altintas, Y., "In-Process Detection of Tool Breakages Using Time Series Monitoring of Cutting Forces", Int. J. of Machine Tool Manuf. 1987 (in press).

MODELLING A BELT CONVEYOR

P.DRAKATOS, E. SOTIROPOULOS, A. DENTSORAS
University of Patras, Mechanical Engineering Dept.
Patras, GREECE

ABSTRACT

The mathematical model of a belt conveyor expresses quantitatively the relation of the materials-machine parameters. Also it expresses the mathematical relation among the size variables, the operation and the capacity of the conveyor. Therefore, there is the ability of comparison among conveyors of various sizes and of experimenting on models in laboratory environment corresponding to real structures which may extend even for kilometers.

The purpose of the present work is the creation of a mathematical model to simulate a belt conveyor, paying attention to various variables and parameters such as size, function and conveying capacity. Some simplifications are used for the relations among variables. For the present case, a belt conveyor is chosen with three equal - size idlers.

The creation of the mathematical model is based on the theorem of Π_i terms or Vaschy-Buckingham theorem and its solution is approached by numerical methods. The coefficients of the equation are determined by a suitable program written in BASIC.

The extent of application of either experimental and real data on mathematical model of simulation is constrained by the relative error which-in order to be acceptable-must not exceed 5%.

1. Introduction

Because of the high cost of the installed conveyors, it is a common practice to study their function on models using computers.

The purpose of the present analysis is the determination of the participation of both functional variables and parameters to the creation of the mathematical model simulating the real structure. For this reason, some simplifications are introduced in the system among variables and useful assumptions are made for the identification of their properties.

NATO ASI Series, Vol. F49
Computer Integrated Manufacturing
Edited by I. Burhan Turksen

A trough belt conveyor is considered with three equal-size idlers. The creation of the model is based on Theorem of Π_i terms of Vaschy-Buckingham theorem and the solution is obtained by numerical methods.

NOMENCLATURE

α	grain size (m)
B	Belt width (m)
C_{lmk}	Coefficient
E	Relative error (%)
F	Load section (m^2)
$K_1, K_2, \ldots$	General exponents of variables
L	Belt length (m)
L	Length notation
M	Mass notation
N	Power (kw)
p	Number of variables
Q	Conveying capacity (t/h)
q	Number of fundamental sizes
T	Time
$X_1, X_2 \ldots$	variables
u	Conveyor speed (m/s)
γ	Bulk weight (N/m^3)
ε	Error
$\Pi_1, \Pi_2, \ldots$	Dimensionless similarity monomial or Π_i terms
Φ	Angle notation
φ_d	Dynamic repose angle (degrees)

2. SIMILARITY MONOMIALS

The function of a belt conveyor is influenced by a great number of variables.

These variables can be listed under the following two major categories:

1. Independent variables
2. Dependent variables

The most important variables which strongly affect the function of a conveyor are listed in the following table.

TABLE 1. Affecting variables

a. Independent variables

Grain size	α	$M^oL^1T^o\Phi^o$
Dynamic repose angle	φ_d	$M^oL^oT^o\Phi^1$
Bulk weight	γ	$M^1L^{-2}T^{-2}\Phi^o$
Belt length	L	$M^oL^1T^o\Phi^o$
Belt speed	u	$M^oL^1T^{-1}\Phi^o$

b. Dependent variables

Conveying capacity	Q	$M^1L^1T^{-3}\Phi^o$
Power	N	$M^1L^2T^{-3}\Phi^o$

For the creation of the similarity monomials the theorem of Vaschy-Buckingham is applied (known also as Theorem of Π_i terms). For this reason the following dimensioned matrix is formulated

TABLE 2. Dimensioned variables matrix

	k_1	k_2	k_3	k_4	k_5	k_6	k_7	k_8
	α	B	u	N	φ_d	γ	Q	L
M	0	0	0	1	0	1	1	0
L	1	1	1	2	0	-2	1	1
T	0	0	-1	-3	0	-2	-3	0
Φ	0	0	0	0	1	0	0	0

The variables were set in the above matrix in such a manner, that the right determinant is non zero as shown below.

$$\begin{vmatrix} 0 & 1 & 1 & 0 \\ 0 & -2 & 1 & 1 \\ 0 & -2 & -3 & 0 \\ 1 & 0 & 0 & 0 \end{vmatrix} = -1 \begin{vmatrix} 1 & 1 & 0 \\ -2 & 1 & 1 \\ -2 & -3 & 0 \end{vmatrix} = -1 \neq 0 \qquad (1)$$

According to Vaschy-Buckingham Theorem, every complete equation describing a process can be written in the following form:

$$f_1(\Pi_1,\Pi_2,\ldots\Pi_{p-q})=0 \qquad (2)$$

Where: Π_i = dimensionless similarity monomials or Π_i terms.

q = number of fundamental units participating in variables

p = number of variables

For the present case p=8 and q=4.

Every monomial is determined by the product:

$$\Pi = X_1^{k_1} \cdot X_2^{k_2} \cdot X_3^{k_3} \cdots \tag{3}$$

where $X_1, X_2, \ldots$ is a general notation for the variables. From the matrix of TABLE 2, four similarity monomials can be obtained for the conveyor under the general form:

$$\Pi = \alpha^{k_1} \cdot B^{k_2} \cdot u^{k_3} \cdot N^{k_4} \cdot \varphi_d^{k_5} \cdot \gamma^{\kappa_6} \cdot Q^{\kappa_7} \cdot L^{k_8} \tag{4}$$

By computing of exponents k_1, k_2, k_3, k_4 the above monomials can be modified as follows:

$$\Pi_1 = \frac{\alpha}{L}, \quad \Pi_2 = \frac{B}{L}, \quad \Pi_3 = \frac{u.\gamma.L^2}{Q}$$

$$\Pi_4 = \frac{N}{Q.L}$$

The term Π_1 can be replaced by the term Π_5:

$$\Pi_5 = \Pi_1^2 . \Pi_3 = \frac{\alpha^2}{L^2} . \frac{u.\gamma.L^2}{Q} = \frac{\alpha^2.u.\gamma}{F.u.\gamma} = \frac{\alpha^2}{F}$$

Where F is the load section on the belt.

According to eqn.(2), the similarity equation of the belt conveyor can be determined by the equation:

$$f_2(\Pi_2, \Pi_3, \Pi_4, \Pi_5) = f_2\left(\frac{B}{L}, \frac{u.\gamma.L^2}{Q}, \frac{\alpha^2}{F}\right) = 0 \tag{5}$$

Every term of the above equation, because of its diverse nature, influences the behaviour of the system in a different way. For the solution of eqn. (5), the best method is to proof the result of influence of every term not separetely, but based on a single parameter which must be related with the system's properties, having also great significance for its operation.

This parameter can be the belt width.

Eqn. (5) can be written as:

$$\frac{B}{L} = f_3\left(\frac{u.\gamma.L^2}{Q}, \frac{N}{Q.L}, \frac{\alpha^2}{F}\right) \tag{6}$$

An attempt to express in analytical way the above relation, drives to a complicated form since:

1. In the similarity monomials B/L, $u\gamma L^2/Q$, N/QL, α^2/F, the variation of B,L,u,γ,α results in variation of Q,N,F so that the monomials Π_3,Π_4,Π_5 get different values. A corresponding set of curves can then be created.
2. The variables B,L,u,γ are also parameters.

Both above facts make impossible the formulation of a simple equation able to interrelate the variables according to their values, which had been experimentally obtained.

3. METHOD OF SOLUTION

According to numerical analysis and, with supposing good approximation eqn.(6) can be written in the following form:

$$\frac{B}{L} = \sum_{l,m,k=0}^{n} C_{lmk}\left(\frac{u.\gamma.L^2}{Q}\right)^l.\left(\frac{N}{Q.L}\right)^m \left(\frac{\alpha^2}{F}\right)^k \qquad (7)$$

where k,l,m=0,1,2,3,...,n

Eqn.(7) contains $(n+1)^3$ C coefficients from C_{ooo} to C_{nnn}.

The computation of C_{lmk} is obtained by a suitable program (see Appendix) and the Least Squares Method so that the deviation of the initial ratio B/L- which results from the true values of B,L -from that computed from eqn. (7) for the same value of n to be minimum.

To simplify the computational process, the first determination of C_{lmk} is carried out by supposing that n=1. The values of C_{lmk} so computed can be considered as final, if the ratio B/L resulting from eqn.(7)-where C_{lmk} is used-differs from the same ratio of the real system by a relative error no more than 5%.

For n=1, there are $(1+1)^3$=8 values C_{ooo}-C_{111}. According to the real conveyors data of experimental data concerning the variables the terms of eqn. (7) are computed and in sequence the eight values of the coefficients C_{ooo}-C_{111} are determined. It is possible to determine many sets of coefficients according to the combinations of the variables of the conveyor to be used. However, this set is chosen that gives a ratio $(B/L)_{theor}$ so that:

$$\sum_{j=1}^{m}[(B/L)_{realj}-(B/L)_{theoryj}]^2 = min \qquad (8)$$

where: $(B/L)_{real}$ = ratio B/L computed for the real values of B,L.

$(B/L)_{theor}$=ratio B/L computed from eqn.(7) according to values of $C_{ooo}-C_{nnn}$.

m = number of ratio B/L measurements in every data set.

As was already mentioned, the relative error must be exceed 5%.

$$\text{Error:} \quad \varepsilon=(\frac{B}{L})_{real}-(\frac{B}{L})_{theory} \tag{9}$$

$$\text{Relative error:} \quad E=\frac{\varepsilon}{(\frac{B}{L})_{real}} \cdot 100\% \tag{10}$$

If, for the chosen set, the error is higher than the permissible, then the process of determination of $C_{ooo}-C_{nnn}$ is repeated setting n=2.

This method usually gives satisfactory results, although it is time -consuming. For this reason, when applied, must be accompanied by a reduction of the number of experimental data.

4. NUMERICAL SOLUTION

The method of solution already mentioned, was entirely applied for the computation of C_{lmk} of eqn. (7) and the creation of the mathematical model of similarity for the conveyor. For this purpose, a computer program was created (see Appendix).

According to the current literature[2], values of size, operation and capacity of real conveyors were used as data presented in the following TABLES 3,4 and 5.

These data were divided in three sets according to the trough angle and the material grain size.

A set: Trough angle B=20 deg., grain size =0.001 (m)

B set: Trough angle B=35 deg., grain size=0.01 (m)

C set: Trough angle B=45 deg., grain size =0.1(m)

Every set was divided in subsets, according to the dynamic repose angle as follows:

1st subset: dynamic repose angle 0 deg.

2nd subset: dynamic repose angle 5 deg.

3rd subset: dynamic repose angle 10 deg.

4th subset: dynamic repose angle 20 deg.

5th subset: dynamic repose angle 25 deg.

6th subset: dynamic repose angle 30 deg.

For every subset a set of data was given consisting of belt width, load section and capacity for the following values of conveyor length: 500, 1000, 1500 m (see Tables 3,4,5). For all cases the speed of the conveyor was regarded as constant and the bulk weight equal to 15715 N/m^3.

Using the data of Tables 3,4,5 as input in the program, the similarity monomials Π_2,Π_3,Π_4,Π_5 were computed. Finally, according to eqn. (7) and setting n=1, the coefficients C_{ooo}-C_{111} were computed.
The results for every data set are the following:

A SET:	B SET:	C SET:
C_{000} = -0,0207732984	C_{000} = -7,1456585.10^{-4}	C_{000}=2,29012288.10^{-3}
C_{001} = 10,0952722	C_{001} = -0,358758801	C_{001}=5,43163077.10^{-3}
C_{010} = 8,03297104.10^{-3}	C_{010} =6,91479608.10^{-4}	C_{010}=-4,04468361.10^{-4}
C_{011} = -4,07829468	C_{011} =0,12713442	C_{011}=-2.03888137
C_{100} = 3,14506489.10^{-9}	C_{100} =-5,44364498.10^{-11}	C_{100}=-4,87428528.10^{-10}
C_{101} = -2,16506272.10^{-5}	C_{101} = 2,05561936.10^{-8}	C_{101}=-1,77513676.10^{-10}
C_{110} = -1,1591274.10^{-9}	C_{110} = 1,15333496.10^{-11}	C_{110}=1,68859361.10^{-10}
C_{111} = 7,96263337.10^{-6}	C_{111} =-7,18819123.10^{-9}	C_{111}=6,9438675.10^{-11}

According to the above coefficients, the following sums for the deviation between $(B/L)_{real}$ and $(B/L)_{theor}$ were computed.

$$\text{A SET:} \quad \sum_{i=1}^{m} \left[(\tfrac{B}{L})_{realj} - (\tfrac{B}{L})_{theoryj} \right]^2 = 5,46435389.10^{-7} \qquad (11)$$

$$\text{B SET:} \quad \sum_{i=1}^{m} \left[(\tfrac{B}{L})_{realj} - (\tfrac{B}{L})_{theoryj} \right]^2 = 3,3372961.10^{-7} \qquad (12)$$

$$\text{C SET:} \quad \sum_{i=1}^{m} \left[(\tfrac{B}{L})_{realj} - (\tfrac{B}{L})_{theoryj} \right]^2 = 2,5934353.10^{-7} \qquad (13)$$

For the determination of the mathematical model, the coefficients C_{000}-C_{111} of the C Set were chosen, since they give the lower squares sum. Using these values in eqn. (7), the values of $(B/L)_{theor}$ deviate from the corresponding $(B/L)_{real}$ and -in some case the appearing error exceeds 15% and even 23%. This obviously means that the model cannot give reliable results.

For the error to become lower than 5%, the number of coefficients C_{000}-C_{nnn} was increased by setting n=2. The computation of $(1+2)^3 = 27$ values of C_{000}-C_{222} was repeated only for the C data set-which corresponds to the lower sum of squares-by gradual decrease of the data subsets number. These 27 values of the coefficients are presented in Table 6, where the gradual decrease and the corresponding error can also be sen.

From Table 6 one can conclude that eqn. (7) can be used with satisfactory approximation as a mathematical model with values l,m,k=0,1,2 for variables which correspond to the data of the 4,5,6 subsets of the C set, because in these cases the relative error does not exceed 4.48%.

5. CONCLUSIONS

According to the preceding analysis, the mathematical model of a conceyor can be obtained.

The model can be applied by success for values of variables and coefficients C_{000}-C_{222} of the subsets 4,5,6 of C data set. (Tables 5,6). The usefulness of eqn. (7) is because of its property to connect two size parameters, namely B (width) and L(length) with the parameters of load and operation of the conveyor. This fact permits the determination of the equivalent width or length of the belt when all remaining variables are known and the function of an equivalent model or real structures in order to facilitate experimenting, since real conveyors usually extend to kilometers of length.

From the results for $(B/L)_{theor}$ deduced from eqn. (7) it can be concluded that the average of the relative error decreases and that the above ratio approaches the corresponding ratio of the real sizes as the order of data sets and subsets values increases.

6. REFERENCES

1. A. Spinakovsky, V. Dyachkov, "Conveyors and Related Equipment", Mir Publishers, Moscow.
2. Drakatos A.P., "Materials Handling Equipment", University of Patras, 1984.
3. Drakatos A.P., "The Application of Similarity Methods During the Experimental Research on Soil Compaction Machines", PH.D., Thesis, Thessaloniki, 1975.

4. Manatakis E. K., Drakatos P.A., "A Statistical Analysis as a Criterion of Soil-Machine System"., Transactions of the ASME vol.107,Sept.,1985.
5. G. Murphy,"Similitude in Engineering", Ronald Press Company, N.Y. 1950, p.36.
6. Roussas, G.G., "A First Course in Mathematical Statistics" (1st ed.), Chapter 17, Addison-Wiley, New York, 1973.

Flow chart

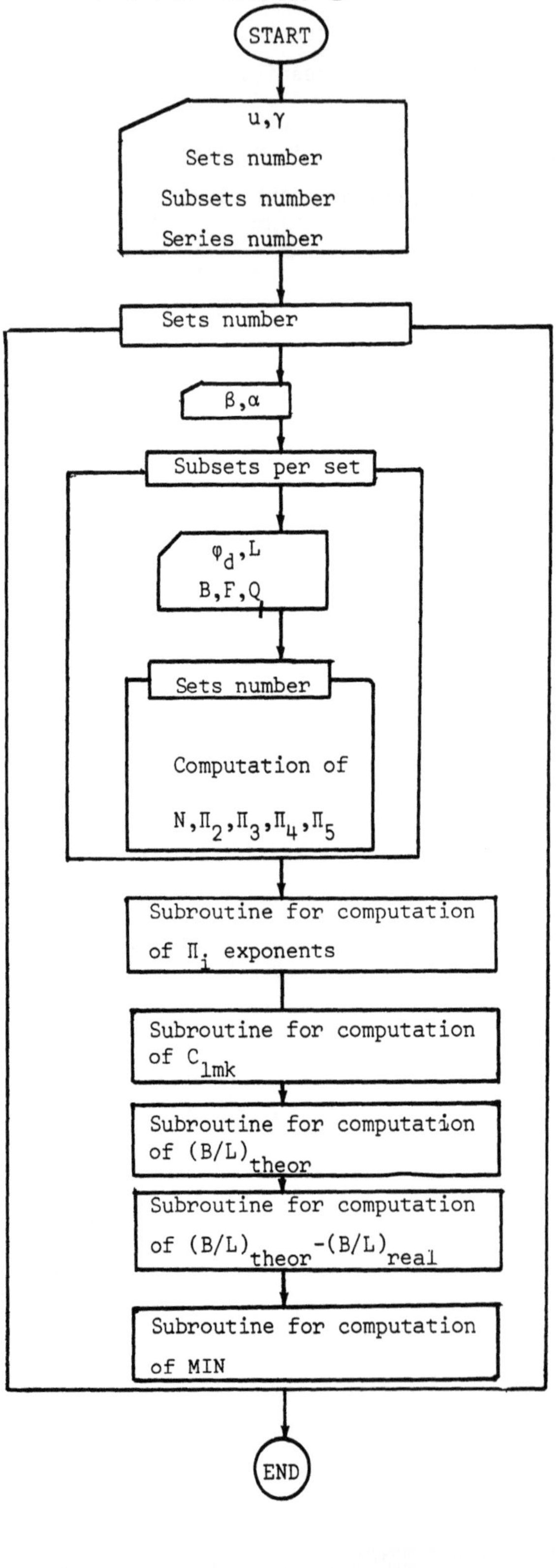
START
u,γ
Sets number
Subsets number
Series number
Sets number
β,α
Subsets per set
φd,L
B,F,Q
Sets number
Computation of
N,Π2,Π3,Π4,Π5
Subroutine for computation of Πi exponents
Subroutine for computation of C1mk
Subroutine for computation of (B/L)theor
Subroutine for computation of (B/L)theor-(B/L)real
Subroutine for computation of MIN
END

Table 3. A set of data. Trough angle $\beta=20^{\circ}$. Grain size $\alpha=0.001$ m
Belt speed $u=0.5$ m/s, Bulk weight, $\gamma=15715$ N/m^3

Belt Length L (m)	Belt Width D (m)	Subsets / Repose angle: 1st Subset 0°		2nd Subset 5°		3rd Subset 10°		4th Subset 20°		5th Subset 25°		6th Subset 30°	
		Area $\times10^{-3}$ (m^2)	Capacity t/h	Area $\times10^{-3}$ (m^2)	Capacity t/h	Area $\times10^{-3}$ (m^2)	Capacity t/h	Area $\times10^{-3}$ (m^2)	Capacity t/h	Area $\times10^{-3}$ (m^2)	Capacity t/h	Area $\times10^{-3}$ (m^2)	Capacity t/h
500	0.35	4.55	14.67	5.57	13.00	6.60	21.60	8.64	27.90	9.75	31.50	10.78	34.80
	0.40	6.32	20.70	7.62	24.60	9.01	29.21	11.80	38.10	13.28	42.90	14.77	47.70
	0.45	8.36	27.05	10.13	32.70	11.89	38.40	15.61	50.40	17.47	56.40	19.42	62.70
	0.50	10.68	34.40	12.91	41.70	15.14	48.90	19.79	63.90	22.20	71.70	24.71	79.80
1000	0.60	16.16	52.10	19.42	62.70	22.85	73.80	29.82	96.30	33.44	108.00	37.07	119.70
	0.70	26.48	85.50	31.77	102.70	37.85	120.60	48.59	156.90	54.44	175.80	60.29	194.70
	0.90	39.27	127.00	47.19	154.20	55.37	178.80	71.90	232.20	60.54	260.10	89.28	289.50
	1.00	54.72	176.70	65.59	211.90	77.01	248.70	99.77	322.20	111.65	123.84	123.84	399.90
1500	1.20	72.65	234.50	87.05	280.50	102.10	329.70	132.29	427.00	148.18	478.50	163.97	529.50
	1.35	93.09	300.50	111.57	360.30	130.71	422.10	169.35	546.90	189.52	612.00	209.86	677.70
	1.50	116.13	375.00	139.07	443.00	162.85	525.90	210.88	681.00	235.97	762.00	261.33	843.90
	1.65	141.67	457.50	169.54	547.80	198.60	641.40	257.05	830.10	287.53	928.50	318.46	1028.00

Table 4. B set of data. Trough angle $B=35^{\circ}$. Grain size $\alpha=0.01$m.
Belt speed $u=0.5$ m/s. Bulk weight $\gamma=15715$ N/m^3

Belt Length L (m)	Belt Width D (m)	1st Subset 0°		2nd Subset 5°		3rd Subset 10°		4th Subset 20°		5th Subset 25°		6th Subset 30°	
		Subsets – Repose angle											
		Area $X10^{-3}(m^2)$	Capacity t/h	Area $X10^{-3}(m^2)$	Capacity t/h	Area $X10^{-3}(m^2)$	Capacity t/h	Area $X10^{-3}(m^2)$	Capacity t/h	Area $X10^{-3}(m^2)$	Capacity t/h	Area $X10^{-3}(m^2)$	Capacity t/h
500	0.35	7.34	23.7	8.18	26.40	9.10	29.40	10.87	35.10	11.89	38.40	12.82	41.40
	0.4	10.13	32.7	11.33	36.60	12.54	40.50	14.96	48.30	16.26	52.50	17.56	56.70
	0.45	13.38	43.2	14.96	48.30	16.54	53.40	19.69	63.60	21.46	69.30	23.13	74.70
	0.5	17.09	55.2	19.04	61.50	20.99	67.80	25.08	81.00	27.22	87.90	29.36	94.80
	0.60	25.83	83.34	28.71	92.70	31.68	102.30	37.72	121.80	40.97	132.30	44.13	142.50
1000	0.75	43.36	136.70	47.01	151.80	51.84	167.40	61.59	198.90	66.70	215.40	71.81	231.90
	0.90	62.80	202.8	69.68	225.00	76.83	248.10	91.13	294.30	98.66	318.20	106.18	342.90
	1.00	87.33	282.0	96.89	312.90	106.25	344.40	126.44	408.30	136.84	441.90	147.25	475.50
1500	1.20	116.03	374.70	128.48	414.90	141.90	456.90	167.54	541.20	181.34	585.60	195.09	630.00
	1.35	148.64	480.00	164.53	531.30	181.16	585.00	214.51	692.70	232.06	749.40	249.53	805.80
	1.50	185.06	597.60	206.12	662.40	215.75	729.00	267.18	862.80	289.01	933.30	310.75	1003.50
	1.65	225.75	729.00	250.09	807.60	275.17	888.60	325.71	1051.80	352.18	1137.30	378.66	1222.80

Table 5. C set of data, Trough angle B=45° .Grain size α=0.1m
Belt speed u 0.5 m/s, Bulk weight γ=15715 N/m3

Belt Length L (m)	Belt Width D (m)	Subsets / Repose angle: 1st Subset 0°		2nd Subset 5°		3rd Subset 10°		4th Subset 20°		5th Subset 25°		6th Subset 30°	
		Area $X10^{-3}(m^2)$	Capacity t/h	Area $X10^{-3}(m^2)$	Capacity t/h	Area $X10^{-3}(m^2)$	Capacity t/h	Area $X10^{-3}(m^2)$	Capacity t/h	Area $X10^{-3}(m^2)$	Capacity t/h	Area $X10^{-3}(m^2)$	Capacity t/h
500	0.35	8.64	27.90	9.48	30.60	10.22	33.00	11.89	38.40	12.91	41.70	13.56	43.80
	0.40	11.98	38.70	13.01	42.00	14.12	45.60	16.26	52.50	17.47	56.40	18.58	60.00
	0.45	15.79	51.00	17.19	55.50	18.58	60.00	21.37	69.00	22.85	73.80	24.34	78.60
	0.50	20.16	65.10	21.92	70.80	23.60	76.20	27.22	87.90	27.68	89.40	30.94	100.00
1000	0.60	30.38	98.10	32.98	106.50	35.58	114.90	40.88	132.00	43.66	141.00	46.36	149.70
	0.75	49.79	160.80	53.88	174.00	58.06	187.50	66.52	214.00	70.98	229.20	75.43	243.60
	0.90	73.06	238.50	79.89	258.00	86.03	277.80	98.47	318.00	105.07	339.30	111.57	360.30
	1.00	102.75	331.80	110.92	358.20	119.47	385.80	136.66	441.30	145.67	470.40	154.68	499.50
1500	1.20	136.28	440.10	147.15	475.20	158.20	511.50	171.68	584.40	192.80	622.80	204.75	661.20
	1.35	174.37	563.10	188.40	608.40	202.71	654.60	221.60	747.90	246.74	786.80	215.44	845.70
	1.50	217.39	702.00	234.76	758.10	252.50	815.40	288.36	931.20	307.22	992.10	326.08	1053.00
	1.65	265.23	856.50	286.13	924.00	307.87	994.20	351.44	1134.90	374.39	1029.00	397.24	1282.00

Table 6. Determination of coefficiency C_{000}-C_{222} (n=2) for C Set with gradual decrease of experimental data

	Subsets 1-6	Subsets 2-6	Subsets 3-6	Subsets 4-6
C_{000}	1.40205521 E-03	3.4361786 E-03	-3.98346114 E-04	-7.10511631 E-03
C_{001}	-1.34074888 E-03	2.72676333 E-03	-1.19911025 E-03	2.98162404 E-03
C_{002}	-1.23190534 E-03	7.77430701 E-04	9.33759556 E-03	0.0406055148
C_{010}	-8.86405842 E-05	-1.01004736 E-03	-6.85399604 E-05	3.43276232 E-03
C_{011}	3.41089121 E-04	-3.33459532 E-04	-9.76317342 E-04	-1.303159 E-03
C_{012}	-3.79038882 E-04	1.86581544 E-03	1.02736711 E-03	-0.0138783585
C_{020}	2.74722977 E-05	9.88556379 E-05	2.78093209 E-04	-9.14345909 E-05
C_{021}	1.04734958 E-05	-3.1939385 E-04	4.33345131 E-04	-7.6899541 E-05
C_{022}	5.29746492 E-04	-5.19103241 E-04	-1.3761547 E-03	-3.95883288 E-05
C_{100}	-3.8293885 E-11	2.02924052 E-10	-1.7780767 E-13	-1.32430548 E-10
C_{101}	-4.58248809 E-11	1.05829518 E-10	2.30953785 E-12	8.74965072
C_{102}	-1.4135746 E-10	5.54064772 E-10	2.1100688 E-10	-5.33208098 E-10
C_{110}	-1.08486747 E-11	-1.10343802 E-10	4.23094083 E-11	2.67781192 E-10
C_{111}	2.59791592 E-11	-8.18090187 E-11	1.3212709 E-10	2.55912411 E-10
C_{112}	3.48226963 E-11	1.32919432 E-10	-8.6695423 E-11	-1.07354177 E-10
C_{120}	1.23994719 E-12	3.66691199 E-12	-2.71305391 E-11	-9.5120217 E-11
C_{121}	-2.46822533 E-12	2.27588395 E-11	-3.59909635 E-11	-1.8166714 E-10
C_{122}	-7.24772063 E-12	-1.45438202 E-10	-2.21450765 E-11	6.2853149 E-11
C_{200}	4.19733834 E-18	-3.37556435 E-18	-6.72517812 E-18	2.4213576 E-17
C_{201}	7.01684682 E-19	-9.82015588 E-18	1.53670114 E-17	-2.20414858 E-17
C_{202}	-7.31185255 E-18	1.16525096 E-17	-6.42565456 E-18	2.35284069 E-17
C_{210}	2.7683523 E-19	-4.73252214 E-18	1.43701531 E-18	8.37658267 E-18
C_{211}	2.6231305 E-18	-4.04599004 E-18	-4.97640258 E-18	6.18089415 E-18
C_{212}	-3.3919237 E-19	1.88682817 E-18	4.7268102 E-18	-6.80902283 E-18
C_{220}	-4.83683363 E-19	2.43062891 E-18	7.08056423 E-19	-5.8762429 E-18
C_{221}	-1.17425182 E-18	2.47134142 E-18	-8.87352547 E-19	-7.10784453 E-19
C_{222}	1.41785159 E-18	-1.65601055 E-18	-1.4951033 E-20	1.2313434 E-18
Maximum relative error E	13.12 %	10.63 %	7.69%	4.48%

Programmable Parts Manipulation And Assembly

E.E. Pickett
Department of Industrial Engineering
University of Toronto
Toronto, Canada M5S 1A4

Abstract

This paper deals with some of the main issues concerning programmable manipulation as used in the assembly of discrete parts. The discussion centers on the planning, specification, programming and execution of the various types of manipulator motions. These motions include free, guarded and compliant motions. The latter two are of particular importance in carrying out tasks in environments with significant uncertainty in object locations. Examples are provided illustrating the use of compliant motion that incorporates sensory feedback to reduce the effect of such uncertainty. The discussion focuses on models that have been proposed in this area, which may serve as the basis for general task planning systems.

1.0 Introduction

Programmable systems offer the ability to implement flexible automation in many manufacturing tasks. The key characteristic to this flexibility is the versatility of physical devices, such as manipulators, in carrying out the more dexterous requirements of these tasks. This versatility derives from the generality of the manipulator's physical structure and sensory capabilities which are exploited through a programmable controller.

Assembly is one step in the manufacturing process which includes: forming of parts, transfer and presentation of parts, joining and bonding of parts and the inspection of parts and subassemblies. The system which transports and manipulates these parts during assembly must successfully integrate the functional requirements of physical dexterity, manipulation control and multisensory data acquisition and interpretation. The performance of such a system is

NATO ASI Series, Vol. F49
Computer Integrated Manufacturing
Edited by I. Burhan Turksen

evaluated in terms of its speed, throughput, reliability, product quality and resulting product yield, as well as its impact on other parts of the manufacturing process including inventory, parts handling, work-in-progress and product design. The rationale for implementing programmable manipulation may involve any combination of these performance parameters as well as other criteria such as the need to gain insight and experience in such a potentially important method for production.

Programmable automation can best be justified for small to medium size batch production of parts of different size and shape. This flexibility is very much a function of the programmable nature of these machines. Numerical control (NC) machines and industrial robots are examples of programmable devices currently used in industry.

NC machine tools were first developed in the early 1950's to aid in the production of advanced aircraft whose parts were designed to be machined rather than riveted. The manufacturing characteristics of this product involved complex part geometry, close tolerances and small lot sizes.

In numerical control a part program specifies the sequence of processing steps to be performed by the NC machine on the work piece. The part program is coded onto punched tape and a servo-controller converts these data into physical actions of the machine tool. The part program is changed to accommodate work pieces of different designs. In more complex machining applications part programming becomes an extremely tedious task and prone to error. An example of this is machining a contour which requires that the tool path be previously defined and continuously controlled along more than one axis.

In computer assisted part programming much of the tedious computational work is transferred to the computer. Special high-level languages such as APT (Automatic Programmed Tool) have been developed which allow the programmer to communicate machining instructions in simple English-like statements. The computer then performs the computations to calculate the tool trajectory.

The present day industrial robot has its origin in both the NC machine and the teleoperator. The teleoperator is an articulated device that allows a task to be performed at a distance. The first programmable industrial robot, developed in the 1960's, combined the programmability of NC machines and the general physical structure (articulated linkage) of the teleoperator.

The ability of a programmable manipulator to perform a variety of tasks follows from the generality of its physical structure and control. At present most manipulator programming systems require the user to specify an execution sequence which consists mostly of motion instructions needed to accomplish the task. In some applications it is necessary to specify additional instructions which permit the manipulator to acquire and respond to sensory input. It is the responsibility of the user to ensure that events such as collision with objects in the workspace during the motion of the manipulator arm and slippage between the payload and gripper device do not occur. The planning of such activities is tedious and places a heavy demand on the time of skilled programmers. As a result the cost of programming may be a significant fraction of the total cost of an application. This has been a major factor in limiting the applications of the industrial robot (Lozano-Perez [1983a]).

A basic problem in manipulation is planning motions to perform the specified task and then controlling the manipulator as it executes the instructions necessary to achieve these motions. The motions of a manipulator are called trajectories and will, in general, refer to the path followed by the end of the manipulator arm. The trajectory is fully specified by a time sequence of arm positions, velocities and accelerations. For trajectory planning it is ordinarily required to transform from the Cartesian space of the hand to the joint space of the manipulator. This is done by first solving the inverse kinematic problem. The direct kinematics consists in part of finding the position of the end of the manipulator, given the joint angles, whereas the inverse kinematics involves finding the joint angles for a given position of the manipulator arm. The inverse kinematic solution also provides joint velocities and accelerations corresponding to the end effector velocities and accelerations.

Given a trajectory as well as the forces to be exerted by the end-effector it is possible to determine a time sequence of torques to be applied to the manipulator joints so as to move the arm in the desired manner. This is accomplished by solving the inverse dynamics problem.

In practice, these preplanned torque values computed for individual joints would not produce the desired trajectory and forces at the end-effector. This is due to errors arising from the inaccuracies of the kinematic and dynamic specifications of the manipulator, computational imprecision and mechanical effects such as friction and vibration. The appropriate way of handling this problem is provided by feedback control which involves measurement of the actual motion during execution of the instruction sequence, computing the error from the desired motion and modifying subsequent torque inputs so as to reduce the error at subsequent times. Feedback control design is concerned with obtaining a rule for deriving the correction from the estimated error. The rule may be regarded as a mapping of past errors into future input corrections. The control system structure is of major importance in determining the overall performance of the manipulator. Here however, we are concerned with the planning and specification of various types of motion strategies and synthesizing these strategies for use in manipulation.

There are three common classes of motions used in manipulation, which are: free, guarded and compliant motions. Free motions take place in free space, that is, the portion of task space that is free of obstacles. The objective during free motion is to reach the goal position without collision. Planning free motions is a problem in obstacle avoidance. Many manipulator tasks needed in assembly cannot be adequately expressed as a sequence of positions. Whenever the manipulator is constrained by positional constraints due to the environment, an alternative mode of control must be used. The two common classes of motions involving external position constraints are guarded and compliant motions.

A guarded move occurs when the motion of the manipulator towards an object is terminated by means of a signal from a contact or force sensor. Guarded moves may be used to locate imprecisely known surfaces, objects or object features by determining the positions of a suitable number of contact points of the object relative to the manipulator. Compliant motion takes place when the manipulator is in continuous contact with a surface or object. Data from force, proximity or visual sensors can be used to modify the motion so as to maintain or achieve a desired relationship with other objects. Compliant motions are designed to maintain contact between objects even in the presence of uncertainty in the object's location. The ability to perform compliant motion implies the combined use of position and force control. This is necessary because compliant motion is accomplished by controlling position in a direction tangent to a sliding surface and controlling force normal to the surface. A strategy to specify any compliant motion task can consist of first identifying all external position constraints and then determining the degrees of freedom which require force control and those that require position control.

Motions involved in assembly tasks may combine all three classes of motions in various ways. For example a typical assembly task such as inserting a peg in a hole might involve a guarded departure from the current position after the peg has been grasped; a free motion toward the destination position of the task step; a guarded approach to contact the hole at the destination; and a compliant motion to insert the peg under positional constraints imposed by the hole.

Much of the difficulty in performing motion synthesis arises from the need to operate under uncertainty regarding the positions of the objects and of the manipulator. Sensing can enable manipulators to perform tasks in the presence of significant uncertainties. Mention has already been made in regard to the use of sensing to terminate motions and comply with external constraints. Other principal uses of sensing include identifying objects and determining their positions, initiating motions, and choosing among alternative actions. An example of the latter might involve checking to determine whether or not an insertion has been properly performed and if not deciding what corrective action to take.

In relatively uncomplicated manipulation, involving low positional accuracy and no sensing or computation, it is quite easy to specify motions by guiding the arm through a series of positions representing the nominal trajectory. This form of motion specification can be implemented without the use of a general purpose computer. However, in most assembly tasks, it is necessary to specify the desired action of the manipulator in response to sensory input, data retrieval, and computation. In these cases the capabilities of a general purpose computer and a programming language with instructions to access sensors, draw inferences and specify manipulator motions is required. These languages are referred to as explicit or manipulator level languages. A major difficulty with these languages is that even when tasks are relatively simple the programming costs of a particular application can be comparable to the cost of the manipulator itself. Also, manipulator level languages are not accessible to the typical assembly worker because of the level of programming expertise required. Faced with this situation some of the more recent approaches to specifying manipulator actions have included attempts to develop task-level languages. This type of language is used to specify how objects should be positioned in relation to one another rather than the motions of the manipulator necessary to achieve those goals. An instruction such as "pick up object A and place on object B" is given without specifying the manipulation path.

The purpose of a task planner is to transform task level specifications into manipulator level specifications. The planner requires information describing the manipulator, objects being manipulated, the environment, and the desired final state. The output from the planner is a program which will perform the task when executed by the manipulator. To enhance the reliability of the synthesized program, the planner must acknowledge any manipulator capabilities for free motion, compliant motion, guarded motion and error checking.

The ideal task planner might need only a description of the initial and final states to plan a task. This has been the goal of much research on automatic problem solving (e.g. Lozano-Perez [1983b]). The problem solving systems proposed in this context typically do not provide output at the level of detailed manipulator motions. In contrast, task planners require more information about intermediate states, but they do provide a more detailed manipulator level program.

The implementation of a comprehensive task level planning system would provide for the manipulator level synthesis of programs to achieve: collision free paths, grasping, error detection and recovery schemes, manipulator motion, and sensory operations. This appears to be unattainable in practice because of the overwhelming complexity of the general problem and the deficiencies of existing methods for geometric reasoning and sensory planning. There has, however, been significant progress towards solving some of the component problems associated with task planning.

The presentation in this report is organized as follows. Section 2 contains a brief review of various approaches to manipulator programming systems. Tradeoffs between flexibility and cost have characterized the evolution of these systems. In Section 3, the topic of manipulator motion control is discussed. Many tasks can be accomplished by relatively simple position control, but more complex tasks in uncertain environments necessitate both position control and compliant motion. Examples of these are provided. Section 4 discusses several topics in task planning, whereby certain tedious problems in manipulator programming are alleviated by automatic planners. Specifically, planning for collision avoidance and fine motion planning are discussed. A fine motion strategy is one that combines force and position control in a way that guarantees achieving the final configuration from all likely initial configurations (Lozano-Perez, Mason and Taylor [1983]). Section 5 offers a brief summary and conclusions.

2.0 Manipulator Programming Systems

Manipulator programming systems fall into three categories (Lozano-Perez [1983b]), and include those that use:

i guiding systems, in which the user leads the manipulator through the motions to be performed. All manipulator programming systems support some form of guiding.

ii explicit programming languages, these are manipulator level languages in which the user writes a program specifying all the manipulator motions and sensing needed to the accomplish the task.

iii task level programming languages, in which the user specifies geometric models of parts and a description of the task in terms of these models. A task planner transforms the task level specifications into detailed manipulator motions.

A common method of programming manipulators to perform tasks is the method of guiding. This involves moving the end-effector to each desired position along a path trajectory and recording the arm configuration (in terms of joint positions) corresponding to that location. Operations not involving joint movements such as closing the gripper or activating a hand-held device occur at the arm configuration where the instruction is encountered. The task is thus described by a sequence of positions and events which can be "played back" at a later time when the task is to be performed.

Programming by guiding offers a very rudimentary control structure. The "program" is simply a single execution sequence, with no loops, branches and no significant amount of computation. These limited capabilities are sufficient to handle well structured pick and place tasks where the position of parts is known with certainty. These tasks exhibit such characteristics as work station features remaining unchanged during execution of the task and the use of fixtures and part feeding devices to ensure accurate part positioning.

The basic guiding systems cannot handle applications which require on-line decision making which may involve sensing, database access and interaction with other machines in the assembly system.

Higher level guiding systems have been introduced that offer simple control structures and allow control to be transferred to different points in the taught position sequence. These extended guiding systems also permit the use of rudimentary sensing operations. The manipulator can wait for an external signal before specified motions are attempted or it can by itself indicate that an event has occurred by moving until a touch sensor on the end-effector encounters an object.

For most situations involving the assembly of parts, manipulator programming requires the capabilities of a general purpose computer programming language. This makes it possible to handle tasks requiring conditional actions and on-line decision making which are necessary as part of carrying out sensing activities.

An explicit programming system consists of a manipulator controller, manipulator level language and a processor (usually an interpreter). In manipulator applications an interpreter provides a number of advantages over a compiler. An interpreter executes code as it is encountered. A compiler passes through the code more than once before it becomes executable. Thus interpreters provide better interactive environments that permit partial execution of sections of code and changes to code as the program executes. Furthermore, their structure leads to enhanced debugging capabilities, they are smaller than compilers and thus easier to construct and are more readily extensible. Compilers can, however, achieve a greater efficiency of execution as well as early detection of some classes of programming errors. The processor accepts program statements written in manipulator level language by the user and interprets them for execution by the controller. The controller generates the trajectories and servos the joints.

Explicit programming systems permit tasks to be performed by carefully specifying the manipulator actions. Carrying out sophisticated tasks such as are involved in assembly operations requires sophisticated programming systems which are able to support: sensing, representation of objects and object features, motion specifications and flow of control. This requirement can sometimes best be achieved by modelling the manipulator programming system on an existing programming system. Thus, an early system, WAVE, designed as a general purpose manipulator language with low level syntax, was modelled after the assembly language of the PDP-10. Although WAVE's syntax was difficult to use, an extensive collection of high level functions was provided, many of which are still not available in commercial systems. WAVE provided the mechanisms to: 1) coordinate joint motions to achieve continuity in velocities and accelerations; 2) position the end-effector according to a description given relative to Cartesian coordinates; 3) specify compliance relative to Cartesian coordinates. A subsequent system, AL, was an attempt to develop a high level manipulator language that provides all the capabilities required for manipulator programming as well as the programming features of high level languages such as ALGOL and Pascal. In addition to the features provided by WAVE it is capable of: concurrent execution of processes, providing the data and control structures of an ALGOL-like language and, support of task modelling by the manipulation of coordinate frames. AL was also designed to support some of the features necessary to handle task level specifications that will be discussed in the following.

To prepare a program for a manipulator level programming system requires expertise in computer programming as well as a knowledge of manipulator kinematics and dynamics. In addition, expertise in the design of sensor based assembly strategies and motion strategies is needed.

Task level programming languages attempt to simplify the human interaction required in manipulator programming. Their development parallels the development of computer programming languages, where the user describes his task in terms of abstract information structures and problem-oriented operations and leaves it to the compiler to provide the details of program implementation such as register management, loop control, translation of expressions into machine instructions, and so on. Task level programming allows the user to describe the task in terms of operations on the object and takes over the function of translating these specifications into explicit instructions which can be executed by the manipulator controller. This translation is carried out by the task planner. The decisions to be made by the task planner include how to grasp objects, selection of object destination as well as the determination of a trajectory that avoids collision with obstacles.

To perform these tasks the planner must consider the geometry of the objects in the environment, the configuration of the objects, the uncertainty associated with the geometry and configuration, the accuracy of the robot itself and the sensory capabilities available. Such information can be obtained from geometric models of objects and of the manipulator. For this reason, task level programming languages are also referred to as model based programming languages.

3.0 Manipulation for assembly

Most industrial manipulators are used as positioning devices. The individual motions are specified by defining configurations corresponding to the beginning and end of the motion. A configuration of a rigid object is a six element vector specifying its position and orientation. The configurations of a multi-linked manipulator are typically equated with the configurations of its end-effector and can be specified in terms of vectors of the joint angles. The greater the difference in the end point configurations of a given path, the less constrained will be the intermediate configurations of the manipulator. Thus, the programmer has less control over the path followed by the end effector and the volume swept out by the rest of the arm. To overcome this problem the programmer normally also selects some additional configurations sufficient to guarantee reaching the destination without collision.

It is often desirable to impose constraints on the velocity and acceleration of the arm motion. This allows for smoother motion of the arm, avoids unnecessary forces on the grasped object and generally reduces mechanism wear. The process of computing a sequence of positions, velocities, and accelerations is referred to as trajectory planning. This sequence of arm configurations forms the input to the arm's feedback control system which servos the joints. There is a minimum rate usually between 20-200 Hz at which configurations must be supplied if the arm is to be capable of smooth and predictable motion. Typically the trajectory planner does not meet the rate required to provide servo-rate demands. As a consequence a post process incorporating a joint space interpolation technique is used to increase the rate at which configurations are supplied to the feedback controller.

Some manipulators incorporate a trajectory planner that allows the grasped object to move along a straight line at constant velocity with constant orientation. This is especially useful for tasks such as insertions, which exhibit linear geometries. Current systems can implement three types of paths: uncoordinated joint motions, straight line motion in joint coordinate space and straight line motion in Cartesian space. Each of these represent a different trade off between speed of execution and the degree of trajectory control. In most assembly applications the binding constraint is control over the trajectory.

Manipulators developed for position control are quite effective for tasks where the positional accuracy requirements are low relative to the manipulator accuracy. Pure position control is, however, inadequate for tasks that involve tight tolerances or motions that are subject to geometric constraints that are not known with certainty.

Examples of tasks that include these situations arise primarily when the desired motion cannot be specified exactly until it is executed. This is typically the case when the position or identity of the part is determined from sensors, where data only becomes available when the motion is about to be executed. Specifying this type of computation can require the facilities of a general purpose programming language.

The simple example of placing a grasped object at a specified position on a table illustrates the situation. An absolute positioning system would attempt to solve this problem by planning and executing a trajectory that would move the object to a location directly above the desired position. This would be followed by a vertical straight line motion to place the object on the table. If the position of the table is known exactly and the control system is sufficiently accurate, absolute positioning might suffice. It is more likely, however, that the object would collide with the table or stop short of the surface and drop as a result of releasing the grasp. The task could best be solved by controlling the force perpendicular to the table rather than position. This type of motion is referred to as a guarded move and can be used in locating an imprecisely known surface by moving toward it and terminating the approach motion when the value of a force sensor exceeds the threshold. Guarded moves are also used to identify features of an imprecisely located object. The contact points can then be used to determine the object's position relative to the manipulator. A limited form of this technique is available in extended guiding systems.

3.1 Programmed positioning

The simplest manipulator operation is programmed positioning of the manipulator arm where the objective is to acquire, move or position parts in task space. Ultimately this involves manipulator kinematics and dynamics, and control of trajectory planning. Our initial concern is with kinematics which deals with the geometry and time dependent aspects of position and motion. For tasks that do not involve sensing, the positioning of the manipulator arm can be specified as a sequence of desired arm configurations. To do this a structure is needed to represent the positions of objects, their features, such as surfaces, edges and holes, and the configuration of the manipulator itself. A coordinate frame imbedded in any object or feature allows its position and orientation to be readily described. Although some of these positions are fixed throughout the task, others will change and may require sensory information to establish their current values. The structure must also be able to express one coordinate frame relative to another. For example, to move the gripper to a grasp position on an object requires knowledge of the gripper position relative to a base reference frame, the object position relative to the same base frame and the grasp position on the object relative to the object frame. The desired gripper position is found by composing the relative grasp position with the object frame.

A common representation for positions of objects, object features and manipulator configuration in task space is the homogeneous transform. These transforms are used in computer graphics and computer vision applications to describe the relationship between objects. A homogeneous transformation is represented by a 4x4 matrix which may be interpreted as a product of rotation and translation matrices. It can be used to express the coordinate frame of one object in terms of the coordinate frame of another by combining the translation of the origin and the rotation of the axes into the 4x4 structure.

Thus the elements of the homogeneous transformation can be interpreted as four vectors describing a second coordinate frame. For example, the 4x4 identity matrix, I, can be considered a coordinate frame with origin at (0,0,0) and three orthogonal coordinate axes (x,y,z). A homogeneous transform, A, composed with I results in a second coordinate frame whose origin and axes orientation are given by the matrix A.

The product of two transforms which relate coordinate frame B to frame A and frame C to frame B is a transform which relates C to A. The inverse of a transform which relates frame B to frame A is another transform which relates A to B.

These homogeneous coordinate transforms when defined symbolically are used to provide a structural description of a task. This is illustrated by means of a simple example. The task consists of picking up several short bars from a bar feeder and inserting them into holes in a block which is held in a fixture, see Fig. 1. The bar feeder presents each bar at a fixed and known position. A special tool with clamping device is used to grasp the bar and is attached to the end of the manipulator arm. The fixture is transported by a conveyor, but is held stationary during the insertion process. Before the bars can be inserted, the exact position of the fixture (and the holes) must be determined. To aid in doing this a marker is located at a corner of each fixture. A fixed overhead camera is used to locate the marker and establish the configuration of the fixture in the horizontal plane. It is assumed that the hole locations are known relative to the fixture.

The transforms CAM, Z, BAR_F represent respectively the positions of the camera, manipulator base, and bar feeder relative to the task reference frame (TASK). T is a transform which describes the configuration of the end of the manipulator arm relative to its base (Z) and is a computable function of the joint coordinates. E is a transform which represents the tool relative to the end of the manipulator arm. The transform FIXTURE relates the position and orientation of the

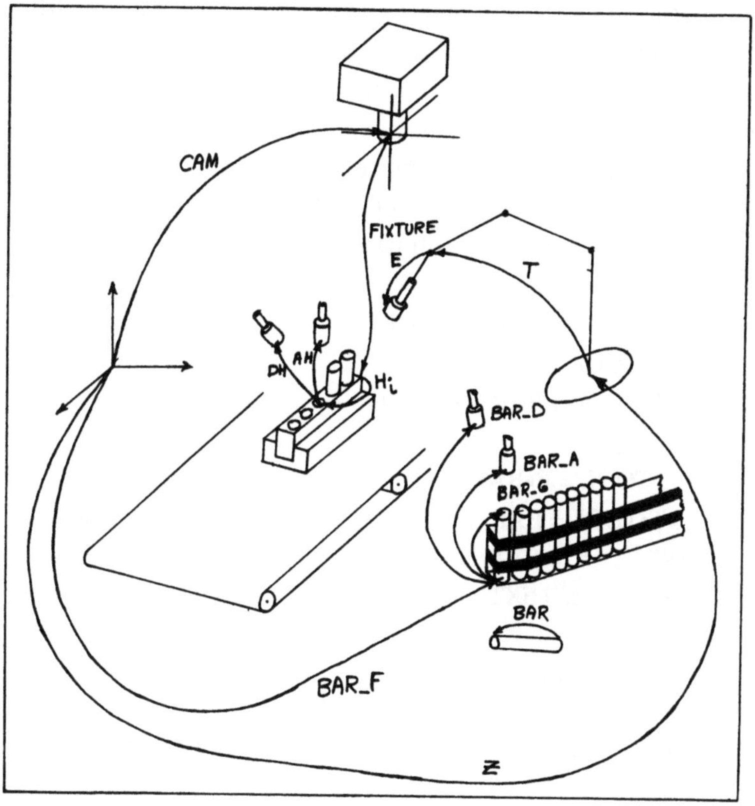

Fig. 1 A manipulator positioning task

fixture to the camera (CAM). There are five transforms denoted by H_i (i=1, ... , 5) which represent the position of the ith hole in the block relative to the fixture (FIXTURE). Additional transformations needed to fully specify the task include: BAR_A, BAR_G, BAR_D, the approach, grasp and depart positions, respectively, relative to the bar feeder; BAR, the bar length; and AH, DH, HD, the approach position, depart position and hole depth, respectively, relative to the ith hole, (i=1,....,5).

To visually obtain the position of the fixture relative to the TASK reference frame the homogeneous transform FIXTURE is constructed from the image. Since the camera is in a fixed position relative to task space (CAM), the position of the fixture relative to the TASK reference frame is given by FXT = CAM•FIXTURE. In a similar manner the position of the ith hole relative to the TASK reference frame is obtained by composing FXT with H_i i.e., $HOLE_i$ = FXT•H_i. To maintain control over the tip of the manipulator a transform, TOOL, is defined. Before the bar is grasped, TOOL is located at the end of the special tool. After the bar is grasped it is located at the end of the bar so that the end of the bar can be brought into coincidence with the bottom of the hole. The configuration of the manipulator, that is the position and orientation of the special tool, is Z•T•E. Thus, when required to move to a particular position, denoted posn, the expression, MOVE posn is replaced by MOVE Z•T•E.

The task can now be represented as a series of transform equations solvable for the set of transformations, {T}, i.e., the manipulator control inputs, as follows:

	TOOL = E	initialize TOOL
	FXT = CAM•FIXTURE	obtain fixture position
pos1	Z•T•TOOL = BAR_F•BAR_A	approach bar feeder
pos2	Z•T•TOOL = BAR_F•BAR_G	bar pick-up position
	grasp	
	TOOL = E•BAR	redefine TOOL
pos3	Z•T•TOOL = BAR_F•BAR_D	depart bar feeder
pos4	Z•T•TOOL = FXT•H_i•AH	approach hole
pos5	Z•T•TOOL = FXT•H_i•AH^{-1}	contact hole
	release	
	TOOL = E	redefine TOOL
pos6	Z•T•TOOL = FXT•H_i•HD	insert bar
pos7	Z•T•TOOL = FXT•H_i•DH	depart fixture

Once the transforms are numerically specified the transform equations can be solved. Some of the transforms can be obtained from knowledge of the task geometry, eg., CAM and BAR_F; others from engineering specifications, eg., BAR and HD and some can best be determined by moving the manipulator arm to a particular position, e.g. BAR_G, obtaining T, for this position and solving for the transform BAR_G. The latter is a form of guiding.

A manipulator program can now be given that will pick-up and insert the bars by moving the arm through the sequence of positions 1 to 7. A "move to position n" instruction, i.e. MOVE posn, will trigger a calculation that determines the corresponding value of T:

T•TOOL = CBD•posn

where CBD is a combined transform specifying the current working coordinate system. For example, at pos1, CBD=Z^{-1}•BAR_F. The program is stated as follows:

```
TOOL = E
READ(CAMERA, FIXTURE)
  FXT = CAM•FIXTURE
FOR I = 1, 5 DO
      CBD = Z-1•BAR_F
      MOVE BAR_A
      MOVE BAR_G
      GRASP
      TOOL = E•BAR
      MOVE BAR_D
      CBD = Z-1•FXT•H[I]
      MOVE AH
      MOVE AH-1
      MOVE HD
      RELEASE
      TOOL = E
      MOVE DH
END
```

To achieve successful insertion of the bar into the hole, it was assumed that the uncertainty in the position of the bar is less than the clearance between the bar and the hole. The positional accuracy is provided by the camera imaging system. This assumption may not be adequate for close fitting tolerances. In this case, the best strategy is to sense the forces generated as the bar is inserted to help guide the manipulator to the final state. This type of manipulator motion is called compliant motion, and is the subject of the following section.

3.2 Compliant motion and control

Mechanical assembly offers significant challenges for the developers of manipulator systems. Many of the operations can only be successfully performed under conditions of high tolerances, but the operations may be specified using geometric features that are located with low tolerances. The variation from one assembly to the next means that the manipulator must sense the location of the features of interest, and adapt its behaviour accordingly. Active compliance is necessary in such situations which require continuous motion in response to continuous sensory input. Using sensory data it is possible to obtain desired relationships between parts or between the manipulator and parts being acquired, such as in grasping objects that are fixed and rigid. Compliant motion is an operation specific to manipulation and special sensor based mechanisms are needed as part of the manipulator system in order to perform such motions. There are several techniques for specifying compliant motions but the most general for use in planning and specifying manipulator tasks is hybrid control (Mason [1983]).

The hybrid controller approach allows the user to control on force/torque in addition to position/orientation. Force is controlled along certain degrees of freedom while position is controlled along the remaining ones. A compliance frame is used to describe the components of translation and orientation of the object. The task degrees of freedom are defined to be the translation along and the rotation about the axes of the time varying compliance frame. At any time during the motion, the degrees of freedom are specified as to which are position controlled and which are force controlled.

In an example involving the insertion of a peg in a hole, the compliance frame is chosen so that the z-axis is coincident with the longitudinal axis of the peg. The origin of the compliance frame is at the tip of the peg Fig. 2. Translation along and rotation about the z-axis are position controlled while the remaining degrees of freedom are force controlled. The position/orientation and force/torque goals can now be specified.

The duality between position and force control can be more formally expressed in terms of natural constraints and artificial constraints (Mason [1981]). Natural constraints in position or force are defined by the geometry of the task. In the previous example the peg is able to translate and rotate along and about the z-axis, but cannot translate or rotate about the other axes. Similarly force and torque along and about the z-axis are zero provided the coefficients of friction are zero, while forces and torques for the other axes are unconstrained. Thus, the natural constraints for this task configuration are

$$v_x = 0 \qquad \omega_x = 0$$
$$v_y = 0 \qquad \omega_y = 0$$
$$f_z = 0 \qquad \tau_z = 0$$

These are homogeneous linear equations expressed in force and velocity components relative to the compliance frame where v, f, ω, and τ are velocity, force, angular velocity, and torque respectively. For each task configuration, a generalized surface with up to six degrees of freedom can be defined in a configuration space (Mason [1981]). This is called a C-surface. Position constraints lie along the normals to this surface and force constraints along the tangents.

Additional constraints called artificial constraints are introduced to specify desired motions or forces. That is, each time a desired position or force is specified, an artificial constraint is defined. Unlike the natural constraints the artificial force constraints are specified along C-surface normals and artificial position constraints along

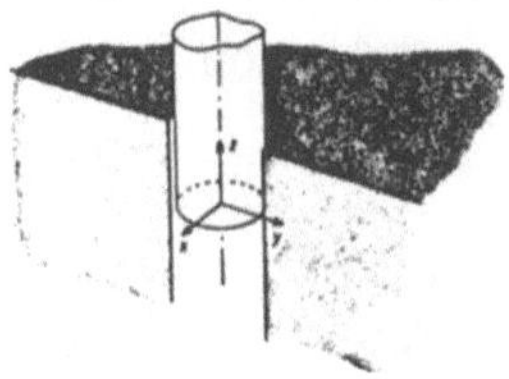

Fig. 2 Compliance frame of peg-in-hole task

tangents so that consistency with the natural constraints is preserved. The artificial constraints for the peg in hole example are

$$f_x = 0 \qquad \tau_x = 0$$

$$f_y = 0 \qquad \tau_y = 0$$

$$v_z = c_1 \qquad \omega_z = c_2$$

where c_1 and c_2 are velocity of the insertion and angular velocity of the peg with respect to the inner surface of the hole. These latter two values depend on the goal strategy and may vary with time.

It is noted that Mason's work provided the basic theoretical framework for structuring compliance tasks. This is especially important from the viewpoint of the control strategy synthesis problem which is to find artificial constraints which will produce the given goal trajectory for a given C-surface. Control strategy synthesis is an important part of the overall manipulator planning problem and is thus related to automatic planning, some aspects of which are dealt with in Section 4. Methods for automatically choosing the natural constraints from knowledge about the task geometry may eventually be developed. Mason has also developed a substantial proportion of the conceptual structure necessary to do this.

The specification of the insertion problem as a sequence of compliant moves is now considered in light of the foregoing discussion. There are four positions in the insertion sequence: approach, contact, alignment and insertion Fig. 3. Associated with each of these positions are restrictions on the peg's position and orientation which determine the number of degrees of freedom available. There are five degrees of freedom associated with motion of the peg in the approach position. On contact with the hole, the degrees of freedom will be reduced to two, i.e. translation in the x, y and z directions are constrained, as well as rotation about the y-axis After rotating the peg to the aligned position the degrees of freedom are again reduced to two: translation and rotation along and about the z-axis. When the peg is fully inserted only one degree of freedom remains: rotation about the z-axis.

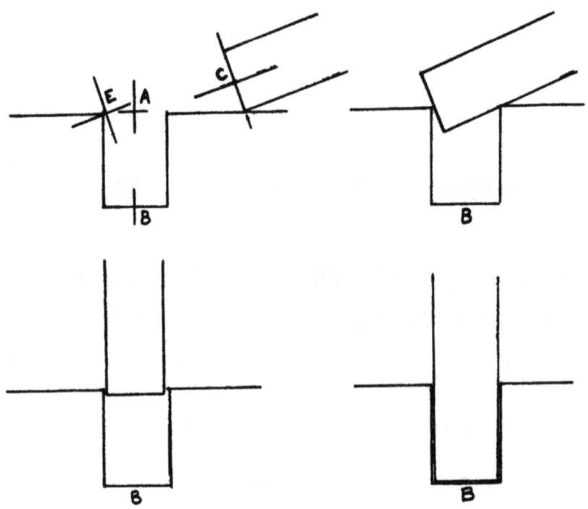

Fig. 3 Pin insertion sequence

In developing a manipulator program to execute the insertion it is assumed that the position of frames A, B and E representing the surface, bottom and edge of the hole respectively are only nominally known. Thus, since there is some uncertainty as to the exact position of the hole, an initial search is implemented in the form of a compliant motion along the surface. This motion is terminated by a force due to encountering the hole.

```
WITH FORCE (X = 0.2)
MOVE E
UNTIL FORCE (Z = -1)
```

The peg is then aligned with frame A by maintaining a negative Z force and a positive X force to ensure contact with the hole. The peg is free to comply in the y direction, centering it in the hole.

```
WITH FORCE (X = 0.2, Z = -1)
MOVE A
UNTIL TORQUE (Y = 5)
```

The final step is to insert the peg, complying in both position and rotation, and stopping when it reaches the bottom.

```
WITH FORCE (X = 0, Y = 0)
WITH TORQUE (X = 0, Y = 0)
MOVE B
UNTIL FORCE (Z = -5)
```

The peg is now inserted in the hole with the coordinate system at the tip of the peg in the center of the bottom of the hole. The uncertainty with which the position of the hole is known need only be within one-half the diameter. Despite this amount of uncertainty a tight fitting tolerance can be accommodated.

4.0 Task planning

Current manipulator systems require that tasks be specified by a sequence of motions rather than specifying the component steps of the operations to be carried out. The objective of task-level programming systems is to specify the task by symbolically described states, leaving it to the system to plan the required detailed motions.

Task planning involves the three classes of motions previously discussed: free, guarded and compliant. Avoiding collisions is an issue in all types of motions, but it is in free motion planning that it assumes central importance where it is termed obstacle avoidance. The problem of planning free motions becomes a geometric one, in which the configuration of the manipulator, the object and obstacles in the workspace must all be considered. In the following, several approaches to obstacles avoidance are described, which serve to illustrate the inherent complexity of obstacle avoidance.

When planning motions that require, or are likely to involve, contact with surfaces, the problem changes from a purely geometric one to one of a dynamic nature. The use of active compliance enables manipulators to carry out such fine motion tasks in the presence of significant sensing and control uncertainty. Sensory feedback is inherent in compliant motion. Because of the strong dependence on surface geometries, planning fine motions must be tailored to each task. The development of a general, flexible system for automatic synthesis of fine motions would be of significant benefit to task planning in assembly operations. In Section 4.2, several approaches to the design of such a system are described.

4.1 Obstacle avoidance

The problem of planning free motions is fundamentally the geometric problem of finding a path for a solid moving object(s) among other solid objects. Collision avoidance is an issue in all types of manipulator motions but it is in free motions where it plays its clearest role. Although many obstacle avoidance algorithms exist none are both general and efficient.

Algorithms for obstacle avoidance can be grouped into the following classes (Lozano-Perez [1983a]): 1) hypothesize and test; 2) penalty function; 3) explicit free space.

The first approach to be proposed was the hypothesize and test method. A candidate path is considered, say a straight line path between the initial and goal positions, and a model of the manipulator is checked for collision as it moves along the path. If a collision occurs the hypothesized path is modified on the basis of the information from the collision. The new path is tested again for collisions. This procedure is continued until a collision free path is found or no feasible path can be found. A drawback with this approach is that a single collision provides very little information to suggest how the path should be changed. As a result the method can lead to dead ends and excessively long paths.

The penalty function approach is based on the determination of a penalty function, defined locally in task space, which takes on very large values for manipulator configurations that cause collisions but drops off sharply as the distance from obstacles increases. The penalty function can be augmented by including a penalty term for deviations from the shortest route and adding penalties due to individual obstacles. The most desirable path might logically be expected to be the one following the local minima of the penalty function. Penalty function approaches are more suitable for paths that require only small modifications, since obtaining globally optimal deviations from a nominal path may be computationally prohibitive.

The explicit free space approach is concerned with the free space between obstacles. Here free space is represented by the union of (possibly overlapping) free regions. Obstacle avoidance is then the problem of finding a path in these free regions that connects the initial and final configurations. An advantage of these methods is that the explicit description of free space allows search methods to be defined that are guaranteed to find paths if, in fact, a path exists. Furthermore, it is feasible to search for short paths rather than determining only a safe path. A drawback is that path computation using explicit free space methods may be expensive. Both the hypothesize and test and penalty function methods may be more efficient for uncluttered spaces, but for relatively cluttered spaces these methods can expend an excessive amount of time in path searching.

Many spatial planning applications contain elements of what are termed the Findspace and Findpath problems. Findspace involves finding a configuration at which to place an object such that it does not intersect other objects. This is referred to as a safe configuration. The objective in Findpath is to move an object from one configuration to another such that it does not collide with any obstacle during transfer. This is called a safe path. The Findpath problem can be viewed as solving a sequence of contiguous Findspace problems.

The concept of configuration space is useful in solving the Findspace and Findpath problems. The configuration space (Cspace) of an object is the set of all its possible configurations. For example, the Cspace of a 2-dimensional object which can assume any location and orientation in a plane is the set of all tuples (x, y, θ), where x, y represents the position of a specified reference point on the object relative to a fixed Cartesian coordinate system, and θ represents the angle made by a reference line on the object with one of the Cartesian axes. Similarly a rigid object allowed to assume any position and orientation in 3-dimensional Cartesian space can be represented by a 6-tuple $(x,y,z,\phi,\theta,\gamma)$ in Cspace.

The Cspace of a manipulator is the set of all manipulator configurations. Free space (Fspace) of a manipulator is a subset of Cspace and is defined by all configurations of the manipulator for which no collisions with obstacles occur. Since any manipulator configuration can be represented by a point in its Cspace, the Fspace contains all the solutions to the Findspace problem. The Findpath problem, by extention, consists of finding a path joining the initial and goal position such that all positions along the path are solutions to the continuous Findspace problem. Thus, in Cspace any continuous path that joins the points representing the initial and final configurations of the manipulator, and is completely contained in Fspace is a collision free path. Once the representation of Fspace is developed, a collision free path can be obtained quite simply. However, obtaining an exact representation of Fspace for six degree of freedom manipulators is a difficult problem. This is because Cspace representation of obstacles with simple shapes can be very complex.

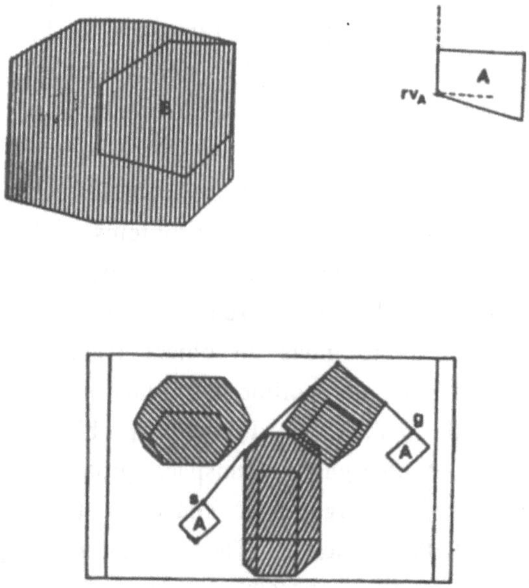

Fig. 4 Cspace obstacles and collision free path

One approach used to obtaining the Cspace representation is to first "grow" obstacles in Cartesian space into obstacles in configuration space Fig. 4. The process of growing an obstacle consists of finding its boundary in Cspace. The boundary is given by the locus of Cspace points (configurations) traced by the manipulator or payload when in contact with the obstacle. In the case of a 2-dimensional polygonal shaped object moving among 2-dimensional polygonal shaped obstacles without rotation, the shape of the grown obstacle is polygonal. When the moving object is allowed to rotate the grown 2-dimensional objects are no longer polygonal but are defined by a number of curved surfaces, each valid over a particular range of object rotation. Growing obstacles is in general, computationally expensive. Donald (1984) has developed equations for surfaces of obstacle in 6-

dimensional Cspace. Though mathematically complete the use of these equations to plan collision free paths has not yet proven practical.

Thus in higher dimensional Cspace, approximations to Cspace obstacles are used. Lozano-Perez [1981] uses a series of lower dimensional "slices" to approximate the higher dimensional Cspace obstacle. This means that some parameters values (joint angles) are fixed (either at a constant value, or over a small interval) and convex polyhedral approximations to the grown obstacle are obtained as the remaining parameters (i.e. joint angles) vary over their complete range.

Once the representation of obstacles in Cspace has been obtained, Fspace is represented as the complement of the space occupied by the Cspace obstacles. Other methods, in distinction to the above, which discretize the set of all points in Cspace into a number of subsets and test each subset for inclusion in Fspace have been proposed. Gouzenes [1984], discretizes Cspace with an n dimensional grid, where n is the number of degrees of freedom of the manipulator. Each cell is tested for being part of Fspace or obstacle space. This involves the use of a collision detector that checks for collision of the volume swept by the manipulator as it moves over the range of configurations represented by the cell and the obstacles.

Instead of dividing the Cspace into a uniform grid and testing each cell, it is more efficient to start with large regions of Cspace and subdivide it further only if the region does not belong to Fspace. A region of Cspace is checked and if no collision is detected then this region is stored as part of Fspace, otherwise it is divided and investigated further.

As is obvious, the development of Fspace representation by the method outlined above requires an efficient collision detection scheme. The use of enclosing volumes (e.g. spheres, cylinders) to model the manipulator link and obstacles can lead to fast collision detection schemes. Further efficiency can be realized by the use of a hierarchy of spheres to model the links and obstacles (Moravec [1981], Sawatsky [1985]). At the highest level objects are represented by only a single sphere. If the sphere is collision free so is the object; otherwise, the object is represented by two smaller spheres. If collision is again detected the number of enclosing spheres is doubled. This process continues until either no collision is detected or the sphere diameter closely approximates the object. Interference detection between spheres is a trivial computation (distance between centers $\geq$ sum of radii of spheres) so that collisions are easily detected.

By detecting collisions in regions of Cspace a graph representation of Fspace may be developed. The nodes of the graph represent regions of Cspace which have been found to be free of obstacles. The edges of the graph link the adjacent regions of Fspace. Finding a collision free path is reduced to locating the regions of Fspace in which the initial and goal configuration belong, and finding a path on the Fspace graph between these regions.

4.2 Fine Motion Planning

Fine motion planning in an assembly environment is characterized by the requirements for control of the grasped object while in contact or in close proximity to obstacle surfaces. The objective of an automatic planning system is to relieve the requirement for detailed specification of manipulator motions, which in the case of compliant motion involves force as well as position control. Lozano-Perez, Mason and Taylor [1983] describe a formal approach to the synthesis of compliant motion strategies from

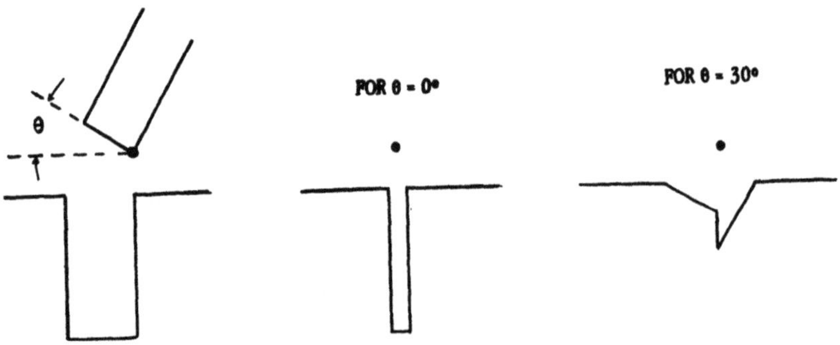

Fig. 5 Cspace representation of peg-in-hole task

geometric descriptions of assembly operations and explicit estimates of errors in sensing and control.

This procedure consists of first transforming the assembly environment into a configuration space representation Fig. 5. It then allows unconstrained motion of the manipulator in Fspace and constrained sliding motion along Cspace surfaces. These two types of motion are incorporated into the planning procedure by means of Coulomb friction and generalized damper models. It should be noted that in most of the work involving fine motion planning, motions are modelled as velocities so that both direction and magnitude of the resulting contact force are accounted for.

The Coulomb friction model assumes that the frictional force, f, due to sliding motion along a constraining surface is given by: $f=-\mu F_N$, where F_N is the normal force of the constraining surface acting on the manipulator and μ is the known coefficient of dynamic friction (assumed equal to the coefficient of static friction).

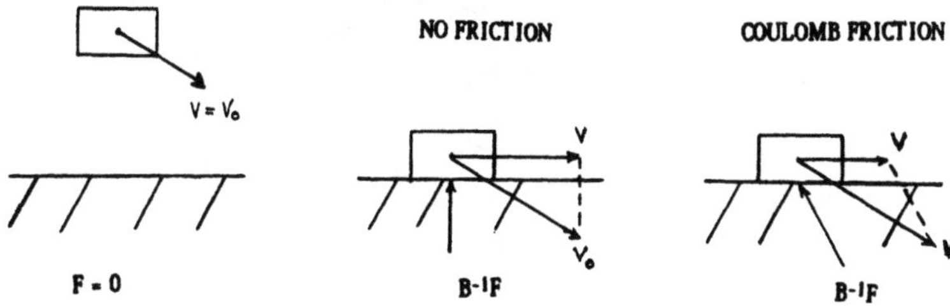

Fig. 6 Generalized damper model

The generalized damper model is used to combine both constrained and unconstrained motions into one algebraic and geometric representation. It assumes that the force, F, acting on the manipulator by the environment is given by: $F=B(V-V_0)$, where V_0 is the commanded velocity, V is the actual velocity, and B is a damping matrix (B is typically diagonal with constant entries). Thus for unconstrained motion F=0 and $V=V_0$, i.e. the actual velocity is the commanded velocity, Fig. 6. For constrained sliding motion, $V=V_0+B^{-1}F$, the actual velocity is tangential to the constraining surface, provided there is no sticking.

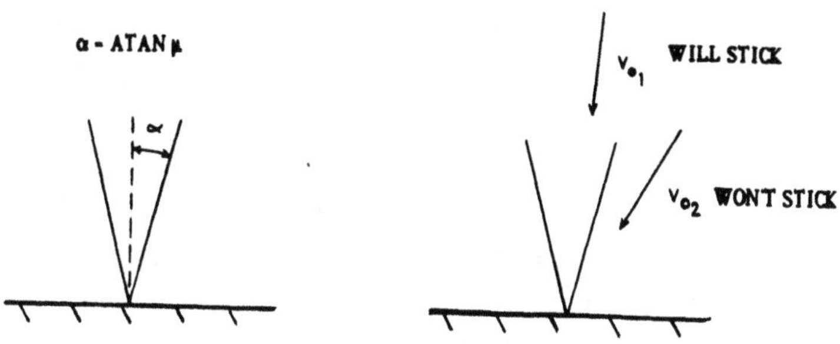

Fig. 7 Friction cone

A simple geometric construct called a friction cone is used to determine when sticking occurs. All manipulator velocities that point into a cone α degrees from the vertical ($\alpha=\arctan\mu$) will stick upon contact with the surface; all velocities that don't point into the cone will slide upon contact Fig. 7.

Now if perfect sensing and perfect control were available given an initial position and command velocity, all possible resultant manipulator configurations could be determined. Given a goal region, it is straightforward to plan a sequence of velocity commands that allow the manipulator to obtain that goal.

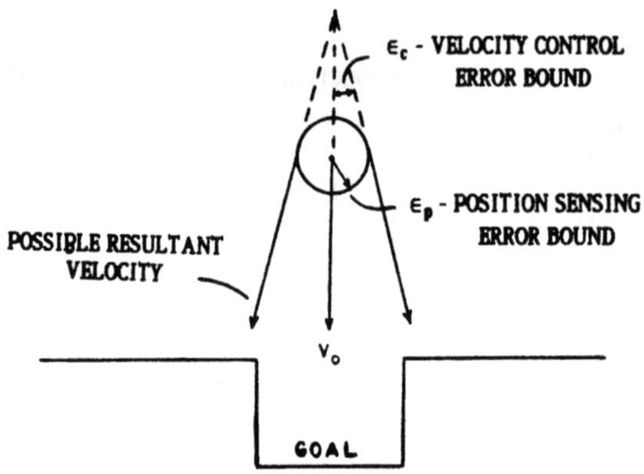

Fig. 8 Effect of sensing and control errors

If perfect sensing and control are not available but known or estimable bounds on the sensing and control errors are, all possible resultant positions can still be determined for a given initial position and velocity. The effect of uncertainty is to greatly expand the number of possible resultant positions for a given initial position and nominal velocity. Included among these may be many in which the manipulator encounters sticking or wedging before reaching the goal, making it much more difficult to plan a sequence of velocity commands, see Fig. 8. Note that in Fig. 8, ϵ_C denotes the known constant bound on velocity control error and ϵ_P is a similar bound on the uncertainty in nominal position.

To plan velocity commands which avoid sticking or wedging, Lozano-Perez, Mason and Taylor [1983] introduce the following strategy: generate a sequence of regions where each region is guaranteed to be attained by at least one velocity command initiated from the previous region. With such a sequence of regions, a sequence of velocity commands can be specified that move the manipulator from one region to the next, and ultimately to the goal.

There are two important problems to resolve regarding this strategy. The first is how to generate a sequence of regions; the second is, having obtained the regions, how are the velocity commands specified, i.e. by what criteria are the velocity commands selected. It is at this point that the fine motion planning procedures diverge, each incorporating different solutions to these two problems.

Pre-Images

In attempting to solve these two problems Lozano-Perez, Mason and Taylor [1983] take an approach that trivially solves problem two if a very non-trivial solution to problem one can be found. This is done by incorporating the concept of pre-images.

A pre-image of a region, R is the set of all positions from which a given velocity command, θ , is guaranteed to recognizably attain R. The concept of recognizability is important and means that the manipulator's sensors have the ability to determine when it has attained a region unambiguously.

The pre-image of the goal is found for all possible velocity commands θ . If the initial position is fully contained within any of these pre-images, then problem two is trivially solved. If not, then one backchains by finding the pre-image of every pre-image, etc., until the initial position is fully contained within some pre-image. The second problem is again trivially solved, as the sequence of velocity commands to be used is uniquely determined by the sequence of pre-images.

The two major limitations to this procedure are 1) that there is an infinite number of pre-images of a region R, indexed by all possible velocity commands, and 2) that there is a very non-trivial question as to when (or even if) pre-images are computable. This question of the computability of pre-images is closely tied to the idea of recognizability. Lozano-Perez, Mason and Taylor [1983], use a decision rule to eliminate potential trajectories based only on sensor readings and elapsed time. Mason [1983] constructs a decision rule based also on all past sensor readings. Neither have demonstrated that their pre-images are computable; in fact, both suspect that they aren't. However, Mason has been able to prove two important results, those of correctness and bounded-completeness (bounded-completeness states that if a solution with a bounded number of velocity commands exists, his procedure will find a solution). The bottom line is, however, that neither of the two procedures are very useful from the viewpoint of implementation.

Backprojections

In trying to solve problems one and two, Erdmann [1984] takes a slightly different approach. Instead of finding pre-images, Erdmann finds backprojections.

A backprojection of a region, R, is the set of all positions from which a given velocity command, θ , is guaranteed to attain the goal (but not necessarily recognizably attain the goal). This procedure, as opposed to the previous procedure is computable. However, it has the drawback that one can't always specify when velocity commands are to terminate, i.e. problem one is now easily solvable but problem two is not.

Erdmann has done substantial work in this area. He has developed algorithms for computing backprojection regions and from an implementation viewpoint his procedure is more useful. He has also undertaken steps in relating pre-images to backprojections in the hope of obtaining a compromise solution.

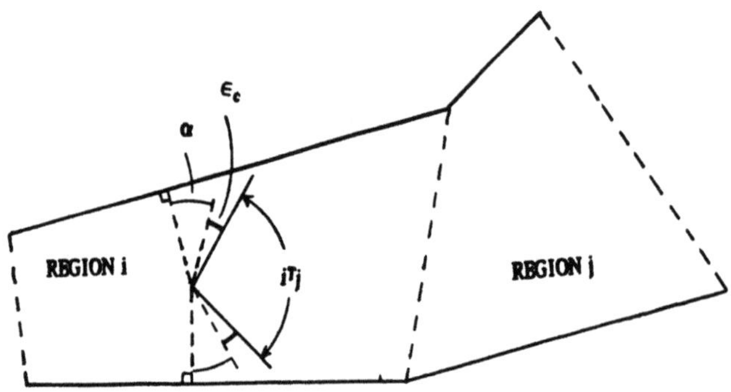

Fig. 9 T-cone

T-cones, task graphs and state graphs

Another course of action is suggested by Turk [1985]. He suggests the first problem be solved by simply partitioning the task configuration space into an arbitrary sequence of connected regions. Done manually, this is trivial for R^2 and R^3: even finding a mechanized method of partitioning R^2 or R^3 would not be too difficult. However, it is not at all clear as to whether or not arbitrary partitioning of (x,y,θ) space or $(x,y,z,\phi,\theta,\gamma)$ space would be feasible.

The solution to problem two can take many directions under this approach. Turk describes a solution that incorporates T-cones, task graphs, and state graphs.

For each region in the sequence, starting with the region containing the initial manipulator position, a T-cone is calculated, which is the set of all velocity commands that will guarantee travel into the next region from every position within the current region Fig. 9. Thus T-cones take into account the given velocity control error bound and the given coefficient of friction. The T-cone from region i to region j is written ${}_iT_j$.

An initial or nominal task graph is constructed, consisting of a node for each region of the task and arcs connecting adjacent nodes representing the T-cones between regions Fig.10. To obtain the final task graph from the nominal task graph, adjacent T-cones are tested for nonzero intersections. For example, if the intersection of ${}_2T_3$ and ${}_3T_4$ is nonzero, it is called ${}_2T_4$, and this is the set of all velocity commands that will guarantee travel from region 2 to region 4, Fig. 11. Continuing in this manner, T-cones that will take the manipulator end-effector from any given region to the furthest possible region are found. These T-cones then represent the range of velocity commands that will take the manipulator from the current region as far as

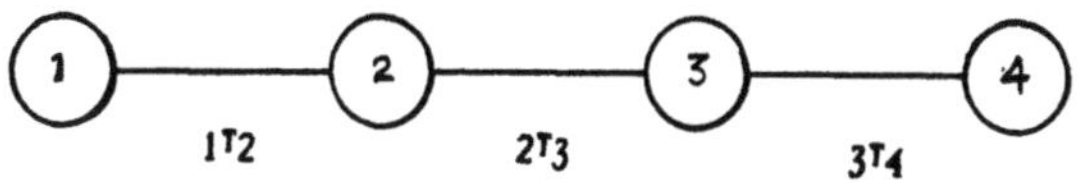

Fig. 10 Nominal task graph

Fig. 11 Intersection of T-cone

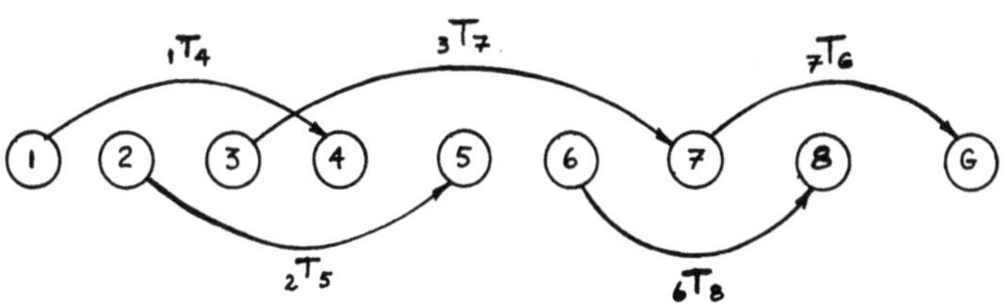

Fig. 12 Final Task graph

possible towards the goal. From these new T-cones the final task graph is constructed.

The final task graph consists of a node for each region and these nodes are connected by arcs that now represent the more general T-cones Fig.12. Regions with overlapping arcs are regions where more than one T-cone are applicable. These regions are called transition zones, in that anywhere in the transition zone the manipulator can switch from a velocity command contained in the first T-cone to a velocity command contained in a second T-cone, i.e. either velocity command will be good while in the transition zone. The larger the transition zone, the better; transition zones larger than the manipulator position uncertainty allow the manipulator to unambiguously switch from one velocity command to another while it is in the transition zone.

From the final task graph the state graph is determined. In the state graph, the T-cones from the task graph are represented as nodes, or states. These states are connected by an arc if the T-cones overlap or meet at one node of the task graph. Thus the state graph is a directed graph representation that makes explicit the alternative fine-motion plans Fig.13.

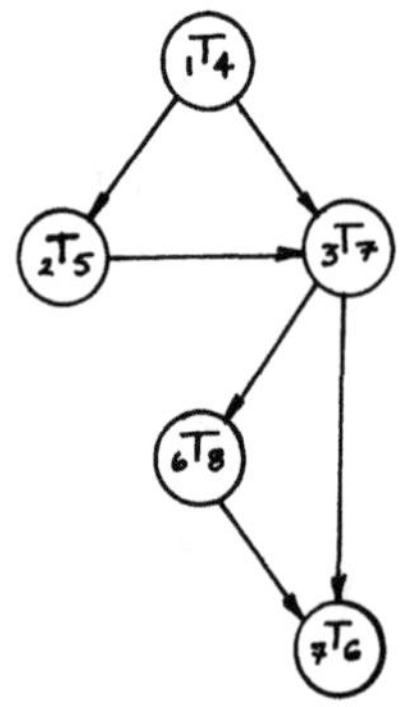

Fig. 13 State graph

Once the state graph is completed, problem two can be solved. The state graph is searched for paths leading from the initial state to the goal. Any such path represents a sequence of T-cones, hence a sequence of velocity commands, that will successfully take the manipulator from its initial position to the goal.

There are two problems remaining: 1) from among all possible paths in the state graph which take the manipulator from the initial state to the goal, which should be chosen? 2) having selected a path, and hence having selected a sequence of T-cones, from within each T-cone which velocity command is chosen?

It is natural to try to solve 1) and 2) in such a way that the chosen velocity commands are in some sense optimal. One approach (the approach taken by Turk) is the following rather simple optimization. From among all possible paths in the state graph, choose the one that gives rise to the fewest number of commands and largest transition zones; having selected this path, choose the velocity command that is in the middle of each T-cone.

There are many other ways of choosing a sequence of velocity commands that are in some sense optimal. This is an area in which there is much further room for research.

5.0 Summary and conclusions

The objective of this paper has been to give the reader an overview of some of the main issues concerning programmable manipulation as used in the assembly of discrete parts. The discussion throughout has centered on planning, specification and execution of the three types of manipulator motions: free, guarded and compliant. Emphasis has been placed on the fact that precise positioning and close tolerancing are typical in many assembly tasks.

Sensing enables manipulators to perform tasks in the presence of significant errors. Vision may be used to determine the approximate shape and position of objects, but generally not with sufficent accuracy for assembly by pure position control. Positioning errors inherently limit the tasks achievable by pure position control.

In mechanical assembly the allowable motions are often tightly constrained by external objects. The basic method for achieving constrained motion in the presence of position uncertainty is in the use of controlled compliance and guarded moves. Using this strategy the manipulator can achieve and maintain contact with surfaces or objects that may vary significantly in shape, position and orientation from expectation.

The problems arising with current manipulator technology concern the adequacy of sensor devices, processing of sensor data and the integration of sensor systems with the control and programming system of the manipulator. In future, it can be expected that there will be increased integration of sensing and motion especially in the case of compliant motion.

Two of the basic components associated with task planning were discussed: planning collision free paths and fine motion planning. Planning free motions is a problem in obstacle avoidance. The type of algorithm that has received the most attention is that which builds an explicit description of the constraints on motion and search for connected regions satisfying those constraints. Many obstacle avoidance algorithms exist but none of them are both general and efficient.

An approach to the automatic synthesis of compliant fine motion strategies for assembly tasks was presented. The approach uses geometric descriptions of parts as well as estimates of measurement and motion errors to produce fine motion strategies. This provides a basis for the development of the programming and control mechanisms best suited for sensor based motion. However, much work remains to be done in this area.

Acknowledgements

The suggestions and comments made by R.G. Whiting and R. Jha of the Department of Industrial Engineering and G.W. Reed of the Department of Electrical Engineering are greatfully acknowledged. Some of the work was partially supported by NSERC Operating Grant No. A4128.

References

1. Donald, B.R., 1984. Motion Planning with six Degrees of Freedom, Technical report AI-TR-791, MIT.

2. Erdmann, M.A., 1984. On Motion Planning with Uncertainty, Technical report AI-TR-810, MIT.

3. Gouzenes, L., 1984. Strategies for Solving Collision Free Trajectory Problems for Mobile and Manipulator Robots, Int. J. of Robotics Res., Vol. 3, No. 4.

4. Lozano-Perez, T., 1981. Automatic Planning of Manipulator Transfer Movements, IEEE Trans. Systems, Man, Cybern., Vol. SMC-11, No. 10, pp.681-698.

5. Lozano-Perez, T., 1983(a). Task Planning in Robot Motion: Planning and Control, M. Brady et al. Eds. Cambridge, MA: MIT Press.

6. Lozano-Perez, T., 1983(b). Robot Programming, Proc. IEEE, Vol. 71, No. 7, pp.821-841.

7. Lozano-Perez, T., M.T. Mason and R.H. Taylor, 1983. Automatic Synthesis of Fine-Motion Strategies for Robots, A.I. Memo 759, MIT.

8. Mason, M.T., 1981. Compliance and Force Control for Computer Controlled Manipulators, IEEE Trans. Systems, Man, Cybern., Vol. SMC-11, No. 6, pp.418-432.

9. Mason, M.T., 1983(a). Compliance, in Robot Motion: Planning and Control, M. Brady et al., Eds. Cambridge, MA: MIT Press.

10. Mason, M.T., 1983(b). Automatic Planning of Fine Motions: Correctness and Completeness, Technical report CMU-RI-TR-83-18. Carnegie-Mellon University.

11. Moravec, H.P., 1981. Robot Rover Visual Navigation, UMI Research Press, Ann Arbor, MI.

12. Sawatzky, G. and H. El-Zorkany, 1985. Using an Efficient Collision Detector in the Solution of the Find-Path Problem of Industrial Robots. Proc. SPIE Vol. 579, Conf. on Intelligent Robots and Computer Vision, Cambridge, MA.

13. Turk, M.A., 1985. A Fine-Motion Planning Algorithm, Proc. SPIE Vol. 579, Conf. on Intelligent Robots and Computer Vision, Cambridge, MA.

Shape Models in Computer-Integrated Manufacture

J.R. Woodwark
IBM UK Scientific Centre
St Clement Street
Winchester
SO23 9DR
UK

Abstract

Existing computer-aided manufacturing systems are based on models of shape tailored to a particular process. Computer-integrated manufacture involves many different processes, and hence requires more complete models of components or assemblies. This can be provided by solid modelling techniques extended to include extra data related to manufacture such as form features and tolerances. These developments are being applied to many manufacturing processes, from casting to assembly, but profound questions remain concerning both the structure and the use of shape models in manufacture.

Introduction

"It is now widely recognised that the foundation for all software in highly automated processes in manufacture has to be a geometrically complete model of the product"
M.A. Sabin, CAE Consultant to the ACME Directorate, UK Science and Engineering Research Council (80).

NATO ASI Series, Vol. F49
Computer Integrated Manufacturing
Edited by I. Burhan Turksen

In most branches of engineering, the geometry of both components and tools has a large influence on manufacturing processes; in order to computerize manufacture, computer-based representations of shape (67, 99) and efficient programs to utilize them are essential. This is a common and accurate assumption, but seldom articulated as clearly as in the quotation above.

Both computer-integrated manufacture and computer-based shape representations have aspects other than those of mutual relevance. CIM includes non-geometric tasks such as stock control, while computer-based shape representations have, perhaps, a much 'higher profile' as the basis for computer graphics **per se** (75). In this paper, however, the aim is to review the places where these technologies intersect or merge.

'Restricting' ourselves to discussing the application of shape models to CIM still leaves an enormous subject area. The size of this paper has been reduced to a reasonable length by the following reductions in scope:

a) We will concentrate on <u>mechanical</u> systems. Other engineering disciplines involve geometry and computerized manufacture, but the problems posed are much more specialized (eg integrated circuit design or terrain mapping).

b) While economic manufacture depends inevitably on good <u>design</u>, we will nevertheless ignore the enormous topic of computer-aided design as far as this close linkage allows.

c) Many practical manufacturing processes, such as punching of sheet metal, very largely involve only two-dimensional geometry. Furthermore, draughting and other two-dimensional systems represent a large proportion of currently-installed computer-aided design capacity. However, we will concentrate on three-dimensional geometric representations, and also on <u>complete</u> representations of objects, or so-called solid models. This is both because the more interesting problems are in this area, and because it is clearly the direction in which the technological imperatives are pointing.

d) Many of the technologies reviewed in this paper are incorporated into CIM systems cutting metal today. However, it is not always easy to obtain technical details about them. While a number of commercial systems are mentioned, academic contributions are more generally cited. Although some of these papers are a few years old, the techniques they describe may only now be making improvements to working CIM systems.

e) Even with the foregoing limitations, it is clearly impossible to review the literature of CIM completely in a paper of acceptable length. In making a more or less representative sample, the author's location has made him particularly aware of UK work in the field, and there is no doubt some bias in this direction, although worldwide contributions are extensively cited.

Additionally, it is emphasized that, while IBM is a major player in CIM, this paper in no way gives an 'IBM position'. The gleanings and views that it contains are, alas, merely those of the author himself.

The rest of the paper is divided into three sections. In the first, we shall look briefly at the shape models of limited capability that were developed with manufacture expressly in mind. They are commonly used to model only those parts of a component which are to be manufactured by a particular process, usually machining. In the second section we shall consider the solid model, and the additions to solid models that are necessary to make it a complete <u>component model</u> for CIM. Finally we will examine the progress that has been made in using such models for a range of specific manufacturing processes.

Geometry Models for Computer-Aided Manufacture

2&1/2D Models

Very simple geometries can be used to construct models of quite complicated components. Heat-exchanger endplates are a good example (69): just holes in a plate, but a lot of them, and their positions are very important. The simplest more general two-dimensional shape model is often considered to be the profile, or closed curve, a construction supported by many draughting systems. Typically, profiles are made up from sequences of straight lines and arcs. They can be used to define components to be manufactured from sheet material, or turned parts. In cutting out such a shape, the important operation to be performed is offsetting, so that a profile can be formed with a cutter of finite size. Offsetting involves deriving a new curve from the profile, in such a way that a cutter travelling along the offset curve produces the outline of the profile.

More complex models may be constructed from many profiles. These can represent a series of apertures in a plate or, if a height value is attached to each curve, a '2&1/2D' component is defined. The 2&1/2D representation is commonly associated with milling operations, as the geometry of a milling cutter makes the machining of 2&1/2D shapes very easy. There are a number of systems built around the 2&1/2D concept. One of the earliest and best known is GNC (34). Profiles are described using a simple language, and then graphics are used to verify the shape and to 'drive' the cutter around it. In milling 2&1/2D shapes, it is insufficient just to cut around each profile. The middle will not simply fall out, like a hole cut in a plate, but must be machined away. In the original version of GNC the necessary extra 'area clearance' cuts were generated interactively, using a graphics cursor, although later systems

(10, 70) provide more or less satisfactory automatic generation of area clearance cutter paths. In addition, GNC provides further facilities associated with the technology of the milling process. For instance, in hard materials or with certain cutters, the cutter cannot descend vertically to the depth of a profile. Instead, it must pursue a sloping path, 'ramping' into the material; GNC allows such motions to be input and verified.

GNC, and systems like it, largely rely on the user, assisted by computer graphics, to avoid making nonsense either of the definition of the shape or of the cutter path. With a stylized two-dimensional view of the machining process, the user is not always in a good position to do this. For instance, on the screen, a particular radius cutter may appear to cut an object correctly; in reality, another part of the milling machine spindle interferes with the workpiece. Advances towards more complete representation of 2&1/2D shapes include techniques for solving this problem. For instance, it is possible (1) to construct an approximate representation of the workpiece from which interferences can be deduced at reasonable computational expense.

Sculptured Surfaces

Parametric surfaces were one of the first computer-based representations of shape to appear; they allow components that could previously only be approximated using drawing board techniques to be represented exactly as a digital model. Over the years, various different formulations have been constructed. Examples are surface patches (31) and ducts (52).

Systems to machine sculptured surfaces, such as NMG (79) and UNISURF (7) were among the first CAM systems. In cutting sculptured surfaces, tooling is somewhat more complex than for 2&1/2D components. The most simple form of cutter is the ball-nosed slot drill, which terminates in a hemispherical swept volume. Offsetting can then be done along normal vectors to the surface. Ball-nosed cutters have the unfortunate property that the surface speed at their tip is zero. More complex cutters have been designed (6), but these require more complex offsetting techniques. Cutters of all sorts repay a careful choice of cutting direction, which may be varied during cutting if a machine tool with sufficient axes is available (63). Another major part of the machining process is the choice of a cutter path. It is easiest to program the cutter to move along iso-parametric lines on the surface, but these may diverge, leaving large cusps in the component. Other techniques, such as contouring or rastering, are sometimes preferred. As with 2&1/2D components, a single offset finishing cut is not adequate to form the component. Roughing may be performed by repeating the finishing cut at decreasing levels above the final surface, but more efficient strategies, usually involving more robust cutters, are much better (6, 99)

Interference problems are again apparent, and measures to overcome them have been taken in many systems. In APT (78), for instance, check surfaces can be defined which bound the cutting of a particular surface. This facility can be used to prevent interference, but also to change the shape being produced. Intersecting surfaces can be machined without the intersection being calculated explicitly, and even a **de facto** blend formed between them.

The nature of sculptured surface models limits the number of processes other than machining for which they can be used. This is a problem, because sculptured surfaces are very suitable for representing cast and moulded components and, as we shall see in the next section, a number of production-oriented analyses are useful on such objects. However, it is not too difficult to perform some analyses on sculptured surfaces. Examples are:

a) Calculating surface area, remembering that this is the area of the surface defined, which may not be the whole object.

b) Calculating curvatures, which can give help in cutter selection.

c) Determining split lines for castings. This is performed using surface normals. It allows different orientations of the component in a split mould or die to be assessed.

Component Models for Computer-Integrated Manufacture

Solid Models

Solid models are complete models of components, rather than of a number of contours or surfaces to be cut in a particular operation. They therefore offer the possibility of more automated planning and control of manufacturing processes (24). The theoretical relationships between the various types of solid model are complex (and are reviewed in detail in (76) and (99)). For present purposes, suffice it to say that the most important types are the set-theoretic and the boundary solid models. The set-theoretic (jargon CSG) solid model consists of simple, or primitive solid shapes added and subtracted together using operators derived from set-theory. The boundary (jargon B-rep) solid model consists of pieces of the model's surface linked together in a topologically consistent way. The set-theoretic model has advantages from the point of view of input, theoretical rigour and numerical stability. The boundary model has advantages in terms of modification and the accessibility of geometric information.

In the present context, the limitations of current modelling technology are of particular interest. The domain of a modeller is the range of shapes that it can represent. Some 'faceted' modellers approximate the surface of objects as planar tiles. Others permit exact representations of the common engineering curved surfaces: the cylinder, cone, sphere and torus. The most advanced systems provide sculptured surfaces or blends, but these are still the subject of research.

A second matter of importance is the efficiency of modelling systems. A universal measure is the order of the algorithms employed by a system. This determines how rapidly processing times increase with model complexity. Many simple algorithms in modelling have worse-than-linear performance in this respect; a system that deals with a nut and bolt in a flash might take hours over a moderate-sized assembly. The order of algorithms affects the speed at which pictures can be drawn but also, and more importantly in the manufacturing context, the rate at which geometric data can be extracted in a useful form.

Tolerances

Determining how a shape model will be dimensioned, and how tolerances will be attached to these dimensions, is important for most manufacturing operations. It is commonly ignored in processes such as the generation of NC cutter paths, where the machine accuracy is assumed to be adequate for the component being manufactured. In considering topics such as clamping strategies and assembly, the effects of tolerances cannot be neglected.

Initial work (9) focused on the computer-aided processing of tolerance chains, divorced from an explicit shape model. Conventional tolerancing practice may be applied to models by simple labelling (20). This is a three-dimensional equivalent to draughting practice, and does not facilitate computer processing of tolerance data. When early boundary modellers became available, a scheme was proposed (47) whereby the elements of the model were considered as parts of a mechanism, and resolved as a problem in mechanics; but it has not been widely favoured. More recently, volume-oriented techniques have been proposed, in the form of tolerance zones (77). All of these approaches have problems. To some extent the difficulty is the informality of 'traditional' engineering techniques and standards. The challenge is to construct a rigorous tolerancing model **ab initio**, and then to link this to computable modelling strategies. This is a topic in which the obstacles are still far from having been overcome.

Form Features

Very few of the specific processes that we would like to apply to models, and that we shall shortly mention, are easy to implement directly onto 'raw' geometry. There is a wide consensus that there are higher-level geometric entities - corresponding more or less closely to engineering terms such as 'boss' or 'keyway' - which can be identified and, when identified, will make processes acting on models easier to construct. Such geometric entities are often called features (see 73 and 74 for reviews of work on features).

There are a number of fundamental problems in generating a representation in terms of features. For example, can the surface or volume of a component be uniquely segmented into feature surfaces or volumes, or should features be allowed to overlap? Can features be clearly identified in a complex geometric domain, for instance the surface of a casting, where geometric elements are all smoothly blended together? Tolerances (which we already have difficulty representing) can affect features. (For instance, a simple circular hole, tightly toleranced, can imply a completely different manufacturing process from the same geometry with a wide, or general, tolerance.) How can we make sure that tolerances are taken into account (22)?

Even if such problems can be overcome or circumvented, there remains the question of whether features must be input, or can be recognized automatically. The first option is the simpler, and experimental systems have been written which either restrict the designer to a certain set of features (commonly related to manufacturing operations, (3)), or permit the manual labelling of features on an extant model (21). The problem with feature input is that it relies on the user to identify an adequate set of features. Even if he can do this reliably, we can only expect a person to have one production process in his mind at once. This is a basic problem with 'design by features', as we would hardly expect the component to be created again every time a different set of features is required. Labelling existing models is more flexible in this respect, but automatic recognition offers several potential advantages:

a) Consistency (if problems of model non-uniqueness can be circumvented).

b) Applicability to different processes (by using a different 'rule base').

c) A saving in manpower (of course).

Initial attempts at feature recognition were restricted to limited - essentially two-dimensional - domains, such as turned parts (50) or single holes (85). More recent progress can be classified by the type of model on which feature recognition is based. Boundary models have been more extensively used for feature recognition (46), employing techniques based on shape grammars ((51) and (56), the latter in a parts-coding application). A criticism that can be levelled at feature recognition based on topological information is that its preoccupation with faces can mean that features are recognized as similar which are very different from a manufacturing point of view (see examples in (100)). On the other hand, set-theoretic models make the recognition of volume characteristics very natural, but the problem in this case is the intractability, especially the non-uniqueness, of the set-theoretic model itself. Heroic attempts at basing feature recognition on set-theoretic models have involved volume decomposition (98) and the reorganization of set-theoretic expressions (11).

Product Databases and Data Exchange Formats

Databases are of course a major component of CIM systems, but usually this means a fairly conventional application to, for instance, inventory control. The possibility of including geometric information as individually retrievable items in a database has often been mooted (44, 66, 97). The relational model is considered the most respectable in database circles, and it has been applied to geometric information in a CIM context (42). However, the word is that performance of relational databases in this environment is inadequate even to maintain the present level of response of CAD or CAM systems. More **ad hoc** databases, such as TORNADO (89), designed especially for geometric information, have been somewhat more successful.

While attempts to adapt shape representation techniques to 'respectable' data structures within modelling programs have not had widespread success, there is currently a great deal of interest in standardized structures for the transmission of shape models. There are a number of different formats in use, but by far the most widely known is the IGES (Initial Graphics

Exchange Specification), which originated at the US National Bureau of Standards. As its name suggests, in its initial embodiment it was concerned with the transfer of draughting system data. Subsequent versions were extended to accommodate parametric surfaces, and the most recent versions contain formats for the transmission of both boundary and set-theoretic solid models. It remains to be seen to what extent these latest (and apparently last) versions of IGES will influence (or constrain) the internal structure of solid modelling systems. (For a good review of this rather confusing field, see (68).)

Applications of Component Models

Cutter Path Verification

Although the logic of manufacture places cutter path generation before verification, the latter is the simpler process, and in many respects a subset of the former; we shall therefore deal with it first. Proving new NC machine tapes, or rather cutter paths, can be a hair-raising business, and some method of verification off the machine is a saving in nerves as well as of machine time.

The 'conventional' way to verify cutter paths is to draw out the cutter centre-line on a graphics display or plotter. Such plots are, however, confusing and not all that helpful. Some effort can be made to show the machined region in more detail in 2&1/2D (65), but the obvious vehicle for verification is the solid modeller. The cutter path can be expressed as a series of swept volumes, which are then subtracted (differenced) from a model of the workpiece blank.

There are inherent limitations of modelling systems for this purpose. The domain of a modelling system may be inadequate, so that it is unable to represent all the shapes that a cutter can produce; or it may not be able to cope at all efficiently with the enormous amount of data present in a cutter path; or it may not have all the structures appropriate to this application (for instance, it is convenient to label regions of the surface of a model with block numbers corresponding to the cutter movement that formed that part of the shape: implying both that faces can be segmented and that labels can be attached to them).

If an adequate modeller is available, then there are a number of ways in which verification may be attempted. In the simplest method, all the swept volumes are subtracted at once from the blank, yielding a model of the machined component results. Pictures of this can be produced (91), and examination of these may be supplemented by software tools to make precise measurements of dimensions (39, 92). Alternatively, the verified model may be compared to a reference model (49), although this begs further questions. For instance, who has created this perfect model? Why not derive a cutter path automatically from that model? And what if certain acceptable deviations from the perfect model (eg cusps) are unavoidable?

More sophisticated verification techniques involve considering each cutter movement separately. One way to do this is to generate a real-time display of the cutting process (94, 95). This is impressive, but it is rather time-consuming to watch, problems can be obscured by the particular viewing direction chosen, and a moment's inattention may easily lead to a problem being missed. An alternative (49) is automatically to examine individual cutter movements. The range of tests that a movement can be exposed to is considerable:

a) Checking that cutters cut in allowable directions and with allowable parts of their volume (ie parts with teeth!).

b) Evaluating speeds, depths of cut and feeds against economic models of the cutting process (88, 103).

c) Looking at the deformations of the cutter and workpiece under cutting forces.

d) Modelling the flow of coolant and swarf.

These processes form an escalating ladder of complexity on which major challenges remain.

Process Planning and Cutter Path Generation

Process planning is generally considered to be the selection of the optimum route through a manufacturing facility for a given component or assembly. Cutter path planning is the much more restricted question of determining how to machine a component once it is on a particular machine tool (not necessarily metal cutting - measuring machines also need probe paths to be planned, see (53)). However, these processes overlap, and we shall consider them together here. In general, we expect to find a CIM system requiring a spectrum of planning activities, from high-level decisions based on approximate models of alternative production processes, to very specific planning of a single machining operation. (See (40) for a good, very much broader, review of automated process planning.)

APT, GNC and similar systems contain facilities to assist with the machining of single shape elements, or small groups of them, if not to plan their cutting automatically. The approaches taken in these systems can be applied to data extracted from more sophisticated shape models. For instance, in the ROMAPT system (26), geometry is extracted from a ROMULUS boundary model in the form of APT geometry statements. In other work, cutter path generation is more or less automatic, but limited to 2&1/2D models or turned parts or regions of models with such geometries. Such systems have been implemented on both boundary (43, 82) and set-theoretic (84) solid models.

More recent projects (see 83 for a fairly up-to-date review) have taken two lines. In the first (4, 23), essentially algorithmic techniques have been implemented which manage to deal with genuinely three-dimensional data, but produce cutter paths which are technologically poor. In the second (33, 35), process planning software has been developed based on expert system techniques, but requiring the geometry to be input in <u>textual</u> form. ('Countersunk hole of diameter x starts from surface y and opens out into face z' gives the flavour.) A number of attempts to bridge this gap are in progress (28) using techniques of feature recognition and expert systems. The aim is both to create process planning software which can obtain geometry direct from a model, and also to achieve more sophisticated cutter paths.

It is worth itemizing what a fully automatic process and path planning software would have to include:

a) A model of the component with tolerance and surface finish information, and also a model of the blank.

b) The ability to select different machines and cutter (25) combinations on both geometric and tolerance bases.

c) The capability to determine appropriate clamping or fixturing systems automatically (and efficiently, because fixtures are often a major element of cost, especially for small batches).

d) An algorithm capable of producing technologically sophisticated cutter paths, which takes into account all the assessment factors already mentioned in discussing verification.

Casting and Similar Processes

Processes involving material flow have as yet been somewhat less affected by solid modelling technology than has metal cutting. An initial problem was that early modelling systems were unable to cope with the smooth shapes of cast, forged and moulded components. The MODCON system (27) provides an example of a special-purpose modeller containing shape elements and transitions between them specially adapted to the casting process, although lacking generality in other respects. This approach has been repeated in more recent work (62). Other projects have accepted some degree of approximation and used more conventional modelling systems (5, 93). Recent work on blend surfaces (for a review see (101)) promises to make the use of more general modelling systems for castings and so forth much easier.

Having constructed a model, even the simple analysis functions of a solid modeller are useful: for instance, calculating the volume (17, 57) of an object, so as to determine the amount of material required. Other relatively simple geometric data, such as length of split line, can be made available in a form to which process design rules can be applied (29, 90). More sophisticated analyses tend to require the shape to be broken down into simple elements of some sort. This may be an approximate decomposition appropriate to a particular analysis (eg material flow in injection mouldings (41), or cooling rate in castings (62)). Alternatively, conventional finite-element analysis may be performed. Finite-element meshes can be generated manually, but full automation of relevant analyses implies automatic mesh generation. This is feasible in two dimensions, and finite-element methods may thus be used to analyse extrusions (58). In three dimensions, finite-element mesh generation is still a research topic. While some simple (eg tetrahedral, see (102)) meshes can be generated automatically (54), they are not normally considered satisfactory for finite-element analysis of stresses and flows (except supersonic ones!). Techniques to generate more complex meshes still need to be developed.

Design for Assembly

Modelling assemblies, mechanisms or whole machines on the computer is a way of examining complex geometric interactions before anything is built. The use of such an approach is currently uncommon because of limitations in today's hardware and software. While this is of obvious importance in design itself, there are also manufacture-oriented aspects of 'design for assembly' which are worth noting here.

It is possible (86) to formulate a set of general rules that should be followed when designing components for automatic assembly. Examples include considerations of symmetry, tangling, and simply reducing the number of different components. Rules of this sort can be assembled into an expert system (86). As with process planning, the next stage is to make the leap from textual to geometric input to the system. This has been achieved in two dimensions, as a link with a draughting system (87) and the feature recognition route is the obvious one to make a further link with solid modelling.

Industrial Robots

Programming robots in terms of axis movements is more difficult than programming NC machine tools in the same way, because of robots' more complex kinematics. Robot languages (such as VAL (16)) have therefore been developed. These are typically based on existing general-purpose programming languages, in which facilities for co-ordinate transformation and statements addressing particular technological features of the robot are embedded.

While robot languages, as such, make the task of programming infinitely easier, they provide no check on the accuracy of programs, for which a geometric simulation of the robots' movements is necessary. Models of the robots and their surroundings then become necessary (2, 71). For instance, collision detection can be performed using either set-theoretic (19, 30) or boundary (15) modelling systems.

Simulation techniques are also popular in modelling manufacturing facilities before construction or modification. (This is not restricted to plant employing robots, but robot kinematics present the same problems to planners as to programmers.) Set-theoretic solid modelling techniques can be used for this sort of facility design (81), but the best-known systems are perhaps SAMMIE (13), and GRASP (14), which are based on boundary models. In the SAMMIE system, a realistic articulated human figure is available, with further special features, such as optical analyses to determine fields of view; GRASP provides a similar environment for simulating robots. It is possible to move simulated robots using a realistic robot programming language (the link between another system, RoboTeach, and the VAL language is detailed in (71)).

In more complex robot programming systems, models are used in the automation of the planning process. The AUTOPASS system (59), for instance, uses the GDP (96) faceted boundary modeller in system modules to plan trajectories (60) and gripping strategies. Early solutions to these problems were neither very general nor very efficient, and robot motion planning has been a topic of continuing research. The trajectory planning problem has received considerable attention from theoreticians (see (32) for a readable review) but algebraic solutions are often too sensitive to complexity for practical implementation. Feasible approaches rely on sensible approximations of the the working space, and a heuristic search among possible routes. A side-issue is that it is not always appropriate to assume that the robot, when moving at high speed, will take the path programmed, and deflections due to robot dynamics (37) must be considered.

At the beginning and end of a robot movement, different problems occur concerned more with contact between gripper and workpiece and between mating parts. Here, different planning techniques must be used (61, 64), often involving movements to be specified in terms of geometric constraints (as in the RAPT system (18, 72)) which must then be resolved numerically, algebraically, or heuristically. Tolerances are again a concern here, and tolerance interactions between components must be considered (38).

Machine Vision

Two-dimensional machine vision techniques, involving the identification of profiles from binary images, are well-established in industry (eg (48)). It is possible to link such systems with two-dimensional shape models, but a training mode of operation, in which the system 'learns' a set of example components, is more common. For each shape the system calculates simple measures such as area, moments of area, and numbers of holes, and this data is used in operation to recognize the different components and their positions and orientations.

Two-dimensional techniques can be applied to three-dimensional components to a limited extent. If, for instance, parts on a conveyor are to be identified by an overhead camera, the silhouettes that will be presented by three-dimensional parts resting in different positions can all be calculated, and used to identify that part. All the different 'stable orientations' of a component can themselves be determined automatically.

This approach (96) shows an impressive integration of techniques, but it is easy to think of different components which would exhibit identical silhouettes in certain orientations.

True three-dimensional vision presents much more profound problems. Software is available that can process images, to extract edges and so on. The difficulty comes in reconstructing the three-dimensional objects in the scene. The most recent thinking (8) is that three-dimensional vision is easier to implement when models of the objects to be identified are available. (The human observer is known to try to make images fit known prototypes.) Model-based three-dimensional vision is a demanding technology. Models must be available and geometric transformations between the model and scene co-ordinates must be made. (Thus far, some techniques can be borrowed from computer graphics.) Matching models with parts of the scene can then be done by the correlation of surfaces (36), edges, or their combination (45). The problem is further complicated when there are many objects in the scene (the 'bin-picking' problem, see (12)) or when one of a number of different objects must be recognized (55), or both. Research to find reliable and fast techniques for these purposes continues.

Conclusions

There are many areas in which the application of shape models has already made a great impact on computer-aided manufacture. Examples include APT and GNC-like systems, current robot languages, simulation tools and two-dimensional vision systems. In working towards computer-integrated manufacture, we seek a higher level of automation based on more sophisticated shape models. In many parts of the CIM spectrum, desirable medium-term goals have been identified: for instance, effective feature recognition and representation techniques, tolerance representations, and extension of the domain of shape models to make the representation of castings and forgings a routine matter. Not all of the objectives may finally prove to have been the most appropriate ones to pursue, but many of them look highly plausible. In a lot of areas, work on these topics has been in progress for years, but increasing capabilities unmask further problems, or advances in other areas cause increasing requirements. (For instance, objects including sculptured surfaces will obviously make feature recognition tasks more difficult.) Despite the glittering visions of automated factories fully integrated with production-oriented design, there are perhaps surprisingly few groups actively pursuing developments in this field, at least as reflected by the published literature. There is still plenty of opportunity to tackle the many and complicated problems that remain in modelling shape for CIM!

Acknowledgements

The author is grateful to the following people who assisted him with the preparation of this review by supplying reprints, information and advice: Dr T.A. Dean, Dr I.D. Faux, Dr P. Hartley, Mr G.E.M. Jared, Dr J.M. Knapman, Prof Dr-Ing F.-L. Krause, Prof M.J. Pratt, Dr M.A. Sabin, Mr D. Saunders and Dr W.-P. Wang.

References

1. R.O. Anderson, 'Detecting and eliminating collisions in numerically controlled machining', Computer-Aided Design 10,4 (231-237), 1978.

2. A.P. Ambler, 'Robotics and solid modelling: a discussion of the requirements robotic applications put on solid modelling', Proceedings of the 2nd International Symposium on Robotics Research, 1985.

3. F. Arbab, D.G. Cantor, L. Lichten and M.A. Melkanoff, 'The MARS CAM-oriented modeling system', Proceedings of a MIT Conference on CAD/CAM Technology in Mechanical Engineering (281-288), 1982.

4. G.T. Armstrong, G.C. Carey and A. de Pennington, 'Numerical code generation from geometric modelling system', in Solid Modelling by Computers (Proceedings of a General Motors Symposium, Detroit, 1983) (M.S. Pickett and J.W. Boyse, eds), Plenum (139-158), 1984.

5. J.T. Berry and J.A.M. Boulet, 'The application of geometric modeling to metal casting technology', in Solid Modelling by Computers (Proceedings of a General Motors Symposium, Detroit, 1983) (M.S. Pickett and J.W. Boyse, eds), Plenum (105-120), 1984.

6. C. Bell, B. Landi and M.A. Sabin, 'The programming and use of numerical control to machine sculptured surfaces', Proceedings of the 14th International MTDR Conference, Manchester (233-238), 1973.

7. P. Bézier 'UNISURF system: principles, programme, language', In Computer Languages for Numerical Control (J. Hatvany, ed) (Proceedings of the PROLAMAT 73 Conference, Budapest), North-Holland (417-426), 1973.

8. T.O Binford, R.A. Brooks and D.G. Lowe, 'Image understanding via geometric models', Proceedings of the Fifth International Conference on Pattern Recognition, (364-369), 1980.

9. Ø. Bjørke, Computer-Aided Tolerancing, Tapir, 1978.

10. M.S. Bloor, A. de Pennington and J.R. Woodwark, 'POISE - pocket and island evaluator', presented at CAM 78 Conference, Glasgow, 1978.

11. J.E. Bobrow, 'NC machine tool path generation from CSG part representations', Computer-Aided Design 17,2 (69-76), 1985.

12. R.C. Bolles, 'The three-dimensional locating of industrial parts', SRI International Technical Note 234 (Proceedings of 8th NSF Grantees' Conference on Production Research and Technology, Stanford), 1981.

13. M.C. Bonney, C.A. Blunsden, K. Case and J.M. Porter, 'Man-machine interaction in work systems', International Journal of Production Research 17,6 (619-629), 1979.

14. M.C. Bonney, P.J. Edwards, J.A. Gleave, J.L. Green, R.J. Marshall and Y.F. Yong, 'The simulation of industrial robot systems', OMEGA International Journal of Management Science 12,3 (273-281), 1984.

15. J.W. Boyse, 'Interference detection among solids and surfaces', Communications of the ACM 22,1 (3-9), 1979.

16. C.A.J. Braganca and P. Sholl 'VAL-II, a language for hierarchical control of a robot-based factory', Robotica, 3 (265-272), 1985.

17. I.C. Braid, 'Designing with volumes: computation of weight, centre of gravity, moment of inertia and principal axes', University of Cambridge CAD Group Document 81, 1974.

18. S.A. Cameron, 'A RAPT picture-scrapbook', University of Edinburgh Department of Artificial Intelligence Working Paper 106, 1982.

19. S.A. Cameron, 'Modelling solids in motion', PhD Thesis, University of Edinburgh, 1984.

20. CAM-I, 'A study of dimensioning and tolerancing of geometric models, final report', CAM-I Report R-84-GM02.2, 1985.

21. CAM-I, 'Requirements for support of form features in a solid modelling system', CAM-I Report R-85-ASPP-01, 1985.

22. CAM-I, 'Reconciliation of design and manufacturing requirements for product design data using functional primitive part features', CAM-I Report R-86-ANC/GM/PPP 01.1, 1986.

23. CAM-I, 'A volume decomposition algorithm', CAM-I Report R-85-ANC-01 (1 - software design, 2 - walk through), 1986.

24. G.C. Carey and A. de Pennington, 'A study of the interface between CAD and CAM using geometric modelling techniques', Proceedings of the 3rd Anglo-Hungarian Seminar on Computer-Aided Geometric Design, Cambridge, 1983.

25. G.C. Carey, 'Development of an automatic tool selection technique', Proceedings of the 4th Anglo-Hungarian Seminar on Computer Aided Design, Budapest (229-246), 1985.

26. B.T.F. Chan, 'ROMAPT: a new link between CAD and CAM', Computer-Aided Design **14**,5 (261-266), 1982.

27. Y.K. Chan and W.A. Knight, 'MODCON: a system for the CAM of dies and moulds', Proceedings of the CAD-80 Conference, Brighton (370-381), 1980.

28. B.K. Choi, M.M. Barash and D.C. Anderson, 'Automatic recognition of machined surfaces from a 3D solid model', Computer-Aided Design **16**,2 (81-86), 1984.

29. S.H. Choi, P. Sims and T.A. Dean, 'A CAD/CAM package for forging hammer dies', Proceedings of 25th International MTDR Conference (451-457), 1985.

30. P.G. Comba, 'A procedure for detecting the intersections of three-dimensional objects', Journal of the ACM **15**,3 (354-366), 1968.

31. S.A. Coons and B. Herzog, 'Surfaces for computer-aided aircraft design', Proceedings of the AIAA 4th Annual Meeting and Technical Display, Anaheim (AIAA Paper 67-895), 1967.

32. J.H. Davenport, 'Robot motion planning', to appear in Proceedings of a IBM Conference on Geometric Reasoning, Winchester, 1986, OUP.

33. B.J. Davies, I.L. Darbyshire and A.J. Wright, 'Expert Systems in Process Planning', in Knowledge Engineering and Expert Systems in CAD (Alison Smith, ed), Butterworths (7-14), 1986.

34. K.J. Davies, GNC - a graphical NC processor', In Computer Languages for Numerical Control (J. Hatvany, ed) (Proceedings of the PROLAMAT 73 Conference, Budapest), North-Holland (51-61), 1973.

35. Y. Descotte and J.C. Latombe, 'GARI - a problem solver that plans how to machine mechanical parts', Proceedings of ICJAI 81, Vancouver (766-771), 1981.

36. O.D. Faugeras and M. Herbert, 'A 3-D recognition and positioning algorithm using geometrical matching between primitive surfaces', Proceedings of the 8th IJCAI, Karlsruhe (997-1002), 1982.

37. R. Featherstone, 'Robot dynamics algorithms', PhD Thesis, University of Edinburgh Department of Artificial Intelligence, 1984.

38. A.D. Fleming 'A representation for geometrically toleranced parts', to appear in Proceedings of a IBM Conference on Geometric Reasoning, Winchester, 1986, OUP.

39. R. Fridshal, K.P. Cheng, D. Duncan and W. Zucker, 'Numerical control part program verification system', Proceedings of a Conference on CAD/CAM Technology in Mechanical Engineering, MIT (236-254), 1982.

40. I. Furth, Automated process planning, to appear in Proceedings of a NATO Advanced Study Institute on CIM, Istanbul, 1987, Springer-Verlag.

41. E. Galli, 'Proving out part moldability by computer', Plastics Machinery and Equipment, June 1980.

42. H. Grabowski and M. Eigner, 'Employing a relational database in a CAD system', Proceedings of ITCAD 78 (367-377), 1978.

43. A.R. Grayer, 'The automatic production of machined components starting from a stored geometric description', in Advances in Computer-Aided Manufacture (D. McPherson, ed), North-Holland (Proceedings of the PROLAMAT 76 Conference, Stirling (137-145), 1977.

44. A.R. Grayer, 'Geometric modelling in production', in Advanced Manufacturing Technology (P. Blake, ed), North-Holland (55-69 and 263-269), 1980.

45. W.E.L. Grimson and T. Lozano-Perez, 'Model based recognition from sparse range or tactile data', International Journal of Robotics Research 3,3, 1984.

46. M.R. Henderson and D.C. Anderson, 'Computer recognition and extraction of form features', Computers in Industry 5 (329-339), 1984.

47. R.C. Hillyard and I.C. Braid 'Analysis of dimensions and tolerances in computer-aided mechanical design', Computer-Aided Design 10,3 (161-166), 1978.

48. S.W. Holland, L. Rossol and M.R. Ward, 'Consight-I: a vision controlled robot system for transferring parts from belt conveyors', General Motors Research Laboratories Research Report GMR-2790, 1978.

49. W.A. Hunt and H.B. Voelcker, 'An exploratory study of automatic verification of programs for numerically controlled machine tools', University of Rochester Production Automation Project Technical Report TM-34, 1982.

50. R. Jakubowski, 'Syntactic characterization of machine parts shapes', Cybernetics and Systems 13 (1-24), 1982.

51. G.E.M. Jared, 'Feature recognition in geometric modelling', Proceedings of the 4th Anglo-Hungarian Seminar on Computer Aided Design, Budapest (23-33), 1985.

52. A.R. Johnson and D.P. Sturge, 'The DUCT system of design and manufacture for patterns moulds and dies', Proceedings of the 20th International MTDR Conference, Birmingham (17-20), 1979.

53. S. Kawabe, F. Kimura and T. Sata, 'Automatic generation of NC commands for 3-coordinate measuring machine', Proceedings of the International Conference on Production Engineering (941-946), 1980.

54. A. Kela, H.B. Voelcker and J.A. Goldak, 'Automatic generation of hierarchical, spatially addressable finite-element meshes from CSG representations of solids', Proceedings of an International Conference on Accuracy Estimation and Adaptive Refinements in Finite Element Computation (ARFEC), Lisbon, 1984.

55. J. Knapman, '3D model identification from stereo data', Proceedings of the 1st International Conference on Computer Vision, London, 1987.

56. L. Kyprianou, 'Shape classification in computer-aided design', PhD Thesis, University of Cambridge, 1980.

57. Y.T. Lee and A.A.G. Requicha, 'Algorithms for computing the volume and other integral properties of solids', Communications of the ACM **25**,9 (Part 1: 635-641, Part 2: 642-650), 1982.

58. E.C. Libardi, J.R. Dixon and M.K. Simmons, 'Designing with features: design and analysis of extrusions as an example', Proceedings of ASME Spring National Design Engineering Conference, Chicago, 1986.

59. L.I. Lieberman and M.A. Wesley, 'AUTOPASS: an automatic programming system for computer controlled mechanical assembly' IBM Journal of Research and Development **21**,4 (321-333), 1977.

60. T. Lozano-Perez and M.A. Wesley, 'An algorithm for planning collision-free paths among polyhedral obstacles', Communications of the ACM **22**,10 (569-570), 1979.

61. T. Lozano-Perez, 'Automatic planning of manipulator transfer movements', IEEE Transactions on Systems, Man and Cybernetics **SMC-11**,10, 1981.

62. S.C. Luby, J.R. Dixon and M.K. Simmons 'Creating and uing a feature data base', Computers in Mechanical Engineering (25-33), November 1986.

63. K. Marciniak, 'Influence of surface shape on admissible tool positions in 5-axis milling', Computer-Aided Design **19**,5 (233-244), 1987.

64. M.T. Mason, 'Mechanics and Planning of Manipulator Pushing Operations', International Journal of Robotics Research 5,1, 1986.

65. P. McGoldrick and R. Gibson, 'NC plotting made simple', Numerical Engineering 1,4, 1980.

66. A. Meier, 'Applying relational database techniques to solid modelling', Computer-Aided Design, **18**,6 (319-326), 1986.

67. M.E. Mortenson, Geometric Modelling, John Wiley, 1985.

68. J. Owen and M.S. Bloor, 'Neutral formats for product data exchange: the current situation', Computer-Aided Design, **19**,8 (436-443), 1987.

69. J. Pejlare, 'Cadric - Calculation of drilling co-ordinates', In Computer languages for Numerical Control (J. Hatvany, ed) (Proceedings of the PROLAMAT 73 Conference, Budapest), North Holland (221-225), 1973.

70. H. Persson, 'NC machining of arbitrarily shaped pockets', Computer-Aided Design **10**,3 (169-174), 1978.

71. M.S. Pickett, R.B. Tilove and V. Shapiro, 'RoboTeach: An off-line robot programming system based on GMSolid', in Solid Modelling by Computers (Proceedings of a General Motors Symposium, Detroit, 1983) (M.S. Pickett and J.W. Boyse, eds), Plenum (159-184), 1984.

72. R.J. Popplestone, A.P. Ambler and I. Bellos, 'RAPT: a language for describing assemblies', The Industrial Robot (131-137), September 1978.

73. M.J. Pratt, 'Solid modeling and the interface between design and manufacture', IEEE Computer Graphics and Applications (52-59), July 1984.

74. M.J. Pratt, 'Current status of form features research in solid modelling', in preparation.

75. P. Quarendon and J.R. Woodwark, 'The model for graphics', in Techniques for Computer Graphics (D.F. Rogers and R.A. Earnshaw, eds) (Proceedings of a NATO Advanced Study, Institute, Stirling, 1986), Springer-Verlag (39-64), 1987.

76. A.A.G. Requicha, 'Representations of rigid solids, theory, methods and systems', ACM Computing Surveys **12**,4 (437-464), 1980.

77. A.A.G. Requicha, 'Representation of tolerances in solid modelling: issues and alternative approaches', in Solid Modelling by Computers (Proceedings of a General Motors Symposium, Detroit, 1983) (M.S. Pickett and J.W. Boyse, eds), Plenum (3-22), 1984.

78. D.T. Ross, 'The design and use of the APT language for automatic programming of numerically controlled machine tools', Proceedings of the Computer Applications Symposium, Chicago (80-99), 1959.

79. M.A. Sabin, 'An existing system in the aircraft industry. The British Aircraft Corporation Numerical Master geometry system', Proceedings of the Royal Society of London **A 321** (197-205), 1971.

80. M.A. Sabin, 'A research programme in CAE', Computer-Aided Engineering Journal **4**,2 (79-82), 1987.

81. P.J. Sackett and J.R. Woodwark, 'Computer aided plant layout appraisal', Proceedings of the 12th International Symposium on Production Management in the Metalworking Industry, Belgrade (353-363), 1983.

82. G. Spur and H.-J. Germer, 'Three-dimensional solid modelling capabilities of the COMPAC system and some applications', Proceedings of the ACM CAE82 Workshop in Geometric Modelling, Milan, 1982.

83. R. Srinivasan and C.R. Lin 'Evolutionary trends in generative process planning', in Methods and Tools for Computer Integrated Manufacturing (Notes for CREST course CIM 83) (U. Rembold and R. Dillman, eds), Springer-Verlag (179-193), 1984.

84. T.W. Stacey and A.E. Middleditch, 'The geometry of machining for computer-aided manufacture', Robotica **4** (83-91), 1986.

85. M. Staley, M.R. Henderson and D.C. Anderson, 'Using syntactic pattern recognition to extract feature information from a solid geometric data base', Computers in Mechanical Engineering (61-66), September 1983.

86. K.G. Swift, 'A computer-based design consultation system', Assembly Automation (151-154), August 1983.

87. K.G. Swift, Knowledge-Based Design for Manufacture, Kogan-Page, 1987.

88. V.A. Tipnis, S.A. Vogel and H.L Gegel, 'Economic models for process planning', Proceedings of the Sixth North American Metalworking Research Conference, Gainesville, Florida (379-387), 1978.

89. S. Ulfsby, S. Meen and J. Oian, 'TORNADO: a DBMS for CAD/CAM systems', in File structures and data bases for CAD (J. Encarnacao and F. Krause, eds), North-Holland (335-350), 1982.

90. M. Vaghul, G.E. Zinsmeister, J.R. Dixon and M.K. Simmons, 'Expert systems in a CAD environment', Proceedings of the ASME Spring Design Engineering Conference, Boston, 1985.

91. A.F. Wallis and J.R. Woodwark, 'Creating large solid models for NC toolpath verification', Proceedings of the CAD-84 Conference, Brighton (236-243), 1984.

92. A.F. Wallis and J.R. Woodwark, 'Interrogating solid models', Proceedings of the CAD-84 Conference, Brighton (236-243), 1984.

93. K.K. Wang and P. Khullar, 'Computer-aided design of injection molds using TIPS-1 system', Proceedings of the CAM-I International Spring Seminar, Denver (35-46), 1980.

94. W.-P. Wang and K.K. Wang, 'Real-time verification of multiaxis NC programs with raster graphics', Proceedings of the IEEE International Conference on Robotics and Automation, San Francisco (166-171), 1986.

95. W.-P. Wang and K.K. Wang, 'Geometric modeling for swept volume of moving solids', IEEE Computer Graphics and Applications, (8-17), December 1986.

96. M.A. Wesley, T. Lozano-Perez, L.I. Lieberman, M.A. Lavin and D.D. Grossman, 'A geometric modelling system for automated mechanical assembly', IBM Journal of Research and Development **24**,1 (64-74), 1980.

97. R.N. Wolfe, '3D geometric databases for mechanical engineering', IBM Thomas J. Watson Research Center Report RC7655, 1979.

98. T.C. Woo, 'Computer-aided recognition of volumetric designs - CARVD', In Computer languages for numerical control, (J. Hatvany, ed) (Proceedings of the PROLAMAT 73 Conference, Budapest), North-Holland (121-135), 1973.

99. J.R. Woodwark, Computing Shape, Butterworths, 1986.

100. J.R. Woodwark, 'Some speculations on feature recognition', to appear in Proceedings of an International Workshop on Geometric Reasoning, Oxford, 1986, (to be published as a special issue of Journal of Artificial Intelligence).

101. J.R. Woodwark, 'Blends in geometric modelling', to appear in Proceedings of 2nd IMA Conference on the Mathematics of surfaces', Cardiff, 1986.

102. B. Wordenweber, 'Finite element mesh generation', Computer-Aided Design **16**,5 (285-291), 1984.

103. I. Yellowley, A. Wong and B. DeSmit, 'The economics of peripheral milling', Proceedings of the Sixth North American Metalworking Research Conference, Gainesville, Florida (388-394), 1978.

Image Acquisition and Processing Techniques for Industrial Inspection

B. G. Batchelor[1] & G. E. Foot[2]

Abstract

The concepts and design of automated visual inspection systems are described. The multidisciplinary, systems-orientated nature of the subject is emphasised; an industrial inspection system inevitably incorporates elements of mechanical handling, lighting, optics, sensors, as well as electronic and software-based image processing. Image acquisition, for example, is regarded as being of equal importance to image processing. There is a detailed discussion of illumination & viewing techniques. Different types of image sensor are compared. Image processing systems may incorporate slow, low-cost, flexible software, or fast, fixed-function, and expensive hardware. Their relative advantages are reviewed. A number of practical applications of automated visual inspection techniques are then discussed.

1. Introduction

The importance of inspection in manufacturing is well understood, since quality and reliability are both prime factors influencing customer choice. Rapid advances in production techniques have increased the need for improved inspection methods. In particular, there is a requirement for increased speed of inspection, to match that of fast, modern production processes. The implicit inspection provided by manual production and assembly are no longer present when full automation is achieved. The quality assurance engineer has many inspection and measurement tools at his disposal, which have been developed to a high level of sophistication over many years. One of these is provided by Automated Visual Inspection (AVI). It offers an attractive, relatively new approach[3] to improving product quality, although the technology has not yet had the benefit of being refined during a long period of evolutionary development. The potential benefits of this young area of industrial automation will be discussed in this article, as will some of the techniques and applications.

Figure 1 shows the organisation of a typical AVI system. Individual AVI installations may differ from this in detail, but this block diagram serves as a useful model on which to base our discussion. **Notice that an AVI system almost invariably contains elements of mechanical handling, lighting, optics, electronics (analogue and digital systems), software and algorithms.All of these must be carefully selected and integrated so that they operate together harmoniously.** These two sentences have been highlighted, to emphasise the fact that they represent the most important single statement in this article. For appropriate applications, a properly designed AVI system offers several advantages over other inspection techniques and provides:

(a) greater reliability than human inspectors who are prone to producing errors, due to fatigue, boredom, distraction, hunger, illness, etc.

(b) greater versatility than all forms of mechanical (contact) sensing and gauging. (It may even

---

[1] Department of Electrical & Electronic Engineering,
University of Wales Institute of Science & Technology,
Cathays Park, Cardiff, CF1 3XE, Wales, U. K.

[2] Department of Electrical & Electronic Engineering,
Polytechnic of Wales,
Pontypridd, Mid Glamorgan, CF37 1DL, Wales, U.K.

[3] Automated visual inspection was first proposed as early as the 1930s but there was no understanding, at that time, of the need for powerful computing techniques, which in any case did not exist

NATO ASI Series, Vol. F49
Computer Integrated Manufacturing
Edited by I. Burhan Turksen

be possible to do certain things using AVI that cannot be achieved in any other way.)
(c) higher speed both human and mechanical inspection systems
(d) reduced unit inspection costs.
The principal benefits of optical sensing arise from the fact that it offers a means of non-contact inspection, which is inherently safe, clean and gentle in its treatment of the artifacts being examined. Apart from the human-eye, there is, for, example, no single alternative means of inspecting a red-hot steel slab, a cream cake, an aerosol spray, a free-falling 'gob' of molten glass, a dollop of unbaked dough or clay, a complete motor vehicle or a micro-electronic circuit. However, it must be emphasised that, to date, AVI has not been able to approach the great potential that was anticipated for it in the 1970s. By 1980, it was realised that the promise was far greater than the achievement. In order to appreciate why, it is necessary to understand that AVI inevitably demands high computation rates, which could not be provided in a cost-effective manner, until very recently. Even today, there are numerous applications that we know how to solve, but for which we cannot provide fast, effective, low-cost hardware. This article develops this theme, by showing that, in a number of laboratory studies, the feasibility of a wide variety of important inspection tasks has been established, even though this work has not yet resulted in the construction of factory-floor inspection systems. In doing so, some of the numerous image acquisition and processing techniques that have been proposed will be encountered. We shall also consider some of the image processing systems that are likely to influence the development of AVI and allow it to reach more areas of manufacturing within the next few years.

2. The Need for Improved Inspection Techniques

The need for improved inspection techniques will be demonstrated by listing a number of outstanding and important industrial applications. Others are described in Section 6. This informal approach is not able, nor is it intended, to demonstrate that there is a viable market for AVI systems, but instead that the potential gains for certain areas of manufacturing industry from the use of this technology are very great indeed. The authors are of the opinion that industrial managers often grossly underestimate the potential value of using AVI. Caution in using a new and "unknown" technology is a major cause of investment inertia. Lack of understanding of the potential of opto-electronic sensing, accompanied by a lack of imagination, also leads to reticence in installing AVI systems. The following list is merely a small subset of the applications that have been studied to date.
(i) Inspecting glassware for a wide variety of a faults, including cracks, loose glass, gross shape deformations, and an unpleasant fault known as a spike. The last of these is like a stalagmite of glass, protruding upwards inside the jar or bottle. It can easily become broken during filling, which may leave a nasty "cocktail" for a drinker to swallow. Another fault, this time in drinking glasses is discussed in Section 6.6.
(ii) Detecting flaws on red-hot steel slabs, thereby saving the cost of reheating. This offers a truly enormous potential pay-off: 1% of the fuel bill in the U.S.A.. [1]
(iii) Detecting foreign bodies in food-stuffs, toiletries and pharmaceuticals. (There are obvious dangers in this type of fault.)
(iv) Finding cracks in structural metal-work, for example in bridges, oil-drilling platforms and air-frames.
(iv) Finding scratches on the surfaces of the bores of hydraulic cylinders. (Failure of such a component in the braking system of a car might result in the death of passengers or other innocent people.)
(v) Locating wires protruding from the cap an electric light bulb. One of the authors (BGB) has experience of three companies who recognise this problem as being serious, since electrocution can result from the use of a such a lamp.[2] Also see Section 6.1.
(vi) Detecting foreign bodies, fingerprints and minute scratches and pits on magnetic recording media for computers. (These faults can lead to data drop-out, or worse to a head crash.)
(vii) Inspecting printed circuit boards, both with and without components. (The high cost of building circuit boards with faults in them is the major justification for this application.)
(vii) Guiding robotic devices to pick objects from a conveyor belt, to load machines. See Section 6.5.

There are three main reasons for installing an AVI system. Unfortunately, the first of these very often dominates all other considerations.

(a) To reduce manufacturing costs.
(b) To improve the reliability, safety and life-time of the product.
(c) To remove undesirable jobs, which may be dangerous, detrimental to the long-term health of the work-force, very boring, or merely inefficiently performed by a human being. (We must, of course, be careful not to automate and thereby eliminate those jobs which are valued by people who could find no other employment and who enjoy what to others may be a very boring task.)

3. Image Acquisition, the Optical Sub-system

Figures 2 to 12 show some of the more significant methods for illuminating and viewing a range of industrial objects. A more complete catalogue is provided by Batchelor [2] . It is important to be aware of the need to consider all aspects of the optical sub-system. In particular, the following points should be borne in mind:
(a) The choice of illumination technique should be made considering the factors listed in Table 1.
(b) Fibre optical devices provide a useful means of guiding light in a well controlled manner to the point where it is needed. (Hill, [2])
(c) It should be questioned whether a standard lens is able to provide the appropriate quality of image, or whether a specially designed lens is required. Very considerable benefits can often be obtained by the use of a specially designed lens, in preference to a standard (i.e. sub-optimal) one. The image contrast and geometric distortion can both be improved.
(d) All optical surfaces must be kept clean. Various techniques exist for this. (Figure 13.)
(e) Stray light, especially from fluorescent tubes and sunlight, must be kept away from the point of inspection. It is almost always cheaper to provide a screen to eliminate the effects of variable ambient light than it is to improve the computer.
(f) The image sensor (camera) must be carefully chosen, bearing in mind the characteristics of the images being produced. Figure 14 and Tables 2 and 3 show how the characteristics of a number of types of image sensor can vary.
(g) If a certain operation can be performed in optics, rather than inside a computer, the cost, speed and efficiency of the overall AVI system will usually all be improved.

4. Image Processing for Inspection

Many authors and industrial design engineers have assumed that a large proportion of inspection algorithms can be divided into two parts:
(a) **preprocessing,** in which fairly simple operations are performed. Typical preprocessing functions are image filtering, edge detection, contrast enhancement etc. They are characterised by being "dumb" compared to those included in (b).
(b) **tail-end processing** which is likely to require some "intelligence" and to be highly dependent upon the inspection task to be performed. For example, we may need to locate the shoulders of a bottle and then measure the radius of curvature. The measured value is then checked against some predefined tolerance parameters. It may be necessary to perform several quite different tasks of this type for each image. A particularly good example of "smart" tail-end processing is provided by the task of inspecting male screw threads. [3]

A number of authors share the opinion that such a division is a useful representation for a broad range of inspection algorithms. A more general model is illustrated in Figure 15. This is proposed in the light of a considerable amount of experience, gained by one of the authors(BGB) using an interactive image processor. [2] Notice that a "smart", master process, which will be implemented in slow software, controls a "dumb" slave process, which will be implemented in fast hardware. It is also important to note that the master must be able perform all of those tasks which the slave is unable to accomplish. To allow the master and slave to communicate effectively, the same data structure must be employed by both of them to represent images.

There are several important groups of image processing operations that can be included under the umbrella of preprocessing. [2] These include:

Monadic, point-by-point operations, characterised by a mapping function of the form

$$c_{i,j} \longleftarrow f(a_{i,j}) \quad \text{(repeated for all i and j in the input image)}$$

where $\{a_{i,j}\}$ and $\{c_{i,j}\}$ represent the input and output images respectively. Monadic functions can be implemented digitally using simple look-up tables, stored in ROM or RAM

Dyadic operations, characterised by a mapping function of the form

$$c_{i,j} \longleftarrow f(a_{i,j}, b_{i,j}) \quad \text{(repeated for all i and j)}$$

In this instance, there are two input images $\{a_{i,j}\}$ and $\{b_{i,j}\}$ and again $\{c_{i,j}\}$ is the output image. Dyadic operations can be performed using a standard ALU.

Local operations, in which the intensity at a given point (i,j) in the output image is calculated from a region in the input image and which is normally centred on the point (i,j). Rather than define a general purpose local operator, which would require considerable space and produce little benefit, we shall merely illustrate them by providing three examples. A simple low-pass filter may be obtained by (unweighted) local averaging:

$$c_{i,j} \longleftarrow [a_{i-1,j-1}+a_{i,j-1}+a_{i+1,j-1}+a_{i-1,j}+a_{i,j}+a_{i+1,j}+a_{i-1,j+1}+a_{i,j+1}+a_{i+1,j+1}]/9$$

Division by 9 is included to normalise the result, so that the output ($c_{i,j}$) lies within the same range as the inputs (the $a_{i,j}$).

The following equation defines the well-known Sobel edge detector

$$c_{i,j} \longleftarrow [|a_{i-1,j-1}+2a_{i,j-1}+a_{i+1,j-1}-a_{i-1,j+1}-2a_{i,j+1}-a_{i+1,j+1}| + |a_{i-1,j-1}+2a_{i-1,j}+a_{i-1,j+1}-a_{i+1,j-1}-2a_{i+1,j}-a_{i+1,j+1}|]/6$$

The third example is a non-linear operator for edge detection

$$c_{i,j} \longleftarrow \mathrm{MAX}[a_{i-1,j-1},a_{i,j-1},a_{i+1,j-1},a_{i-1,j},a_{i,j},a_{i+1,j},a_{i-1,j+1},a_{i,j+1},a_{i+1,j+1}]-a_{i,j}$$

Local operators can be defined using larger windows than the 3*3 neighbourhoods which these three examples employ. Operators like these can be performed in real time upon a digitised video signal using quite simple circuitry. ([4,5], Figure 16) By generalising the concept of a local operator, we can include such operations as erosion, dilation, object counting (Euler number) and connectivity analysis in this category. Even such seemingly unlikely candidates as thinning and chain coding can be performed using local operators, in conjunction with other easily implemented functions.

Histogram modification operations, such as histogram equalisation can also be performed in real time, albeit with some delay. However, the circuitry to do this is rather more complicated than that required for the other three classes discussed so far. [6]

Of course, there are numerous other image processing functions which do not fall within any of these four categories. However, any low-cost machine which could perform a broad variety of computations from these four classes at high speed would find ready application in AVI. It is important to note that it is often possible to perform certain calculations at moderately high speed using a serial processor, once the image has been recoded, using for example run-length or chain code. This illustrates a general point: it is commonly found that seemingly important functions, like region labelling, can be implemented in an indirect way, or their use avoided altogether, by completely reformulating the algorithm. There are nearly always many algorithmic paths which can reach a given goal. (Figure 17) However, despite all our guile, there will always be certain functions which cannot be implemented easily using fast hardware, and for which we must resort to the use of a software-based processor.

Now, it is all too easy to lose sight of the fact that none of the image processing operations mentioned above represents anything more than an "atom", for possible inclusion in an AVI algorithm. The latter are inevitably much more complex than the individual atoms; a "typical" inspection algorithm might include several dozen such atoms. As a result, a much more complex machine than that suggested in Figure 16 is needed for all but the most trivial AVI applications. In other words, we must take a high level, "systems" approach to the design of image processing equipment for inspection.

5. Architectures for Preprocessing

It is not possible, in the small space available here, to review all of the specialised architectures that have been proposed for image processing. [7] We shall discuss just a few of them.

5.1 Linear Array Processor [5,8]

The LAP was developed at the U.K. National Physical Laboratory and has the architecture shown in Figure 18 It can perform a range of local operations at high speed. Notice that it would appear that the M68000 system controller should be able to perform some calculations concurrently with the processing within the array. We have found this to be difficult to achieve in practice, because the array requires a lot of supervision by the system controller while the LAP is running. As a result, any software-based image processing is best left until the array has completed its assigned task. It is possible to interleave software processing with operations in the array, data being moved around the system via the VME bus.

5.2 Kiwivision pipe-line processor [5,9]

This is a pipe-line processor and has the system organisation illustrated in Figure 19. Notice that the pipe-line structure can be varied to suit individual application requirements and that data can be recirculated through it several times. This keeps the cost of the system down but reduces its speed. On the other hand, high speed processing can be achieved for quite complex algorithms, by simply adding more units into the pipe-line. Notice too that the M68000 processor, which acts as the system controller, is able to perform certain image processing operations concurrently with the flow of data through the pipe-line. Moreover, images generated by the M68000 processor can be reprocessed using the pipe-line. The interleaving of hardware and software processing is possible because image data can be moved around via the VME bus. The basic architecture of Kiwivision was conceived and initially developed by one of the author's (BGB) colleagues (Dr. C.C.Bowman). It is being developed further and should soon become available commercially.

5.3 Maxvideo pipe-line processor

The Maxvideo family of image processing boards is sold by Datacube, Inc. [5,10] These boards can be controlled via the VME bus. A group of Maxvideo boards provides a flexible image processing system, comparable in power and versatility with Kiwivision. The manufacturer suggests the organisation shown in Figure 20 as one possible means of interconnecting them.

5.4 VCS

VCS is an interactive image processor and consists of a large amount of software, coded in 'C', running on a conventional computer which contains a frame-store.[4] Such an arrangement is rather slow in operation, but this is not normally important, because interactive image processing is used primarily as an algorithm development tool, rather than for factory floor inspection systems. VCS boasts a very rich repertoire of image processing commands and is sold by Vision Dynamics Ltd. (Hemel Hempstead U.K.). A number of other companies have similar products inspired by earlier work by one of the authors(BGB). [2]

---

[4] A frame store comprises a RAM which has special input and output circuitry to allow a video signal from a camera to be digitised and processed images to be displayed.

5.5 Comparison of the different architectures

Table 4 lists the execution times of the LAP, Kiwivision, Maxvideo and software systems for a range of operations.

Notice that the execution times listed in Table 4 relate only to algorithmic "atoms". They do not give any indication about the speed of complete AVI algorithms, since it is very difficult indeed to provide realistic figures for a representative set of "typical" AVI algorithms[5]. Rather than present a few such results, which would we feel provide an unrealistic picture, we shall instead discuss the different modes of operation of the array and pipe-line processors. This affects their speed when applied to real AVI tasks and ultimately their utility and cost.

The LAP, being a fixed architecture SIMD machine cannot be tailored for each particular application, as the pipe-line processors can. One of the main advantages of the LAP is that it can be programmed almost as easily as a conventional serial machine. Notice that, the processing times for an array are additive; an algorithm containing many atoms is performed in sequence. On the other hand, the pipe-line processors permit some improvement in processing speed to be achieved by simply adding more units into the pipe-line. It is not always possible to interchange cost and speed as freely as one might wish, because the pipe-line hardware and architecture may well impose certain limitations upon the order in which the algorithmic "atoms" can be performed. This does not in any way restrict what can be achieved by recirculating data through the pipe-line, but it may result in a reduction in the processing speed.

One of the major benefits of the pipe-line processors is their modularity. This means that it is possible to tailor such systems to suit each particular application, achieving an appropriate balance between processing speed and cost, Moreover, novel types of processing module can be built and incorporated into the pipe-line, thereby forming the basis for a constantly improving system. The authors' colleagues are designing an integrated circuit systolic array to allow the median, rank and range filters [11] to be implemented. Datacube Inc. have announced further modules for their Maxvideo family, in addition to those mentioned in Figure 20.

In one of the author's (BGB) experience, the Maxvideo boards were found to be far from easy to use. The hardware has been designed to provide great flexibility, but it is sometimes difficult for relatively inexperienced users to exploit this to good effect. The provision of a software controller for these boards seems to be an essential first step for anyone contemplating using them seriously. Kiwivision, though less ambitious in the flexibility of its pipe-line modules, is more systematic in its structure than the Maxvideo system; Kiwivision may be likened to a RISC machine.

One important fact emerges from our experimental work: none of the hardware systems that we have examined is able to implement all of the image processing functions that we should like to perform at high speed. For example, the LAP cannot efficiently acquire an image histogram. Neither the LAP nor the pipe-line processors can easily label regions in a binary image. The Kiwivision and Maxvideo systems cannot readily implement certain filters, such as local-area histogram equalisation. In summary, the use of some software-based image processing to augment that implemented in hardware is inevitable.

In order to avoid creating a new processing bottleneck, we propose using concurrent computing techniques. We shall describe how this can be integrated with Kiwivision, although it would be possible to combine concurrent processing with other image processing hardware, such as the Maxvideo pipe-line or LAP.

---

[5] One example is provided in Figure 29, where it is evident that the complete algorithm could not easily be executed in fewer than 5 television frame periods. A combination of concurrent and pipe-line processing techniques could be used in a system with a throughput rate of 1 object per frame-period and latency of 5 frame periods.

5.6 Concurrent processing techniques[6]

It has been pointed out elsewhere that the **throughput rate** (i.e. number of parts to be inspected per second) is an important feature of an inspection system, whereas the **latency** (delay between receiving data from a camera and computing a decision based upon that data) is not normally of great significance. [12,2] When objects are to inspected on a conveyor belt, it is often possible to separate the camera and reject/accept device in order to accommodate a larger latency. (Figure 21) Noting this fact, one of the authors (BGB) is planning an enhanced pipe-line processor, which will contain several slave processors, each able to perform the same, complex software-based operations. (Figure 22) One way of viewing this system is that some of the image stores in Figure 19 have been given their own local processors. Although each of these slaves is a serial (von Neumann) processor, high throughput rate can be obtained by the simple expedient of using several of them. Notice that this does not reduce the latency, a fact which renders this architecture unsuitable for robot control, in-process inspection and other applications where real-time control is needed. A concurrent processor is surprisingly economical. [12,2] Notice the following points:

(a) The software is identical for each of the slaves.

(b) The slave hardware should ideally consist of a standard single-board processor with a small amount of additional circuitry, which is required to provide an interface for the digital video bus.

(c) System development is straightforward. The control software required by the master is straightforward since the slaves operate autonomously, once they have been initiated. The simple procedure of plugging more slaves into the system is all that is needed to increase the throughput rate.

(d) The provision of spares, servicing and fault-finding are all simplified, because a concurrent system is highly modular.

It is possible to anticipate the performance of such a system with confidence, since it is only a very minor extension of our current capability. The system shown in Figure 22 is capable of efficiently interleaving hardware and software-based processing, thereby allowing as much of the algorithm as possible to be implemented in the fast hardware. (See Figure 23) Notice that partially processed image data can be recirculated through the pipe-line, even after some software processing has been performed on it.

An obvious question to consider is whether the concurrent processing approach could be extended to implement the whole of the inspection algorithm, without using any special purpose preprocessing hardware. While this is technically feasible, no general-purpose, commercial, AVI system has yet been based upon concurrent computing techniques. The viability of this approach is doubtful, if we are to use conventional microprocessors; even a fast modern processor, such the M68000, takes much too long to implement certain operations, such as image filtering. (Table 4)

One approach which offers the potential for high processing speed in a program-controlled machine, at a reasonable cost, is offered by the transputer. [13] The same network of transputers can operate as a concurrent processor, pipe-line, SIMD or MIMD machine, simply by changing its software. A 67-element array of transputers is capable of performing a 3*3 linear convolution operation, on an image of 256*256 pixels, in 16 milliseconds. [14] If greater speed than this is required, more transputers may be added to the array. To a first approximation, the throughput rate of a transputer network, operating as an SIMD machine (i.e

---

[6] Parallel and concurrent processing techniques were in use in Roman times. Since photocopiers had not been invented, the mass production of documents was often performed by an army of slaves who could write, while their slave-master, who was noted for having a strong voice, read aloud from the original document. This is parallel processing. When the slavemaster became hoarse, concurrent processing was used instead. In this eventuality, each slave copied one page at a time and, when he had finished it, passed it to a colleague and received a new page for copying from another scribe. Naturally a document containing N pages could most effectively be copied by N slaves. Notice that a well disciplined army of scribes, working concurrently, does not need any supervision, other than an initial order to copy a given book.

implementing the monadic, dyadic or local operators) is proportional to the number of processors in the array. The latency may not decrease as fast as this, for complete AVI algorithms, which frequently contain a significant "tail-end" component which is serial in nature.

The authors are convinced that a promising long-term approach to providing modular image processing techniques for AVI, is offered by the transputer. However, this possibility is some distance into the future, as there is a great deal of engineering work to be done before a cost-effective inspection system based upon the transputer can be produced. No doubt it will, one day, be possible to produce a fully programmable array of transputers or similar devices, which can implement the whole inspection algorithm. What is not yet clear is what network architecture will be most effective. The principal difficulty seems to be transferring the video data in/out of the transputer array at high speed.

6. Applications

The following applications were selected principally to illustrate some of the general points which have been made above.

6.1 Detecting loose wires on the cap of an electric light bulb

Application requirements: Check that the cap has two eyelets and that these are well formed. In particular, the presence of wires protruding from the eyelets must be detected.
Commercial importance: Badly made electric light bulbs are dangerous, especially in Europe where the electrical supply voltage is high. One particularly dangerous fault is the presence of loose wires which protrude from the eyelets. There is a danger that these wires, which connect internally to the filament, may cause a fatal electric shock to a person fitting the bulb into a "live" socket; deaths have resulted from such faults in the past.
Lighting: Either omni-directional illumination (Figure 4) or polarisers (Figure 6) are needed in order to avoid excessive glinting from the brightly reflecting metal and metallised glass surfaces. Front diffuse reflection (Figure 7) may cause shadows on the rough, ill formed solder eyelets.
Optics: Not critical.
Camera: The camera should be capable of withstanding optical overload due to glinting, without either becoming damaged or blooming. Blooming is a characteristic of some cameras, which become saturated (i.e. the camera seems to indicate the presence of large "white" regions) around a small very bright region, arising from, say, glinting. (See Tables 2 and 3.)
Image processing philosophy: There are several satisfactory methods of processing the image. The simplest and cheapest is based upon simple thresholding, but this is far from robust, because it is critically dependent upon the lighting remaining constant. A far better procedure is based upon a non-linear filter[7], and is able to cope with poor quality images. (Notice, however, we should never employ sophisticated image processing algorithms to compensate for poor lighting.) After filtering, simple fixed-parameter thresholding, can be applied. The resulting binary image must then be interpreted, so that the presence of a wire can be detected. This procedure is trivial, if the orientation of the bulb can be fixed by mechanical means beforehand. A more complicated image processing technique has been devised which is able to operate successfully, even if there is no such constraint. (However, we should not accept the presence of a robust algorithm as providing a license to tolerate sloppy design of the image acquisition or mechanical sub-systems.)
Illustrations: Figure 24

6.2 Verifying the existence of a chamfer

Application requirements: A certain hole in a component should be chamfered. Verifying the existence and depth of a chamfer is required.
Commercial importance: Many components possess chamfered holes. Failure to produce

[7] The filter is closely related to the non-linear edge detector outlined in Section 4.

the correct chamfer could result in a pin, which is to be inserted into the hole, becoming jammed, bent or broken. The absence of a chamfer can thus lead to costly damage.
Lighting: The chamfer has a vee-shaped cross section. Thus, if we can arrange for specular reflection from the drilled front face of the component, we can obtain high image contrast. To achieve this we need to be able to provide illumination from the camera. This can be accomplished using either a beam-splitter or a fibre-optic device, such as a ring light or endoscope. (Figures 9, 10, 11b) obtain illustrations given in Figure 25, an endoscope was used.
Optics: A certain amount of magnification may be needed for some holes. This can also be achieved using an endoscope.
Image processing philosophy: An edge detector is required to locate the outer edge of the chamfer. Once the edges of the chamfer have been found, the interpretation is straightforward.
Remarks: The images shown above were derived from an electrical connector. The glinting is due to the bright metal spring contacts inside the hole.
Illustrations: Figure 25.

6.3 Reading punched lettering on a metal surface

Application requirements: Punched labels on a nominally flat aluminum surface are to be read.
Commercial importance: There are numerous instances where punched or embossed labels are to be read or verified. There are many other instances where alphanumeric data is to be read in a factory environment, including: the reading of conventional printing, ink-jet printing, stampings, pressings, etchings, mouldings etc.
Lighting: In this instance, considerable care was taken to obtain a poor quality image, in order to create a difficult problem, and thereby show off the power of the image processing. In practice, a better quality image (i.e higher contrast and less surface "noise") would be desirable. Notice that there is a good deal of glinting evident in Figure 26. In practice, much of this can and should be avoided, possibly by using polarising filters. (Figure 6)
Optics: The punched lettering may be quite small. Hence, a high magnification lens may be needed.
Camera: The camera should be one capable of withstanding a high local optical overload, due to glinting.
Image processing philosophy: The aim is to produce a "clean" image of the lettering which can then be represented by a skeleton and analysed using AI techniques. The noisy image resulting from the glinting on the rough bright metallic surface can be processed in a number of ways. It was found that the most popular methods did not, on this occasion, offer any realistic solution. In this instance, one of the authors (BGB) had to experiment for a considerable time with an interactive image processor (VCS), before he found a likely candidate.There is really no rationale as to why the particular algorithms demonstrated here were chosen; they occurred to him as he was operating the interactive image processor as it was originally intended to be used, as a problem solving tool. Once the skeleton has been produced, it may be characterised by detecting the limb ends and joints. The distance between these points and the geometric relationship to one another may be used in a fairly obvious way as a basis for the recognition of the alphanumeric characters.
Remarks: Commercial document readers do not employ such sophisticated techniques as are illustrated here, since they have to operate at high speed. A complicated algorithm was needed in this instance, because the image quality is so poor. It is possible to obtain a very much better quality image by altering the lighting. Clearly, this would be preferable in practice.
Illustrations: Figure 26.

6.4 Shape sorting

Applications requirements: The value (i.e. number of "pips") and suit of a playing card is to be determined. "Picture cards" are excluded from this simple demonstration. It will not be assumed that the card is in any particular orientation.

Commercial importance: No direct significance. This is a model of certain industrial problems, such as identifying packages, cartons, etc.but is included here to illustrate the kind of indirect approach which is sometimes used to achieve a certain goal.
Lighting: Not critical
Optics: Not critical. Colour (i.e. red/green) filters can be used to distinguish Hearts and Diamonds (both red) from Clubs and Spades (black).
Camera: Not critical
Image processing philosophy: The "value" of a card can be determined using the Euler number. (Assuming that all "holes" have been eliminated, the "value" of a card, such as the 5 of Clubs, is equal to the Euler number, minus 4; there are two small suit symbols in the corners of the card and two numerals.) The suit is identified by counting the number of (large) blobs in the convex deficiency of one of the suit symbols. (The convex deficiency is defined as the difference between the convex hull of a blob-like shape and the original blob itself, while the convex hull is the shape equivalent to that enclosed by a rubber band, stretched around the blob.) This process defines a number of areas, known as concavities, which can be of two kinds:

(a) holes, also called "lakes"
(b) indentations, also called "bays".

(To see why we use the terms "lakes" and "bays", think of the silhouette as an island. See Figures 27 and 28) The suits produce the following results:

Suit	Number of blobs
Diamonds	0
Hearts	1
Spades	2
"Impossible"	3
Clubs	4
"Impossible"	5 or more

Remarks: The algorithm outlined above was used some years ago in a "fun" demonstration of the use of vision to control a robot sorting playing cards into their suits. [15]
Illustrations: Figure 27

6.5 Guiding a robot to pick up a con-rod

Application requirements: A con-rod is to be picked up from a flat, featureless platform.
Commercial importance: Robotic plucking of con-rods from a moving conveyor belt is of considerable economic importance and has already reached the factory floor.
Lighting and optics: The Consight system, developed by General Motors, used a variant of the structured-lighting approach outlined in Figure 12a. Details are presented elsewhere. [16]
Camera: The Consight system uses a line-scan camera and produces very high contrast, effectively binary images.
Image Processing Philosophy: Once a binary image has been obtained, there are several possible means of analysis. The one presented here is related to the shape-sorting algorithm discussed immediately above. (Figure 28) The centroid of the silhouette is first computed. Following this, the convex hull of the silhouette is obtained. There are three major bays and one very small bay defined by the convex hull of a con-rod. It happens that the line joining the centroid of the third largest bay to that of the silhouette defines a conveniently computed axis, which is close to the principal axis[8] of the con-rod. The important point, however, is not that these axes are aligned, but that either of them is able to define the orientation of the component. Thus, the coordinates of the centroid and that of the third largest "bay" yield enough

---

[8] The principal axis is defined as that line about which the second moment is a minimum. The principal axis can also be computed easily. The "long axis" of a con-rod, recognised intuitively by the human eye, is also close to these two lines.

information for us to determine the exact position and orientation of the con-rod, so that a robot can be directed to pick it up.

Remarks: A related technique can be derived for determining the "handed-ness" of a blob-like shape, or whether it is "heads up" or "tails up". By scanning in a clockwise direction around the centroid of a con-rod, we encounter the centroids of the "bays" in the order:

1 4,2,3
or 4,2,3,1
or 2,3,1,4
or 3,1,4,2

depending upon where we start. However, by travelling in the opposite direction, we encounter the centroids in a different order:

1,3,2,4
or 3,2,4,1
or 2,4,1,3
or 4,1,3,2.

Illustrations: Figure 28

6.6 Detecting flakes of glass on the rim of a drinking glass

Application requirements: A common fault occurring in mass-produced drinking glasses is the presence of loosely attached glass, in the form of a very fine filament or flake. These artifacts arise as a result of the method of manufacture: a thin-walled bottle is first blown and then its neck is removed, using a gas flame. As the neck falls away, a filament or small blob of glass may be formed on the rim, giving rise to these very undesirable faults.

Commercial importance: The importance of the problem lies in the need to protect the drinker from being cut or swallowing the loose glass. It is difficult to inspect glass-ware by any means other than vision. Human inspectors are unreliable and unfortunately allow too many defective glasses to be passed as "safe". There are numerous situations where loose glass can contaminate food-stuffs, pharmaceuticals or toiletries.

Lighting: Back lighting (Figure 2) or dark-ground illumination (Figure 3) is suitable. The former technique was used to obtain the pictures in Figure 29.

Optics: A close focus lens is needed as the filaments/flakes can be very fine.

Camera: A newvicon camera was used for Figure 29. In practice, the tumbler would probably need to be rotated in front of a line-scan camera, in order to obtain an all-round view.

Image processing philosophy: We can use conventional linear, or nonlinear, low-pass filtering or the crack detector, followed by simple, fixed-value thresholding. Following this, we have to eliminate small specks, which may be due to camera, dirt or other forms of "noise". (In this instance, we have actually done so, even though the "noise" is due to undesirable faults, namely speck-like pieces of glass.) Finally, a decision is made on a simple count of the white pixels.

Image Processing Algorithm, defined using the VCS language:

```
DIG             / Digitise an image
WRI TEMPY       / Save image on the disc
LNB        / Expand bright regions          I Together, these
LNB        / Expand bright regions          I six lines implement
LNB        / Expand bright regions          I the crack-detection
NEG        / Negate                         I algorithm, defined
LNB        / Expand bright regions          I by Batchelor [2].
LNB        / Expand bright regions          I Normally, they would
LNB        / Expand bright regions          I be implemented in
NEG        / Negate                         I a single macro.
SWI / Switch images
REA TEMPY       / Restore the original image
SUB             / Subtract images
THR 220,255     / Fixed value thresholding is OK after filtering
NEG             / Negate
EDGE_MASK       / Macro: Remove edge effects by masking
```

```
REM_SPOT      / Macro: Remove isolated white points ("noise")
WRI TEMPY     / Save the image for later
INT / Integrate along image rows
COPY_RHS      / Macro: copy the right-most column into all others
THR 100,255   / Remove all but the biggest
EXW           / Expand it
NEG           / Negate
SWI / Swap images ready for getting image from disc
REA TEMPY     / Read the image obtained above
MIN / Use the stripe picture as a mask
LPF           / Low-pass filter          | Makes small blobs
LPF           / Low-pass filter          | very dark compared
LPF           / Low-pass filter          | to big ones
TH 85,255     / Delete the small (i.e. dark) blobs
WP COUNT      / Count white points that remain.
```

The drinking-glass is rejected if COUNT is greater than some predefined tolerance value, determined experimentally. The user of the AVI system is able to adjust this threshold parameter in order to adjust its performance, i.e. give more/fewer false alarms.

Implementation: The algorithm defined above is typical of many inspection procedures. It consists of a serial arrangement comprising several processing stages, each of which could be implemented in a single television frame period. These steps could be performed simultaneously, albeit on different images, using an advanced pipe-line structure.

Illustrations: Figures 29 and 30

7. Conclusion: the need for a Systems Approach

Automated Visual Inspection systems comprise several distinct parts, involving different technologies. These must all be made to operate harmoniously together. Failure to ensure that they do will almost invariably result in a grossly over-elaborate machine, whose performance is not as robust as it should be, is more expensive than it need be and which may not produce the same accuracy as a well-designed and well-balanced system. Each part of an AVI machine should be individually optimised, bearing in mind that they all have to work together in one system. It cannot be overemphasised that the illumination is just as important as the image processing. Unfortunately, some companies, particularly those which have good computers for image processing, are frequently cavalier when it comes to designing the optical sub-system.

There is peculiar tendency for would-be customers of inspection machines to try and impose their will about the design of the system, in a way which is all too frequently unhelpful. There are unfortunately too few good systems engineers who can properly balance the design of inspection devices and it is certainly not a task for people with inflexible ideas or a narrow range of expertise. Production engineers do not necessarily make good vision engineers!

AVI systems invariably require a considerable amount of individual design effort; there is no such thing as a truly general purpose AVI machine. Even if there were such a device it would require an expert to install it.

The design and installation of AVI systems must never be left to people who are amateurs in the field. The worst offenders are experts from one of the component technologies; they frequently try to impose a solution which accords with their own narrow range of knowledge. For example, computer engineers have achieved notoriety by emphasising the computing aspects of AVI system design, at the expense of the image acquisition (Desk lamps have been used for lighting, while the most advanced software techniques are being employed.) A good design team contains representatives of all of the component technologies that comprise AVI.

A serious problem will arise in the future if the market for AVI continues to grow. Consider just one large company for example, which has about 45000 different products. Not only is there a huge potential market here. The company changes its product range such that the "half-life" is under 5 years. That corresponds to a rate of about 30 new products per week. Suppose that only 1% of these were to be suitable for AVI, then there would be a requirement for one new industrial vision system every 3 weeks. Considering the remarks just made, it is clear that there is likely to be a considerable bottleneck caused by the demands for highly specialised and relatively rare expertise.

For this reason, it is essential that AVI systems engineers prepare themselves for an explosion of interest in the subject, by developing suitable design aids. [17,18,19] The interactive image processor is just one such aid. It is essential that we develop techniques which, for example, will allow the factory-floor image processing system (target system) to be designed directly from the algorithm, specified as a command sequence such as that in Section 6.6. Other useful design aids would be:

(a) A computerised database of illumination devices, lamps, reflectors, lenses, etc.

(b) An expert system which could offer advice about the most appropriate image acquisition techniques, given a descripition of the object to be viewed, its likely faults, how it is to be moved, etc.

(c) An expert system for giving advice about the most appropriate image processing technique, given a description of the appearance of the images.

Finally, let us briefly consider some things that Automated Visual Inspection cannot do. The technology is only appropriate, at the moment, for use in those situations where large numbers (millions) of objects are made, or there is a continuous flow of material (e.g. in a steel mill, or in the manufacture of adhesive tape). It is far too expensive a technology for industries where objects are made in small numbers. Now, it has been established that about 75% of industrial artifacts are made in batches of 50 or fewer items. If we are ever to automate the inspection of items such as these, then we must provide a facility, which the authors have termed a Flexible Inspection Cell, in which a wide range of objects can be examined, This would require software switching of inspection function, as new products are being produced. The physical structure and computing system organisation of a Flexible Inspection Cell is shown in Figure 31. One day, we may even see automated manufacturing and inspection of one-off components. When that happens, the day of the label "Hand made" will be numbered.

Acknowledgements

The authors wish to thank Dr. C. C. Bowman for kindly allowing them to reproduce the material contained in Table 4.

References

1. Suresh B. R., Fundakowski R. A., Levitt T. S., Overland J. E. & Beckering T. L. , "The automated surface inspection of hot steel slabs", proc. conf on Industrial Applications of Machine Vision, Research Triangle Park, North Carolina, U. S. A., May 1982, pub. I.E.E. New York, 1982, E. Computer Society, pp. 112-118.

2. Batchelor, B. G., Hill D. A., Hodgson, D. C. (eds) "Automated Visual Inspection", IFS (Publications) Ltd., Bedford, U.K. & North Holland (a division of Elsevier Science Publications), Amsterdam, Netherlands, 1985.

3. Batchelor, B. G., "Preliminary Note on Automatic Inspection of Male Screw Threads", Proc. 3rd Int. Conf. on Automated Visual Inspection & Product Control, pp. 139-156, IFS (Publications) Ltd., Bedford, U.K.

4. ZR33481 Data Sheet, Zoran Corporation, Inc., Santa Clara, California, U.S.A.. 1985.

5. Batchelor, B. G., "Systems Considerations for Automated Visual Inspection", Proc. 1st IFAC Workshop on Digital Image Processing in Industrial Applications: Vision Control, Helsinki, June 1986.

6. McCollum, A. J., "High Speed image enhancement using Histogram modification", M.Sc. Dissertation, University of Wales, 1984.

7. Batchelor, B. G. & Braggins, D. W., "Commercial Vision Systems", in press.

8. Plummer, P., National Physical Laboratory, Teddington, U. K., Linear Array Processor, Private Communication

9. Bowman, C. C., "High Speed Image Processing for Machine Vision", Ph.D. thesis, University of Wales, 1986.

10. Maxvideo Family User's Manual, Peabody, Mass., U.S.A., 1986.

11. Bailey, D. G., "Hardware & Software Developments for Applied Digital Image Processing", Ph.D. thesis, University of Canterbury, New Zealand, 1985.

12. Batchelor B. G., "A laboratory based approach to designing automated inspection systems", proc. conf on Industrial Applications of Machine Vision, Research Triangle Park, North Carolina, U. S. A., May 1982, pub. I.E.E. New York, 1982, E. Computer Society, pp. 80-86

13. Transputer Reference Manual, Inmos Ltd., Bristol, England, U.K., 1986.

14. Harp, J. G., Roberts, J. B. G., Ward, J. S., "Signal Processing with Transputer Arrays (TRAPS)", Computer Physics Communications, vol., 37, pp. 77-86, 1985.

15. Cotter, S. M., "Industrial Vision Systems", Ph.D. thesis, University of Wales, 1985.

16. Holland, S. W., Rossol, L., Ward, M. R., "Consight-1: A Vision controlled Robot system for transferring parts from belt conveyors" , G.M.R Symposium on Computer Vision and Sensor-Based Robots , General Motors Research Labs., USA . Sept. 1978.

17. Jones, W. E. D., Batchelor, B. G., Bowman, Using AI to Assist in the Selection of Optical, Illumination and Algorithmic Techniques for Industrial Inspection", Proc. conf. on Automated Inspection & Measurement, Cambridge Mass., October 1986, to be published by Society of Photo-Optical Instrumentation Engineers.

18. Kelly, D. E., Batchelor, B. G., "KWIK, A Software Environment for Developing Code for Factory-floor Image Processing Systems", Proc. conf. on Automated Inspection & Measurement, Cambridge Mass., October 1986, to be published by Society of Photo-Optical Instrumentation Engineers.

19. Batchelor B. G., "An appraisal of artificial intelligence techniques for industrial vision systems", proc. Int. Conf on Automated Optical Inspection, Innsbruck, Austria, April, 1986, pub. Society of Photo-optical Instumentation Engineers, vol, 654, 1986, pp. 250-263.

Table 1 Factors to be considered when choosing lighting for inspection.

Feature	Remarks
Spectrum	Choose colour for separation of critical features
	Broad-band or monochromic illumination required?
	Infra red content - IR may lead to overheating of object under examination. Many cameras are sensitive to IR and optics (e.g. polarisers) may not function properly in IR wave-band
	Ultra-violet content - UV may damage certain materials and it causes fluorescence in others. UV can lead to personal injury (eyes and skin). Many cameras are sensitive to UV and optics may not function properly in UV wave-band.
Intensity	Overall intensity level
	Variation over field of illumination (Should uniform or patterned illumination be used?)
	Should intensity be kept constant or varied to obtain special effects?
	Should stroboscopic illumination be used?
	How does brightness vary during age of lamps?
	Does light output vary significiantly during mains cycle?
	Should light output be controlled by automatic feedback loop? If so, where do we place the sensor for the best effect?
System Design	Choose number and type of lamp and optics
	Choose appropriate optics/fibre optics
	Is coherent lighting to be used?
	Should light be polarised?
Environmental	Optics must be kept clean, free from dust, fumes, etc.
	Does atmosphere in the factory affect lamp life or performance? (Chemicals may attack the glass envelope.)
	Optics and lamps must be protected from abuse by the work force
	Is heat output from lamps excessive?
	Protect work-force from stroboscopic illumination
	Eliminate ambient light, especially sunlight and from fluorescent tubes
	Protect products from broken glass, resulting from lamp breakage
Maintenance	Regular cleaning of all optical surfaces by skilled personnel
	Regular checking and/or replacement of lamps

Table 2 Characteristics of various photoconductive camera tubes

Property	Plumbicon & saticon	Newvicon	Si-vidicon	Standard	Chalnicon vidicon
Resolution	medium	good	adequate	good	good
Sensitivity	medium	very high	high	adequate	good
Linear (gamma)	yes	yes	yes	no (about 0.7)	yes
Dark curent	very low	medium	medium	adequate	low
Blooming	little	little	noticeable	very little	little
Sensitivity -to burn-in	poor/ medium	good	very good	poor/ adequate	good
Lag	very little	medium	little	noticeable	little/ medium
Sensitivity to near UV	adequate	adequate	adequate	adequate	medium/ good
Sensitivity	very little	medium	good	adequate	little

Table 3 Comparing several types of solid state image sensor.

Properties	CCD	SSPD	CID
Resolution	high	high	lower
Noise	low	higher than CCD	
Dark current	highest	usually lower than CCD	
Uniformity	medium	better than CCD & fewer spikes	
Blooming	poor	good	good
Pattern noise	some	good	some

Table 4 Comparison of the execution times of the Maxvideo, Kiwivision and LAP systems. The figures given below show the execution times, measured in milliseconds. Key:

{.} indicates times which are dependent on the picture characteristics. Typical execution times are given.

[.] indicates times which could be obtained by extending the Kiwivision system by adding new boards, whose overall design structure is already known.

Asterisks (***) indicate that a command cannot easily be performed by the hardware.

(.) indicates hardware execution times for calculations which do not completely implement a command but greatly assist the software. Operations which can be performed by the Maxvideo boards in real time, on a CCIR or RS-170 video signal, are indicated by bold type.

Function	PDP11	68000	LAP	Kiwivision Stand	Kiwivision Fast
Adjust aspect ratio	4800	?	***	***	***
Invert X axis	6800	220	***	[11]	[11]
Cartesian-radial coord. transform.	20000	?	***	***	***
Radial-Cartesian coord. transform.	32000	?	***	***	***
Zoom *2	12800	?	100	***	***
Shift image	7100	260	16	[11]	[11]
Halve resolution	1400	?	***	[11]	[11]
Rotate image, arbitrary angle	23000	5500	***	***	***
Rotate by 90 degrees	7000	730	***	[11]	[11]
"Modulus" of int'y	8600	410	8	11	11
Bit complement of int'y	5200	320	6	11	11
Mutiply int'y by constant	12000	410	13	11	11
Double intensity	7600	410	8	11	11
Negate	6400	300	6	11	11
Delete LSBs of int'y	5600	230	6	11	11
Add const. to int'y	8400	410	10	11	11
Halve intensity	5300	330	7	11	11
Square intensity	13000	410	22	11	11
Set image to black	4000	220	6	11	11
Linear rescaling of int'y	21000	700	130	55	22
Histogram equalisation	15000	980	***	***	23
Highlight defined int'y band	7700	410	9	11	11
Percent threshold	13000	?	110	88	24
Simple threshold	7100	290	7	11	11
Apply LUT to int'y	7700	360	***	11	11
Add two images	7300	490	11	11	11
Minimum of two images	7000	530	11	11	11
Multiply two images	14000	1700	26	11	11
Max imum of two images	7000	500	11	11	11
Subtract images	8400	570	10	11	11
Exclusive OR of two images	8800	380	10	11	11
Copy image	4600	220	6	11	11
Copy RH column to all cols.	3900	270	88	11	11
Integrate along rows of image	11000	1400	97	11	11
Maximum along rows of image	6900	410	100	11	11
General 3*3 linear convolution	52000	6800	35	11	11
Laplacian	32000	3100	22	11	11
3*3 low pass filter	48000	2800	36	11	11
Horizontal gradient	14000	3300	10	11	11
Vertical gradient	14000	3300	10	11	11
Largest 8-neighbor	26300	2300	24	55	11

Non-linear filter (crack detector)	14000	5800	62	110	69
Simple gradient operator	18000	1200	15	33	11
Roberts edge operators	22500	1400	17	33	11
Sobel edge operator	46000	2900	34	33	11
Region labelling	{11000}	?	{350}	{4200}	{2700}
Convex hull	26000	?	{1000}	700	510
Fill holes	{8800}	?	***	99	77
Count white 8-neighbours	24000	{1500}	10	11	11
Expand white blobs	{20000}	{1500}	8	22	11
Remove isolated white points	24000	{1500}	21	22	11
Shrink white blobs	{17000}	{1500}	8	22	11
Binary edge operator	18000	{1500}	9	22	11
Critical points for connectivity	13000	{1500}	20	22	11
Average int'y	7200	250	90	11	11
Int'y lower limit	7600	540	180	22	11
Centroid	4800	?	***	[11]	[11]
Highlight sharp corners on curve	11000	?	***	***	***
Euler number	13000	?	50	77	11
Generate chain code	{12000}	?	(35)	(11)	(11)
Count white points in image	4700	370	74	11	11
Average speed-up compared to PDP 11	1.0	15.2	741	789	1034

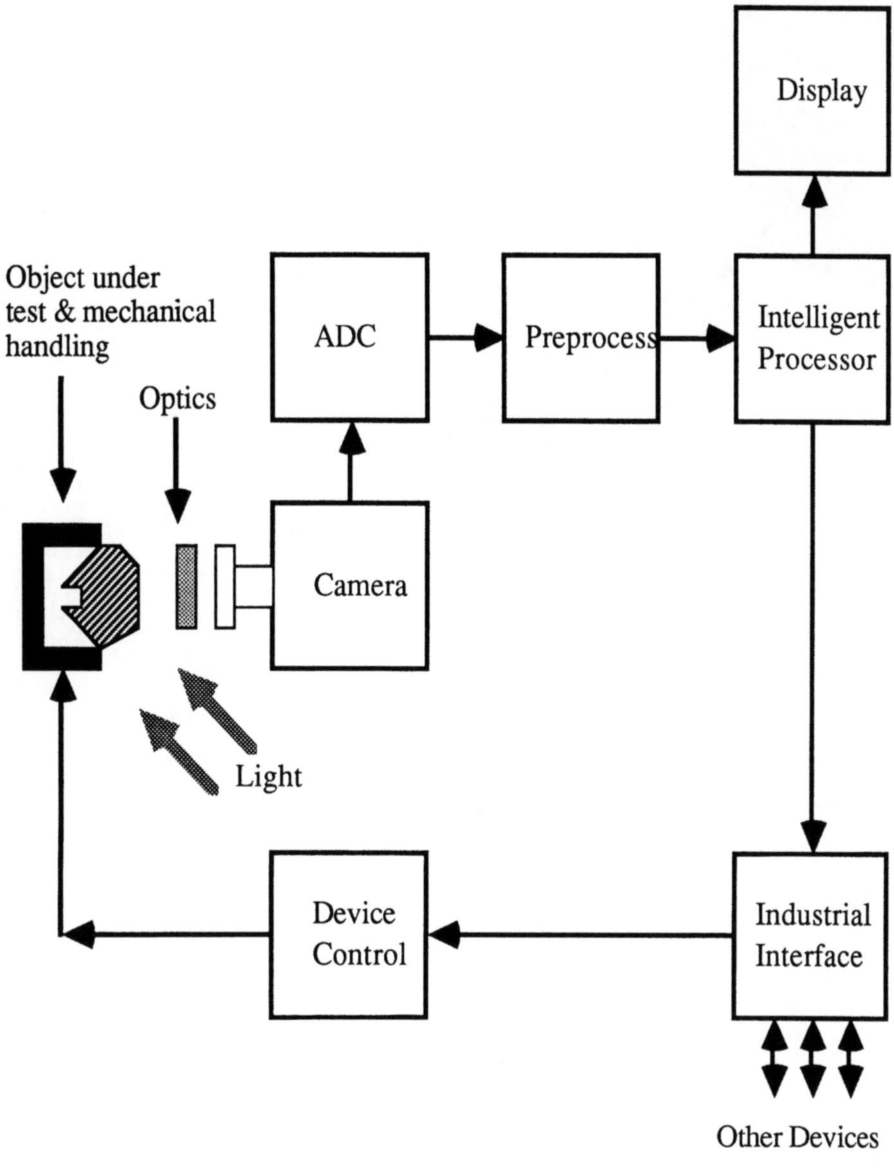

Figure 1 Archetypal automated visual inspection system.

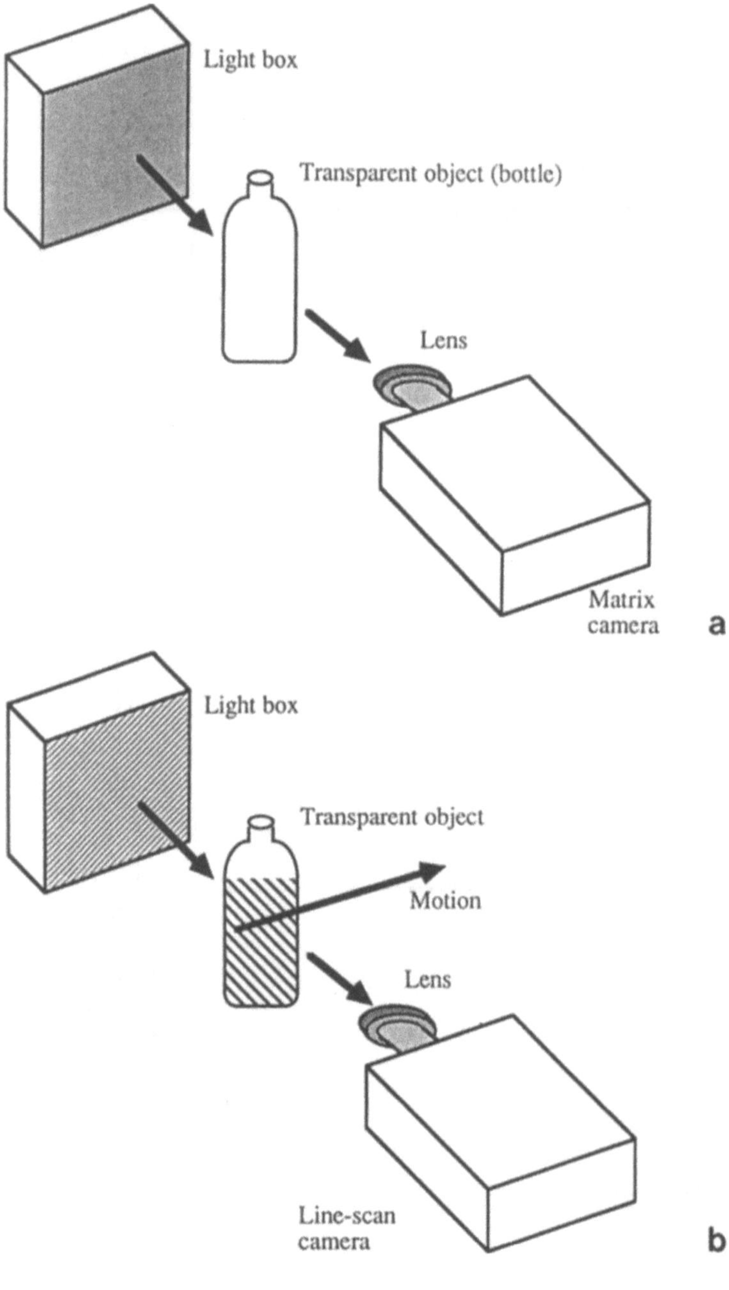

Figure 2 Back illumination
(a) Using a standard (matrix) camera.
(b) Using a line-scan camera. Notice that the object to be inspected is in continuous motion on a conveyor. Such a camera is ideally suited for this situation. The alternative would be to use a matrix camera, together with stroboscopic illumination, but flashing lights present a health hazard to people working nearby.

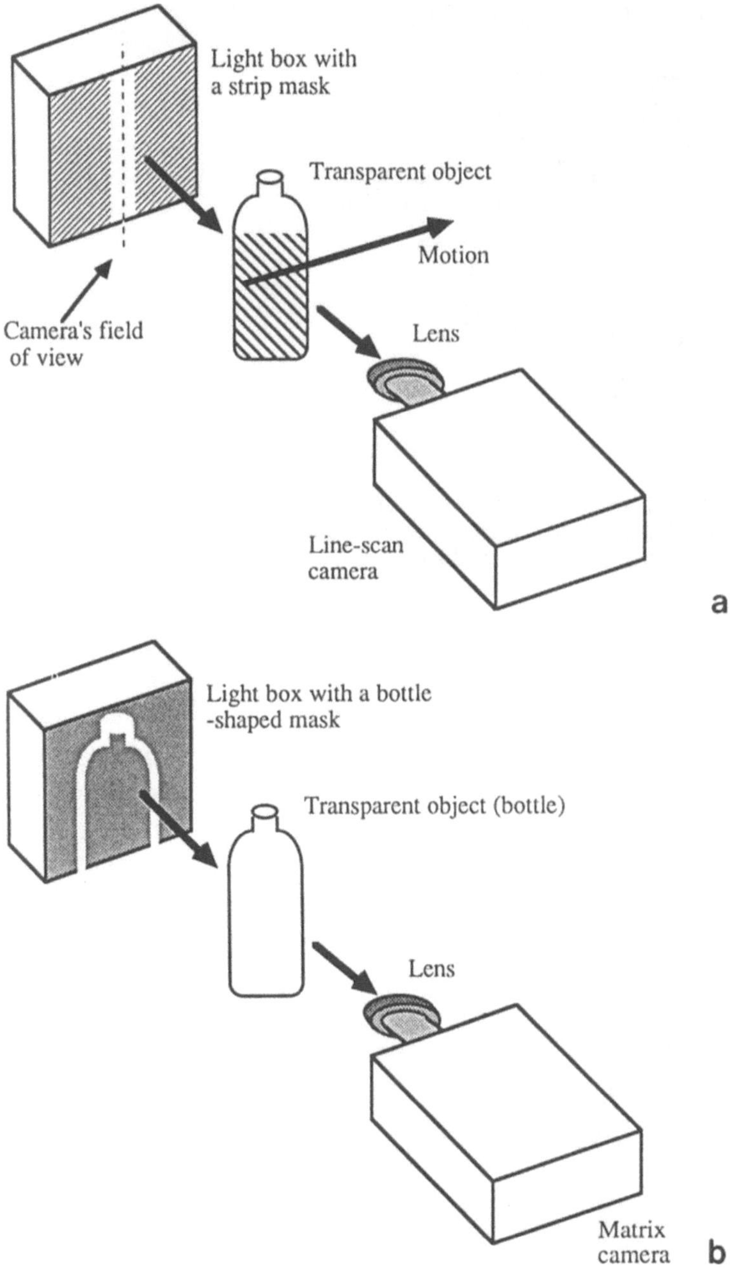

Figure 3 Dark ground (or dark field) illumination
(a) Using a line scan camera. Notice the black mask on the light source.
(b) Using a matrix camera. Notice that the mask has the same shape as the bottle and the light source or possibly the camera must be strobed.

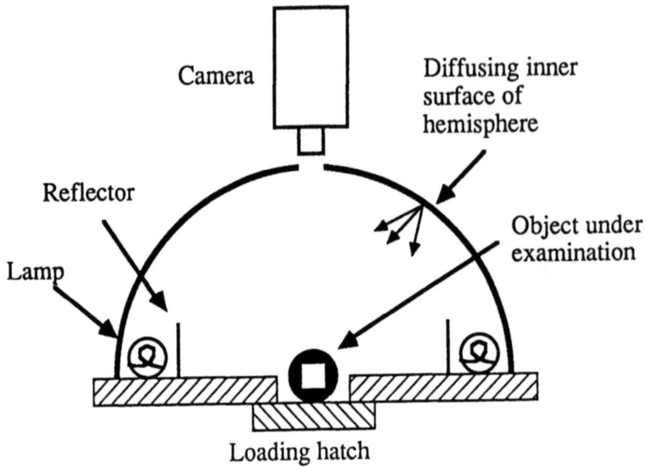

Figure 4 Omni-directional light source. This shows the cross-section through a hemispherical diffusing dome and that the lamps cannot directly illuminate the object under examination. Such an arrangement does not cast shadows. Nor does it cause highlights, due to glinting, from bright metallic surfaces.

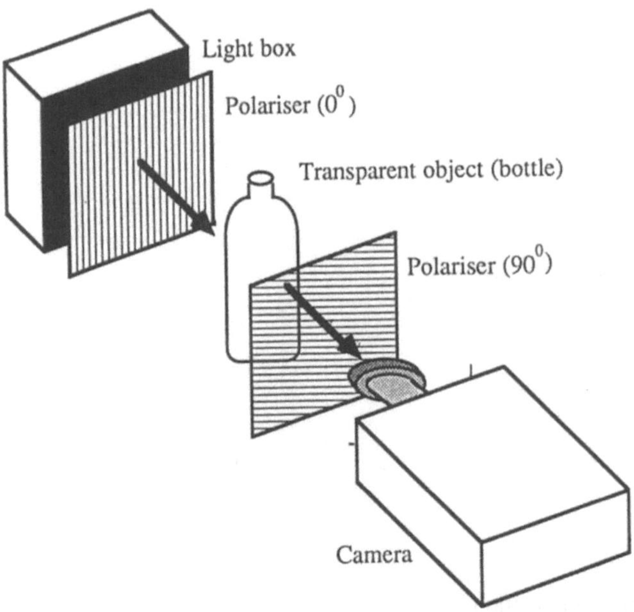

Figure 5 Using crossed linear polarisers to show the stress in transparent materials.

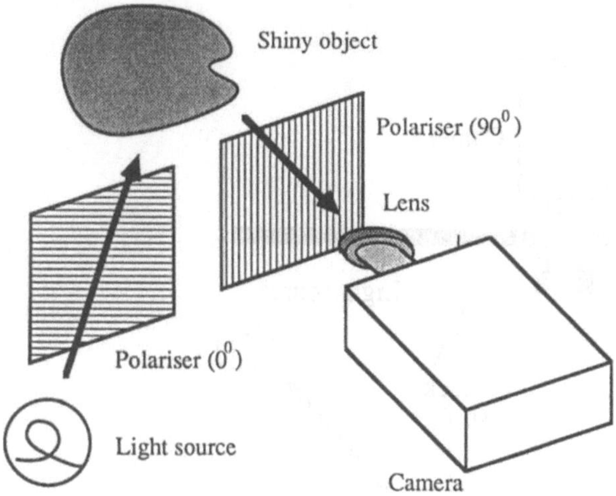

Figure 6 Using crossed linear polarisers to illuminate bright metallic objects, without creating highlights, due to specular reflection. Many cameras present serious difficulties when facing a glinting metallic surface. (See Tables 2 and 3.) As an alternative, a single circularly polarising filter may be used instead; illumination and viewing both use the same polariser.

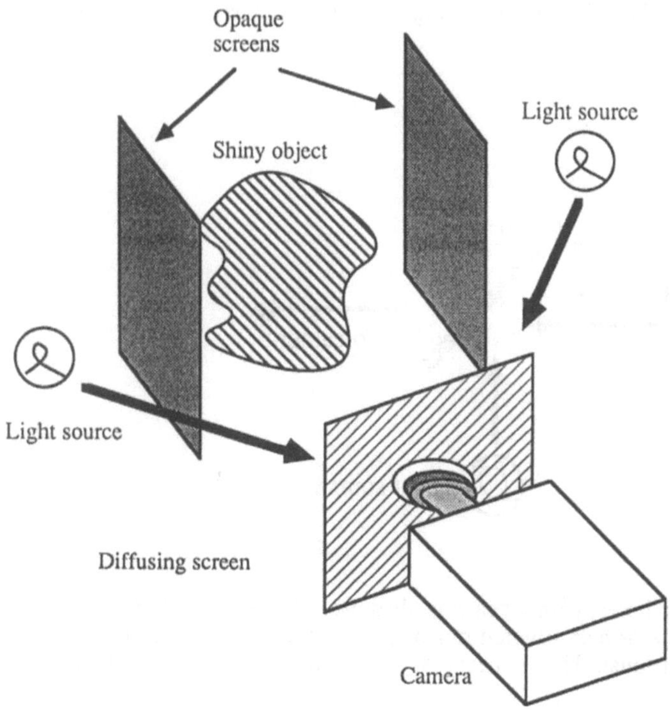

Figure 7 Using diffusing screen, acting as a broad illumination source, to avoid problems caused by glinting.

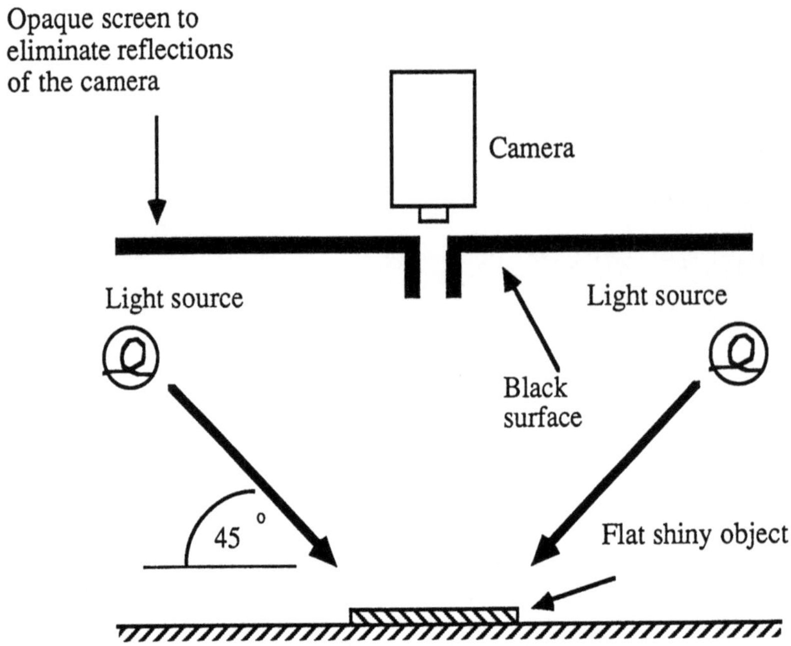

Figure 8 Using carefully controlled directional lighting to avoid problems due to glinting.

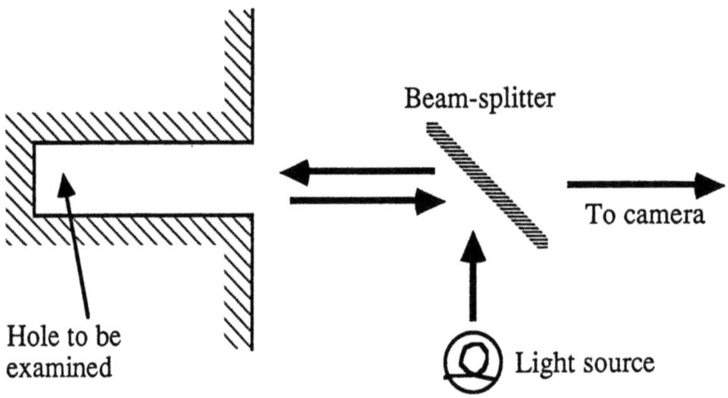

Figure 9 Viewing inside deep holes, without casting shadows. The beam-splitter should be of the thin-film type (i.e. pellicle or half-silvered mirror), rather than the double-prism variety, although the latter is more robust. This is necessary to avoid creating bright spots in the camera's field of view.

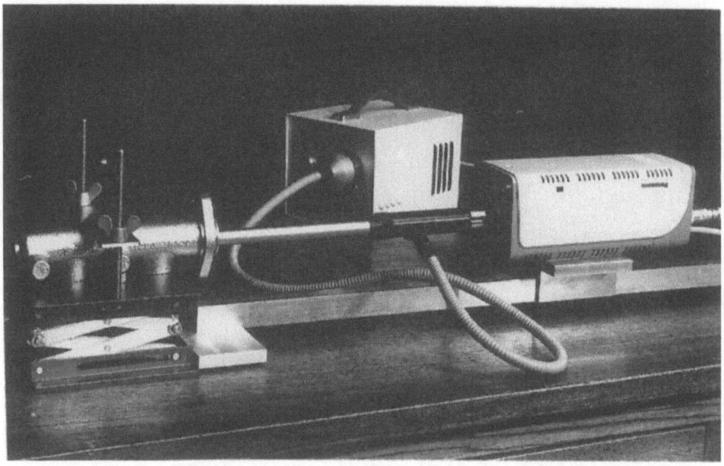

Figure 10 Endoscope (or intrascope), fitted to the C-mount of a standard closed-circuit television camera and being used to examine the internal surface of an hydraulics cylinder. An endoscope can be fitted with a small mirror, which allows it to "look" sideways. High magnification can also be achieved by choosing an endoscope which has an objective lens with a short focal length. Illumination is provided by a ring of optical fibres surrounding the objective lens. (c.f. Figure 11c)

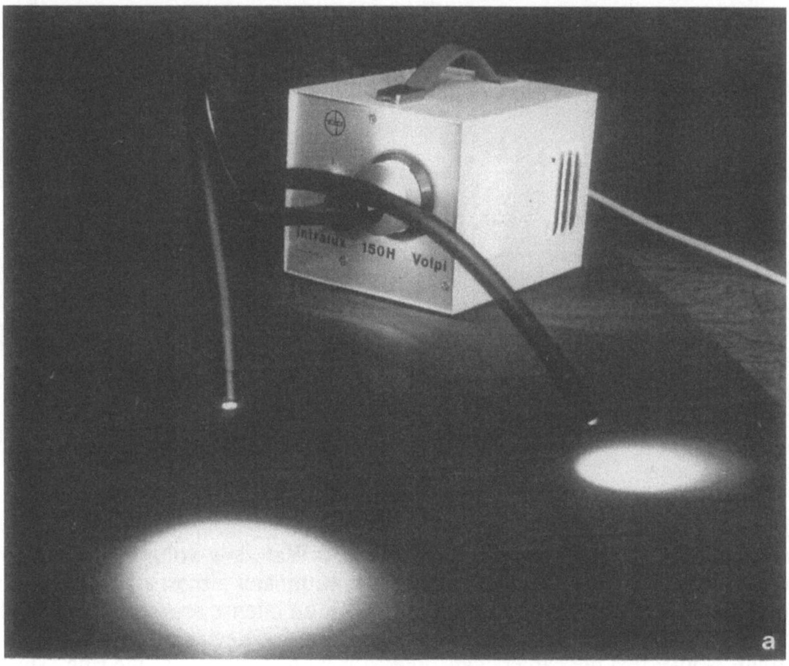

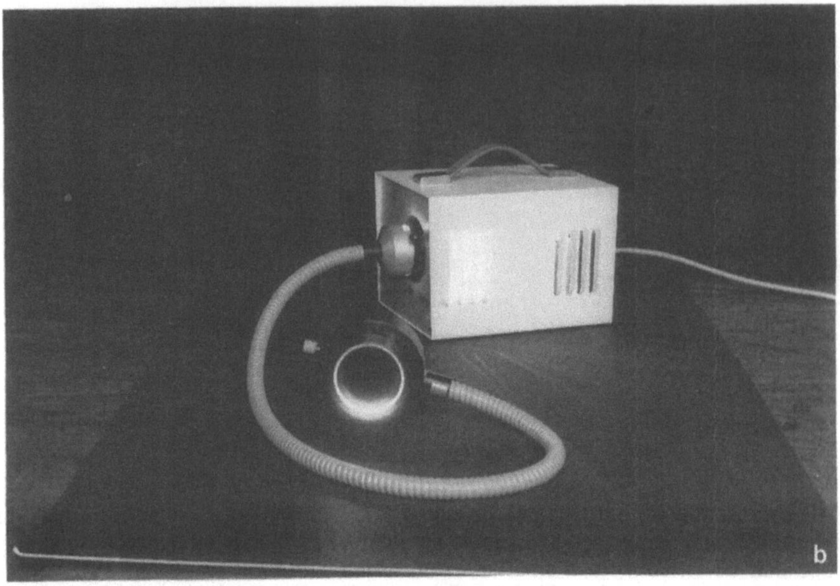

Figure 11 Fibre optic illumination devices.
(a) Twin swan neck. The lamp housing contains a 150-Watt, low-voltage lamp, which is fitted with a dichroic reflector. This, together with a filter, eliminates almost all of the infra-red light. Glass optical fibres also absorb infra-red. The result is an intense source of "cold" light, which would not, for example, melt chocolate, or damage electronic devices.
(b) Ring light. Notice that the camera can "look" through the centre of the ring. This provides an ideal device for illuminating a circular object, and also provides an alternative to the schemes, illustrated in Figures 9 and 10, for viewing inside deep holes, without casting shadows.
(c) Cat-o'-nine-tails (actually there are 45 tails in this bundle) By holding the fibres in a grid, it is possible to use this type of bundle to provide a very flexible illumination source.

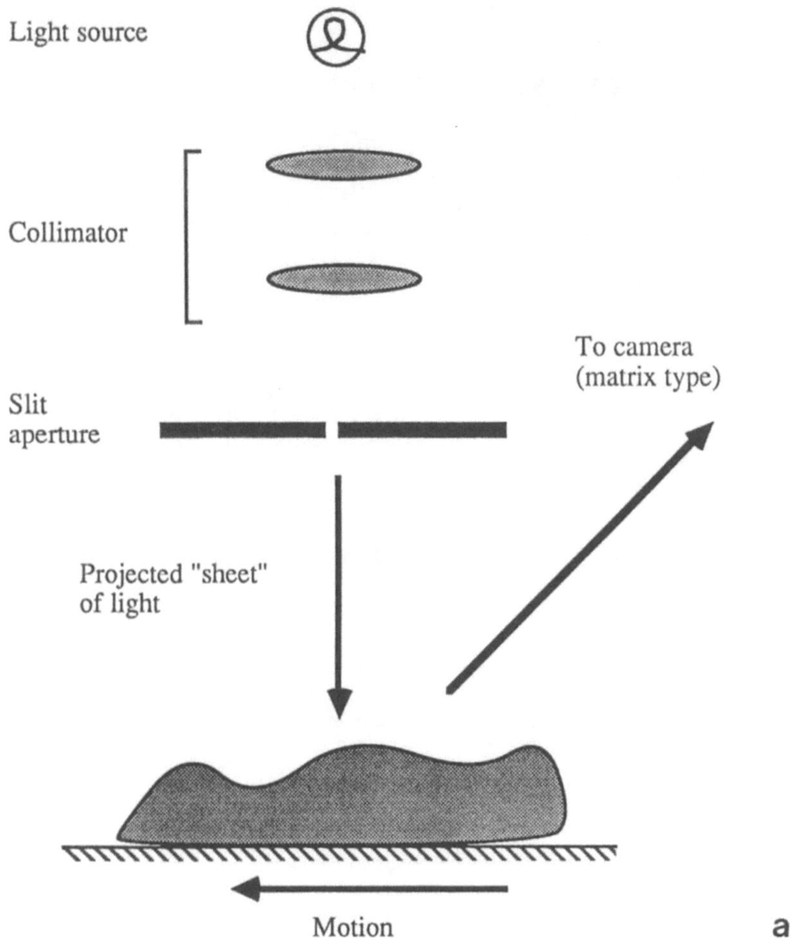

Figure 12 Structured lighting to obtain data about the 3-dimensional structure of an object.
(a) The layout of the optical system. This is a different scheme from that published by various other authors (including the present one!), in that the positions of the camera and light source have been interchanged. The reason for doing this is that the resulting data is easier to interpret. Notice that the object is physically moved across the camera's field of view, to obtain height data over the whole (top-surface) of the object. In addition, it is important to appreciate that glinting can cause problems, as can shadowing. The former can be overcome, using polarisers, while the latter requires the use of two structured-light systems, whose outputs are combined together.
(b) Showing the projected light stripe across a simple box-like object on a flat table.
(c) The view as seen by the camera. Notice that, for each frame of video data, the height information in one plane can be calculated. This presents a fundamental limitation upon the speed of this technique.
(d) Geometry of the optical system, assuming that a pin-hole camera is used.
(e) A depth map obtained from a plastic moulding (the base of a U.K. 13-Amp mains plug). In this type of image, the intensity indicates the height of the surface, rather than its luminance.

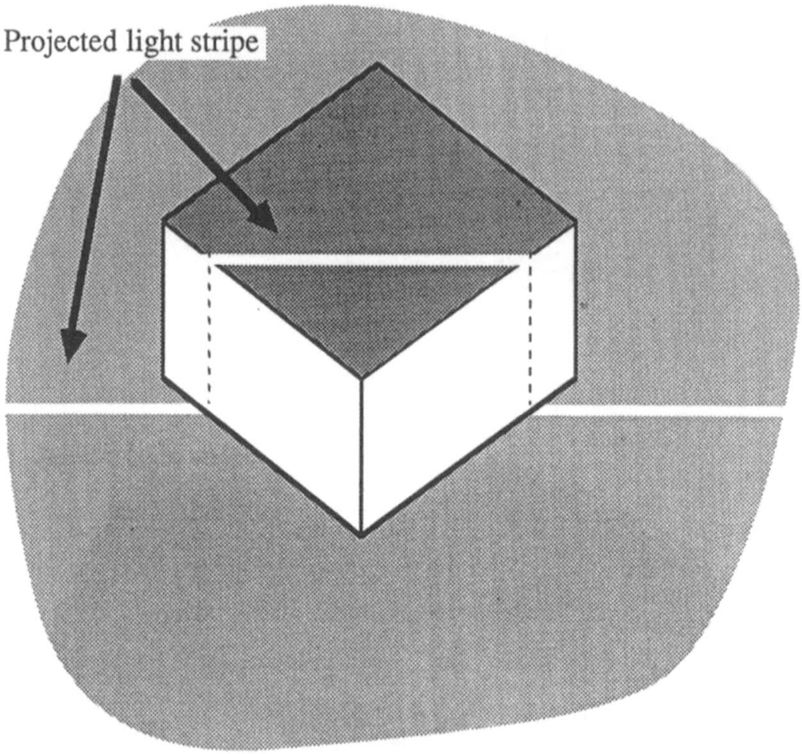

b

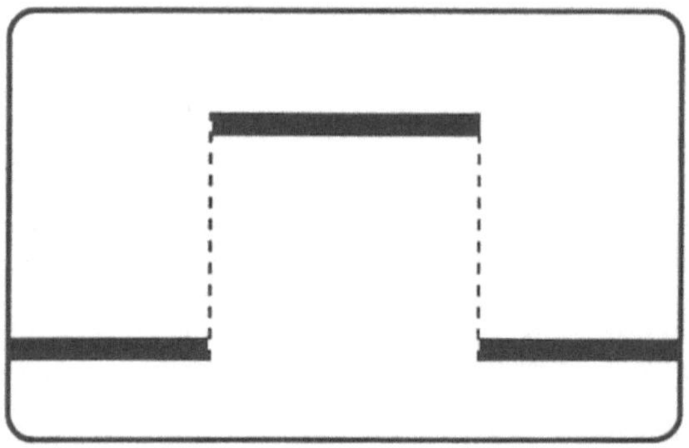

c

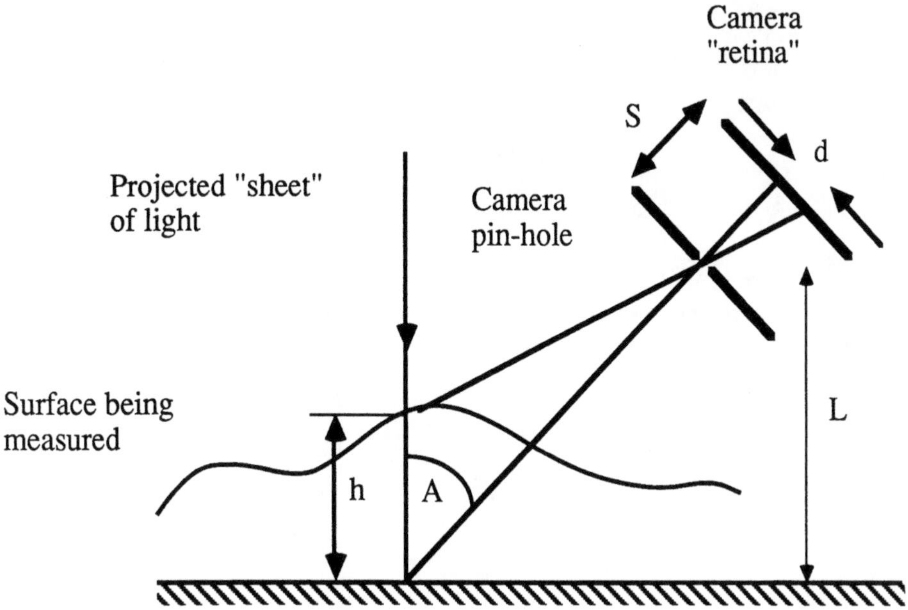

$$h = \frac{d.L.(1 + \tan^2 A)}{S.\tan A + d}$$

d

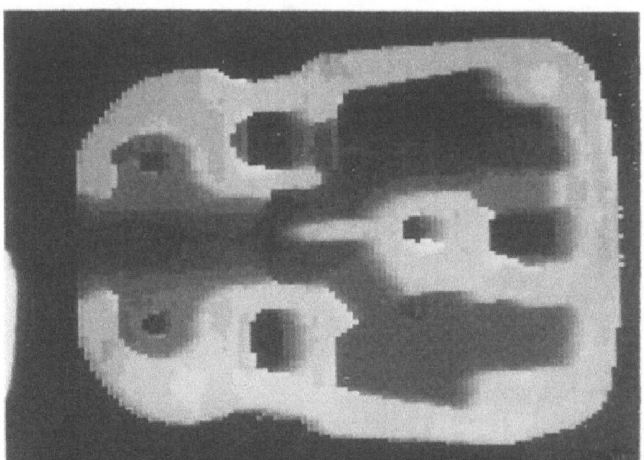

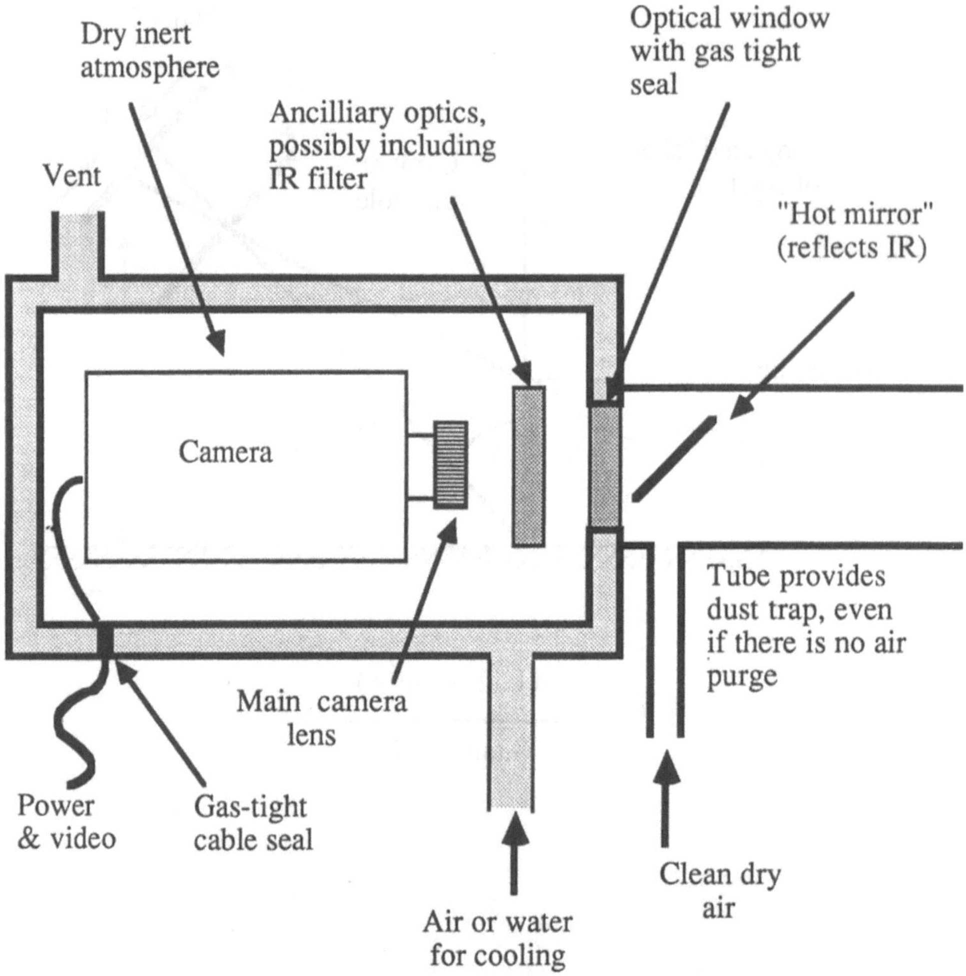

Figure 13 Showing a number of techniques for protecting a camera and keeping its optical surfaces clean.

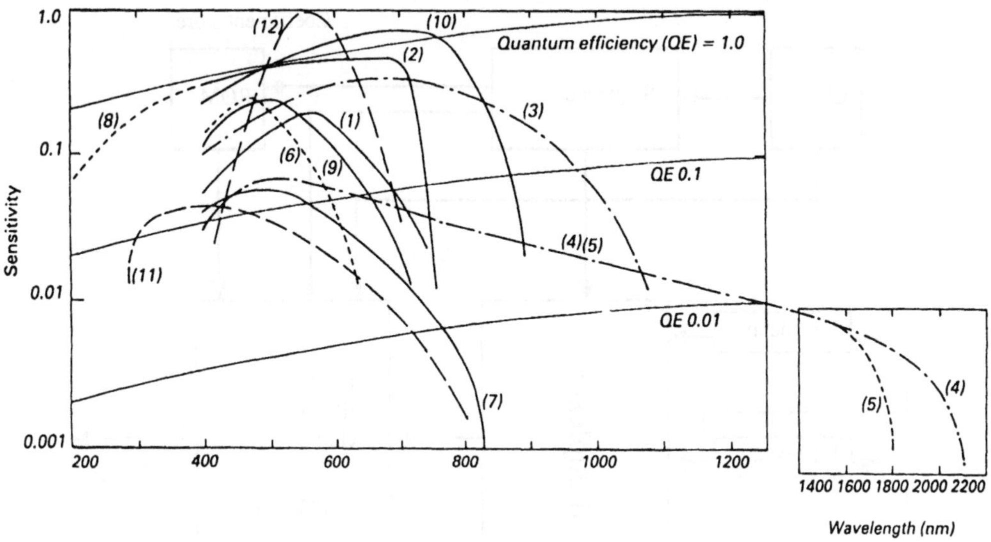

Figure 14 Spectral response curves of various cameras and the human eye.

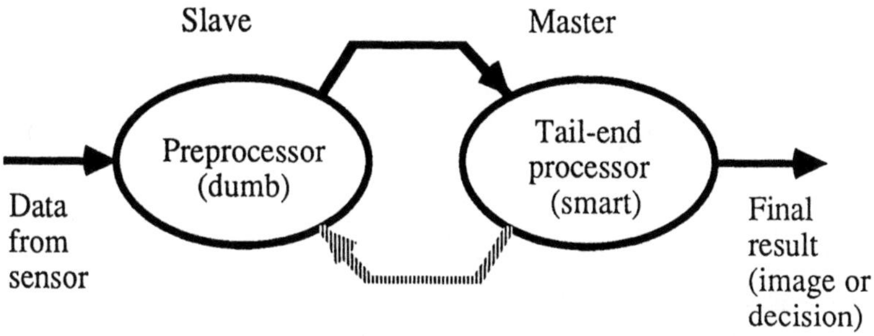

Figure 15 In many inspection algorithms, the data flow is linear, from the data source (i.e. the camera) through the dumb processor to the smart processor, and then to its final destination (i.e. some kind of mechanical device for accepting/rejecting the parts). In this situation, we can refer to the *preprocessor* and *tail-end* processor. However, it is sometimes found that it is necessary to provide a feed-back loop (grey), so that the image processing sub-system can perform hypothesise-and-test algorithms. The smart process acts as a master controlling a dumb slave process. This model anticipates the architecture of the image processing sub-system proposed in Figure 22.

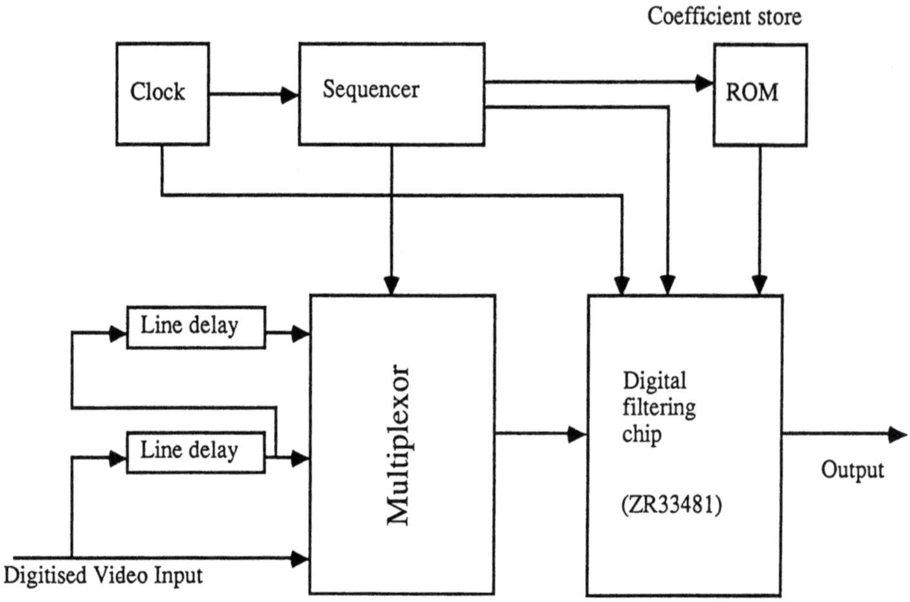

Figure 16 Implementation of a 3*3 linear local operator using the Zoran ZR33481 digital filtering chip. Such a circuit operates in real time and is of modest cost.

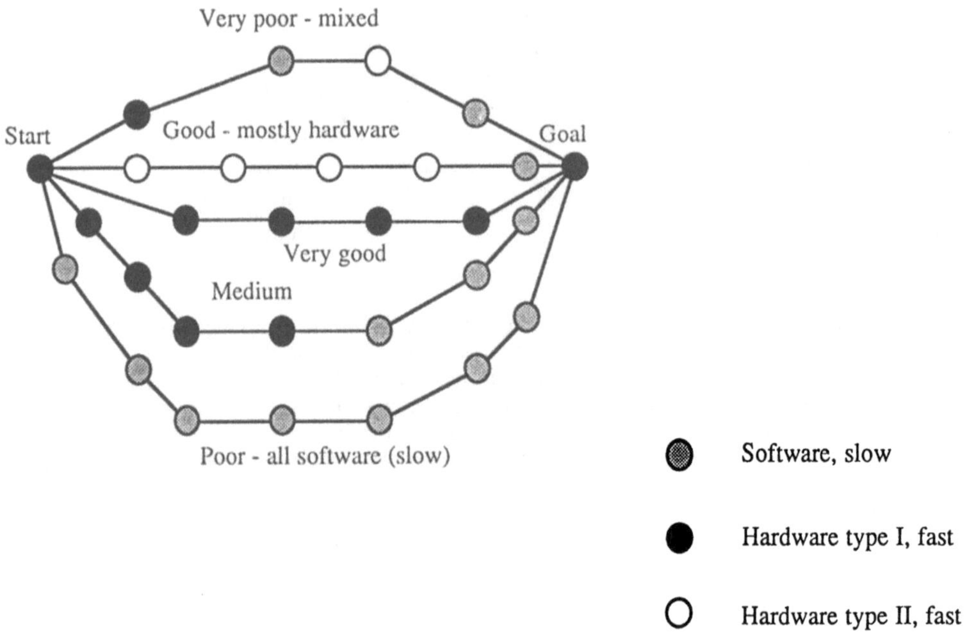

Figure 17 There are many algorithmic paths which can achieve a given result. They should be judged, not only by their performance, but also by the ease of implementation. Notice that some algorithms are preferable to others for the simple reason that they allow a nearly "homogenous" implementation.

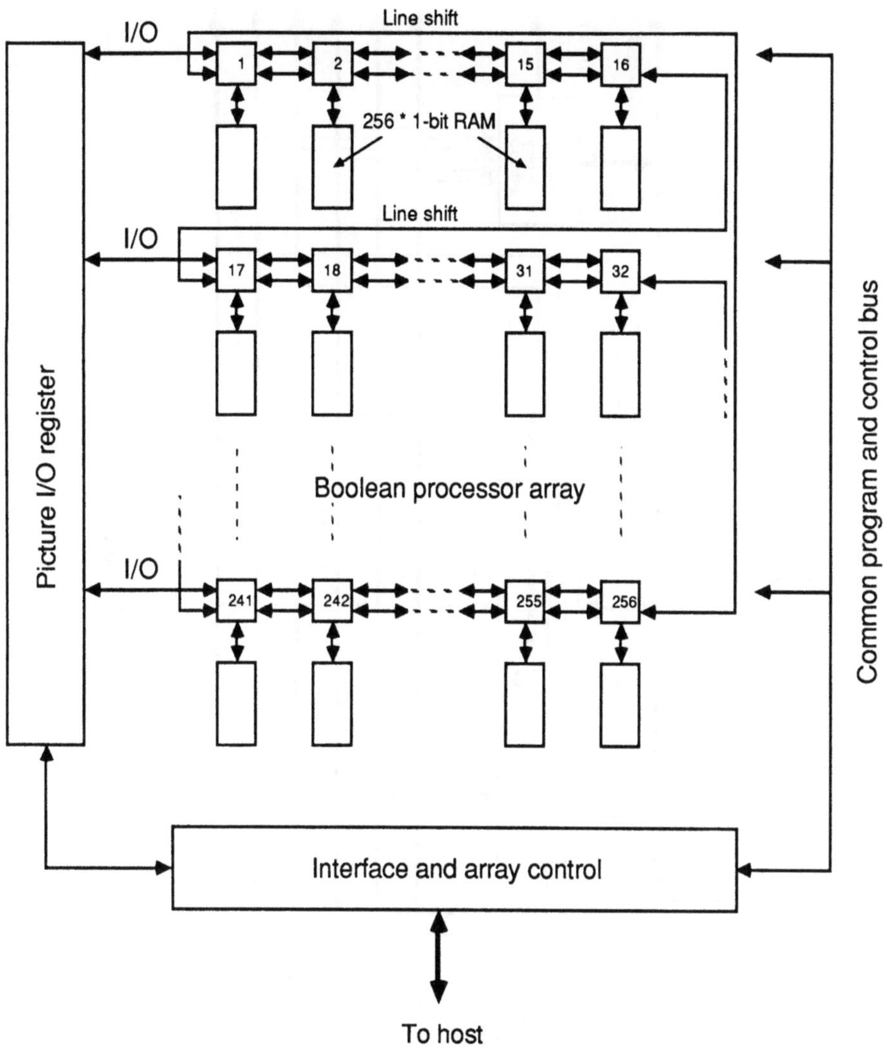

Figure 18 Linear array processor

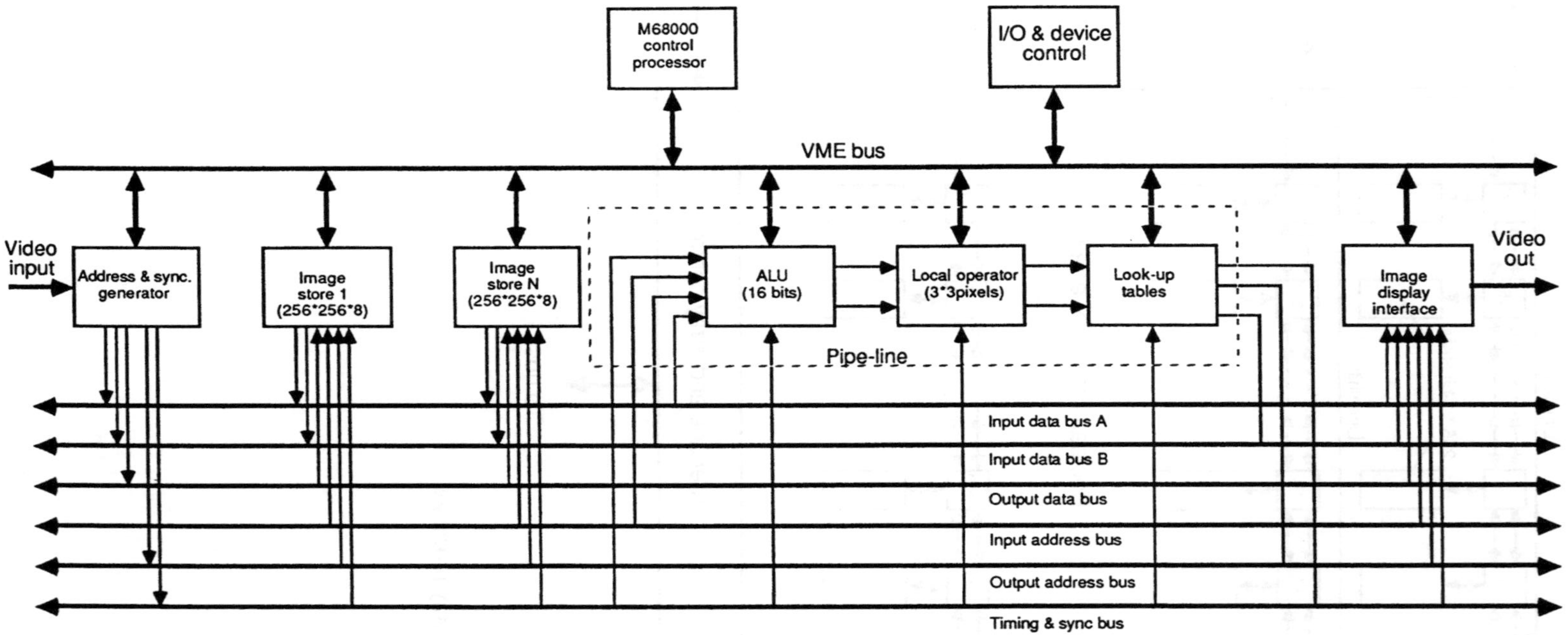

Figure 19 Kiwivision pipe-line processor. This uses the standard pipe-line structure referred to in Table 4, but it is not restricted to this form.

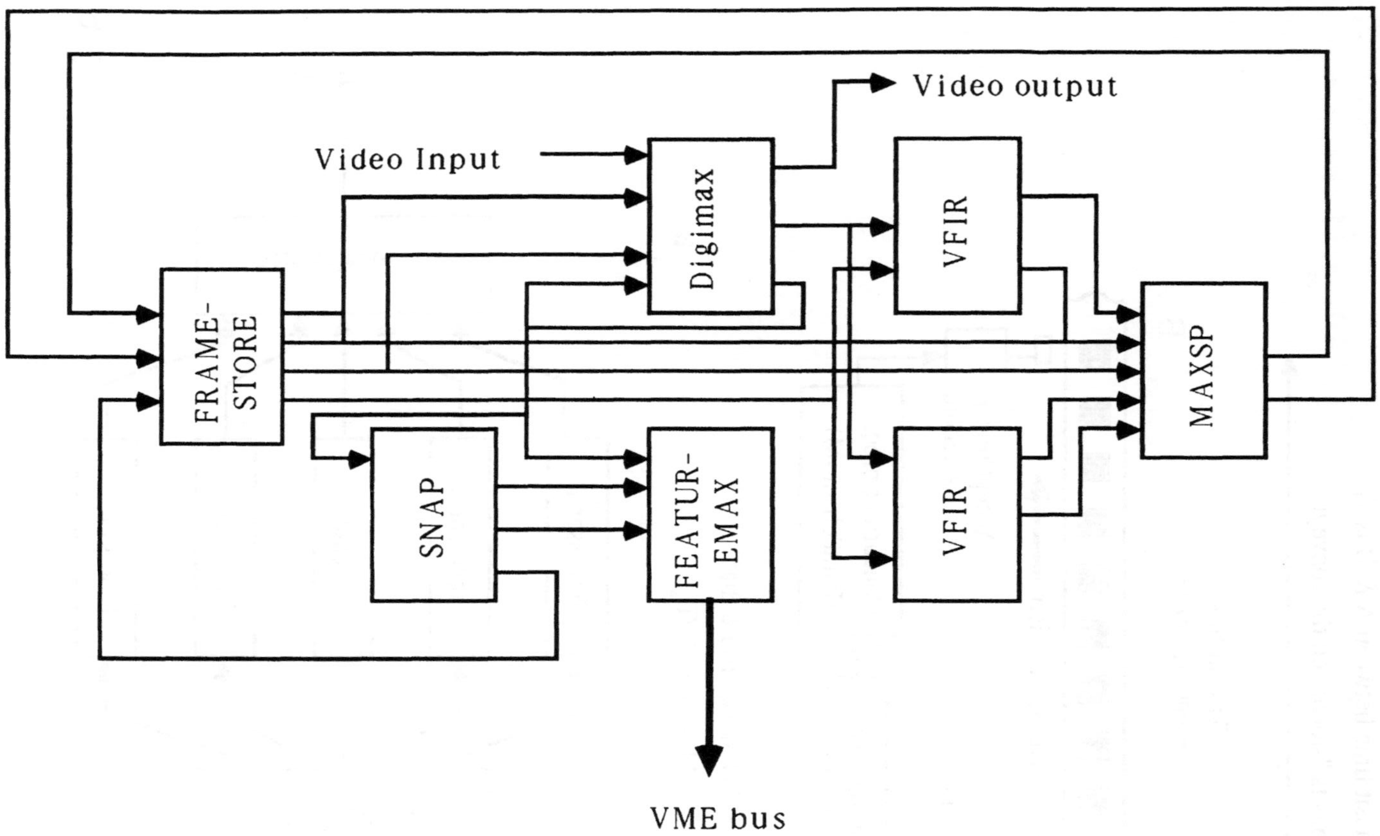

Figure 20 Maxvideo pipe-line processor. This is the standard configuration suggested by Datacube, Inc. It can implement the Sobel edge detector, thresholding, feature (e.g. corner) detection, feature location and counting in near real time.

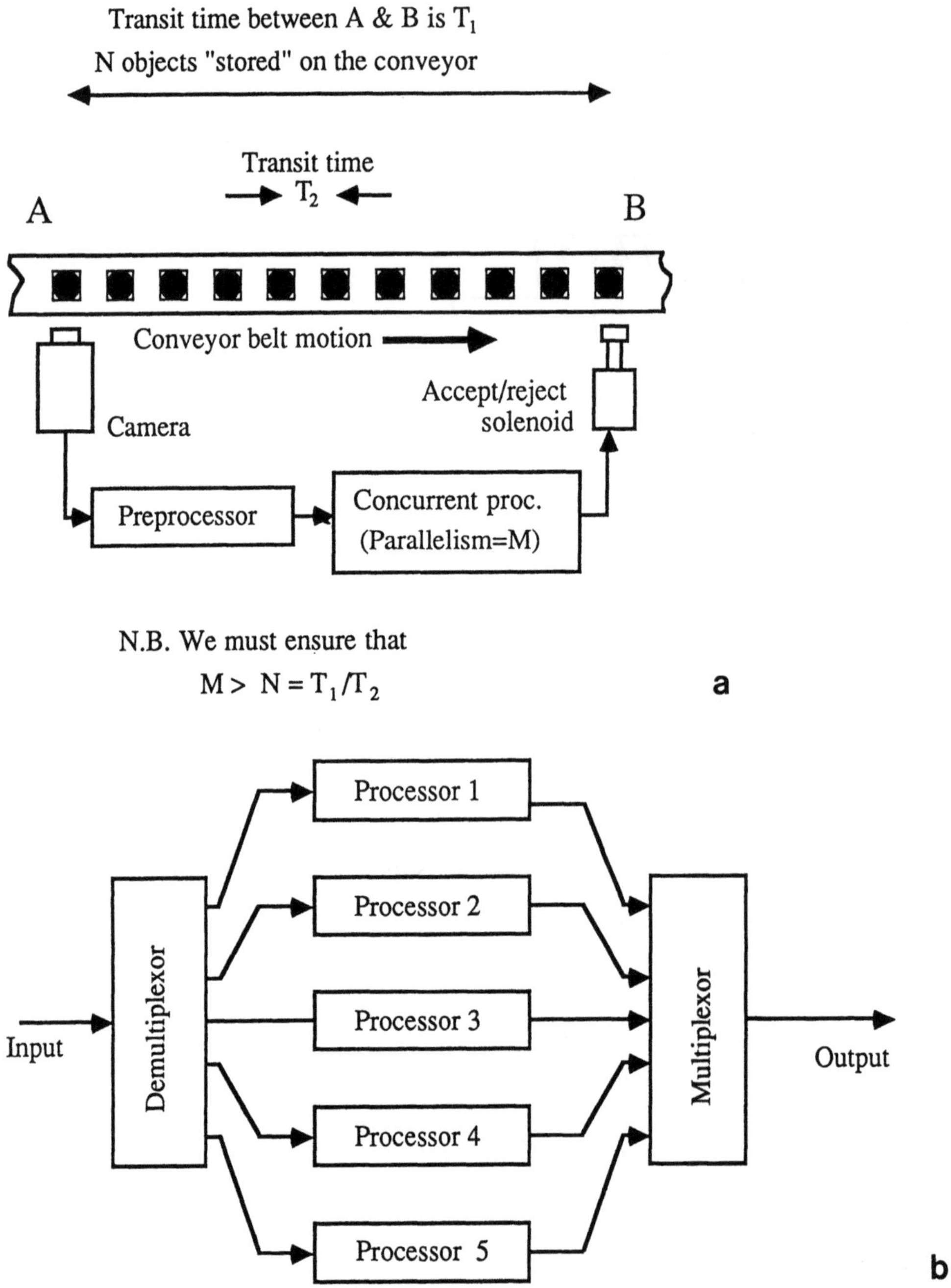

Figure 21 Concurrent processing applied to inspection on a conveyor belt.
(a) Physical layout of the conveyor belt and inspection system.
(b) Organisation of a concurrent processor.

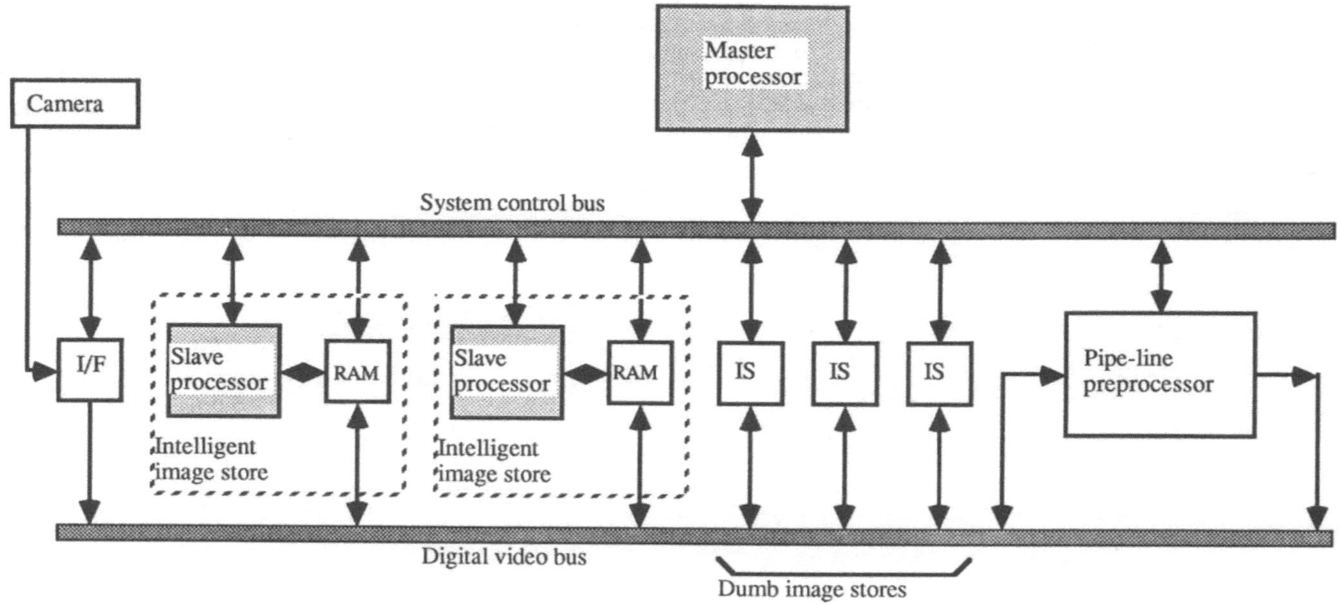

Figure 22 Outline architecture for a system capable of performing hardware and software image processing operations concurrently. There may be several dumb image stores and several intelligent image stores in a practical system. The latter would require some firmware; the authors are currently developing a system, based upon Kiwivision and a software package akin to Autoview Viking.

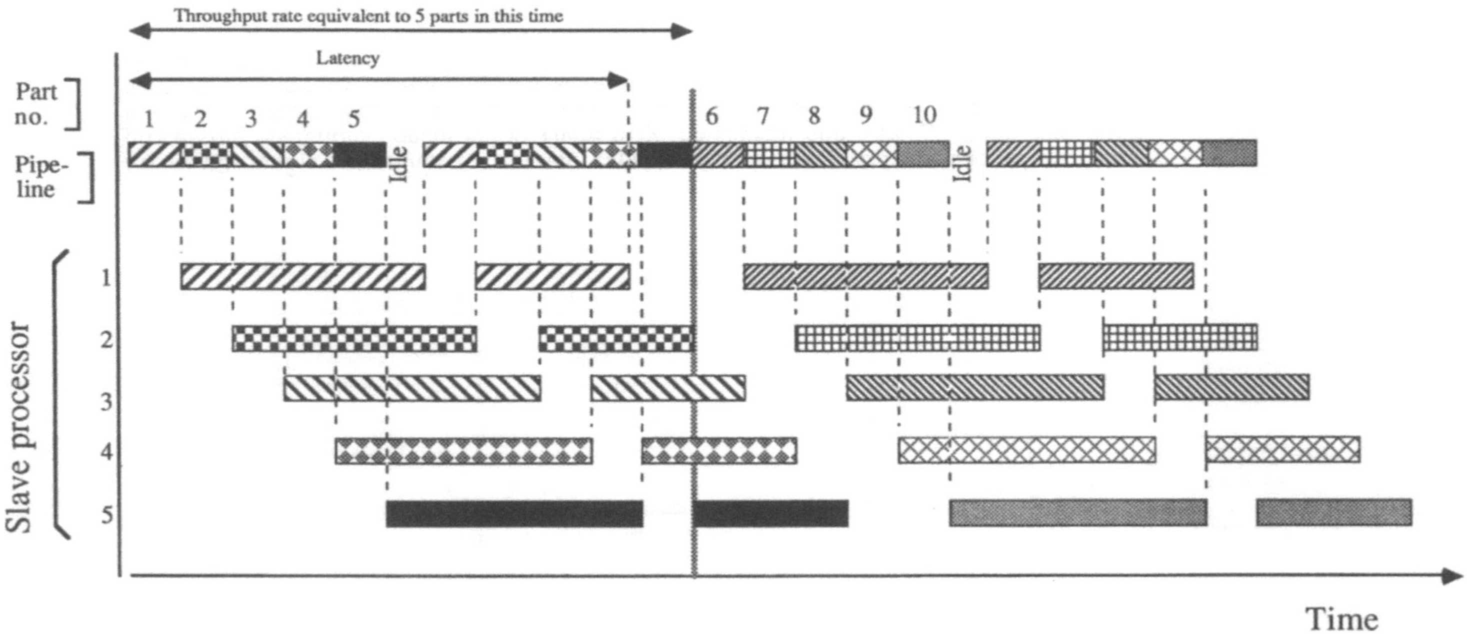

Figure 23 Sequence chart for the multi-processor whose organisation is shown in shown in Figure 22. The pipe-line is effectively multiplexed, as it serves the five slave processors, in turn. Each object to be inspected is represented by a different type of shading. Notice that the pipe-line and a slave each require *two* periods for processing a single image. The arrows show the data flow for part number 1. In this example, the slave processors limit the speed of operation, since they are unable to complete the first part of their calculations in under five times the period taken by the pipe-line processor.

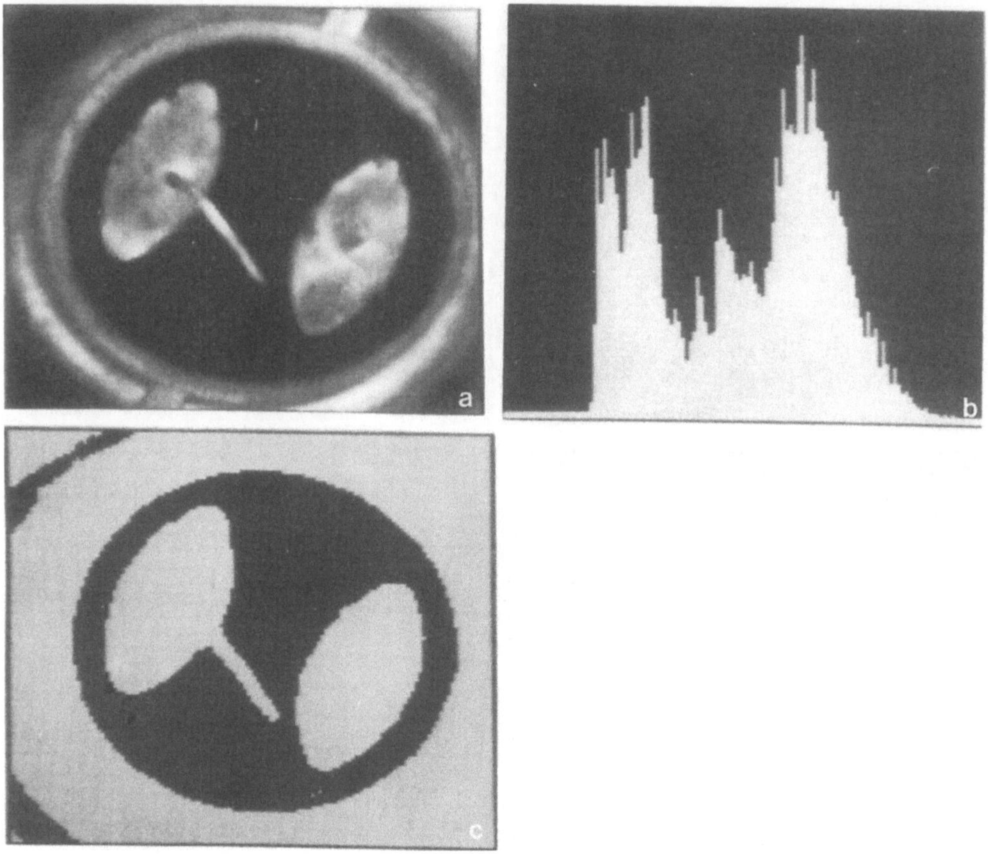

Figure 24 Detecting wires on the cap of a light bulb
(a) Original unprocessed image, obtained using an omni-directional light source. (Figure 4) Notice that there is a wire protruding from the left-hand eyelet.
(b) The intensity histogram of image (a). Notice the strongly bimodal nature of this histogram. This is typical of an image showing a bright object against a dark background, or vice versa. The "valley" between the two peaks often indicates a useful intensity value at which to perform a threshold operation.
(c) Simple thresholding applied to image (a). The threshold parameter was calculated automatically from the intensity histogram. Although thresholding works well in thias particular instance, it is not nearly as reliable as the algorithm discussed in the text.

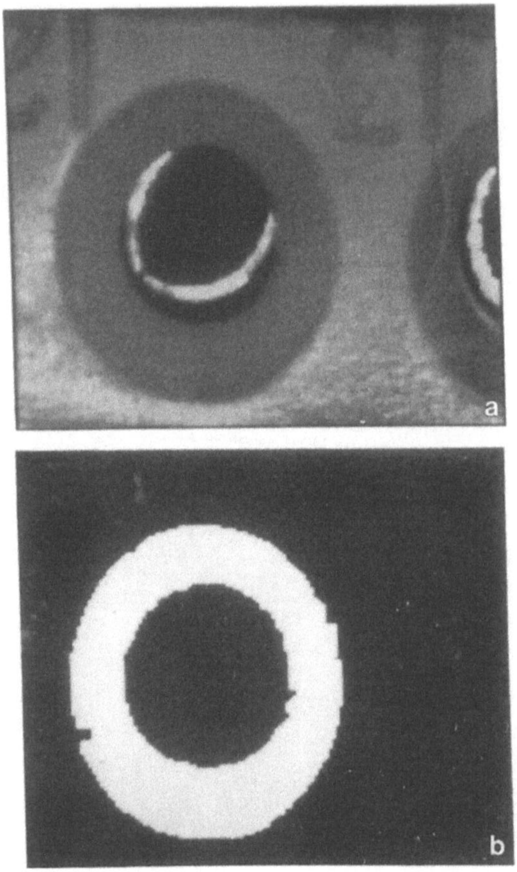

Figure 25 Detecting the presence of a chamfer
(a) Original unprocessed image. Notice the presence of glinting, which could have been avoided by using a circular polariser. The processing is well able to cope with this.
(b) The chamfer detected.

Figure 26 Reading punched lettering on a metal surface
(a) Original unprocessed image. Notice the presence of glinting and that the lettering is indistinct.
(b) Linear highpass filtering
(c) Fixed value thresholding applied to image (a). There is no threshold value which properly separates the lettering from the background. No algorithm was discovered which would suppress the background "noise" in this image, whilst leaving the outline of the lettering intact.
(d) A non-linear noise-reduction filter applied to image (a). Although this eventually lead to a "blind alley" in the processing, this result seemed, at first, a little more promising than that shown in images (b) and (c). As a result, the search was then concentrated on algorithms which are generically related to this one.
(e) One of the author's(BGB) favourite commands did not, on this occasion, yield any promise of providing satisfactory results. As a result, this avenue was soon abandonned.
(f) Another noise-reduction algorithm. The results were felt to be sufficiently promising to warrant further investigation.
(g) Simple thresholding applied to image (f). Although this is distincly promising, it does rely upon fixed value thresholding, which is an undesirable procedure to use in a situation, where the lighting can directly affect the results.
(h) The skeleton derived from image (g).
(i) A non-linear "low-pass" filter applied to image (f). This is a rather more robust algorithm, than simple thresholding used to obtain image (g).
(j) The skeleton derived from image (i) contains a number of undesirable spurs, which make the subsequent processing more difficult.
(k) An alternative processing scheme results in a much cleaner skeleton.
(l) The limb ends and joints of the skeleton in image (k). It is a fairly simple task to define rules which can recognise the alphanumeric characters using features such as these.

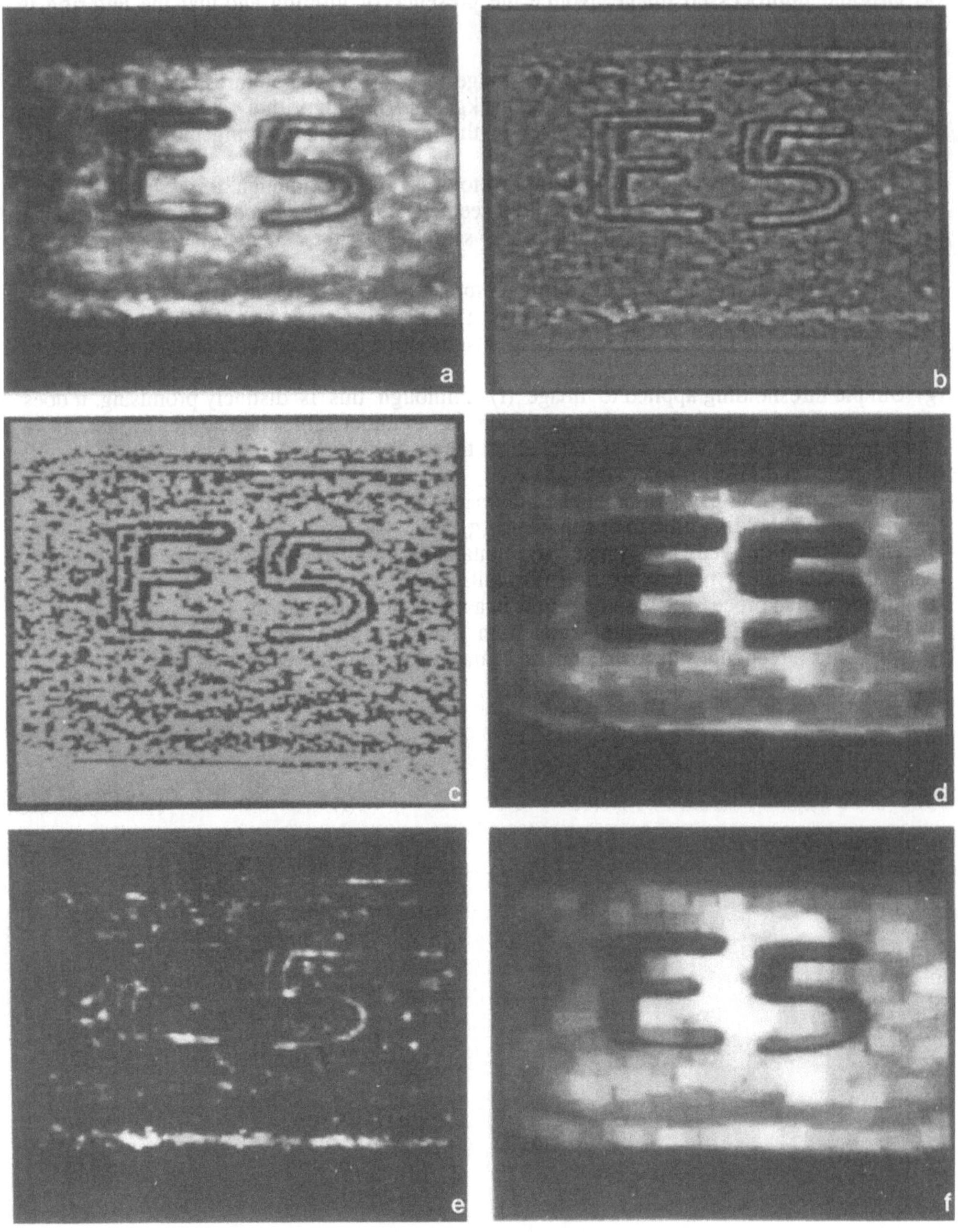
a
b
c
d
e
f

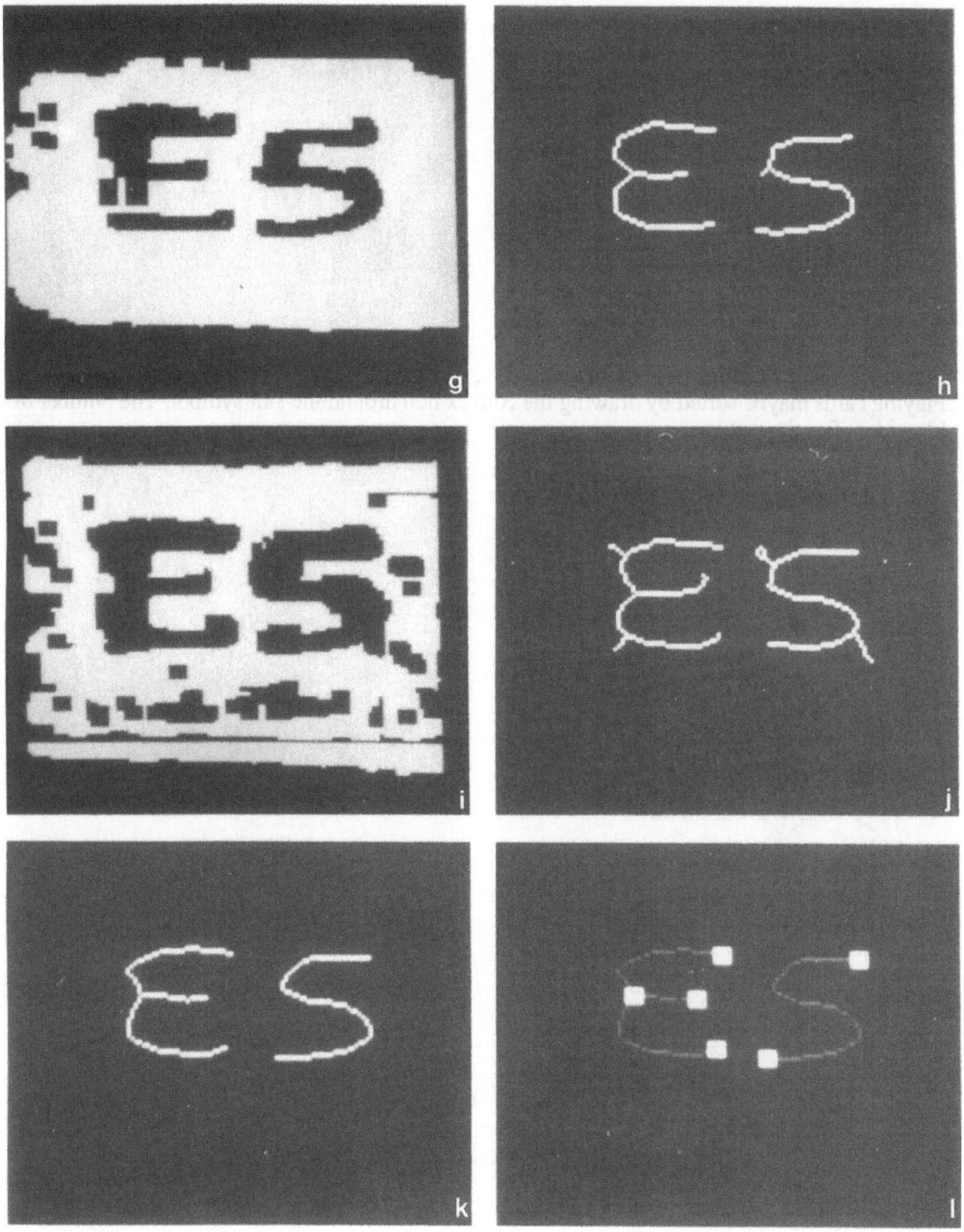
g
h
i
j
k
l

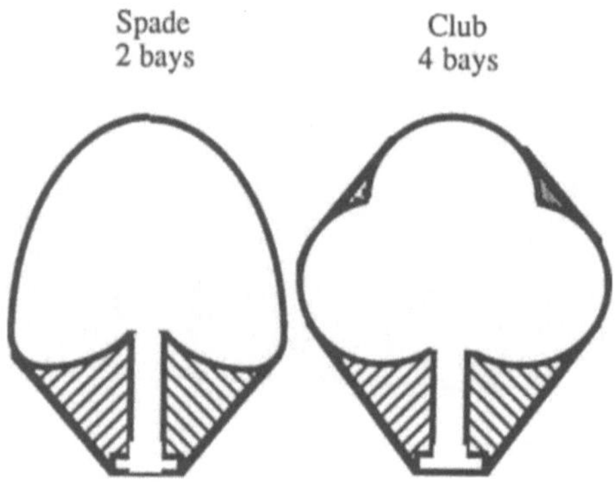

Figure 27 Shape Sorting.
Playing cards maybe sorted by drawing the convex hull around the suit symbol. The number of bays identifies the suit.

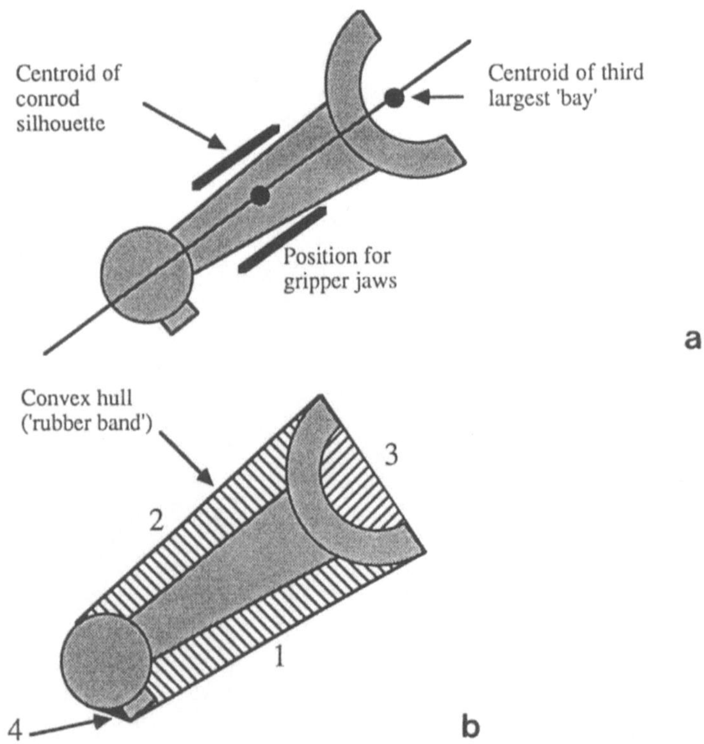

Figure 28 Picking up a con-rod.
(a) The centroid of the silhouette and that of its third largest bay defines a conveneient axis to determine the orientation of the con-rod. The gripping points can be calculated by a simple procedure, using the position of the centroid of the silhouette and the orientation of this line.
(b) The con-rod silhouette, its convex hull and the four "bays", which together comprise the convex deficiency.

Figure 29 Detecting glass flakes on the rim of a drinking glass.
(a) Original unprocessed image. (The flake is about 6mm x 3mm. Notice the presence of some very small bubbles and specks of loose glass.
(b) A non-linear low-pass filter applied to image (a).
(c) Simple thresholding applied to image (b).
(d) The histogram of the original image suggests that it should be possible to apply fixed value thresholding.
(e) Simple thresholding applied to image (a). This corresponds to a point near the foot of the "hill" in the histogram shown in (d). Notice that the flake is only partially detected and that the background on the left is fragmented; it is not possible to threshold the image in (a) reliably. It is commonly found that simple thresholding is not as good as we expect, having examined the image by eye.
(f) The intensity profile along a vertical line shows clearly that the background intensity is changing and that what appears to be very good contrast is, in fact, rather worse than the eye can easily detect. This is a common phenomenon and is due entirely to the lack of sensitivity of the eye to low frequency intensity changes compared to local intensity anomolies and sharp edges.

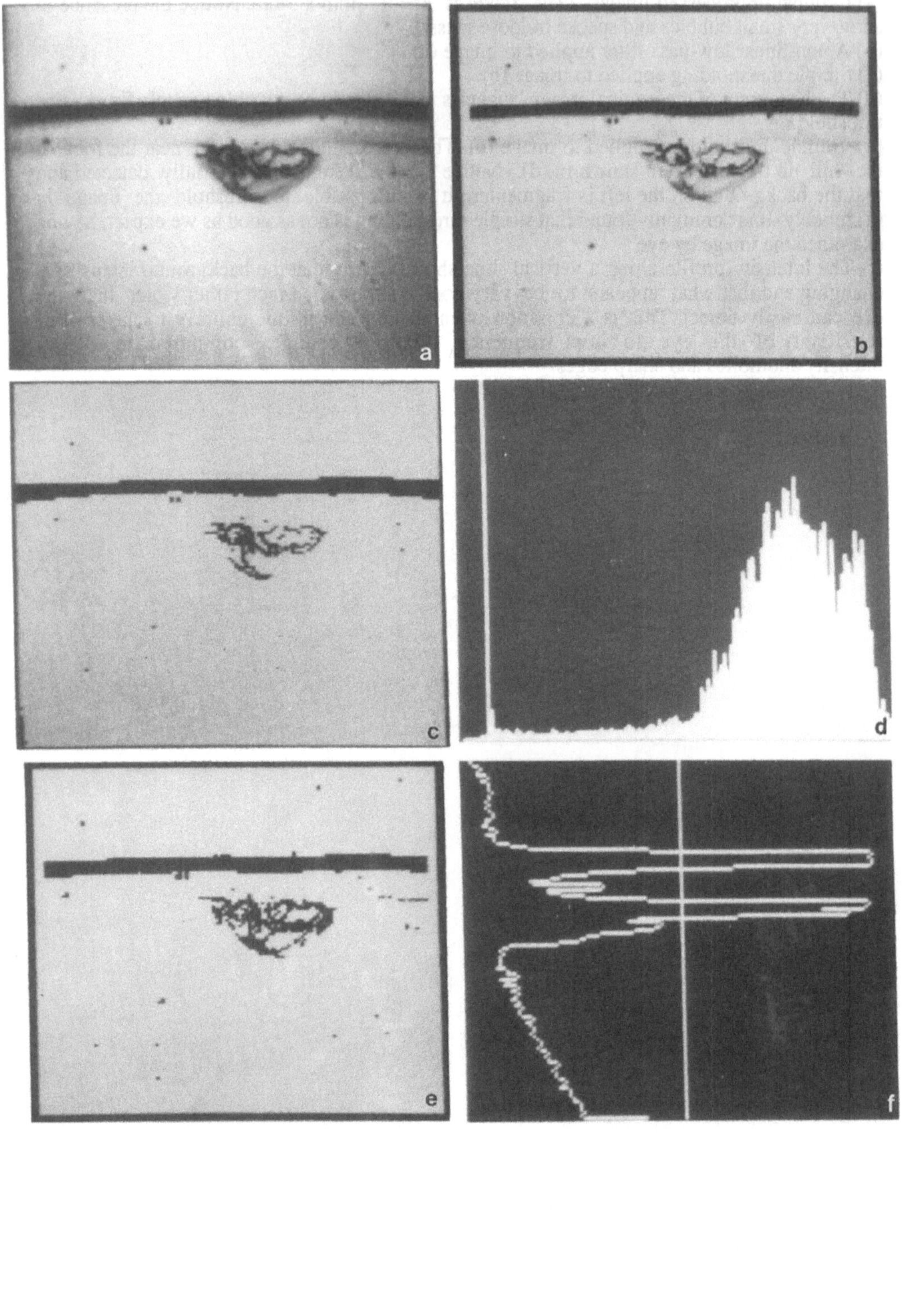
a
b
c
d
e
f

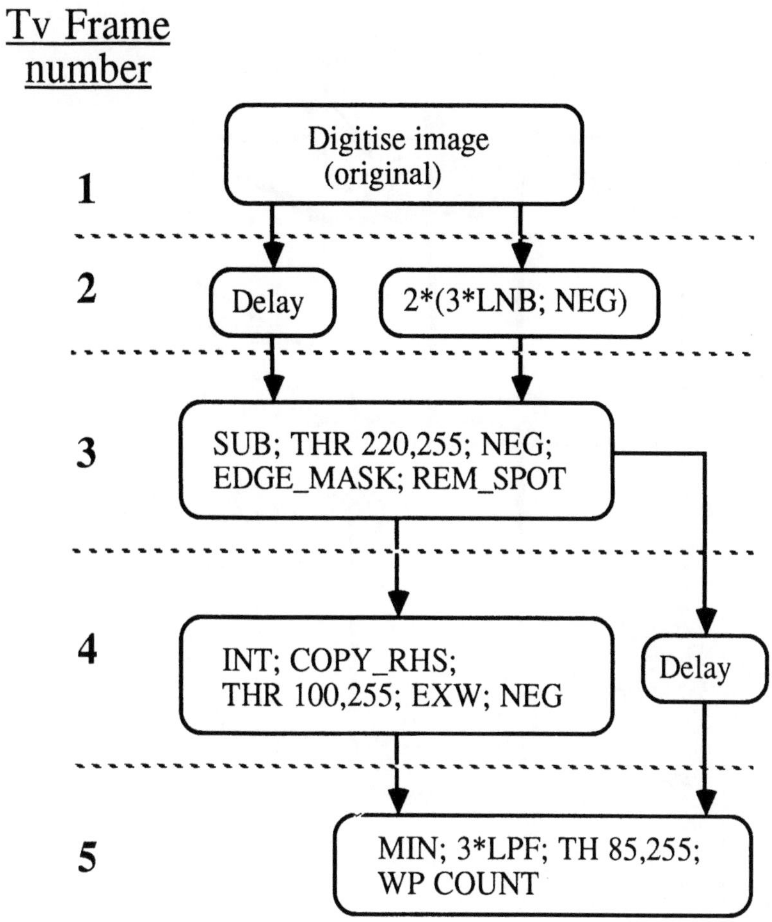

Figure 30 The algorithm for detecting flakes of glass on the rim of a drinking glass, cannot be performed within a single television frame period; a pipe-line processor, such as Kiwivision, requires a number of passes throught the image.

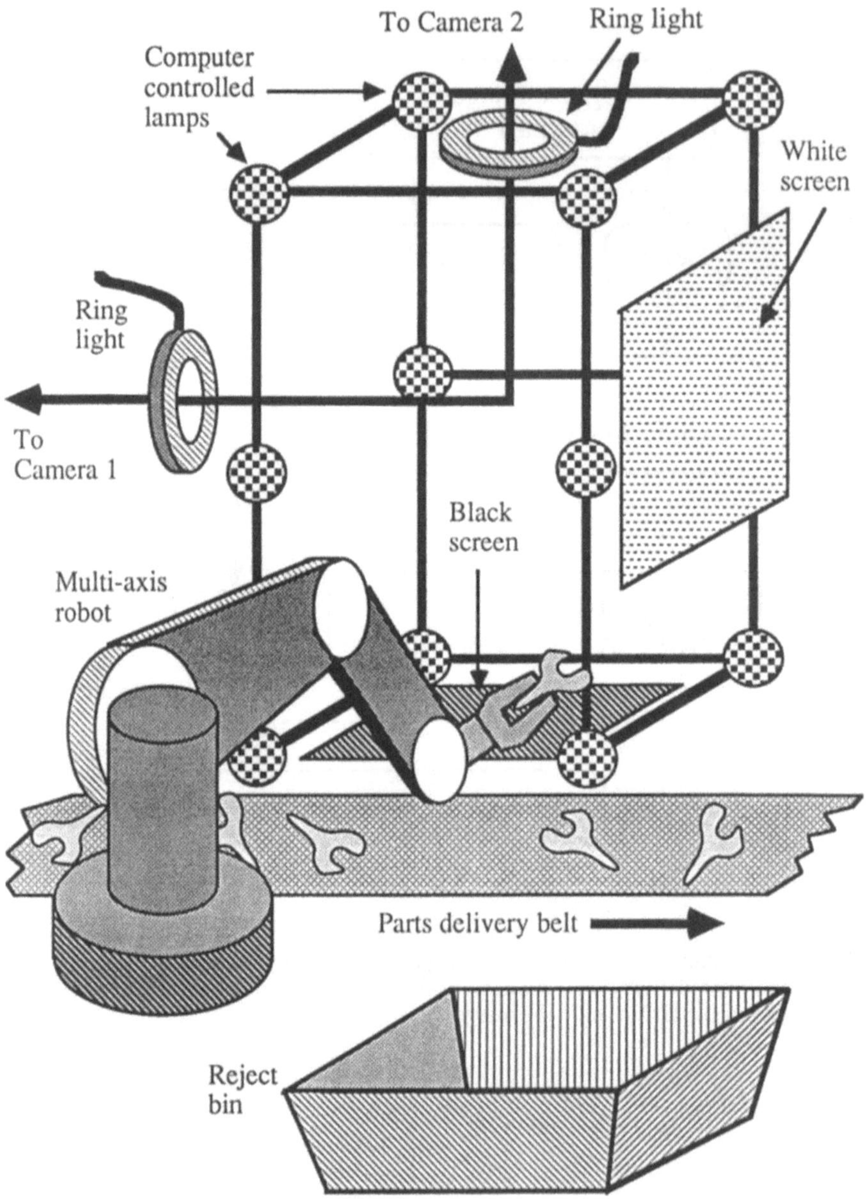
To Camera 2
Ring light
Computer controlled lamps
White screen
Ring light
To Camera 1
Black screen
Multi-axis robot
Parts delivery belt
Reject bin

a

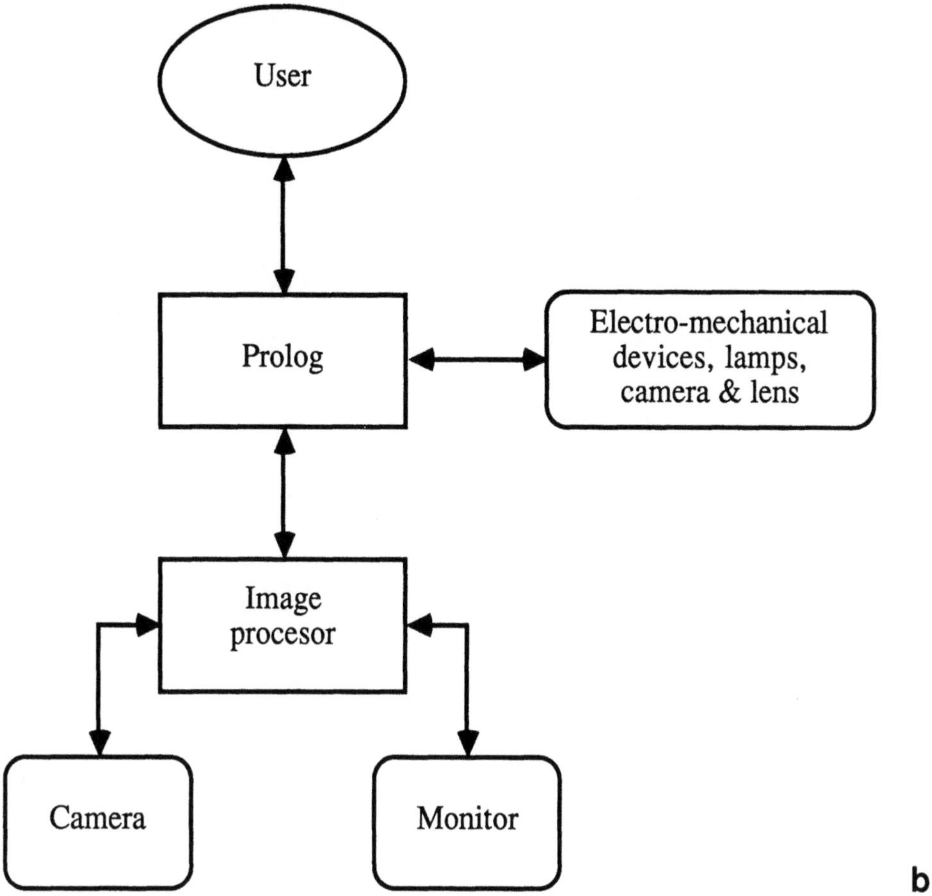

Figure 31 Flexible inspection cell (proposed)
(a) Physical arrangement.
(b) Organisation of the computer system

NATO CIM AUTHOR INDEX

NATO CIM CONFERENCE

SUBJECT INDEX

NATO ASI Series F

Vol. 1: Issues in Acoustic Signal – Image Processing and Recognition. Edited by C. H. Chen. VIII, 333 pages. 1983.

Vol. 2: Image Sequence Processing and Dynamic Scene Analysis. Edited by T. S. Huang. IX, 749 pages. 1983.

Vol. 3: Electronic Systems Effectiveness and Life Cycle Costing. Edited by J. K. Skwirzynski. XVII, 732 pages. 1983.

Vol. 4: Pictorial Data Analysis. Edited by R. M. Haralick. VIII, 468 pages. 1983.

Vol. 5: International Calibration Study of Traffic Conflict Techniques. Edited by E. Asmussen. VII, 229 pages. 1984.

Vol. 6: Information Technology and the Computer Network. Edited by K. G. Beauchamp. VIII, 271 pages. 1984.

Vol. 7: High-Speed Computation. Edited by J. S. Kowalik. IX, 441 pages. 1984.

Vol. 8: Program Transformation and Programming Environments. Report on an Workshop directed by F. L. Bauer and H. Remus. Edited by P. Pepper. XIV, 378 pages. 1984.

Vol. 9: Computer Aided Analysis and Optimization of Mechanical System Dynamics. Edited by E. J. Haug. XXII, 700 pages. 1984.

Vol. 10: Simulation and Model-Based Methodologies: An Integrative View. Edited by T. I. Ören, B. P. Zeigler, M. S. Elzas. XIII, 651 pages. 1984.

Vol. 11: Robotics and Artificial Intelligence. Edited by M. Brady, L. A. Gerhardt, H. F. Davidson. XVII, 693 pages. 1984.

Vol. 12: Combinatorial Algorithms on Words. Edited by A. Apostolico, Z. Galil. VIII, 361 pages. 1985.

Vol. 13: Logics and Models of Concurrent Systems. Edited by K. R. Apt. VIII, 498 pages. 1985.

Vol. 14: Control Flow and Data Flow: Concepts of Distributed Programming. Edited by M. Broy. VIII, 525 pages. 1985.

Vol. 15: Computational Mathematical Programming. Edited by K. Schittkowski. VIII, 451 pages. 1985.

Vol. 16: New Systems and Architectures for Automatic Speech Recognition and Synthesis. Edited by R. De Mori, C.Y. Suen. XIII, 630 pages. 1985.

Vol. 17: Fundamental Algorithms for Computer Graphics. Edited by R.A. Earnshaw. XVI, 1042 pages. 1985.

Vol. 18: Computer Architectures for Spatially Distributed Data. Edited by H. Freeman and G. G. Pieroni. VIII, 391 pages. 1985.

Vol. 19: Pictorial Information Systems in Medicine. Edited by K. H. Höhne. XII, 525 pages. 1986.

Vol. 20: Disordered Systems and Biological Organization. Edited by E. Bienenstock, F. Fogelman Soulié, G. Weisbuch. XXI, 405 pages. 1986.

Vol. 21: Intelligent Decision Support in Process Environments. Edited by E. Hollnagel, G. Mancini, D.D. Woods. XV, 524 pages. 1986.

Vol. 22: Software System Design Methods. The Challenge of Advanced Computing Technology. Edited by J.K. Skwirzynski. XIII, 747 pages. 1986.

NATO ASI Series F

Vol. 23: Designing Computer-Based Learning Materials. Edited by H. Weinstock and A. Bork. IX, 285 pages. 1986.

Vol. 24: Database Machines. Modern Trends and Applications. Edited by A.K. Sood and A.H. Qureshi. VIII, 570 pages. 1986.

Vol. 25: Pyramidal Systems for Computer Vision. Edited by V. Cantoni and S. Levialdi. VIII, 392 pages. 1986.

Vol. 26: Modelling and Analysis in Arms Control. Edited by R. Avenhaus, R.K. Huber and J.D. Kettelle. VIII, 488 pages. 1986.

Vol. 27: Computer Aided Optimal Design: Structural and Mechanical Systems. Edited by C.A. Mota Soares. XIII, 1029 pages. 1987.

Vol. 28: Distributed Operating Systems. Theory und Practice. Edited by Y. Paker, J.-P. Banatre and M. Bozyiğit. X, 379 pages. 1987.

Vol. 29: Languages for Sensor-Based Control in Robotics. Edited by U. Rembold and K. Hörmann. IX, 625 pages. 1987.

Vol. 30: Pattern Recognition Theory and Applications. Edited by P.A. Devijver and J. Kittler. XI, 543 pages. 1987.

Vol. 31: Decision Support Systems: Theory and Application. Edited by C.W. Holsapple and A.B. Whinston. X, 500 pages. 1987.

Vol. 32: Information Systems: Failure Analysis. Edited by J.A. Wise and A. Debons. XV, 338 pages. 1987.

Vol. 33: Machine Intelligence and Knowledge Engineering for Robotic Applications. Edited by A.K.C. Wong and A. Pugh. XIV, 486 pages. 1987.

Vol. 34: Modelling, Robustness and Sensitivity Reduction in Control Systems. Edited by R.F. Curtain. IX, 492 pages. 1987.

Vol. 35: Expert Judgment and Expert Systems. Edited by J.L. Mumpower, L.D. Phillips, O. Renn and V.R.R. Uppuluri. VIII, 361 pages. 1987.

Vol. 36: Logic of Programming and Calculi of Discrete Design. Edited by M. Broy. VII, 415 pages. 1987.

Vol. 37: Dynamics of Infinite Dimensional Systems. Edited by S.-N. Chow and J.K. Hale. IX, 514 pages. 1987.

Vol. 38: Flow Control of Congested Networks. Edited by A.R. Odoni, L. Bianco and G. Szegö. XII, 355 pages. 1987.

Vol. 39: Mathematics and Computer Science in Medical Imaging. Edited by M.A. Viergever and A. Todd-Pokropek. VIII, 546 pages. 1988.

Vol. 40: Theoretical Foundations of Computer Graphics and CAD. Edited by R.A. Earnshaw. XX, 1246 pages. 1988.

Vol. 41: Neural Computers. Edited by R. Eckmiller and Ch. v. d. Malsburg. XIII, 566 pages. 1988.

Vol. 42: Real-Time Object Measurement and Classification. Edited by A.K. Jain. VIII, 407 pages 1988.

NATO ASI Series F

Vol. 43: Sensors and Sensory Systems for Advanced Robots. Edited by P. Dario. XI, 597 pages. 1988.

Vol. 44: Signal Processing and Pattern Recognition in Nondestructive Evaluation of Materials. Edited by C.H. Chen. VIII, 344 pages. 1988.

Vol. 45: Syntactic and Structural Pattern Recognition. Edited by G. Ferraté, T. Pavlidis, A. Sanfeliu, H. Bunke. XVI, 467 pages. 1988.

Vol. 46: Recent Advances in Speech Understanding and Dialog Systems. Edited by H. Niemann, M. Lang, G. Sagerer. X, 521 pages. 1988.

Vol. 47: Advanced Computing Concepts and Techniques in Control Engineering. Edited by M.J. Denham, A.J. Laub. XI, 518 pages. 1988.

Vol. 48: Mathematical Models for Decision Support. Edited by Gautam Mitra. IX, 762 pages. 1988.

Vol. 49: Computer Integrated Manufacturing. Edited by I. Burhan Turksen. VIII, 568 pages. 1988.

NATO ASI Series F

MIX
Papier aus verantwortungsvollen Quellen
Paper from responsible sources
FSC® C105338
FSC
www.fsc.org

If you have any concerns about our products,
you can contact us on
ProductSafety@springernature.com

In case Publisher is established outside the EU,
the EU authorized representative is:
Springer Nature Customer Service Center GmbH
Europaplatz 3, 69115 Heidelberg, Germany

Printed by Libri Plureos GmbH
in Hamburg, Germany